Introduction to Forest and Shade Tree Insects

Introduction to Forest and Shade Tree Insects

Pedro Barbosa
Department of Entomology
University of Maryland
College Park, Maryland

Michael R. Wagner
School of Forestry
Northern Arizona University
Flagstaff, Arizona

ACADEMIC PRESS, INC.
Harcourt Brace Jovanovich, Publishers
San Diego   New York   Berkeley   Boston
London   Sydney   Tokyo   Toronto

ACADEMIC PRESS, INC.
San Diego, California 92101

United Kingdom Edition published by
ACADEMIC PRESS LIMITED
24-28 Oval Road, London NW1 7DX

Library of Congress Cataloging-in-Publication Data

Barbosa, Pedro Date
    Introduction to forest and shade tree insects / Pedro Barbosa and
Michael R. Wagner.
        p.   cm.
    Includes index.
    ISBN 0-12-078146-8   (hardcover) (alk. paper)
    1. Forest insects.   2. Forest insects—Control.   3. Shade trees-
-Diseases and pests.   4. Insect pests—Control.   5. Insect-plant
relationships.   I. Wagner, Michael R.   II. Title.
QL463.B34   1988
634.9′67—dc19   88-4006
                                                         CIP

PRINTED IN THE UNITED STATES OF AMERICA
89 90 91 92   9 8 7 6 5 4 3 2 1

We dedicate this book to our teachers and students.

To teachers such as Daniel M. Benjamin, Harry C. Coppel, and T. Michael Peters, who guided and encouraged our basic curiosity about insects in their world.

To students such as JoAnn Bentz, Liz Blake, Nancy Breisch, John Capinera, Kevin Carlin, Garry Domis, Dave Frantz, Jane Greenblatt, Eugene Harrington, Karen Kester, David Lance, Yigun Lin, David Long, Nikhil Mallampalli, Debbie McCullough, Ken Miller, Gina Rizzo-Orr, Alejandro Segarra, Bob Tisdale, Fernando Vega, and Zhao-Yi Zhang, who are the most important purpose for our professional activities. We hope this book will help inspire their professional careers.

CONTENTS

PREFACE

This book is intended for use as an introductory text on the biology, ecology, and control of forest and shade tree insects. Although we assume the users will be upper-level undergraduates and graduate students, we have written it for those who may have no more than a general background in biology, zoology, or ecology. Courses in forest entomology, forest protection, or other related courses are often taught from widely divergent viewpoints. Not only do these courses cover different topics, but students in these courses range from lower division undergraduates to doctoral students. Thus, no one book can provide the exact range of topics, appropriate level of detail, or the precise sequence of subject matter for all these courses. Alternatively, we have provided what we hope is the best compromise—the major subjects needed for most courses. Ultimately, each instructor must select those topics and sequences of topics that are most appropriate for his or her course. We have in most cases attempted to provide considerable detail on the assumption that it is easier for an instructor to selectively delete rather than supplement a superficial discussion.

The underlying assumption of most of the book and of Section I ("Insect–Tree Interactions") in particular is that one cannot simply examine the influence of insects on trees but must also consider the plant (tree) characteristics. The interrelationships between insects and trees are dynamic and inextricably linked associations resulting from years of coevolution or at least coexistence. Section II ("Forest and Shade Tree Insects: Their Impact on Trees") reviews the nature and effect of major groups that live on, feed on, or injure trees. In the same fashion that the tree and its insect associates are linked, so too are the survival of tree insects and the physical and biotic factors of the environment. Thus, Section III ("Ecology of Forest and Shade Tree Insects") covers the primary ecological forces affecting forest and shade tree insects. Finally, Section IV ("Control of Forest and Shade Tree Insects") present the underlying principles and the array of control strategies available for the management of forest and shade tree pests. This section includes not only the types of control, but also the procedures that influence effective control of forest and shade tree insects, i.e., evaluation, sampling, and other elements of the decision-making process. Finally, an appendix on general entomology is provided to highlight major aspects of entomology.

Instructors who teach an undergraduate forest entomology course may require students to begin with the general entomology appendix, followed by Section II ("Forest and Shade Tree Insects: Their Impact on Trees") and complete the course with Section IV ("Control of Forest and Shade Tree Insects"). Graduate or upper-level undergraduate courses might ignore the appendix and de-emphasize the applied Section IV, focusing on the general concepts outlined in Sections I, II, and III. Graduate level courses in insect ecology and host–plant interactions might focus in detail on Section I ("Insect–Tree Interactions") and Section III ("Ecology of Forest and Shade Tree Insects"), while minimizing the topics covered in Sections II and IV. In general, our purpose has been to provide rigorous discussion of topics that can be presented in a variety of sequences to serve the needs of courses in forest entomology, forest pest management, ecology of forest insects, and host–plant interactions.

In summary, this text is designed to go beyond a discussion of a series of forest and shade tree pests and their control. We hope to present the basic underlying principles of biology and ecology that allow an understanding of various insect/tree associations within the self-perpetuating ecosystem known as a forest.

Pedro Barbosa
Michael R. Wagner

ACKNOWLEDGEMENTS

We would like to acknowledge the assistance of the many people who contributed to the production of this book. Numerous individuals allowed us to utilize their photographs and figures to illustrate important concepts. We have tried to acknowledge these people in the text when we presented their work. Amy Bartlett-Wright and Emily Mead produced numerous figures and line drawings for the text. The Northern Arizona University Ralph M. Bilby Research Center provided excellent typing and photographic services. The following individuals provided the essential critical reviews of various sections of the book: D. L. Dahlsten, J. L. Foltz, A. E. Hajek, W. T. Johnson, W. P. Kemp, A. M. Lynch, W. J. Mattson, D. G. McCullough, M. W. McFadden, T. L. Payne, T. D. Schowalter, and all the anomymous reviewers for Academic Press. Many others provided encouragement and suggestions, and we apologize for not remembering all of them in this acknowledgement.

Finally, we would like to acknowledge the tremendous effort of Elizabeth A. Blake of the Northern Arizona University School of Forestry. Liz managed many of the technical aspects of the book production. This book would not have been possible without her continued persistence. Our sincere thanks!

Pedro Barbosa
Michael R. Wagner

PART I

Insect–Tree Interactions

CHAPTER 1

Introduction to Forest and Shade Tree Entomology

Introduction

A multitude of fascinating relationships exists among insects and the other life forms of a forest. Interactions between insects and plants, insect parasites, predators, pathogens, birds, mammals, fungi, and other forest organisms are discussed throughout this book. The discussions and examples in introductory textbooks such as this too often make assumptions that influence students' understanding of basic principles but often are not clearly stated. Examples of such assumptions of relevance to this text are the definition of a forest insect pest and the importance of human activity on the development and solution of problems in forest insect pest management. These and other issues discussed below provide an important basis for the choice of topics and for the nature of the discussions in this book.

What Is Forest and Shade Tree Entomology?

A forest is a highly organized group of communities whose plant species diversity and structural complexity provide numerous niches. The ecological dynamics of the forest are the sum total of a complex series of interactions among animals, plants (including microorganisms), and physical factors. Those interactions that lead to loss of yield, reduction in quality, or alteration of trees' aesthetic value are of concern in forest entomology. Indeed, the objective of this discipline is to understand insect–tree interactions sufficiently to prevent or retard economic or socioeconomic damage. The interweaving relationships of the forest environment require the thorough understanding of insects other than those that damage trees and of other organisms associated with forests or forest insects. Although soil insects, soil microflora, symbiotic fungi, vertebrate predators, and so on, may not directly damage trees, their activities may affect the behavior, physiology, and ecology of dendrophagous species.

When Is a Dendrophagous Species a Pest?

A dendrophagous species is considered a pest when it interferes with the intended use of a tree, forest, or forest product. The relationship between intended use and type of injury determines the significance of inflicted damage and the appropriate strategy for control.

For example, an insect defoliator that reduces the growth of trees, and therefore the production of wood fiber, may be considered a pest in a plantation but not in a wilderness area. Growing wood fiber is an objective or intended use for trees in a plantation. Because an insect defoliator may reduce the amount of wood fiber, it would be considered a pest. In a wilderness area the production of wood fiber is not an intended use; therefore, reduction of tree growth does not interfere with the intended use, and the insect defoliator would not be considered a pest.

The multiple uses of forests and forest products make it difficult to sort out the many ways that insects can become pests. Forests are really multiproduct factories. Wood, paper, recreation, grazing, watershed, and aesthetics are just a few of the many products that are derived from the forest factory. All intended uses of the forest must be considered when determining whether an insect species is a pest.

In general, the variety of uses of forests and forest products can provide overall categories of forest insect pests. Wood utilization in the United States includes traditional uses for the interior and exterior of houses and for furniture. Wood for railroad crossties and powerline poles has been an important contribution to the transportation and communication industry. The per capita use of paper products per year is approximately 135 lb in Europe, 23 lb in Asia, and over 600 lb in the United States. In the latter use of forest products, an insect that interferes with the production of wood intended for sawmills or pulp plants would clearly be a pest.

In addition to the uses of wood products, an incredible variety of compounds is derived from extracts of tree tissues, such as bark and foliage. Valuable compounds are also obtained from the chemical modification or conversion of cellulose, lignin, and other extractives. Although in some countries these extractives are important commercial products, the primary types of chemically modified wood products are pulp and paper, cellulose textile fibers (i.e., rayon), and cellulose plastics (Champion, 1975). An enormous variety of chemicals is derived from foliage, particularly from the essential oil fraction. These chemicals are obtainable in varying amounts and are used to various degrees around the world in medicine, veterinary practices, and cosmetics (Table 1.1). In some countries such as

Table 1.1. Present uses, production (in the USSR), and future potential of chemicals extracted from tree foliage[a]

Foliage Derivative	Production in the USSR	Uses
Chlorophyll-carotene	250 tons/yr[b] produced in 10 plants (approximate production capacity); estimated present Soviet demand about 700–800 tons/yr	In medicine, externally in a variety of salves and ointments for the treatment of burns; highly purified preparations given intravenously for blood disorders In cosmetics, toothpastes, bath additives, shaving soaps, shampoos, skin lotions, and perfumes In cattle breeding and poultry raising for faster growth, greater weight, and healthier animals
Sodium chlorophyll	Still in the development and marketing stage; present production about 20 kg/yr; estimated potential market 100 tons/yr	Same as for chlorophyll-carotene paste
Provitamin concentrate	In early stages of development; no estimate of potential markets given	In cattle breeding to increase growth rate and weight In animal and poultry nutrition as a vitamin supplement In cosmetic preparations
Conifer wax	By-product of the manufacture of chlorophyll-carotene paste	As a lubricant In cosmetics and pharmaceuticals In veterinary practice
Medium and heavy fractions of essential oils	In the early research and marketing stage	In perfumes and cosmetics In insect control
Conifer-salt residues	In the early research and marketing stage	In cosmetics for foot and body baths

[a] From Barton (1976).
[b] Ton, as used here, refers to metric ton = 1000 kg = 2200 lb.

Russia, forest foliage is converted to muka, a vitamin supplement for livestock fodder (Keays, 1976; Barton, 1976). Currently the United States and Canada have several projects for evaluating the potential of muka and other foliage derivatives (Dickson and Larson, 1977; Gerry *et al.*, 1977; Hunt and Barton, 1978). Other wood extractives and tree products include Canada balsam (a resin used on microscopic slides), tannin for the leather industry, maple syrup and sugar, nuts, fruits, and adhesive extenders used in wood bonding.

Insects that alter the aesthetic qualities of a tree or a forest also are important, though their economic impact is often difficult to categorize or assess. The importance of large, healthy shade trees to property values and people's positive emotional response to trees and forests, or to the insect pests themselves, justify concern about pests that damage the aesthetic resource (Fig. 1.1). Social inputs are also important in forest entomology because strong public concern about the relationship between environ-

Figure 1.1.   Late instar dispersal within gypsy moth infestations can sometimes disturb urban and suburban areas. Larvae often cover sidewalks and homes and get into swimming pools and reservoirs. (Courtesy USDA Forest Service.)

mental protection and forest pest management is often voiced and can influence management decisions.

Forest trees help protect watersheds and regulate stream flow by reducing the impact of rain on the soil and slowing the flow of accumulated water. In the southwestern United States, where water is scarce, National Forests have been established with the primary goal of protecting watersheds. Forest insect damage to overstory trees in managed watersheds might increase water yields. Again, the use of the word *pest* must be based on interference with achievement of a specific goal. A forest insect that kills trees in a watershed in the northeastern United States may have a very different impact on management goals than a similar insect in the arid southwest.

In summary, a forest or individual tree can serve a multitude of diverse functions. The diversity of reasons for protecting a tree or a forest complicates the process of determining which insect is a pest. Perhaps the best approach to understanding forest communities is to understand the interactions that exist within the forest.

Why Is an Insect a Tree Pest?

The population level reached by a given species may determine its status as a pest; indeed, the populations of many insects of concern to forest managers often reach high numbers (Fig. 1.2). However, although the population density of some insect species may be useful for evaluating potential tree damage, density by itself does not always provide a valid index for estimating the impact of dendrophagous species. Some species must be present in extremely high numbers to damage a tree, whereas others need increase only several fold to make a substantial impact. Ultimately, it is the effect that an insect has on a tree, irrespective of its population density, that best determines when a species is a pest. For example, although diameter increment loss may result from 75% or greater defoliation, defoliation of less than 50% usually causes insignificant growth loss.

Increment loss has traditionally provided forest managers with a measure of the economic consequence of tree damage, but increment loss due to dendrophagous insects is often an inadequate measure of the impact of insect activity. Although such measures are often adequate for single shade trees, it is difficult to translate the growth-increment loss of a single tree to the survival and health of a stand or an entire forest. An alternative would be a holistic approach that attempts to interpret the significance of insect activity in relation to other dynamic components of the ecosystem over time. What often appears to cause losses in individual trees or over short periods of time may result in gains to the forest stand or

Figure 1.2.   Masses of redhumped oakworm caterpillars. (Courtesy USDA Forest Service.)

long-term beneficial changes. If one places less emphasis on the importance of single trees and considers the influence on the forest, the result of defoliation may take on a different perspective. Moderate to severe defoliation may contribute a 20 to 200% increase in the nitrogen, phosphorus, and potassium content of the litter. Insect excrement and cadavers, together with fragmented leaves and twigs, enrich the litter, enhance conditions for soil organisms, and provide a greater concentration of nutrients for growth (Rafes, 1970, 1971; Mattson and Addy, 1975) (Table 1.2).

An economic assessment of the activities of dendrophagous species may differ significantly from an ecological assessment. For example, feeding by bark beetles or defoliators on weakened or overmature trees helps eliminate the least productive members of a forest community and may be ecologically advantageous. However, similar damage to commercial timber requires specific control strategies before feeding and brood formation can reduce the quality of timber.

The appropriateness of resource utilization by dendrophilous (tree-loving) species is manifested in the differential effect of their feeding. For example, suppressed trees often succumb more readily to defoliation than dominant individuals. Many forest insects are so-called secondary species because they primarily attack weakened, dying, or dead trees. These

Table 1.2 Mean nutrient concentrations (%) of leaves, frass, and tissues of gypsy moth (*Lymantria dispar*) larvae feeding on various tree types[a,b,c]

	N	P	K	Ca	Mg
Leaves					
Oak	2.59	0.05	1.04	1.15	0.13
Ash	1.79	0.18	1.50	2.12	0.33
Maple	1.33	0.14	1.38	1.87	0.38
Lime	2.65	0.11	1.79	1.65	0.21
Hazel	1.38	0.14	1.49	1.92	0.38
Frass	3.26	0.20	3.21	1.87	0.76
Larvae	9.62	1.53	2.98	0.39	0.27

[a] From Rafes (1971).
[b] Similar trend also occurs in *Gilpinia hercyniae* (European spruce sawfly) feeding on spruce (Lunderstadt *et al.*, 1977).
[c] The concentration of nutrients in excretory products of other types of insects may not increase but decrease. The willow aphid (*Tuberolachnus salignus*) when feeding on stems of willow (*Salix acutifolia*) produces honeydew that has the same amino acid and amine composition as the host plants but at lower concentrations (Mittler, 1958).

insects are scavenger species, hastening the death and/or decomposition of overmature, weakened individuals, thereby liberating resources for growth of young trees. Species such as the spruce bark beetle (*Dendroctonus rufipennis*) have been likened to predatory mammals that prey only on the surplus (weakened and old individuals) of their prey population (Gobeil, 1941).

New growth may also result from opening the canopy to light and heat since understory trees typically respond with enhanced growth as the dominant trees are defoliated. In theory, the thinning of older, weakened trees and the enhancement of young growth cause stand production to increase because density becomes optimal and growth more vigorous. In summary, the biological impact of the activity of dendrophagous species can be viewed from a number of perspectives. The effects of insect activity on individual trees may be distinct from those on the forest stand.

Successional and Land-Use Patterns

Changes in forests, whether biotic or abiotic, natural or mediated by human activity, have a significant role in determining which insects become pests and why they become pests. Forests throughout the world can be characterized by their dominant tree species, and to a great degree these species shape the nature of interactions among the life-forms of the forests. The forests of North America vary considerably, but there are approximately five principal forest regions. These regions contain forest communities that often provide conditions appropriate to the survival and numerical increase of various dendrophagous species. Thus, dendrophagous pest species are often closely associated with each forest region, and the various regions often have different forest insects.

The northernmost states have spruce–fir or northern coniferous forests, which extend into Canada. This forest type is vulnerable to defoliation by the spruce budworm (*Choristoneura fumiferana*), spruce sawfly (*Neodiprion abietis*), larch sawfly (*Pristiphora erichsonii*), and casebearers. South of the spruce–fir forest is a region of hardwood forest. The northeastern uplands form a transition zone dominated by beech (*Fagus* spp.), birch (*Betula* spp.), maples (*Acer* spp.), white pine (*Pinus strobus*), and hemlock (*Tsuga* spp.). The region south of New England is dominated by oak-hickory forests, which eventually grade into the southerly oak–pine forests. The hardwood region is also characterized by defoliators, such as the gypsy moth (*Lymantria dispar*), various oakworms, cankerworms, and tent caterpillars (*Malacosoma* spp.). Tent caterpillar defoliation, for example, is characteristic of the aspen–cottonwood–oak forest of the northern central hardwood areas.

Major changes in insect-pest complexes occur in the southern forest region. The dominance of pines [e.g., longleaf, slash, shortleaf, loblolly (*Pinus palustris, P. elliottii, P. echinata*, and *P. taeda*, respectively)], gives the southern pine beetle (*Dendroctonus frontalis*) and other bark beetles prominence as major dendrophagous species. In the far west, the western pine forests [which are primarily ponderosa, sugar, and lodgepole pines (*Pinus ponderosa, P. lambertiana*, and *P. contorta*, respectively, and Douglas-fir (*Pseudotsuga menziesii*)], the Rocky Mountain region forests [including Douglas-fir, pines, Engelmann spruce (*Picea engelmannii*), and other spruce species], and the Pacific coast forests [including western white pine *Pinus monticola*), firs (*Abies* spp.), cedars (*Juniperus* spp.), and Douglas-fir] are dominated by the Douglas-fir beetle (*D. pseudotsugae*), western pine beetle (*D. brevicomis*), mountain pine beetle (*D. ponderosae*), and spruce beetle. A few defoliators also are characteristic of these forests, such as the Douglas-fir tussock moth (*Orgyia pseudotsugata*), the western spruce budworm (*Choristoneura occidentalis*), the pine butterfly (*Neophasia menapia*) (Clement and Nisbet, 1977; Randall and Heisley, 1933), and the pandora moth (*Coloradia pandora*) (Patterson, 1929; Wygant, 1941).

Perhaps because the life span of most trees in the United States (from 150 to 300 years) is much greater than that of humans, we are inclined to think of forests as unchanging. However, within the relatively recent history of the northeastern forests, significant changes have occurred that, ultimately may influence the nature and severity of pest problems; these changes have been due, in part, to the effect of dendrophagous insects on trees and the tree species composition of the forests. In other situations, changes induced by other agents have altered the ongoing interactions between insects and trees.

One of these agents has been human intervention. Past land-use patterns have a significant effect on which forest insects become pests. The history of forest use in the Northeast, even with the usual historic uncertainty, is the best known of the forest regions. Simple forest management was undertaken by Native American tribes in this area when they periodically burned forested areas to clear camp sites and to remove undergrowth from hunting areas (Clement and Nisbet, 1977). Windstorms and lightning-initiated fires also affected the character of the forest (Oliver and Stephens, 1977). In these very early forests, insects such as the spruce budworm and the hemlock looper (*Lambdina fiscellaria fiscellaria*) no doubt defoliated portions of forests, inducing changes in many areas. When European colonists arrived, change became more rapid and the types of change more diversified as the agriculturally oriented colonists cut forests for farms and pastures and for building material. By the 1800s almost all of southern New England was an open landscape. During the initial development of railroads in the middle of the 19th century, wood for locomotives and charcoal put additional demands on timber resources, involving millions of cords of second growth timber.

The railroads also initiated a major reversal in land-use patterns in New England by transporting people to the farm areas of the midwest, leaving the farms in New England, which grew back to grasses, shrubs, and a new cycle of trees (Randall and Heisley, 1933; Clement and Nisbet, 1977). These and other changes paralleled changes in animal life, particularly that of insects. Although most of the changes in insect species composition and numbers went unnoticed, some species grew in prominence. The white pine weevil (*Pissodes strobi*) did not become a problem in the eastern United States until the mid-1800s, when cleared agricultural fields were abandoned and subsequently became occupied primarily by white pine. These poorly stocked stands were repeatedly weevilled. In addition, the many acres of plantations established in the early 1900s provided extensive breeding areas for the weevil (Belyea and Sullivan, 1956).

Insects have also played a major role in the reduction and modification of the early forest in the northeast. Widespread outbreaks of spruce budworm, larch sawfly, hemlock looper, balsam woolly adelgid (*Adelges piceae*), European spruce sawfly (*Gilpinia hercyniae*), and beech scale (*Cryptococcus fagisuga*) killed a large proportion of their hosts, altered the age distribution of host tree populations, and thus helped change the species composition of forests (Swaine, 1933). Changes in the character of the forests induced by agents such as fungi have also occurred; however, the associated changes in insect fauna have been poorly documented.

Dendrophagous Insects in the Urban Forest

The urban forest is still a rather ill-defined concept: It can range from a single, sparse row of trees lining a street in a large city to green spaces or parklike areas in small towns and city suburbs. Shade trees may grow anywhere, from major urban centers to house lots surrounded by forested land. The public concern for urban forests and shade trees is focused on (a) the importance of the aesthetics of the tree, (b) the low density of trees per unit area, (c) the importance of trees as a source of recreation in areas with a high density of people, (d) the role of trees in climate modulation, and (e) the role of trees in environmental modification (e.g., as wind and sound barriers). These functions of shade trees and urban forests translate into higher property values for well-landscaped land. It has been suggested by the International Society of Aboriculture that well-landscaped property has a real estate value approximately 7% higher than similar nonlandscaped property. Because of the higher value of trees in the urban environment, insects often are perceived as pests at lower densities than in natural forests.

Chapter Summary

Forest and shade tree entomology is the science of understanding the complex interaction between insects and trees and how this interaction leads to losses of the multiple products of the forest. Not all insects that inhabit forests are pests. The concept of "pest" requires a clear specification of desired forest tree products and the potential of an insect to reduce their quantity and quality. An insect species can be a pest in one situation and not in another. Insects have become notorious pests because they can quickly reach high population levels under suitable environmental conditions, and because they have adapted to numerous forest niches. In some situations, insects can play an ecological role as they cause shifts of resources from slow-growing to more-productive trees. Humans, through their activities in the forest, can have profound effects on the nature of forests and the subsequent complex of potential pest insects. Because of the high value we place on trees in the urban environment, insects that become pests can be important factors in urban forest management.

CHAPTER 2

The Influence of Food on Dendrophagous Species

Introduction

What and how much an insect eats determines, in part, its significance as an economically important insect. Forest insects eat a tremendous variety of foods, from dead wood and plant detritus to living tissues. Living food consumed by forest and shade tree insects includes bacteria, fungi, root tissues, leaves, wood, and seeds. Parasitism and predation add to the diversity of feeding habits among forest and shade tree inhabitants, making an insect's role in a forest ecosystem food web difficult to assess. Whereas some wood borers subsist on wood, others such as carpenter ants and some bark beetles live on food found outside their tunnels, and still others live on symbiotic fungi in their tunnels. In still other species, the bulk of the food consumed may not be the prime source of nutrition. For example, the diet of an insect in the forest litter may not be the decaying litter it ingests, but the fungi growing on the litter. Some wood-eating powder-post beetles in the genus *Lyctus* are unable to digest cellulose, an abundant component of wood, but instead use the starch, sugars, and nitrogenous compounds in the wood (Trager, 1953).

In a behavioral and ecological context, the phrase "you are what you eat" accurately describes the effects of the quantity and quality of food on the life of all insects. The food consumed by dendrophagous species primarily provides nutrition but also provides protective shelter and a breeding site. Thus, it affects the behavior, physiology, survival, fecundity, and the distribution of insects.

How Much Do Insects Eat?

Consumption by dendrophagous species can be extensive, particularly when the rate of consumption is considered. The rate of food consumption by most leaf-chewing, dendrophagous herbivores ranges from approximately 50 to 150% of their body weight per day. Piercing–sucking insects, feeding for example on xylem, probably have the highest rates of consumption: 100–1000 times their weight per day (Mattson, 1980). The tree cricket (*Oecanthus* sp.), in a tulip poplar forest averages approximately 7.35 mg dry weight of foliage per day or 81% of its body weight. Similarly, geometrids can consume approximately 56% of their body weight. Consumption by these two defoliators accounted for 60% of leaf consumption by insect herbivores in a tulip poplar (*Liriodendron tulipifera*) forest (Reichle and Crossley, 1967).

Although normal insect grazing ranges from 5 to 30% of annual foliage production in most temperate forest ecosystems, foliage-feeding insects more typically consume 3–8% of annual foliage production (Mattson and Addy, 1975). Current estimates fall within this range. Consumption by insects on maple (*Acer*), ash (*Fraxinus*), and beech (*Fagus*) in southern Canadian forests ranges from 5.0 to 7.7% (Bray, 1964). Primary consumers of a tulip-poplar forest consume approximately 5.6% of the total available foliage (Reichle and Crossley, 1967). Leaf area consumption by insects at endemic levels may be somewhat less than indicated above.

End-of-the-season measurements of consumption may often overestimate leaf area consumption because early feeding damage may appear greater after the leaf grows. Thus, estimates of leaf area consumption by dendrophagous associates of tulip tree forests should be adjusted to approximately 1.9% from the previous estimate by Reichle and Crossley (1967) of 5–8% (Reichle *et al.*, 1973). Nevertheless, the impact of even this level of consumption is considered significant since nutrients concentrated in the frass return 2.5% or 9 g/m^2 of primary production to the forest floor (Reichle *et al.*, 1973).

Although rarely studied, the interactions among various types of leaf feeders in allocating their resources (foliage) may increase the amount of damage to trees. A survey of dendrophagous species of birch, aspen, and willow has revealed that approximately 20 species of foliage-feeding insects attack each tree type. These insects can damage 60–70% of the leaves of birch (*Betula*), aspen (*Populus*), and willow. Damage of only one type (such as chewing, sucking, mining, or galling) is usually encountered on an infested leaf. Only about 20% of leaves have more than one type of damage (Rafes and Sokolov, 1976). This phenomenon may be due to the apparent overabundance of leaves. However, the lack of new attacks on already damaged leaves may be a result of sign recognition by other insects. The observed pattern in some leaf miners of one mine per leaf may be the result of chemical or physical signaling or the result of competitive interaction within a leaf.

The amount of available foliage can, in part, determine the potential extent of damage that a foliage-feeding dendrophagous insect can inflict. One might ask if there is a difference in the relative amount of foliage in coniferous forests compared with deciduous forests, and whether this difference might help determine the potential impact of a pest species. Foliage production by angiosperm forest trees ranges from 23 to 30% of the total net aboveground production over the life of the forest (Bray, 1964). Leaf dry weight per hectare of Japanese deciduous broadleaved forests ranges between 1 and 5 metric tons. Differences in leaf biomass between European and Japanese species are minor (Satoo, 1970; Mattson and Addy, 1975). In general, deciduous forests have less leaf biomass than evergreen forests. In fact, evergreen forests may have three to seven times as much foliage as deciduous forests (Tadaki, 1966; Mattson and Addy, 1975). Although we know that the amount of foliage differs with forest type, we still do not know whether a given level of defoliation is as damaging to coniferous forests, for example, as to deciduous forests.

Consumption by dendrophagous species must be kept in perspective. Although forest and shade trees support a wide variety of insect species, the insects are typically low in abundance and generally remain near steady-state levels. Indeed, population density may be so low that it is nearly impossible to clearly demonstrate a steady-state level with any statistical significance. Populations of relatively few species build to dramatic epidemic levels capable of significant defoliation. However, those that do reach epidemic levels can consume 100% of current foliage production and previous growth (e.g., in conifers) and ultimately may be responsible for substantial mortality. It is this latter group of defoliators that often acquires the status of true pests.

The feeding insect can have an important role in the health and dynamics of the forest. It has been suggested that the rate of nutrient recycling by leaf fragments, frass, and cadavers is enhanced by defoliation. Indeed, some researchers believe that insect feeding acts as a regulator of forest productivity and of tree species abundance and distribution. This hypothesis is partially based on the facts that (a) insects feed on a plant's key sites of energy and biochemical synthesis (i.e., buds, leaves, phloem, and cambium), (b) the variety of insect species insures consistent feeding pressure, (c) the insect–tree relationship may be the result of a long-standing evolutionary interaction, (d) insects respond to variations in the physiological state of their tree hosts, and (e) trees respond to the feeding activity of insect populations (Mattson and Addy, 1975).

Patterns of Feeding Preferences

Insect species feed on such a wide variety of tree tissues that it would be difficult to designate any single feeding type as representative of an order. However, generalizations can be useful in representing the most significant damage caused by representatives of an order. Overall, Lepidoptera, Hymenoptera, and Coleoptera are the most important dendrophagous groups in the forest. Bud, twig, and terminal feeders are principally in the

Lepidoptera and Coleoptera. These two orders and the Hymenoptera and Diptera contain the most important seed feeders. Foliage-feeding species are spread across the Lepidoptera, Hymenoptera, Coleoptera, Diptera, and Orthoptera. Species of Coleoptera, Lepidoptera, and some Diptera represent the major bark-boring species. Wood borers are found primarily among the Coleoptera, Hymenoptera, and Lepidoptera, whereas root-feeding species occur mainly among the Coleoptera. Finally, the Hemiptera are common in most tree tissues except the heartwood (Franklin, 1973).

The preferences exhibited by forest and shade tree insects vary from those of insects that feed on specific tissues of a limited number of hosts to those that feed on virtually every tree in a habitat (see section on host plant range of forest and shade tree insects). The variety in the size of host range is due, in great part, to the responses of insects to chemical and physical stimuli.

The foliar concentration of amino acids is greater in balsam fir (*Abies balsamea*), one of the hosts of the spruce budworm, than it is in white spruce (*Picea glauca*), another host (Kimmins, 1971). This and other differences are responsible for the insect's preference for balsam fir. The choice of hosts made by a dendrophagous species is not always based on a plant's chemical composition but can also be based on physical cues. It is often difficult to separate the influence of physical and chemical factors. Nevertheless, physical characteristics of the host may affect preferences as well as the survival of herbivores. Feeding-site selection by the green spruce aphid (*Elatobium abietinum*) is correlated to the frequency of stomata on the needle surface. Most individuals are found on the underside of the needles of Sitka spruce (*Picea sitchensis*), where most stomata occur. The penetration of needles of both Sitka and Norway spruce (*Picea abies*) is influenced by biochemical and morphological characteristics of the stomata. These traits appear to explain differences in the degree of needle damage (chlorotic banding and needle death) between the two spruce species (Parry, 1971).

Microhabitats created by the physical texture of host trees also are important in determining the survival of some forest insects. In endemic populations of the gypsy moth (*Lymantria dispar*), for example, larvae cease feeding during the day and seek resting sites. Sites such as bark flaps or crevices may provide a protective microclimate where larvae are less susceptible to predation than those larvae resting in the litter (Campbell *et al.*, 1975a,b). Thus, what appears to be a host-plant preference based on observations of the accumulation of larvae may result from differential mortality of larvae. Tree species that have an abundance of resting sites retain larvae, making them less susceptible to predation by small vertebrates (Campbell and Sloan, 1976; Campbell *et al.*, 1975a,b).

For many species, selection of food is not accomplished by larvae but is determined by parental females; that is, wherever eggs are laid is where larvae feed. The anobiid beetle (*Xyletinus peltatus*) shows a strong oviposi-

Table 2.1. Oviposition choice of *Xyletinus peltatus* beetles exposed to various host woods[a]

Wood	Average Eggs/Female	Average Eggs/Wood	Percentage[b]
Yellow-popular	16.3 a[c]	402.3	64.1
Slash pine	2.1 b	58.1	9.2
Sweetgum	3.4 b	72.1	11.5
Douglas-fir plywood	1.4 b	36.6	5.8
Shortleaf pine	1.2 b	24.5	3.9
Loblolly pine	0.5 b	12.8	2.0
Longleaf pine	0.5 b	11.3	1.8
Southern pine plywood	0.4 b	9.6	1.5

[a] From Williams and Mauldin (1974).
[b] Percentage of total eggs laid in test that were laid on each kind of wood.
[c] Means followed by different letters differ significantly.

tion preference for specific hosts (Table 2.1) which appears to be based, at least in part, on the roughness of the wood surface (Williams and Mauldin, 1974). The psyllid (*Cardiaspina densitexta*) is a species that reaches epidemic levels in parts of southern Australia, thereby causing serious defoliation of pink gum (*Eucalyptus fasciculosa*). The oviposition by its females is affected by physical characteristics of host foliage. Eggs are preferentially laid on the undersurface of the leaf: 81% on lower-leaf surfaces versus 19% of upper-leaf surfaces examined. Virtually all eggs are laid in dense groups on the basal third of mature leaves. The presence and thickness of the midrib appears to provide a strong stimulus for oviposition (White, 1970). Ovipositing females also show a preference for mature leaves (Table 2.2). On a surface-area basis (eggs/cm^2), more than 60 and 120 times as many eggs are deposited on mature leaves compared with those laid on young and old leaves, respectively (White, 1970).

Table 2.2. Distribution of eggs among leaves of different age classes and the number of eggs laid by *Cardiaspina densitexta* females confined to one of three age classes of leaves[a,b]

Leaf Type	No. of Eggs, Test I (% of total)	Eggs per cm^2	No. of Eggs, Test II (% of total)	Mean No. of Eggs per Female, Test II
Old	78 (0.4)	0.05	121 (3.0)	1.3
Mature	19,526 (98.8)	6.24	3,773 (93.0)	40.6
Young	155 (0.8)	0.10	163 (4.0)	1.8

[a] Modified from White (1970).
[b] Age classes are old leaves (greater than 1 yr old, situated at base of twigs, thick, rigid, and often necrotic), mature leaves (current season's growth, fully expanded, flexible, and free of blemishes), and young leaves (soft, flaccid, or incompletely expanded new leaves).

Damage: External Appearances

The mouthparts of forest insects vary from the strong mandibles of wood-boring insects to the slender, threadlike mouthparts of aphids and scales (see Appendix). Even more varied are the responses of tissues to feeding punctures on leaves, stems, and buds. These signs and symptoms of feeding damage sometimes may be nondescript but are often specific and characteristic of a particular insect species. Thus, feeding damage, along with other information (e.g., host species, time of the year, and geographical region), can be useful in identifying the cause of damage.

The four primary categories of insect damage to foliage are free-feeding, skeletonizing, mining, and stippling. Free-feeding, perhaps the most obvious damage, consists of feeding without regard for leaf veins (Fig. 2.1). Large pieces of leaves are usually missing as a result of this type of feeding. A variation of this feeding pattern is found in species that chew many small holes in a leaf. Skeletonizers usually feed on leaves without destroying the veins, leaving the damaged leaf characteristically lacelike in appearance (Fig. 2.2). Small insects may consume only the lower or upper layer of leaf tissue without removing the opposing leaf layer, making small patches that appear light brown. Many species, such as the cottonwood leaf beetle (*Chrysomela scripta*), the red-humped oakworm (*Symmerista canicosta*), the orange-striped oakworm (*Anisota senatoria*), and the hickory tussock moth (*Halisidota caryae*) begin as a leaf skeletonizer and progress to devouring most of the leaf, with the exception of larger veins. In still other species, larvae skeletonize leaves, and adults free-feed.

Mining is a feeding activity that damages tissues of leaves, needles, stems or buds. Leaf miners, for example, consume the mesophyll layer of the leaf without feeding on the upper or lower leaf layer. Mines become visible as upper and lower leaf tissue die. Leaf mines are often morphologically distinct. Linear and serpentine mines (Fig. 2.3) are straight or wavy tunnels, respectively, within the leaf. Blotch mines are formed by feeding in circular or irregular patches. Mines often coalesce, or begin as one type and change as the insect ages. Some leaf miners also mine the stem cambium, causing long, thin mines just below the bark. Finally, the minute circular wounds made by the stylets of insects such as leafhoppers, lacebugs, and plant bugs accumulate to form a characteristic stippling pattern of injury (Fig. 2.4). These aggregations of minute dots may appear as a discoloration of the leaf.

Wood and bark borers feed on roots, seeds, stems, branches, and twigs. Their feeding results not only in scoring the wood but also in peppering the surface with emergence holes. Inner-bark (phloem) feeders, such as bark beetles and some wood borers, form galleries that etch the xylem and are visible when the outer bark is removed (Fig. 2.5). Other external signs of the activity of these insects can include wound responses of trees or associated damage (e.g., pitch tubes, wilting, leaf or needle

Figure 2.1. Free-feeding by larvae of the fall cankerworm (*Alsophila pometaria*). (Courtesy, Shade Tree Laboratory, University of Massachusetts. Inserted drawing courtesy of John A. Davidson, University of Maryland.)

Figure 2.2.   Skeletonizing damage (center and apex of bottom leaf) caused by the cottonwood leaf beetle (*Chrysomela scripta*). (Courtesy Shade Tree Laboratory, University of Massachusetts.)

discoloration, and twig breakage). Ambrosia beetle damage is distinct because of the blue or black color of their tunnels in wood, which is due to the symbiotic fungus that lines their tunnels (Fig. 2.6).

Internal examination of wood can provide evidence of damage that is often characteristic and easily associated with a particular insect group. For example, a type of damage known as honeycombing is characterized by extensive irregular galleries in sapwood and heartwood. The most com-

Leaf Mine

Figure 2.3.   Illustration of one of the many serpentine
leaf mines found on the foliage of forest and
shade trees.

mon honeycombing species are carpenter ants and termites. Unlike the
carpenter ants, termites feed on the wood they excavate, and their tunnels
are often caked with a mudlike material. In general, gallery walls produced
by carpenter ants are thicker than those produced by termites and lack the
mudlike material. In both cases, the structural integrity of the wood is
severely impaired (see chapter on wood borers).

Internal abnormalities can indicate either feeding damage or a tree's
response to feeding. Ring distortions include separations of growth zones
by resin flow (resinosis) or bark formation (healing responses). These may
result from the activity of insects such as bark beetles. Smaller separations,
so-called gum spots, are often made in response to bark miners or flatheaded borers. Cambium-mining maggots (Diptera) and feeding weevils
form scars in the wood that result in defects called checks, flecks, or pith
flecks (Figs. 2.7 and 2.8).

Another type of damage that undermines the physical integrity of
seasoned wood results from the feeding of various beetles and is referred
to as powder posting. Powder-post beetles (Anobiids, Bostrichids, powder-post weevils, etc.) often infest wood, rendering it a powdery and
fragile material.

Figure 2.4.   Example of stippling damage on leaf on American sycamore (*Platanus occidentalis*) caused by feeding punctures of *Erythroneura* spp. leafhoppers. (Courtesy of M. McClure.)

Wood borers produce and extrude *frass*, a combination of wood shavings and fecal pellets, through holes in the wood surface. The presence of frass, which ranges from the fine powder produced by the powder-post beetles to the sawdustlike material produced by pine sawyer beetles, can serve as a "fingerprint" in species identification (Solomon, 1977) (see Fig. 16.11; chapter on detection, survey, and surveillance).

The shelters constructed or induced by insects are often related to their feeding. To varying degrees they may also provide useful information for the identification of the inhabitants. Larval webs or tents are often constructed of silk and foliage (Figs. 2.9a and 2.9b). Other easily recognizable, protective structures are formed by bagworms (*Thyridopteryx ephemeraeformis*) and casebearers (Fig. 2.10). Leaf tiers form simple

Figure 2.5. *Ips* sp. galleries etched on the wood surface. (Courtesy Shade Tree Laboratory, University of Massachusetts.)

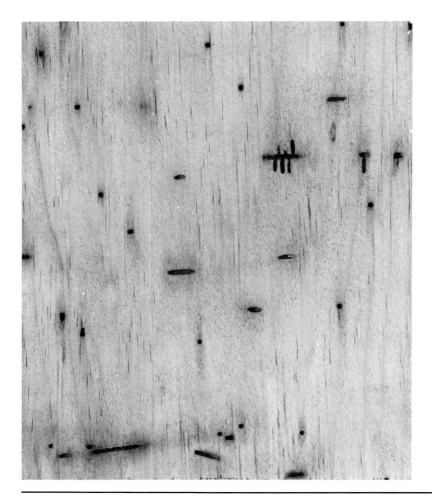

Figure 2.6.   Portions of fungus-lined tunnels of ambrosia beetles as they occur on the surface of cut lumber. (Courtesy Shade Tree Laboratory, University of Massachusetts.)

structures, within which they feed, by either joining two single leaves (Fig. 2.11), or two or more needles, together with silk. Similarly, leaf rollers fold or roll leaves together to form a shelter (Fig. 2.12). Finally, in some species, secretions form what appear to be protective coverings, for example, spittle masses and treehopper egg mass coverings. Spittle masses are frothy secretions produced by species such as the saratoga spittlebug (*Aphrophora saratogensis*), the pine spittlebug (*A. parallela*), and others (Fig. 2.13).

Figure 2.7.   The appearance of injury due to attack by cambium miners in the cambium region. (Courtesy USDA Forest Service.)

Host Range of Forest and Shade Tree Insects

Dendrophagous species can be specific (monophagous) in their choice of host tree. For example, the white pine weevil feeds on the meristematic tissue of its host, eastern white pine (*Pinus strobus*); the elm bark borer (*Saperda tridentata*) breeds in corky bark of living elms (*Ulmus americana* and *U. rubra*); the cherry scallop shell moth feeds on wild and choke cherries (*Prunus serotina* and *P. virginiana*, respectively); and the larch casebearer (*Coleoptera larcella*) forms a case from and feeds on larch needles (*Larix laricina* and *L. occidentalis*). Conversely, the variation in the number of host tree species fed upon by any one dendrophagous species can be extensive

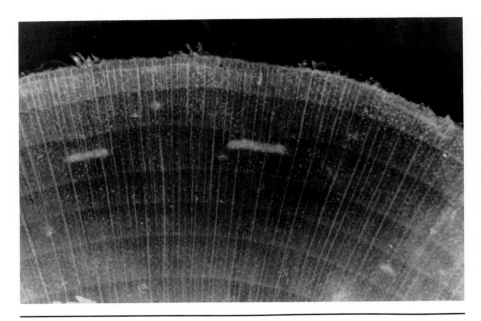

Figure 2.8.   Pith-ray flecks in a cross section of a log represent injury by cambium miners. (Courtesy USDA Forest Service.)

(a polyphagous species). Approximately 458 species of plants are acceptable to the gypsy moth (Forbush and Fernald, 1896). The fall webworm (*Hyphantria cunea*) feeds on well over 100 species of forest and shade trees, and the red turpentine beetle (*Dendroctonus valens*) attacks all pines and occasionally spruces, larches, and firs (Baker, 1972).

Some tree species seem attractive or suitable to more types of insects than others. Oak species are fed upon by over 1000 insect species, of which at least 50 are leaf miners (Frost, 1942). In Britain alone, oak species have 284 insect species associated with them (Table 2.3).

Several species may feed on similar tree tissues. These are often referred to as feeding guilds. White pine, for example, is affected by several feeding guilds that include at least 30 foliage-feeding species, 9 bud-and-twig feeders, 3 cone borers, 19 wood borers, 2 root-boring species, 13 cambium and phloem feeders, and at least 1 bark feeder (Baker, 1972).

Why do some tree species have more insect species associated with them than other tree species? An important hypothesis suggests that the number of insect species associated with a tree is proportional to the tree's recent abundance or its evolutionary history (Table 2.4). Dominant native trees would have the most insect species, whereas recently introduced trees would have the least. A newly introduced tree that is grown

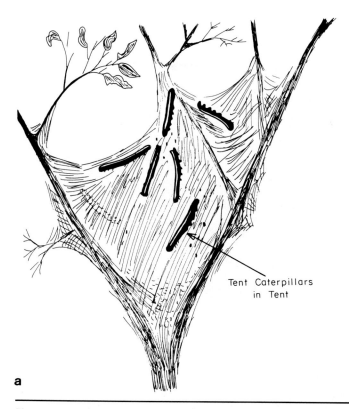

Tent Caterpillars
in Tent

a

Figure 2.9a. Illustration of the tent of the eastern tent caterpillar (*Malacosoma americanum*). (Courtesy of John A. Davidson.)

extensively is likely to be attacked, sooner or later, by several native insects; on the other hand, introduced trees that remain relatively rare, without closely allied species in the native flora, would have few, if any, native insects feeding on them (Southwood, 1960, 1961). For example, widespread plantings of apple (*Malus*) appear to have led to host plant switching by species of mirids and tephritids that adapted from hosts such as willow (*Salix*) and hawthorn (*Crataegus*). These transfers require changes not only in host abundance but also in the insect's tolerance of a new host.

Other researchers argue that the time factor is unimportant, suggesting that extensive geologic or evolutionary time and habitat age are not required for insect species richness (Strong, 1974a,b; Strong *et al.*, 1977). Instead, the leveling-off point of rapid asymptotic species accumulation is set by the range of the host tree in a region, and most pest species are

Figure 2.9b. Nest of pine webworm (*Tetralopha robustella*), which incorporates the insect's frass. (Courtesy Shade Tree Laboratory, University of Massachusetts.)

recruited from the native fauna. These researchers point out that although native trees of Britain have been associated with their insects for the duration of the Quaternary, the number of insects associated with some tree species introduced only 300–400 yr ago is indistinguishable from that of the native trees. Finally, Blaustein *et al.* (1983) indicate that (a) the

Figure 2.10.   Larval cases of the evergreen bagworm, developed after feeding on various host plants: (A) cypress and beech; (B) honey locust (*Gleditsia triacanthos*); (C) cypress and linden (note use of linden petioles rather than leaves); (D) cypress, linden, and eastern white pine (note linden fruit); (E) cypress and plane tree. The lower portion of the bag is ornamented by material from the first host and upper portions by succeeding host plants. (From Santamour, 1980.)

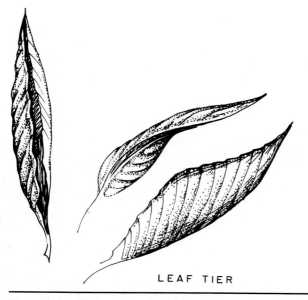

LEAF TIER

Figure 2.11.   Illustrations of one type of leaf tier. (Courtesy A. Bartlett.)

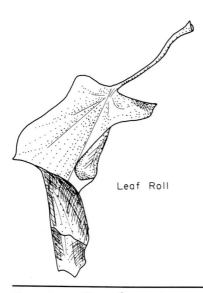

Leaf Roll

Figure 2.12.   Illustration of a simple leaf roll. (Courtesy A. Bartlett.)

Figure 2.13.   Typical spittle mass produced by the pine spittle bug (*Aphrophora parallela*). (Courtesy Shade Tree Laboratory, University of Massachusetts.)

Table 2.3. Comparison of the numbers of
insect species associated with deciduous
and coniferous forest trees in Britain and
European Russia [a]

Tree	Britain	Russia
Oak (*Quercus*)	284	150
Willow (*Salix*)	266	147
Birch (*Betula*)	229	101
Hawthorn (*Crataegus*)	149	59
Poplars (*Populus*)	97	122
Apple (*Malus*)	93	77
Pine (*Pinus*)[b]	91	190
Alder (*Alnus*)	90	63
Elm (*Ulmus*)	82	81
Hazel (*Corylus*)	73	26
Beech (*Fagus*)	64	79
Ash (*Fraxinus*)	41	41
Spruce (*Picea*)[b]	37	117
Lime (*Tilia*)	31	37
Hornbeam (*Carpinus*)	28	53
Larch (*Larix*)[b]	17	44
Fir (*Abies*)[b]	16	42
Holly (*Ilex*)	7	8

[a] From Southwood (1961).
[b] Coniferous; otherwise, trees are deciduous.

geographical range of an introduced host plant can be an important
determinant of the number of pest species it has acquired; (b) native trees
have significantly more insect pests than introduced trees and, thus the
time since introduction may also be an important influence on pest
accumulation rates; (c) generalist pests are probably acquired more quickly
than specialists; and (d) although rapid accumulation of pest species on
some host plants occurs, such a pattern of acquisition is not a general
phenomenon.

Food and Survival

Major shifts in the numerical levels of forest insect populations can result from changes in the quantity and quality of food. The most obvious
changes are those caused by insufficient food. These shortages can be
absolute, for example, in the starvation of slower-developing individuals of a epidemic population whose early-hatching individuals cause
100% defoliation. Or it can be an "effective" shortage, in which seasonal

Table 2.4. Prevalent British trees, their history, and the number of species of associated insects[a]

Tree	History in Britain since Pleistocene Period	Heteroptera	Homoptera (Part)[b]	Macrolepidoptera	Microlepidoptera	Coleoptera	Total
Oak (Quercus robur and Q. petraea)	Native	37	10	106	81	50	284
Birch (Betula spp.)	Native	12	4	94	84	35	229
Hazel (Corylus avellana)	Native	16	2	18	28	9	73
Willow (Salix spp.)	Native	22	20	100	73	51	266
Alder (Alnus glutinosa)	Native	14	8	28	27	13	90
Hawthorn (Crataegus spp.)	Native	17	1	64	53	14	149
Ash (Fraxinus excelsior)	Native	10	2	16	9	4	41
Pine (Pinus sylvestris)	Native	15	3	10	28	35	91
Holly (Ilex aquifolium)	Native	0	0	2	2	3	7
Yew (Taxus baccata)	Native	0	0	1	0	0	1
Sloe (Prunus spinosa)	Native	4	2	48	43	12	109
Poplars (Populus spp.)	Native	8	11	33	26	19	97
Elm (Ulmus spp.)	Native	11	4	33	26	10	82
Beech (Fagus sylvatica)	Native	4	3	24	16	17	64
Common maple (Acer campestre)	Native	2	2	8	12	2	26
Hornbeam (Carpinus betulus)	Native	1	2	7	16	4	28
Juniper (Juniperus communis)	Native	6	0	4	8	2	20
Spruce (Picea abies)	Native in interglacial, reintroduced c. 1500	9	1	6	13	8	37
Lime (Tilia spp.)	Native and introduced	7	2	15	5	2	31
Mountain ash (Sorbus aucuparia)	Native	0	1	2	17	8	28
Fir (Abies spp.)	Native in interglacial, reintroduced c. 1600	5	0	2	1	8	16
Sweet chestnut (Castanea sativa)	Introduced A.D. 100	0	0	0	5	0	5
Apple (Malus spp.)	Native and introduced	18	3	21	42	9	93
Walnut (Juglans regia)	Introduced c. 1400	0	0	0	1	2	3
Holm oak (Quercus ilex)	Introduced 1580	0	0	0	2	0	2
Larch (Larix decidua)	Introduced 1629	3	0	6	6	2	17
Sycamore (Acer pseudoplatanus)	Introduced c. 1250	1	0	5	8	0	15
Horse chestnut (Aesculus hippocastanum)	Introduced c. 1600	0	2	1	1	0	4
Acacia (Robinia pseudoacacia)	Introduced 1601	0	0	0	1	0	1
Plane (Platanus orientalis)	Introduced c. 1520	0	0	0	0	0	0

[a] From Southwood (1961).
[b] Includes the suborder Auchenorrhyncha and the subfamily Psylloidea.

biochemical or physical changes of host tissues result in a shortage of appropriate food. Both circumstances cause similar results. Partially starved individuals produce small and premature pupae. Similarly, individuals forced to feed on qualitatively inferior food tend to produce small pupae; adults from small pupae often exhibit reduced fecundity. Thus, the density of forest insect populations can be affected by the quantity and quality of food.

Mortality of many dendrophagous species is determined by the food they eat. Mortality of the nun moth (*Lymantria monacha*) on apple, oak, spruce, and larch is only 2%; but it is 18% on beech, 61% on pine, and 89% on alder (Sattler, 1939). These host-related effects on survival are due to nutritional differences among the various host species and the toxic compounds they contain.

Differences in survival can be attributed to variation in the nutritional value of host tree tissues from species to species. However, differences among herbivores may be responsible for differential host utilization. Larvae of the diprionid sawflies, European sawfly (*Neodiprion sertifer*), redheaded pine sawfly (*N. nigroscutum* or *N. lecontei*), and introduced pine sawfly (*Diprion similis*) feed and develop on *Pinus* species such as jack pine (*P. banksiana*), red pine (*P. resinosa*), Scots pine (*P. sylvestris*), and white pine. However, a large proportion of their food (78–88%) is not utilized even though only 33–44% of the dry foliage is fiber. Another factor that contributes to the indigestibility of food is starch content, which comprises up to 14% of 1-yr-old foliage (Fogal, 1974). The degree to which host-tree foliage can be utilized varies within a host taxon and depends on the herbivore (Table 2.5).

In addition to overall differences among host species and herbivores, host tissues vary in their quality and their value to insect herbivores. Second instar spruce budworms mine needles, feed on staminate cones, or enter compact vegetative buds. When buds begin to swell, larvae move to old or new foliage or flowers, depending on population numbers, tree phenology, and so on. The behavioral feeding sequence is a result of tissue suitability and availability. Second instars prefer to feed on the newly opened staminate flowers of balsam fir. If these are not available, they mine either into 1- or 2-yr-old needles or into expanding vegetative buds. Consumption of pollen speeds larval development. Larvae that feed on flowers have a developmental advantage of approximately 3 days over larvae feeding only on foliage (Blais, 1952). This can have an impact on numerical change since the average fecundity is four to six eggs lower for each day that development is extended (Greenbank, 1956). This developmental advantage as well as oviposition and overwintering preferences for flowering trees, lead to the enhancement of population numbers and thus contribute to the initiation of outbreaks (Blais, 1952).

Nonnutritional chemicals produced by the tree also affect the survival, growth, vigor, behavior, or population biology of herbivore species

Table 2.5. The digestion and assimilation efficiencies of tree-feeding Lepidoptera[a]

	AD		ECI		ECD	
	High	**Low**	**High**	**Low**	**High**	**Low**
White oak (*Quercus alba*)[b]	46.14	22.87	18.51	6.03	83.38	17.14
	$\bar{X} = 36.02$		$\bar{X} = 11.75$		$\bar{X} = 36.48$	
Red oak (*Q. rubra*)	36.67	25.97	22.20	3.59	81.56	11.66
	$\bar{X} = 31.26$		$\bar{X} = 9.76$		$\bar{X} = 34.42$	

[a] From Barbosa and Greenblatt (1979a). Based on data from Scriber (1975). AD, approximate digestability; ECI, efficiency of conversion of ingested food; ECD, efficiency of conversion of digested food.
[b] Calculations based on the feeding of the penultimate and last instars of *Antheraea pernyi*, *Eacles imperialis* (imperial moth), and *A. polyphemus* (polyphemus moth).

feeding on it. These chemicals are sometimes referred to as secondary plant substances or allelochemicals. They probably play a role in determining the number of insect herbivores feeding or ovipositing on a tree and/or how successfully its various tissues (e.g., leaves, bark, and roots) can be utilized by the herbivore. The biological principles explaining the limits of host range and host tissue utilization are only partially understood but surely include a number of the above factors. Not only are nutritional biochemicals and allelochemicals (repellants, inhibitors, attractants, and toxins) involved, but visual stimuli, physical traits, and microclimatic conditions also determine the character of these insect–tree interactions.

Effects of Nitrogen and Water Content on Host Plant Utilization

Many researchers have emphasized nitrogen and water content in host-tree tissues as the critical determinants of herbivory. Nitrogen is a limiting factor for insects because protein, the structural base of insect tissues, is 7–14% nitrogen, whereas their food averages only 2% nitrogen and rarely reaches 7%. Thus, a dendrophagous species' ability to ingest, digest, and assimilate plant nitrogen is a key component in its survival and potential impact on a forest stand or shade tree.

Available data suggest that woody plants have significantly lower moisture content and nitrogen levels than nonwoody plants. Associated with these deficiencies are high concentrations of tannins, fiber, lignins, resins, and so on. Low leaf-water and/or nitrogen content, along with such digestibility reducers as tannins, lignin, and resins, may be the fundamental factors limiting the extent of herbivory by dendrophagous species. Indeed, although forb-feeding species may be capable of digesting an average of 53% of their food, dendrophagous species digest only an average 39% of their food. Similarly, the efficiency of conversion of ingested forbs (ECI) is generally greater than 20%, compared with an ECI of less than 20% for those species feeding on woody plants. Current research is focusing on the effects of changes in nitrogen and water content on the population dynamics of insects as reflected in larval growth rates, survival, and fecundity (Mattson, 1980; Scriber, 1982).

The larval food source can have a dramatic impact not only on subsequent adult individuals but even on progeny of the following generation. For example, oviposition, fecundity, and eclosion (egg hatch) of the spear-marked black moth (*Rheumaptera hastata*) is affected by the food source of the larvae (Table 2.6). Finally, satin moth (*Leucoma salicis*) larvae from parents fed on eastern cottonwood (*Populus deltoides*) weigh more at eclosion, develop more rapidly, have higher survival, and gain more weight than those from parents fed on big tooth aspen (*P. grandidentata*) (Wagner and Leonard, 1979).

Table 2.6. Influence of larval food source on oviposition, fecundity, and eclosion of the spear-marked black moth (*Rheumaptera hasta*)

| Food Source | Female Survival (%) | Mean No. of Eggs Produced per Female (±SD)[a] | | Mean No. of Eggs Deposited per Female (±SD)[a] | | | Larval Hatch[b] (%) | | |
		Laid	Unlaid	Birch	Alder	Willow	Birch	Alder	Willow
Birch	80.0(60)a[c]	83 ± 16.4a	15 ± 4.5d	65 ± 10.3a	12 ± 3.2d	6 ± 1.6d	91	67	62
Alder	48.0(36)b	71 ± 17.8a	21 ± 5.2c	38 ± 11.4b	21 ± 6.1c	12 ± 4.1d	66	31	28
Willow	44.0(33)b	45 ± 9.6b	33 ± 7.4bc	31 ± 8.6b	11 ± 2.2d	3 ± 1.1d	49	38	34

[a] Values followed by the same letter do not significantly differ at the 5% level of probability. From Werner (1979).
[b] A measure of egg viability.
[c] Number in parentheses is the number of female moths observed during oviposition.

Nutritional Quality and Herbivore Survival

As noted in previous discussions, host tissues available to dendrophagous species are rarely comparable in their nutritional quality or in their suitability. Such variation is an important aspect of the interaction between dendrophagous species and their tree hosts. Even different portions of the wood of an individual tree can differ nutritionally. Changes in nutrient concentrations occur as a tree conserves nutrients by translocating them out of wood as it becomes heartwood (dead wood) or out of leaves about to fall.

Table 2.7.   Chemical composition of tissues of various plants in the Brookhaven oak–pine forest [a,b]

Element and Species	Stem Heartwood	Stem Sapwood	Stem Bark	Stem Inner Bark	Stem Outer Bark	Stemwood and Bark	Branch Live Wood and Bark
Nitrogen							
Quercus coccinea[c]	1.40	1.00	3.19	—	3.15	4.62	—
Q. alba[d]	1.88	1.53	4.15	—	2.80	5.59	—
Pinus rigida[e]	0.97	0.87	2.28	—	1.83	5.76	—
Quercus ilicifolia[f]	—	1.18	5.64	—	3.98	7.74	—
Gaylussacia baccata[g]	—	2.50	4.68	—	4.65	7.06	—
Vaccinium vacillans[h]	—	2.64	5.53	—	4.72	6.30	—
V. angustifolium[i]	—	—	—	—	5.38	8.06	—
Kalmia angustifolia[j]	—	3.38	4.30	—	4.70	6.00	—
Tree sprouts	—	1.27	3.67	—	2.97	5.10	—
Potassium							
Quercus coccinea	0.61	1.36	1.47	2.60	0.59	—	2.07
Q. alba	0.73	1.16	1.82	2.40	1.09	—	2.43
Pinus rigida	0.24	0.49	1.23	4.84	0.11	—	1.13
Quercus ilicifolia	—	1.28	1.96	—	—	—	1.69
Gaylussacia baccata	—	1.34	1.49	—	—	—	1.45
Vaccinium vacillans	—	2.47	2.18	—	—	2.70	2.45
V. angustifolium	—	1.82	—	—	—	1.82	1.82
Kalmia angustifolia	—	1.13	1.09	—	—	1.12	1.72
Tree sprouts	—	1.26	1.65	2.50	0.84	—	2.25
Calcium							
Quercus coccinea	0.50	0.52	13.4	18.7	14.7	—	5.20
Q. alba	1.02	0.85	25.1	28.1	35.6	—	5.70
Pinus rigida	1.04	0.81	1.1	1.6	1.4	—	1.89
Quercus ilicifolia	—	0.55	12.1	—	—	—	5.15
Gaylussacia baccata	—	0.62	2.3	—	—	—	2.33
Vaccinium vacillans	—	0.67	5.6	—	—	3.13	5.94
V. angustifolium	—	3.26	—	—	—	3.26	3.26
Kalmia angustifolia	—	0.53	3.2	—	—	1.13	1.77
Tree sprouts	—	0.69	19.3	23.4	25.2	—	5.45

[a] Modified from: Nutrient concentrations in plants in the Brookhaven oak–pine forest. G. M. Woodwell, R. H. Whittaker, R. A. Houghton, *Ecology* **56**: 318–332 (1975). Copyright © 1975 by Duke University Press. Reprinted by permission.
[b] Data are in mg/g oven dry tissue.

[c] Scarlet oak	[g] Huckleberry
[d] White oak	[h] Blueberry
[e] Pitch pine	[i] Dwarf blueberry
[f] Bear oak	[j] Laurel

The patterns of nutrient concentration are similar for most tree species. Leaves, flowers, and fruits have the highest nutrient concentrations, and heartwood has the lowest. Concentrations in branches, twigs, bark, roots, and sapwood are intermediate (Table 2.7). Many insects as well as some of their relatives (mites, for example) are negatively affected by mineral deficiencies in their host plants. Parameters that can be affected include survival, growth, fecundity, and behavior (Taylor *et al.*, 1953; Allen and Selman, 1955; Cannon and Terriere, 1966; Maddox and Rhyne, 1975). Thus, shifts in the concentration of nutrients in trees must affect the numbers and types of insects found on trees.

Even the differences in the nutritional quality of tissues adjacent to those subjected to herbivory may be important. For example, a number of

Current Twigs	Leaves	Fruits	Flowers	Root Wood	Root Bark	Small Whole Roots	Taproot	Roots: Mean
—	7.90 ± 2.40	4.97 ± 0.95	14.50 ± 3.1	—	—	—	—	2.02
—	7.90 ± 2.10	6.43	26.32	—	—	—	—	3.42
—	9.00 ± 0.53	1.30 ± 0.16	11.30	—	—	—	—	2.00
—	14.00 ± 1.24	—	—	—	—	—	—	1.88
—	77.50 ± 1.35	5.97	17.33	—	—	—	—	3.68
—	5.99	5.97	17.33	—	—	—	—	4.03
—	6.16	—	—	—	—	—	—	4.20
—	9.28 ± 0.62	—	—	—	—	—	—	6.10
—	7.90	—	—	—	—	—	—	—
2.53	5.15	6.60	19.71	2.43	2.28	2.74	—	2.74
3.95	7.04	7.77	17.85	1.83	2.41	2.42	—	2.42
3.61	3.19	0.50	10.86	1.35	0.82	2.79	0.66	2.79
2.66	4.19	—	—	—	—	2.18	—	2.18
2.47	6.72	7.19	16.10	—	—	1.51	—	1.51
2.85	4.72	7.19	16.10	—	—	1.34	1.34	1.34
2.58	4.26	—	—	—	—	1.28	—	1.28
3.25	3.23	—	—	—	—	2.26	—	2.26
3.24	6.10	—	—	—	—	—	—	—
5.15	3.88	6.60	5.55	0.48	8.76	2.60	—	2.60
6.44	4.91	1.83	1.61	1.33	16.26	3.38	—	3.38
1.16	2.18	0.12	0.30	0.42	0.53	0.98	0.30	0.98
5.78	5.09	—	—	—	—	2.90	—	2.90
7.90	8.45	4.22	2.49	—	—	2.11	—	2.11
7.98	8.45	4.22	2.49	—	—	0.88	1.28	0.88
4.81	6.88	—	—	—	—	2.05	—	2.05
3.85	8.28	—	—	—	—	1.39	—	1.39
5.80	4.40	—	—	—	—	—	—	—

gall-producing aphid species attack adjacent foliar tissues of poplar. Three closely related species induce galls on *Populus nigra* (black poplar): *Pemphigus bursarius* (lettuce root aphid) and *P. spirothecae* (spiral gall maker) induce galls on the petiole, and *P. filaginus* induces a pouchlike gall on the leaf blade. No study has been made of the possible role of nutrient concentration in site selection or aphid development, even though in other tree species four- to five-fold differences in nitrogen concentration occur between the blades and the petioles of leaves (Auchmoody, 1974). Although an alternative explanation might be differential concentrations of allelochemicals (see later discussion), in many species a correlation also exists between the distribution of nitrogen and larval development. Such differences may result in different responses of plant tissues to gall makers.

The distribution of nitrogen may differ with tree age, height, and radial area, and time of year. Analyses of nitrogen content of a tree at any time or sampling height may vary from species to species. Generally, however, in softwood species such as pine or spruce, the protein content (which is correlated to nitrogen content) is highest near the bark, decreases toward the center of the tree, and is lowest in the outer heartwood. Larval development and weight gain of the common furniture beetle (*Anobium punctatum*) is best in the outer zone of its host (Sitka spruce) and decreases toward the pith. The gradient in nitrogen content parallels and is significantly correlated to the differences in the suitability of wood as a food source (Bletchly and Taylor, 1964).

The duration of the life cycle of wood-boring species is often correlated to the nutritional value of the wood they feed on. Species feeding on phloem and mining the inner bark usually have shorter life cycles than those that bore into and feed on inner (sapwood and heartwood) tissues. Heartwood feeders often have life cycles with larval development periods of 1, 2, or more years.

One must always keep in mind that the utility of tree tissues is not necessarily determined by the concentration of one element or nutritional compound. The suitability of food is affected, to varying degrees, by the quantity and quality of fats, mineral salts, resin, pentose and hexose sugars, tannins, terpenes, and various other nonnutritional chemical compounds. Thus, the removal of resin from Scots pine and tannin from a European oak (*Quercus* sp.), for example, is beneficial to the survival and development of larval *Anobium punctatum* (Bletchly and Farmer, 1959).

The quality of a host tissue (such as leaves or bark) often varies within a single living tree. The suitability of tissues and feeding sites in the crown, for example, can vary significantly. The small European elm bark beetle (*Scolytus multistriatus*) exhibits feeding preferences for twig crotches (Fig. 2.14) in the outer and upper crown region of elms (Table 2.8). Although no data are available on the nutritional quality of twig tissues, beetle preferences probably reflect twig quality because as available twig

Figure 2.14.   Twig crotch-feeding by the smaller European elm bark beetle (*Scolytus multistriatus*). (Courtesy Shade Tree Laboratory, University of Massachusetts.)

Table 2.8.   Feeding attack of the European elm bark beetle (*Scolytus multistriatus*) at three heights of six mature elm trees[a]

	Height Level (m)[b]		
Tree	5	10	15
A	0	2.9	4.5
B	9.3	34.8	62.1
C	24.0	84.2	92.6
D	17.8	81.7	83.0
E	27.2	65.2	81.7
F	0.4	27.8	63.0

[a] Modified from Riedl and Butcher (1975).
[b] Average percent attack.

crotches in the upper levels are used, new arrivals are forced to move down into the middle crown region (Riedl and Butcher, 1975).

Crown position of foliage can also affect the population dynamics of many insect species. Significant differences in head capsule width, pupal weight, and potential fecundity are found in Douglas-fir tussock moths

Table 2.9. Pupal weights and egg potential of laboratory-reared Douglas-fir tussock, moth (*Orgyia pseudotsugata*)[a]

Species	Foliage Source	Stress[b]	Pupal Weight (mg) ♂	♀	No. of Eggs
Grand fir	Top	None	171.2	597.2	346.0
		Stress	155.0	445.7	270.4
	Bottom	None	158.7	487.8	297.2
		Stress	151.0	423.3	248.7
Douglas-fir	Top	None	177.8	593.5	334.2
		Stress	167.7	483.9	292.0
	Bottom	None	170.6	538.0	323.7
		Stress	144.6	334.8	187.0
Subalpine fir	Top	None	175.2	572.2	335.0
		Stress	147.2	472.2	292.0
	Bottom	None	162.0	509.7	271.3
		Stress	128.2	350.0	214.3

[a] From Beckwith (1976).
[b] Simulated by forcing larvae to feed on old growth of foliage.

feeding on foliage from the top of the crown compared with those feeding on lower crown foliage (Table 2.9). These differences may reflect variation in the chemical profile of foliage from various points in the crown.

Foliage of quaking aspen (*Populus tremuloides*) shows differences in element concentration due to crown position. The concentrations of calcium and potassium in the upper third of the crown can be consistently lower than those in the middle or lower third of the crown. Variation in elemental composition is presumed to reflect differences in the nutritional quality of foliage for herbivores. Analyses of elemental concentrations in red oak (*Quercus rubra*), white oak (*Quercus alba*), chestnut oak (*Q. prinus*), and scarlet oak (*Q. coccinea*) provide other illustrations. Leaf dry weight generally decreases from terminal through interior positions and lower crown for all species. Concentrations of potassium, calcium, and magnesium differ among crown positions; generally, higher concentrations occur in interior positions compared with upper crown and terminal locations. No appreciable change due to crown position occurs in nitrogen, phosphorus, manganese, or iron concentration. Differences can be found in foliar weight and nutrient composition among the four species. Red oak, for example, has the heaviest leaves and the highest nutrient content (for all seven elements) and exhibits the greatest positional differences. On the other hand, white and scarlet oak have low foliage weights and nutrient content compared with red and chestnut oaks (Auchmoody and Hammack, 1975).

Seasonal Changes in Nutritional Quality

Seasonal changes in host trees often consist of chemical changes associated with aging (Table 2.10). For example, leaf starch content of sugar maple, (*Acer saccharum*) decreases from the time that leaves are fully expanded to the time they are becoming senescent (in New England, approximately late June to mid-September). During the same period, leaf sugar and the starch content of the current year's stems increases. These seasonal changes incorporate and overlap patterns of daily changes. For example, sugar concentrations of the maple leaves are at a minimum in early morning and reach their maximum in late afternoon or early evening (Donnelly, 1976). These and other changes represent the storage and mobilization of compounds during growth and nongrowth periods as well as qualitative changes in photosynthetic tissues and hormone-target tissues. Patterns of seasonal change for different tissues (i.e., buds, shoot apices, leaves, and phloem) vary in any given species (Figs. 2.15–2.17).

The leaves of trees increase in toughness as they age (Fig. 2.18). In addition, age-related biochemical and water content changes parallel those in leaf texture and may be an important reason for the preponderance of early season lepidopterous defoliators (Table 2.11). Many insects manifest different survival, size, and fecundity on juvenile versus mature leaves (Fig. 2.19). The elm spanworm (*Ennomos subsignarius*) prefers and exhibits enhanced survival on juvenile hickory and oak leaves (Drooz, 1970a,b; Clark *et al.*, 1975). Similarly, spruce budworm females reared as larvae on late season current-year foliage of balsam fir produce fewer eggs than larvae reared on early season foliage (Table 2.12). Larvae feeding on old foliage also have a higher mortality rate and slower development. Field

Table 2.10. Seasonal variations in carbohydrate (CHO) and crude fat (CF) content in needles of balsam fir (*Abies balsamea*)

Needle Age	Period		Net Change in CHO and CF Content[b]
	Start	**End**	
Current	28 May	5 Aug	+6.4
	5 Aug	9 Sept	−1.9
	9 Sept	2 Dec	+6.6
1 yr	20 Mar	28 May	+12.9
	28 May	22 July	−8.2
	22 July	5 Aug	+2.7
	5 Aug	9 Sept	−4.0
	9 Sept	2 Dec	+3.0

[a] Modified from Little (1970).
[b] Change in oven-dry weight, Δ DW, in percent.

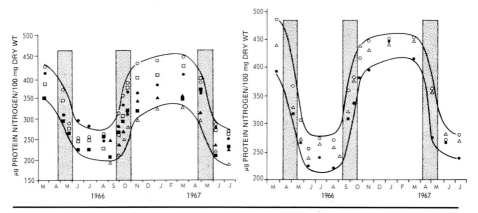

Figure 2.15. Seasonal variations in protein nitogen of extracts of bark (left) and needles (right) of red pine (*Pinus resinosa*). Each symbol represents an average of results from one tree. (The stippled regions are not relevant.) (From Pomeroy *et al.*, 1970.)

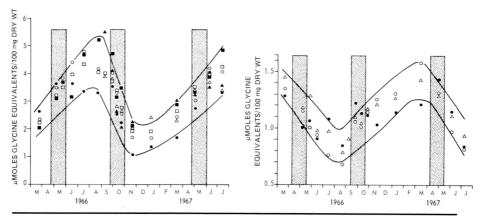

Figure 2.16. Seasonal variations in amino acids of bark (left) and needles (right) of red pine (*Pinus resinosa*). Each symbol represents an average of results from one tree. (The stippled regions are not relevant.) (From Pomeroy *et al.*, 1970.)

populations that reach epidemic proportions and destroy all the current-year growth prior to the completion of larval development produce females with reduced reproductive potential (Blais, 1952). Food consumption by larvae feeding on younger current-year foliage is greater than that of larvae feeding on older current-year foliage. In other species, such as aphids, increased consumption is a common behavior of individuals feeding on nutritionally poor food (Dixon, 1963).

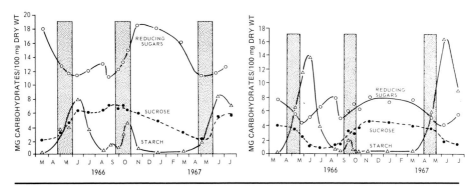

Figure 2.17.   Seasonal variation in carbohydrates of bark (left) and needles
(right) of red pine (*Pinus resinosa*). Each curve represents an average of results
from at least three trees. (The stripled regions are not relevant.) (From
Pomeroy *et al.*, 1970.)

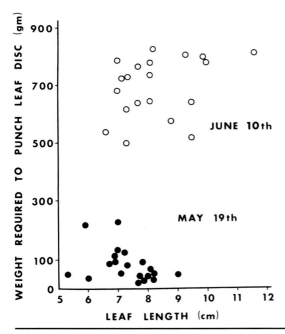

Figure 2.18.   Seasonal changes in the toughness of oak leaves: ○, June 10; ●,
May 19. (From Feeny, 1970.)

Analysis of new and old foliage of balsam fir demonstrates some
important variation in nutrient content. The amino acid concentration
of new foliage of both flowering and nonflowering trees of balsam fir
and white spruce was greater than in old foliage of comparable trees
(Table 2.13). Current-year foliage of flowering balsam fir had the highest

Table 2.11.   Comparison of larval feeding habits of early-feeding (June or earlier) and late-feeding (July or later) Lepidoptera species on oak leaves[a]

Feeding Habit (Not Mutually Exclusive)	Percentage of Early-Feeding Species (Total 111)	Percentage of Late-Feeding Species (Total 90)
Larva completes growth on oak leaves in one season (excluding leaf miners)	92	42
Larva completes growth on low herbage, after initial feeding on oak leaves	3	11
Larva overwinters and completes growth in following season	4	38
Larva bores into parenchyma (leaf miner)	3	26

[a] From: Seasonal Changes in oak leaf tannins and nutrients as a cause of spring feeding by winter moth caterpillars. P. Feeny, *Ecology*, **51** : 565–582 (1970). Copyright © 1970 by Duke University Press. Reprinted by permission.

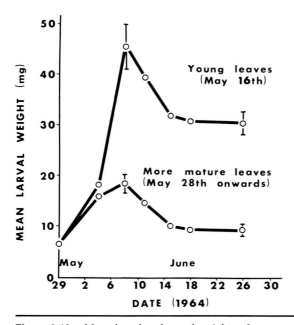

Figure 2.19.   Mean larval and pupal weights of groups of 25 fourth instar winter moths reared on young and on more mature oak leaves: upper curve, young leaves (May 16); lower curve, more mature leaves (May 28 onward). (From Feeny, 1970.)

Table 2.12.   Comparison of fecundities of spruce budworm reared on early and late current foliage[a]

Diet	No. of Budworm Females	Mean Fecundity	Mean Difference
Early foliage	39	168.9	
Late foliage	28	102.9	66.0

[a] From Greenbank (1956).

levels of the majority of the amino acids examined (Kimmins, 1971). The importance of flowering trees to the budworm has already been discussed.

Foliage also changes as the season progresses, and these changes may have an effect on insects. Fall webworms reared on early-, middle-, and late-season apple (*Malus sylvestris*) foliage exhibit significant differences in pupal weight and adult fecundity. Larval and pupal survival of larvae fed on middle-season foliage is one-fourth that of larvae fed on early-season foliage (Morris, 1967). The major effect of aging foliage is in the mean number of eggs laid per female (Table 2.14). In addition to the developmental differences shown, larvae reared on late-season foliage are disoriented and tend to wander off their food. Not only are larvae affected by changes in foliage quality, but the effects of improper nutrition are transmitted to subsequent generations, particularly if they also are nutritionally stressed. The effects of parental nutrition are reflected in the reduced viability of eggs and the reduced ability of first instars to become established on food.

Seasonal changes are also found in the nutritional quality of two elm spanworm hosts in the southern Appalachian region of the United States: pignut hickory (*Carya glabra*) and southern red oak (*Quercus falcata*). The majority of lipids in the foliage of both species undergo significant changes in concentration as leaves mature. Fatty acid and sterol concentrations in hickory foliage decrease with maturation. In oak foliage, although most lipids initially increase in concentration, lipid concentrations decrease during the period of senescence. Unsaturated fatty acids in both species comprise approximately 68–70% of the total fatty acids. There can be an overall decrease of approximately 3% in hickory and 12% in oak in unsaturated fatty acids with leaf maturation (Clark et al., 1975). Developmental changes associated with leaf maturity already have been discussed.

Russian studies show a connection between reserves of fat and nitrogenous matter in insect tissues and the biochemical composition of the food consumed. Fat content in the green tortrix moth (*Tortrix viridana*), the winter moth, the gypsy moth (Fig. 2.20), the lackey moth (*Malacosoma neustria*), the northern birch sawfly (*Croesus sepentrionalis*), the satin moth

Table 2.13. Comparison of mean amino acid concentrations in current and old foliage of flowering and nonflowering balsam fir and white spruce, two hosts of the spruce budworm[a,b]

Amino Acids	Balsam fir				White spruce			
	Flowering Trees		Non flowering Trees		Flowering Trees		Non flowering Trees	
	Current[c]	Old[d]	Current	Old	Current	Old	Current	Old
Alanine	29.62	8.56	26.13	10.06	19.75	4.50	20.06	6.56
Arginine	26.38	11.75	23.50	12.19	18.25	7.38	18.13	7.97
Aspartic acid	6.00	3.63	5.69	2.44	5.00	1.88	4.37	2.00
Cystine	10.38	0.06	9.63	Trace	5.38	0.02	3.38	Trace
Glutamic acid	1.00	0.00	1.13	Trace	0.38	Trace	0.13	Trace
Glycine	19.00	5.13	15.00	3.19	11.88	2.69	10.75	2.78
Histidine	12.38	3.19	9.13	2.78	6.88	1.31	6.63	1.44
Leucine[e]	22.13	9.19	25.56	12.53	16.50	5.63	19.00	8.99
Lysine	23.75	9.56	24.00	10.22	16.25	5.81	16.25	6.81
Valine[f]	7.00	3.88	7.44	4.69	6.00	2.88	5.94	3.41
Proline	17.75	8.38	17.06	8.31	14.13	5.81	15.75	6.84
Serine	15.00	5.31	11.75	3.97	9.75	2.88	8.69	3.06
Threonine	18.38	5.50	13.19	4.19	11.13	3.00	9.50	3.59
Tyrosine	10.50	3.75	8.94	3.56	6.00	2.25	5.63	2.56

[a] Modified from Kimmins (1971).
[b] Data are in mg/g oven dry needles.
[c] 1968 foliage.
[d] 1967 and 1966 foliage combined.
[e] Includes isoleucine and phenylalanine.
[f] Includes methionine and tryptophan.

Table 2.14.   Weight, longevity, and reproductive potential of fall
webworm (*Hyphantria cunea*) individuals reared on early-, middle-, and
late-season foliage of apple[a]

Series	Foliage Type		
	Early	Middle	Late
Weight per male (mg)	140	121	90
Weight per female (mg)	169	135	110
Longevity per male (days)	8.3	8.9	8.0
Longevity per female (days)	8.9	9.2	7.2
Eggs per female	604	372	128

[a] From Morris (1967).

(*Leucoma salicis*), and *Diphtera alpium* coincides with fluctuations in carbohydrate content of leaves. The accumulation of total nitrogen is also directly
related to nitrogen content of their food. The most significant changes
occur in those species that feed on hosts in which sugar content increases
sharply during tree growth, with an equally sharp drop in nitrogen
content. *D. alpium*, which develops in the second half of the summer,
exhibits little change in the amount of body fat and nitrogenous matter
because the mature leaves on which it feeds have a constant level of sugars
and nitrogenous material. The satin moth, which feeds on foliage with a
constant nitrogen level and an increasing sugar content, exhibits little
change in nitrogen content with age, but does exhibit large increases in fat
reserves. The birch sawfly (*Arge pectoralis*) has two generations, one in
early summer and one in late summer. Although there is no significant
qualitative difference among individuals of the first or the second generation, there are major differences between the generations; the larvae and
food of the first generation contain more nitrogen and less sugar, respectively, than those of the second generation (Edelman, 1963).

Simple and complex sugars are translocated in the phloem tissue.
Other compounds may occur in host trees, including terpenes, vitamins,
growth regulators, amino acids, and other organic materials. These compounds vary seasonally and probably diurnally. Seasonal changes in
attraction of bark beetles to their hosts may be related to seasonal changes
in phloem (Pitman, 1966). Some evidence exists that soluble sugar concentration of the phloem in species such as ponderosa pine are important
variables in pheromone production by males of species such as the
California fivespined *Ips* (*Ips paraconfusus*).

Amino acid and overall nitrogen levels in leaves generally exhibit a
seasonal decline in many tree species. Similar declines in white spruce
(Durzan, 1968), "sycamore'" (*Acer pseudoplatanus*) (Fig. 2.21), and Sitka

Figure 2.20. Female gypsy moth (*Lymantria dispar*) on bark surface. (Courtesy Shade Tree Laboratory, University of Massachusetts.)

spruce (Parry, 1974) result in the decreasing suitability of food for their insect associates, including the sycamore aphid (*Drepanosiphum platanoides*) and the green spruce aphid (*Cinara fornacula*). The limiting factor in the food of aphids is often the level of amino-nitrogen. Most aphids feed on phloem sap which contains high concentrations of sugars but low concentrations of amino-nitrogen. The quantity and quality of amino-nitrogen changes with the growth and maturation of leaves and shoots and thus can significantly influence these dendrophagous species (Dixon, 1970).

The number of generations per year of some insect species is determined by the onset of the leaf aging process which, in turn, is affected

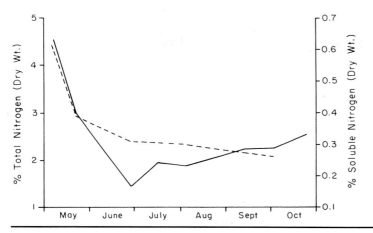

Figure 2.21.   The soluble (——) and total (-----) nitrogen content of sycamore leaves collected in the field during 1959. (From Dixon, 1963.)

by a complex of edaphic and other factors such as plant biochemistry. The Chinese oak silkworm (*Antheraea pernyi*) is a multivoltine species. In southern Ontario, Canada, it has only a single generation each year; but if larvae are fed on leaves from an area 250 miles to the south, this saturniid completes two to three successive generations between May and mid-October (Mansingh, 1972).

Finally, successional shifts in nutrient concentration may result in host quality changes appropriate for exploitation by various types of insects. Analyses of nutrient concentrations in forests of the northeastern United States indicate that high nutrient concentrations occur in early successional herbs and shrubs (Woodwell *et al.*, 1975). Decreasing nutrient concentrations through the shrub stage reach the lowest overall levels in the early succession trees (mainly *Pinus*), and intermediate concentrations occur in late succession trees (i.e., in *Quercus* species). More research is needed in a greater variety of forest types before valid hypotheses can be formulated relating successional stage, nutritional composition, and susceptibility to colonization by dendrophagous species.

Although there are numerous sources of variation in nutritional concentrations, only a few have been illustrated here. Trends in variation in nutrients due to various parameters are summarized in Table 2.15.

Alleochemicals and Herbivory

Trees, like other plants, contain a variety of chemical compounds in addition to the nutrients used by dendrophagous species. Many of these are by-products of the synthesis of primary metabolic products and thus are often called secondary plant substances or allelochemicals. Although they can serve a multitude of functions, allelochemicals often affect the behavior, physiology, and survival of dendrophagous species. Among the more important groups are the terpenoids, flavonoids, alkaloids, cyanogenic glycosides, and phenolics. Considerably more research is needed on the identity of secondary substances in various tree tissues and their effects on herbivores. The effects of secondary substances cannot be considered in isolation because these substances may act in concert with physical and nutritional characteristics of the tree and affect a variety of organisms other than insects.

Studies of defoliation by the winter moth and other defoliators of oaks in Britain provide a good illustration of the potential role of allelochemicals in the interaction between tree and feeding insect. Most lepidopterous defoliators in Britain occur in the spring (Table 2.11). The differential activity of defoliators is believed to be due, in part, to the changes in phenolic content of oak leaves as they age. Condensed tannins appear in late May and increase in quantity until leaf fall. They rise from 0.19% of fresh leaf weight (0.66% dry weight) in April to 2.40% fresh weight (5.50% dry weight) in September (Feeny and Bostok,1968; Feeney, 1968). The addition of as little as 1% tannin in synthetic diet is sufficient to cause significant reduction in larval growth and pupal weight (Table 2.16). However, changes in tannin content of leaves are also associated with other changes, the effects of which are essentially unknown. Tannin or associated compounds bind available nitrogen, making it unavailable to the feeding insect. Thus, what appears to be an abundance of foliage is in reality a limited food supply. Trees also contain a variety of toxins (Barbosa and Krischik, 1987) which, depending on the herbivore, may defend the plant against injury. The concurrence of the major period of larval feeding and the highest concentrations of nitrogen in foliage during the early part of the season lends support to the hypothesis that nitrogen content may also be an important factor governing early season feeding (Feeny, 1970).

Allelochemical as well as nutritional factors may be involved in the selection of leaves of certain ages. The Swaine jack pine sawfly (*Neodiprion swainei*) and the redheaded jack pine sawfly (*N. rugifrons*) are monophagous sawflies on jack pine. Unlike the elm spanworm, spruce budworm, winter moth (*Operophtera brumata*), and many others, both sawfly species preferentially feed and have their highest survival rate on mature foliage (one or more years old). Current season (juvenile) foliage contains extractable components that act as larval anti feedants and are detrimental to development and survival (All and Benjamin, 1976). Likewise, specific terpenes are responsible for preferential feeding on larch (Wagner *et al.*, 1979; Ohigashi *et al.*, 1981). Anti-feedants probably occur in tree hosts of a wide number of forest insects and may very well be a pest management tool of the future.

The type, number, concentration, and biological significance of allelochemicals is quite variable (Table 2.17). Preceding discussions have

Table 2.15. Selected examples of types and sources of nutrient variation with emphasis on foliar nutrients

Type of Variation	Source of Variation and General Trend
Time and Age	**Between Years**
	Variation between years is significant and often related to climatic parameters (e.g., precipitation and temperature) but varies with species, location, and experimental treatment.
	During the Year (Seasonal)
	Seasonal variation exhibits certain trends that hold true for most tree species (i.e., Ca, Fe, and Mn concentrations increase during the growing season whereas N, P, and K generally decrease). There is some indication that protein amino acids and carbohydrates (e.g., sugars) decrease over time within the growing season.
	During the Day
	Very few data are available. There is some indication that seasonal and daily variation of K is related to transpiration patterns and is thus influenced by extremes of water stress.
	Due to Tree Age
	Variation due to tree age does appear to occur; but rather than being a physiological effect, the changes appear to be due to increasing availability of nutrients in the soil as the stand ages.
	Due to Foliage (or Other Tissue) Age
	Variation due to foliage age can be viewed as an integration of seasonal trends. N, P, and K appear to decrease with increasing age while Ca and Mn increase.
Tree Physiology and Morphology	**Among Tree Species**
	Nutrient content requirements and content vary significantly among tree species. No consistent or ordered patterns can be found. One of the few generalizations that can be made is that conifers are lower in alkali metals than hardwoods.

(continued)

Table 2.15. *(continued)*

Type of Variation	Source of Variation and General Trend
Tree Physiology and Morphology	Due to Crown Class
	Both *no differences* and *significant differences* have been found between dominant and suppressed trees. However, although differences in nutrient concentration exist between crown classes within a stand, differences also exist between trees within the same crown class.
	Due to Position on Tree
	Generalizations are difficult, but foliar Ca frequently decreases in concentration with increasing height in the tree, whereas the reverse appears to be true for N.
	Due to Branch Aspect
	No consistent trend exists in the influence of foliage aspect (side of the tree) on the nutrient concentration of that tree foliage.
	Due to Foliage Exposure
	Variation between sun and shade foliage may result from different photosynthetic activity or to plant moisture relations associated with different light intensity and transpirational stress. Data are too few for generalizations.
	Due to Presence of Fruit
	Lower elemental concentrations are observed in foliage associated with fruit compared with similar foliage without fruit.
	Due to Pathogens and/or Herbivory
	The relation between disease or herbivory and nutrient variation has not been well investigated. Evidence indicates that defoliation does change the quality of foliage and its ability to support herbivore growth. This may be due to changes in nutrient concentration and/or allelochemicals. Carbohydrate stored in roots decreases with defoliation. Total phloem sugars and starch in the first and second year of defoliation, respectively, similarly decreases.

Due to Genetic Variability and Provenance

Differences in nutrient concentrations in different provenances or clones have been reported and are based on the hypothesis that different provenances have different nutrient uptake potentials and thus different nutrient concentrations.

Due to Geographical Location

Influence varies and thus no generalizations are possible.

Due to Differences in Soils

Variation tends to reflect the dissimilarities in nutrient concentration caused by soil differences.

Due to Air Quality (Pollutants)

Data are limited, but some studies indicate reduction in protein, carbohydrates, and lipids on exposure to ozone and sulphur dioxide.

Due to Fertilization or Thinning

Both can cause either increases in nutritional quality due to greater availability of a nutrient (or nutrients) or an apparent lowering due to dilution from increased growth.

Due to Burning

Potentially can volatilize and reduce available nutrients. However there is no significant evidence of a decline in foliage nutrients.

Due to Pesticides

Pesticides reduce photosynthetic efficiency. This reduction is often greatest when the lower leaf surface is sprayed. Effects vary depending on formulation, plant species and physiology, methods of application, and edaphic and environmental conditions.

Tree Location and Environment

Management Practices

Table 2.16.   Comparison of pupal weights attained by
winter moth (*Operophtera brumata*) larvae on diets containing
September oak leaf tannin and on control diets[a]

Diet	Mean Pupal Weight (Day 24) with S.E.M. (mg)[c]
Plain artificial diet (control)	30.3 ± 2.1
Young oak leaves (early May)	30.8 ± 1.2
Diet containing 0.1% oak leaf tannin	26.5 ± 1.8
Diet containing 1.0% oak leaf tannin	22.4 ± 1.1[b]
Diet containing 10.0% oak leaf tannin	21.2 ± 1.6[b]

[a] From Feeny (1968). Reprinted with permission of *Journal of Insect
Physiology*, Pergamon Press, Ltd.
[b] Difference from control diet statistically significant.
[c] S.E.M., standard error of the mean.

focused on the influences of allelochemicals on the survival and growth of
dendrophagous species. Many of the same compounds act as attractants,
repellants, and stimulants, thus influence the behavior that controls
important aspects of feeding, mating, oviposition, and reproduction (See
section on host selection). Thus, allelochemicals are important not just as
toxins but also as behavior modifiers.

It should be noted that the presence of allelochemicals does not
necessarily imply a defense mechanism against insect injury. These
chemical substances are also effective against other stresses including
wood-rotting fungi and other pathogens (Flodin and Fries, 1978; Hare,
1966) as well as other plants (allelopathy) (Lodhi, 1975, 1976, 1978).

Food and Reproduction

The fecundity of an adult insect is, in part or in whole, dependent on its
nutrition during immature stages. Adults of many species of the Hymenoptera, Diptera, and Lepidoptera require only some carbohydrate as an
energy source. Other adult insects require no food at all because the
nutritional resources for activity and egg production are stored during
preimaginal stages. Thus, the impact of qualitative and quantitative
changes in food vary, depending on the dendrophagous species and the
developmental stage exposed to those changes. Numerous examples of
the influence of food on reproduction already have been noted. Most
common among these are host plant differences that affect larval growth
and subsequent pupal size (Tables 2.18 and 2.19). Pupal size, in turn, is
strongly correlated to reproductive potential (Fig. 2.22), particularly in

Table 2.17. Selected examples of the types and sources of variation in allelochemicals in trees

Type of Variation	Nature of Variation	Tree Hosts	Source of Information
Organ to Organ (tissue)	Quantitative and qualitative differences in various tissues of the tree	*Abies sibirica* (Siberian fir)	Stepanov (1976a)
	Kaemferol-3- (p-coumarylglucoside), a main needle flavonoid that is absent or at very low concentration in buds	*Larix leptolepis*	Niemann (1976)
	Variation in cortical essential oils (terpenoids), which is greatest in tree tops	*Abies* spp.	Zavarin (1975)
Age of Foliage	Difference in 19 of 36 terpenoid compounds analyzed in young and mature foliage	*Juniperus scopulorum* v. *platinum* (rocky mountain juniper)	Adams and Hagerman (1976)
	Differences in terpenoid concentration	*Sequoiadendron giganteum* (giant sequoia)	Levinson *et al.* (1971)
		Picea glauca (white spruce)	Von Rudloff (1972)
		Pinus ponderosa (ponderosa pine)	Zavarin *et al.* (1971)
	Occurrence of kaempferol-3-glucoside and quercetin-3-glucoside (flavonoids) in young needles but not in old needles	*Picea abies*	Niemann (1979)
Among Trees of a Species	Difference in 37 terpenoids among trees	*Juniperus scopulorum*	Adams and Hagerman (1977)
	High variation of picein in needles and as much as a 60-fold difference among trees	*Picea abies*	Esterbauer *et al.* (1975)

(continued)

Table 2.17. (*continued*)

Type of Variation	Nature of Variation	Tree Hosts	Source of Information
	Significant differences in tanning coefficients (a measure of protein binding activity expressed as tannic acid equivalents) between individual trees	*Betula alleghaniensis* and *Acer saccharum* (yellow birch and sugar maple)	Schultz *et al.* (1982)
Species Differences	Qualitative and quantitative differences between species	*Pinus sibirica* (Siberian pine)	Stepanov (1976b)
		Pinus silvestris (Scots pine)	
		Larix sibirica (Siberian larch)	
		Abies sibirica (Siberian fir)	
		Picea obovata (Siberian spruce)	
Geographic Variation (Differences due to population location within its range)	Relative population differences in 152 foliar terpenoids	*Juniperus ashei* (ashe juniper)	Adams (1977)
	Terpenoid types and concentration in seeds (seedcoats) from trees within the range	*Abies concolor, A. grandis, A. magnifera, A. procera,* and *Pseudotsuga menziesii* (white, grand, California red and noble, and Douglas-firs, respectively)	Zavarin *et al.* (1979)

Daily Variation (Diurnal as well as daily, or day-to-day)	Significant differences in 37 terpenoids between consecutive days; significant diurnal differences in some of the 37 compounds	*Juniperus scopulorum*	Adams and Hagerman (1977)
	Diurnal differences in terpenoids	*Juniperus californica* (california juniper)	Tatro et al. (1973)
	Variation of 5 flavonoid concentrations	*Larix sibirica*	Niemann (1979)
Seasonal Variation	Differences in terpenoids over the growing season	*Picea glauca P. pungen* (blue spruce)	Von Rudloff (1962, 1972)
		Juniperus ashei	Adams (1969)
		J. pinchotii (Pinchot juniper)	Adams (1970)
		J. scopulorum	Powell and Adams (1973)
	Differences in current-year foliage only	*Pinus ponderosa*	Zavarin et al. (1971)
	Changes in concentration and type of various flavonoids and aromatic hydroxy acids	*Larix leptolepis*	Niemann (1979), and refs. therein
	Seasonal concentration peaks in pungenin and picein (acetophenones)	*Picea pungens*	Niemann (1979)
		P. abies	
	Increases in leaf tannins and catechin from spring to autumn	*Quercus robur* (English oak)	Feeny (1970)
	Changes in bark tannins		Hathway (1952)
	Long-term and short-term seasonal changes in tanning coefficients (tannic acid equivalents) of foliage	*Betula allegheniensis Acer saccharum*	Schultz et al. (1982)

Table 2.18.   The influence of the principal host
plants of the fall webworm (*Hyphantria cunea*) in
New Brunswick and Nova Scotia, Canada, on
pupal size[a],[b]

Host Species	Pupal Weight (mg)
Speckled alder	225
Cherry	210
Manitoba maple	204
Apple	184
Elm	180
Poplar	146
Birch	108

[a] From Morris (1971).
[b] *Hyphantria cunea* (Lepidoptera: Arctiidae).

those species whose adults do not feed. In species whose larvae feed on
several hosts, the size and reproductive potential of individuals feeding
on each host differs (Table 2.19). Elm spanworm females reared on
pignut hickory have a reproductive capacity that is 109% greater than
that of oak-fed females (Drooz, 1965). The implication of such differences for forest insect population dynamics in sites dominated by hosts
favoring high reproductive potential is obvious.

Differences in the quality of tree tissues within or among individual
trees of a host species also can lead to differences in reproductive potential. The average number of eggs laid by the mountain pine beetle in
lodgepole pine is affected by phloem thickness. The differences in fecundity and oviposition rate may be related to nutrition: Thin phloem and
excavated xylem appear to provide less food value than thick phloem
(Amman, 1972). Average egg production can range from 3.5 per in. of
gallery in phloem that is 0.06 in. thick to 9.6 per in. in phloem that is
0.18 in. thick. Oviposition rate is similarly affected. The average number
of eggs laid per day can range from 1.7 in phloem 0.06 in. thick to 7.7 in
phloem 0.20 in. thick. Other factors such as temperature, beetle size,
and density also are important.

Similarly, total sugar and L-proline concentrations are greatest in the
staminate flowers and least in the mature needles of balsam fir (the least
preferred tissues fed upon by spruce budworm). Sugar and L-proline
concentrations influence feeding , with the highest concentrations producing the greatest feeding activity by larvae. Sugars in all tissues, including
needles and buds, are readily used and may have a significant impact on
development and reproduction as reflected in pupal weights (Table 2.20).
The preference for certain tissues is reinforced by an adult female oviposition preference for flowering trees (Table 2.21). This preference appears to

Table 2.19. Pupal weights (mg) of elm spanworms (*Ennomos subsignarius*) reared on three food plants[a]

Food Plant	First-day Pupae			Fifth-day Pupae			Tenth-day Pupae		
	No.	Mean	Range	No.	Mean	Range	No.	Mean	Range
					Males				
Northern red oak	16	145.7	101.8–176.6	16	138.7	94.2–166.4	14	132.2	87.5–154.6
White oak	17	131.4	78.2–174.1	17	123.9	76.6–166.7	16	115.1	70.4–150.4
Pignut hickory	16	166.7	124.5–227.1	16	158.4	119.8–220.0	14	149.4	110.0–207.3
					Females				
Northern red oak	17	189.2	92.3–270.2	17	173.5	87.0–215.7	13	162.6	80.3–200.8
White oak	11	174.1	108.6–231.5	11	164.2	93.0–221.5	5	161.9	103.6–210.8
Pignut hickory	15	290.0	241.4–359.2	15	277.0	226.3–344.6	13	258.7	209.5–322.5

[a] From Drooz (1965).

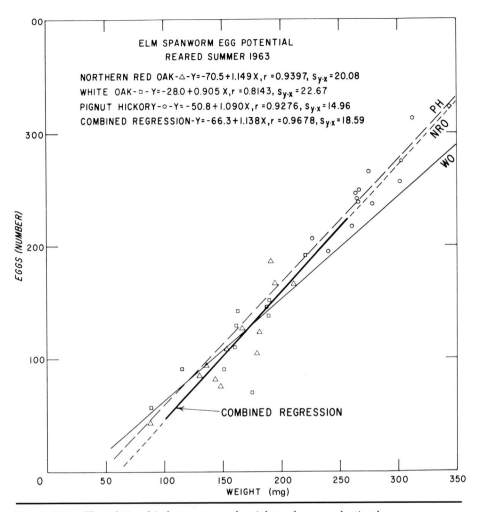

Figure 2.22. The relationship between pupal weight and egg production by elm spanworm females (*Ennomos subsignarius*) whose larvae were reared on various host species, summer 1963. Northern red oak ($\triangle$, NRO): $Y = -70.5 + 1.149X$, $r = 0.9397$, $S_{y \cdot x} = 20.08$; white oak ($\square$, WO): $Y = -28.0 + 0.905X$, $r = 0.8143$, $S_{y \cdot x} = 22.67$; pignut hickory ($\bigcirc$, PH): $Y = 50.8 + 1.090X$, $r = 0.9276$, $S_{y \cdot x} = 14.96$; and combined regression (heavy line, CR): $Y = -66.3 + 1.138X$, $r = 0.9678$, $S_{y \cdot x} = 18.59$; where $r =$ correlation coefficient and $S_{y \cdot x} =$ standard error of the estimate. (From Drooz, 1965.)

be based on the growth pattern of flowering trees. These trees, typically dominant or codominant, are taller, have greater crown lengths, and grow in open areas (Blais, 1952).

Asynchrony of egg hatch can be sufficient to expose feeding stages to qualitatively different food, particularly if environmental stresses cause

Table 2.20. Performance of the spruce budworm (*Choristoneura fumiferana*) on diets containing various levels of added sucrose[a]

Diet	Sugar[b] (%)	Females			Males	
		Survival (%)	Pupal Weight (mg)	Development Time (half days)	Pupal Weight (mg)	Development Time (half days)
Control (=C)	0.5	29.6	69.4	52 ± 0.7	56.2	52
C + sucrose	0.9	46.6	71.0	54 ± 1.6	69.5[c]	44[c]
C + sucrose	1.6	38.2	91.8[c]	51 ± 1.0	72.1	44
C + sucrose	2.2	38.0	100.2[d]	48 ± 0.7[d]	70.8	44
C + sucrose	4.0	40.0	114.9[c]	46 ± 1.4	71.9	42

[a] Modified from Harvey (1974).
[b] Calculated total soluble sugars, including 0.5% in wheat germ
[c] Where t-test with next lower sugar concentration indicated significant differences at the 1% level.
[d] Where t-test with next lower sugar concentration indicated significant differences at the 5% level.

Table 2.21. Ovipositional preferences of
female moths of the spruce budworm
(*Choristoneura fumiferana*)[a]

Plot	Tree Type[b]	No. of Egg Clusters per Pupal Case
A	F	5.7
	NF	1.6
B	F	4.0
	NF	1.2
C	F	6.5
	NF	0.8
D	F	8.5
	NF	2.3

[a] From Blais (1952).
[b] F flowering, NF = nonflowering.

rapid changes in the host tree. However, even in the absence of asynchronous hatch there may be large developmental discrepancies. In any given area, the earliest spruce budworm pupae are larger and the first females more fecund, even though all individuals feed on current foliage (Greenbank, 1956). Reproductive potentials of dendrophagous species feeding on any given host rarely remain constant during a growing season. Instead, the level of progeny production may change as the season progresses and host quality changes (Fig. 2.23). Thus, the effects on fecundity due to consumption of tissues of the same tree species may vary from year to year. Similarly, incorporation of additional plant species into the diet may dramatically affect fedundity (Table 2.22).

Plant Hosts and Protection against Pathogens

Antibacterial substances, extractable from the leaves of various coniferous and deciduous trees, actively interact with pathogens of larval defoliators. Juglone, a natural constituent of shagbark hickory (*Carya ovata*) which affects feeding by insects, is also antagonistic to microorganisms (Brooks, 1963). Although anti bacterial action is observed in extracts of various tree species, conifer extracts appear to produce the most inhibition (Kushner and Harvey, 1962). The mechanisms by which these substances affect the development of microorganisms, such as *Bacillus cereus* and *B. thuringiensis*, remain unclear. The high acidity of certain foliage can lower the usually high alkalinity of the midgut of many Lepidoptera and thus may enable the toxic crystals of certain pathogens to dissolve and to kill the herbivore. However, other modes of action may occur.

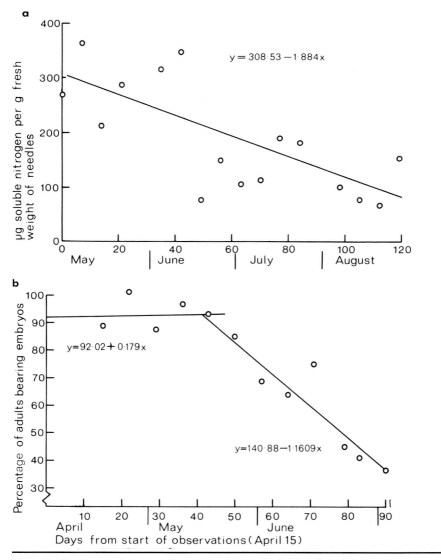

Figure 2.23. Seasonal changes in soluble nitrogen in the needles of Sitka spruce (*Picea sitchensis*); (b) comparable seasonal changes in the proportion of adult green spruce aphids (*Elatobium abietinum*) bearing embryos. (From Parry, 1974.)

Antibiosis induced by plant compounds may be detrimental to symbiotic microorganisms which can play an important role in the assimilation of an herbivore's food. Thus, the antimicrobial properties of conifer volatile oils and terpenoids can drastically disrupt beneficial symbiotic relationships (Table 2.23).

Table 2.22.   Egg mass characteristics of gypsy moths (*Lymantria dispar*) reared on several natural diets[a]

Diet	Egg Masses Examined	Reproductive Index[b]	No. Eggs/Mass (mean ± SE)[c]
Red maple, 1974	13	4.37	108 ± 7a
Red maple, 1975	20	3.15	218 ± 12b
Red oak	16	3.15	292 ± 25c
Red oak–red maple	10	3.36	151 ± 16a
Red oak–white pine	16	2.82	347 ± 25d

[a] From Capinera and Barbosa (1977).
[b] Reproductive index = pupal wt. (mg)/no. eggs.
[c] Means followed by the same letter are not significantly different at the 5% level by Duncan's New Multiple Range Test.

Nutrition and Population Quality

The effects of poor nutrition, although sometimes very subtle, can have a significant long-range impact on forest and shade tree insect populations. The food reserves accumulated during larval development of female western tent caterpillars (*Malacosoma californiacum pluviale*) influence the behavior of their progeny. Food reserves must be used for both the maintenance of adult tissues and the production of eggs. Nutrient limitations are reflected in the yolk content (energy reserves) of eggs, particularly in the last eggs laid, which are the smallest. These eggs give rise to less active individuals compared with the larvae hatching from the first eggs oviposited. These sluggish larvae (from small eggs) are one component of an activity and vigor gradient exhibited by all colonies. A large proportion of active individuals occurs among the offspring of active type females. Colonies with many active members feed frequently, develop rapidly, and have greater survival than their sluggish colony counterpart. The nutritionally mediated range of behaviors allows for survival in a mosaic environment of temporary habitats. The importance of population quality (phenotypic variation that influences population dynamics) is illustrated by many forest and shade tree insects including the fall webworm, the gypsy moth, and others (Barbosa and Baltensweiler, 1987).

Food and Speciation

Population of dendrophagous species in different geographical localities often have significantly different host preferences. The Douglas-fir tussock moth favors white fir (*Abies concolor*) in California, grand fir (*Abies grandis*) and Douglas-fir in Oregon and Washington, and subalpine fir (*Abies lasiocarpa*) in Nevada. These differences may have important implications for numerical change or the long-range process of speciation.

Table 2.23. Influence of volatile oils and terpenoids of Scots pine on insect microflora[a]

Culture	Tested Substances and Effect of Their Action (in %)[b]					
	Volatile oil from needle +/-	Volatile oil from shoots +/-	α-Pinene +/-	β-Pinene +/-	Limonene +/-	α³-Carene +/-
Paecilomyces aspergiloides	12/88	30/70	41/50	55/45	48/52	45/55
Control	69/31	81/19	60/40	51/49	66/34	51/49
Cladosporium brevicompactum	24/76	41/59	25/75	70/30	58/42	26/74
Control	70/30	80/20	67/33	75/25	72/28	78/22
Cladosporium herbarum	29/71	33/67	20/80	34/66	39/61	23/77
Control	90/10	85/15	80/20	70/21	70/30	68/32
Scopulariopsis sp.	50/50	39/61	72/28	63/37	71/29	66/34
Control	64/36	54/46	80/20	59/41	52/48	63/37
Candida sp.[c]	92	88	149	115	98	19
Control	21	30	81	41	39	16

[a] From Smelyanets (1977b).
[b] + = % germinated spores; − = % ungerminated spores.
[c] Number of colonies.

Although still a matter of controversy, isolation imposed by variations in host selection, in association with other prerequisites, may lead to sympatric speciation (Bush, 1975). This hypothesis suggests that hosts often serve as the location of courtship and mating. Thus, mate selection in these species depends on host selection. The principal isolating mechanism would thus be ecological as well as behavioral. Shifts to new hosts could profoundly affect mate selection and provide strong barriers to gene exchange between parental and progeny populations. The appearance of new host races of insects such as the apple maggot (*Rhagoletis pomonella*), which moved from hawthorn to apple and cherries, is presented by proponents of the hypothesis as one example.

The complex taxonomic status of some sawflies may represent another illustration of sympatric speciation. The balsam fir sawfly (*Neodiprion abietis*) complex is an example in which larvae feed on a number of tree types and are subdivided into populations based on host species. All the populations of the *N. abietis* complex in various localities have different seasonal appearances. Phenological variations have developed in other *Neodiprion* races adapted to particular food hosts (Knerer and Atwood, 1973).

Finally, the obscure scale (*Melanaspis obscura*) has been recorded on a number of oak species. A study of the differences in its life history on white oak and pin oak (*Q. palustris*) has shown that nearly a one month lag occurs in the development of most stages on white oak compared with those on pin oak. Obscure scales on white oak overwinter as settled crawlers, whereas on pin oak the larvae overwinter in the second instar. Because adult males live for less than 24h, mating among adults of different host varieties is essentially impossible. Although no morphological differences were found between adult males or females of each type, the bionomic differences between the populations lead the investigators to suggest the presence of two species (Stoetzel and Davidson, 1973).

Chapter Summary

The amount and pattern of insect feeding behavior has many varied implications concerning their economic importance. Feeding by the majority of insect species results in moderate reduction of total foliage areas. Knowledge about feeding patterns is important in understanding impacts. Because of the great diversity of mouthpart adaptations in insects, feeding damage takes on a variety of external appearances. Four primary categories of insect damage include free-feeding, skeletonizing, mining, and stippling.

The quality of food available to insects is a fundamental and extremely important aspect of population dynamics. Food quality in part determines host range, insect population survival, and seasonal patterns of abundance. Levels of nitrogen, amino acids, carbohydrates, minerals, water, and allelochemicals have all been implicated as factors influencing patterns of insect abundance. Considerable evidence exists for the importance of food quality, but much work is still needed. Many complex interactions, such as the effect of foliage quality on pathogenic microorganisms, probably exist and may alter the quality of food for dendrophagous insects.

CHAPTER 3

Tree Vigor and Insect Herbivores

Introduction

This chapter deals with the influence of host vigor on the development and survival of dendrophagous insects and with the factors that alter host vigor. Forest managers believe that a vigorous, healthy forest is less susceptible to damage by insects and diseases than other forests. This belief arises from years of field observations by practicing foresters and parallels the generally accepted notion for animals that stress predisposes individuals to disease.

Although the term *host vigor* has been used in many ways and is often poorly quantified, it usually describes a variety of physiological states that enhance or inhibit herbivore attraction, development, and survival or that influence behavior.

The reasons why unstressed trees may be more resistant to insect attack are not well understood, but several possible mechanisms have been suggested. Perhaps the most popular is that vigorous trees have more energy available to expend on defensive strategies. A good example of this is the ability of many pine trees to "pitch out" bark beetles. A vigorous tree when attacked by a beetle produces large amounts of resin that physically interfere with beetle attack. A stressed tree is not able to produce sufficient resin to resist beetle attack.

It is also possible that vigorous trees are able to grow more quickly past susceptible stages. A predictable succession of forest insects occurs in forests as trees develop from seedlings to pole-sized to mature trees. As trees move through the various stages, they are no longer susceptible to certain insect pests. A fast-growing tree spends a shorter period of time in any susceptible stage than a slow-growing tree. A third possible reason why a vigorous tree may be less susceptible is related to the ability of the tree to recover from insect attack. A healthy tree may sustain as much feeding by a dendrophagous insect as an unhealthy tree, but the net impact is less because the healthy tree recovers more quickly.

The Nature of Tree Stress

Perhaps one of the more difficult aspects of understanding the relationship between tree stress and susceptibility to forest insects is the lack of a consistent definition of stress. Kozlowski (1985) states, "the predominant concept of stress is based on performance, with the tree in a state of stress when some measure of its performance falls below par." This definition indicates that most trees become stressed at some point in their development. The stressed condition of a tree can also change quickly as the result of a changing environment. Generally, stress is due to the effects of biotic factors (insects, disease, competition) or abiotic factors (temperature, moisture, fire). Both biotic and abiotic factors can induce stress when they are above or below the optimum range for a given tree species. For example, both drought (lack of water) and flooding (excess water) can cause serious and sudden increases in tree stress. The relationship between stress and susceptibility may also become confusing when one considers that insect damage can be both a cause and a result of tree stress. For purposes of this discussion the term *vigor* refers to a variety of physiological states in a host tree that enhance or inhibit herbivore attraction, development, and survival, and behavior: low vigor being associated with stress and high vigor with the lack of stress.

Indices of Tree Vigor

Judging the vigor of trees is often an important and, perhaps, confusing responsibility of foresters. Vigor can affect growth rates, competitive abilities, and susceptibility to insects and disease, as well as the aesthetic appearance of high-value urban trees. Although vigor usually refers to the physiological condition of the tree, tree vitality may actually be manifested in several other characteristics or attributes of the tree. Many of these have become the bases for measuring vigor. Methods of vigor estimation can be divided into qualitative (or subjective) and quantitative measures. The best estimate of tree condition often involves both types of assessments. Familiarity with various vigor indices can be a useful asset in forest and urban situations.

Qualitative Estimates

Most qualitative estimates of tree vigor depend on the judgment and experience of the observer and his or her familiarity with local problems. Several attempts have been made in the past to categorize trees according to various physical characteristics. Keen (1936, 1943) devised an elaborate classification scheme based on tree age, tree size, crown form, and crown class. Observations of such attributes can be helpful in providing information about the long-term health of the tree. Crown ratio, the proportion of the height of the crown to the total tree height, is valuable as a quick indication of tree vigor. This method is most often used as a guide for thinning. Trees with crown ratios of less than 30 to 40% are usually removed.

Needle retention can also be used to evaluate tree vigor. Most healthy, vigorously growing ponderosa pines shed 3- to 4-yr-old needles in the fall, retaining 2 to 3 years' worth of needles, but stressed trees may shed all but the current-year foliage, leaving branches with short tufts of needles. In both the above methods, estimates are approximations and rarely entail counts or direct measurement.

Quantitative Estimates

The amount of sapwood produced per unit area of foliage (Waring vigor index) is usually considered to be an effective indication of long-term tree vigor (Grier and Waring, 1974; Mitchell *et al.*, 1983; Waring *et al.*, 1980). The index is the ratio of the radius of the current year's diameter growth to the radius of the sapwood. Wood production has lower priority than root and shoot growth, and increases only as the energy budget of the tree increases. Tree vigor is reflected by the amount of current diameter growth per unit of crown leaf surface. Crown leaf surface is obtained using the relationship between sapwood area at breast height and leaf surface. These ratios have been calculated for several species. Diameter at breast height (dbh), the width of the total sapwood area, and the width of the last annual increment must be measured to obtain a ratio for each tree. Low ratios are indicative of low vigor, and high ratios indicate high vigor.

Starch content is a visual technique that indicates broad vigor classes based on relative amounts of starch in the stem tissue (Wargo, 1975). The starch content reflects the photosynthetic capacity of the tree because it is the most predominant form of reserve energy. Thin cross sections of twig or root material are stained with a solution of potassium iodide and then mounted on slides for viewing or photographing (Wargo, 1975). The degree of staining indicates the amount of starch and the general health of the tree. The stained sections can be compared to standardized levels for the specific species or can be used to rank a group of trees relative to one another. This technique has been used to determine the impact of defoliation and the effects of site quality on Douglas fir (Webb and Karchesy, 1977) and to determine relative vigor classes in several eastern hardwood species.

Plant moisture stress (PMS), probably the most commonly used measure of plant stress, is determined with a Scholander pressure chamber

(Scholander *et al.*, 1965; Ritchie and Hinckley, 1975; Waring and Cleary, 1967) (Fig. 3.1). A fascicle or twig is placed in the chamber, and the pressure is gradually increased until the sap is forced to the cut surface on the xylem vessels. This pressure is equal to the tension on the water column in the xylem at the time the sample was cut. Water status of a plant can be affected by edaphic and atmospheric conditions and by regulators such as stomata. Measurements are usually taken before dawn (predawn) because PMS is minimal and least variable at this time of day. Midday measurements are used to determine maximum moisture stress. High PMS readings indicate that processes such as photosynthesis and growth are

Figure 3.1. The Scholander pressure chamber is probably the most widely used instrument to measure tree moisture stress.

slowing or shutting down, which may lead to physiological damage and eventual mortality. For example, in ponderosa pine seedlings a predawn reading of less than −10 bars indicates a vigorous plant, whereas readings of −20 bars indicate the photosynthetic process is being negatively affected (Cleary, 1970). Consistently high readings usually indicate impending mortality.

Electrical resistance is a quick method for vigor determination (Wargo and Skutt, 1975; Kostaka and Sherald, 1982). Electrical resistance is usually measured with a shigometer (a modified ohmmeter named after its founder, Alex Shigo) (Fig. 3.2). A pulsed electrical current is passed between two stainless steel probes that are inserted vertically into bark fissures. The electrical resistance between the probes is inversely proportional to the concentration of cations, particularly potassium ions, which accumulate in actively growing regions. A high reading indicates low vigor. Electrical resistance has been related to periodic growth, dwarf mistletoe (*Arceuthobium* spp.) ratings, and spruce budworm infestation levels. Readings may vary seasonally and between species and sites. This technique is probably most applicable to localized areas and requires experience on the part of the operator.

All of the vigor indices have positive and negative aspects. It is important to consider the nature of the stress when attempting to select a

Figure 3.2.   Shigometer inserted into a ponderosa pine to measure electrical resistance in the cambium.

suitable vigor index. Considerable research in the area of tree vigor assessment will likely continue in the future.

Moisture Stress and Insect Activity

The association of defoliator and bark beetle outbreaks with drought (i.e., soil moisture deficit) has been observed for decades (Felt, 1914; Blackman, 1924; Craighead, 1925; St. George, 1930; Beal, 1942; Hetrick, 1949; Lorio, 1968; Berryman, 1973; Ferrell, 1974). Moisture deficit is one of the most important stresses on trees. It weakens trees, makes them more susceptible to insect attack, enables them to support high herbivore densities, and thus contributes to insect outbreaks (Mattson and Haack, 1987).

The ability of trees to resist or recover from injury is a function of soil moisture because a deficit can affect (a) resin quantity and quality, (b) nutritional quality of tree tissues, and (c) the growth rate of trees. Several studies have shown that the water supply has a profound effect on the growth and yield of trees (Polster, 1963; Brown *et al.*, 1965). Changes in tree tissues occur simply because the cells of trees can grow only if they take up water. Movement of water usually proceeds from sites of high water potential (high energy content) to sites of low water potential. As soil dries out, the soil water potential (i.e., the tendency for water to move) decreases. The only way water will move into roots is if the water potential of the roots is lower than that of surrounding soil. Under drought conditions, the water potential of soil is lower than that of the roots; thus water is not taken up by the roots. Evapotranspiration produces a tension that reduces water potential in the tree and thus around cells. This lowered potential prevents water from moving into cells of tree tissues and prevents growth because cells are unable to reduce their water potential. Tree growth is reduced, leaves may wilt, the quality of the tree's chemical composition changes, and the physiological state of the tree is altered.

Drought disrupts normal cell metabolism and may result in (a) an increased concentration of certain forms of nitrogen, (b) breakdown of carbohydrates and protein, (c) accelerated translocation of soluble leaf nitrogen and phosphorus compounds from the leaves to the stem, and (d) increased concentrations of compounds that are detrimental to insects and may affect the nutritional quality of the tissues. The concentration of sugars in the foliage and other tissues may increase. Moisture stress can also have an impact on nutrient uptake (Johnson, 1968; Schwenke, 1968; Hodges and Lorio, 1969; Parker, 1977; van den Driessche and Webber, 1975). In addition, retardation of transpiration due to the closing of stomata leads to high leaf temperatures and low photosynthetic rate. For example, net photosynthesis in balsam fir seedlings and other species decreases rapidly after soil moisture is reduced below field capacity and ceases when the permanent wilting percentage of the soil is reached (Satoo and Negisi, 1961; Johnson, 1968). Depletion of soil moisture reduces growth and yield substantially before the permanent wilting percentage of the soil is reached (Kozlowski, 1949). Thus, quite significant changes occur in a tree as a result of moisture stress. These changes, in turn, can affect a variety of aspects of the survival, development, behavior, and ecology of dendrophagous species.

Perhaps the best understood relationship between stress and increased insect activity relates to bark beetles that feed on conifers. The first line of defense against these insects is the ability to produce resin (oleoresin exudation pressure). Stress reduces the trees' ability to produce resin (Hodges and Lorio, 1975). Stressed trees, because of their reduced ability to pitch out beetles, are more susceptible and frequently succumb to beetle attacks. However, an important relationship exists between tree stress and beetle population (Fig. 3.3). Even a very healthy (resistant) tree can be killed if beetles attack *en masse*. Conversely, a very susceptible tree can be killed by a small population of beetles.

Moisture content of wood governs the invasion of host trees by the adult ambrosia beetle and the development and survival of the larvae because of its influence on the beetle's food, ambrosia fungus. *Scolytus multistriatus* prefers freshly cut logs with moisture and phloem conditions similar to those of healthy trees (Rudinsky, 1962). Finally, although certain loblolly pines offer favorable conditions for attack and colonization by southern pine and *Ips* beetles, the favorable changes may be the result of

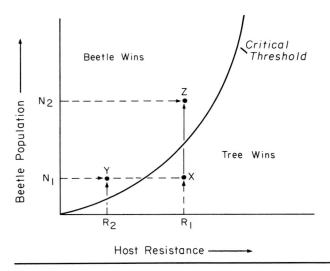

Figure 3.3.   Theoretical relationship between host resistance and the beetle population required to overcome that resistance. (From Berryman, 1982. Copyright © 1982 by the University of Texas Press.)

the effects of moisture stress on wood-staining fungi and subsequent changes in tree tissues as the fungi grow. In trees struck by lightning, for example, the water content, oleoresin exudation pressure, and sucrose content of the inner bark are reduced while reducing sugars increase. After attack by southern pine beetles, both nonreducing and reducing sugars decline. These changes may be due to the activity of the wood-staining fungi which are often carried by the beetles (Hodges and Pickard, 1971).

Not only are low-vigor hosts, such as those under moisture stress, required for the initiation of attack and subsequent pheromone production by many species, but the qualitative variation in the pheromones may also be associated with variation in moisture stress of host trees. Bark beetle pheromones are multicomponent compounds that include various host tree substances. Production of secondary attractant pheromone by pioneer male *Ips* beetles occurs shortly after penetration into living phloem. Specific host factors are necessary for the generation of species-specific attractants. Recent reports have shown conclusively that myrcene, a chemical constituent of the host tree, is converted in male *Ips paraconfusus* to the pheromones ipsenol and ipsdienol (Hendry *et al.*, 1980). Since the vigor of trees varies over time and space when they are affected by moisture stress, it is conceivable that the amount or quality of host terpenes and pheromones is also affected.

Aging and Host Quality

As a tree ages, physical and biochemical changes occur that affect tree vigor. These changes represent differences that may influence the success of insect invasion, colonization, and survival. Often the relationship between aging and susceptibility is not simply that the greater the age, the greater the susceptibility. Often, changes that occur only during a limited age interval are those that enhance susceptibility. For example, infestation of pine by the pine root collar weevil (*Hylobius radicis*) (Fig. 3.4) is greatest in 10- to 13-year-old trees, declines slightly in 14- to 15-year-old hosts and drops sharply in trees 16 years or older (Kennedy and Wilson, 1971).

The physiological differences between old and young trees can be quite substantial. The changes that occur as a result of the aging process vary with species, site, and climate. Nevertheless, as a tree ages, the changes observed include (a) a decrease in metabolism, (b) death of inner wood cells, (c) reduction in wound healing capacities, and (d) reduction in shoot diameter growth. These and other changes may predispose trees to attack due to altered threshold levels of flowering, cambial activity, resin flow, critical sapwood moisture, and/or production of biochemical products by physiologically altered tissues which may attract insects or enhance their survival (Kozlowski, 1969).

Among many beetles species the selection of hosts is based on cues that reflect the host tree's age and size. For example, an ample supply of

Figure 3.4. Stages of *Hylobius radicis*: (a) egg, (b) larva, (c) pupa in cell, and (d) adult. (From Finnegan, 1962.)

phloem and resin seem to be of prime importance in the attraction, oviposition, and larval development of the white pine weevil. Bark thickness can be a sensitive index of host suitability. The few larvae that develop in thin-bark terminals are much smaller than larvae of the same age in vigorous terminals (of trees within a specific age class); the latter have an ample supply of phloem tissue and other critical substances (Kriebel, 1954).

Locust borer attack is most severe in trees in older age classes. Approximately 24% of trees in the 1-in. dbh class are susceptible to attack, 44% in the 1.5-in. class, 60% in the 2-inch class, and almost 100% of the

≥2.5-in. class. As the tree grows older, the bark becomes rougher. The roughness of the bark is a critical stimulus for the oviposition of the borer (Berry, 1945).

Similarly, groups of large trees are foci of attack by the southern pine beetle, whereas smaller trees are less attractive. In a study of the influence of tree size, the two largest pines (11 to 12-in. dbh) attracted 66% of an aggregating beetle population, two 8-in. dbh pines attracted 24%, and the smaller trees attracted only 10% of the population (Vite and Crozier, 1968). In stands of lodgepole pine the number of mountain pine beetles emerging per unit area of host increases with the size (age) of the tree. Beetles prefer trees of large diameter during any given year and over the life of the infestation, although these differences are more directly related to bark (or phloem) thickness, which influences colonization and survival (Amman, 1972).

Flowering is often related to tree age. Young trees do not flower, whereas older trees flower periodically at the expense of vegetative growth. Flowering balsam fir trees are usually in the dominant or codominant crown classes and are larger in diameter, total height, and crown length than nonflowering trees. Nonflowering trees are in the intermediate and suppressed classes. Mortality due to defoliation by spruce budworm tends to be considerably higher in flowering balsam fir than in the intermediate or nonflowering categories (Table 3.1). This differential results from the slow growth rate of mature trees and enhanced larval survival due to feeding on pollen of staminate flowers. These pollen-fed larvae develop more rapidly than those feeding solely on foliage (Blais, 1952). In addition, because of their flight patterns, adult females prefer to oviposit on the tallest and most exposed trees in a stand, which tend to be mature trees (Mott, 1963). Thus, maturity of balsam fir stands aids in the development of spruce budworm outbreaks.

Table 3.1. Number of stems per acre of living and dead balsam fir by diameter classes in a nonflowering and an adjoining flowering stand defoliated by spruce budworms (*Choristoneura fumiferana*)[a]

Type of Stand	Condition of Trees	dbh (in.)					Total	%
		1–2	3–4	5–6	7–8	9–13		
Nonflowering	Living	526	122	54	4	2	708	99
	Dead	0	4	0	2	0	6	1
	Total	526	126	54	6	2	714	100
Flowering	Living	74	68	20	8	2	172	20
	Dead	286	240	126	30	18	700	80
	Total	360	308	146	38	20	872	100

[a] Modified from Blais (1958a).

Recovery from Injury

Susceptibility to attack and injury, as well as the inability of trees to recover easily from injury, is often associated with relatively slow growth rates in trees. Locust borers deposit eggs on vigorous as well as slow-growing trees. However, usually only the vigorous, fast-growing trees can grow over wounds. Wounds in older or less vigorous trees often remain open, resulting in more severe injury (Berry, 1945). The growth rate of trees prior to defoliation is often a factor influencing their recovery from injury (Table 3.2).

The western pine tip moth (*Rhyacionia bushnelli*) attacks the seedling and sapling stages of various pine species. Ponderosa pine is a susceptible host species and Austrian pine (*Pinus nigra* (austriaca)) is relatively resistant. Differences in susceptibility among trees are generally based on the trees' ability to outgrow injury and to develop adventitious buds. The faster-growing trees are better able to outgrow tip moth injury. For each species some trees are more successful than others in achieving some growth in spite of attack. The degree of injury to leader shoots reaches a peak at 40 to 49 in. and decreases above this height. Injury to lateral shoots reaches a maximum from 60 to 100 in. Plantations with closed crowns show a marked decline in tip moth injury (Graham and Baumhofer, 1930). Many other forest insects prefer host trees within specific size ranges (Tables 3.3 and 3.4).

Vigor is usually associated with resistance to or recovery from insect injury. However, there are exceptions, Several insect species, such as *Pissodes strobi*, tend to select vigorous host trees. Similarly, the balsam woolly aphid induces abnormal swelling at the nodes and buds of vigorous branches (Balch, 1952).

Soil Richness and Site Quality

The availability of adequate nutritional requirements obviously affects tree vigor, although the effects of inadequate nutrition are often ambiguous.

Table 3.2.   The mean diameter growth of trees during a 5-yr period prior to defoliation by the pine butterfly (*Neophasia menapia*)[a]

Ultimate Outcome of Trees in Each Group	Mean Annual Diameter Growth (mm)	Significant Difference (mm)
Trees recovered	1.88	
Trees subsequently attacked by beetles	1.32	0.56[b]
Trees that died from defoliation alone	1.26	0.62[c]

[a] Modified from Evenden (1940).
[b] Significant difference between averages of trees recovered and trees attacked.
[c] Significant difference between averages of trees recovered and trees dead from defoliation.

Table 3.3.   The influence of tree height on the susceptibility of Sitka spruce to attack by the spruce weevil (*Pissodes sitchensis-Pissodes strobi*) during a 5-yr period[a]

Height Class (ft in 1964)	Total Trees	Total Attacks per Tree									Trees Attacked (%)
		0	1	2	3	4	5	6	7	10	
2.0– 3.9	15	15	—	—	—	—	—	—	—	—	0
4.0– 5.0	33	31	2	—	—	—	—	—	—	—	6.0
6.0– 7.9	54	34	13	4	3	—	—	—	—	—	37.0
8.0– 9.9	33	12	7	5	4	3	2	—	—	—	63.6
10.0–11.9	47	8	13	5	12	6	3	—	—	—	83.0
12.0–13.9	18	3	5	2	3	2	2	—	1	—	83.3
14.0–15.9	12	—	1	4	4	1	1	—	—	1	100
16.0–17.9	7	—	—	3	1	—	2	1	—	—	100
18.0–19.9	4	—	1	1	2	—	—	—	—	—	100
Totals	223	104	41	24	29	12	10	1	1	1	

[a] From Silver (1968).

Table 3.4.   The influence of length and diameter of leaders of Sitka spruce on the proportion attacked by the spruce weevil (*Pissodes sitchensis-Pissodes strobi*)[a]

Leader Size	1960		1961	
	No. Leaders	% Attacked	No. Leaders	% Attacked
4–7	15	7	10	0
8–10	22	32	17	0
11–16	36	39	27	11
17–23	13	69	31	58
23–34	0	—	8	100
Diameter (in.)				
0.10–0.17	3	0	18	0
0.18–0.24	28	14	38	13
0.25–0.30	41	36	26	50
0.31–0.40	14	86	9	78
0.41–0.46	0	—	2	100

[a] From Silver (1968).

However, the recurrent association between trees growing on poor sites and the numerical increase of injurious insect species cannot be ignored. The Saratoga spittlebug (*Aphrophora saratogensis*) (Fig. 3.5) is most common on hosts growing in sites having poor, sandy, or barren soil (Kozlowski, 1949). Scale insect outbreaks are associated with similar sites (Johnson,

Figure 3.5. Spittle mass of the Saratoga spittlebug (*Aphrophora saratogensis*). (Courtesy USDA Forest Service.)

1968). Outbreaks of the lodgepole needle miner *Coleotechnites milleri* tend to occur in suboptimal sites, that is, those having nutrient-poor soils (Mason and Tigner, 1972). Similarly, outbreaks of the spruce budworm and the red-headed pine sawfly are associated with poor sites (Fauss and Pierce, 1969; Shepherd, 1959). Moisture deficits and competitive interactions among trees are concurrent pressures usually present in sites with nutrient-poor, poorly developed, or disturbed soil profiles.

Both biotic and abiotic factors can be used to describe poor (susceptible) sites. For example, edaphic (soil) characteristics of poor sites that are susceptible to southern pine beetle attack are those that exhibit little slope, poor drainage, and low soil pH. Biotic characteristics of tree stands on poor sites include high total basal area, high proportion of pines, high stand density, low radial growth rate, extensive understory vegetation, and thick bark on potential hosts (Lorio, 1968, Kushmaul *et al.*, 1979).

Black locust grown in areas ranging from sites without much top soil to sites with fertile soil differ in the extent of locust borer (*Megacyllene robiniae*) damage. Most of the susceptible size-class trees on poor sites are heavily infested. Their growth is so slow that they seldom recover. On good sites the proportion of vigorous, rapidly growing trees is high, and the damage to such trees is low (Berry, 1945). Evaluations of damage by the golden or pit-making oak scale (*Asterolecanium variolosum*) to chestnut oak have indicated that on poor sites, particularly those along ridges, many heavily infested 3 to 6-in. diameter trees are severely injured or killed. On good sites the effects of infestation are less evident. Poor sites are characterized by infertile and rocky soil which causes stunted and scrubby growth (Parr, 1937). Similarly, the most severe infestations of the Swaine jack pine sawfly occur in nearly pure stands of jack pine on kalmia-vaccinium or cladonia-vaccinium sites. On these sites jack pine grows densely and poorly (McLeod, 1970).

Fertilization of trees is a method of manipulating site quality that may indirectly influence the number, sex ratio, and size of insects or the extent of insect damage. Although available data are far from consistent, some researchers have concluded that nitrogen fertilization tends to reduce damage by defoliators and phloem miners but favors sucking insects and mites (Stark, 1965). One mechanism that may explain the effects of fertilization involves an inverse relationship between nitrogen content and the content of antiherbivore phenolic compounds in trees (Shigo, 1973).

Feeding by some insect species may aggravate stresses associated with marginal or unfavorable site conditions. For example, an infestation of the balsam woolly aphid can cause changes in wood structure that interfere with water transport. Aphid attack is most serious on sites where water stresses occur; severe damage may develop rapidly on these sites (Page, 1975).

Larches that are physiologically disturbed or weakened due to attack by needle-feeding, needle-sucking, or phloem-feeding insects show major anatomical changes. Hypertrophial cell-clusters appear in the cambium; and, even in slightly damaged trees, the number of parenchyma cells in the phloem increase. In addition, more and more cells of the medullary rays become empty, sieve-cells become compressed, and wound resin canals (or resin canal chains) begin to appear in the xylem (Schimitschek and Wienke, 1966) (see other chapters on the effects of various insect groups). These changes make the consequences of additional stress, such as that imposed by poor site conditions, quite deleterious.

Induction of Changes in Host Trees: The Role of the Insect in Host Quality

The chemical and physical defenses of trees have evolved to protect trees against a variety of mortality agents. Although microorganisms, pollutants, and physical climatic phenomena may induce defensive responses, our focus will be on responses induced by insects. Researchers are just beginning to uncover these phenomena though similar responses involving microorganisms have been known for some time (Kuc and Caruso, 1977; Kuc, 1982).

Studies conducted in the last few years show or strongly suggest that defoliation and other dendrophagous activity result in significant changes in the quality of tree tissues as food (Edwards *et al.*, 1986; Paine *et al.*, 1985). These changes protect the tree tissues by reducing the survival or potential impact of herbivores. This process has been termed the *induction of defensive responses*. Although most studies of induction of defensive responses describe changes due to feeding activities, some induced changes may result from oviposition activity. Gum deposition by cherry trees in response to oviposition by periodical cicadas can be effective in reducing successful egg hatch (Karban, 1983). Flagging may also reduce the success of hatching (White, 1981), although this may be viewed as a simple change due to mechanical damage.

In general, changes induced by defoliators, piercing-sucking insects, borers, and so on may simply be shifts from the normal state to some altered or abnormal physiological state, or they may cause the production of a specific defensive compound. Whatever the mechanism, several illustrations of this phenomenon have been reported (Rhoades, 1979; Wagner, 1988).

Feeding by one insect species can negatively affect the quality of leaves or other tree tissues and may detrimentally affect subsequent herbivores. Carbohydrate metabolism in primary leaves reflects the defoliation experienced the previous season. For example, the carbohydrate content of red maple foliage produced after 2 and 3 yr of leaf removal declines with severity of defoliation. Reductions can be as severe as 58% in some circumstances (Heichel and Turner, 1976). Needles produced after defoliation differ in quality from the original needle or leaves. Regenerated larch needles have a shortage of nutrients (i.e., less protein, fats, and carbohydrates) compared with original needles. These deficiencies have a negative impact on a subsequent defoliator, the Siberian pine moth (Rafes *et al.*, 1972). Two years of defoliation by the Douglas fir tussock moth can result in a reduction in monoterpenes, a reduction in total sugars the first year after defoliation, and a reduction in starch the second year. The trees that produced the least amount of monoterpenes were successfully attacked by a second species, the fir engraver (*Scolytus ventralis*) (Wright *et al.*, 1979).

Studies on the spear-marked black moth and the gypsy moth have shown that these species suffer negative effects from feeding on host foliage of trees previously defoliated for 1 to 3 yr. Feeding on such foliage results in detrimental changes (reduced pupal weights, longer development time, and reduced survival) (Fig. 3.6). A number of other examples of induction by insects of what are presumed to be defensive mechanisms in plants have been reported in a variety of insect–plant associations (Rhoades, 1979; Haukioja and Neuvonen, 1987; Wagner, 1988). Perhaps entomologists should not be surprised by this phenomenon since changes in secondary substances are known to result from stresses due to physical and biotic factors. Some plant pathogens, for example, induce defensive chemicals in plants.

One of the most interesting examples of the induction of defense is demonstrated by the birch-feeding insect *Oporinia autumata*. The growth of this insect is retarded when larvae are raised on birch (*Betula pubescens* spp.

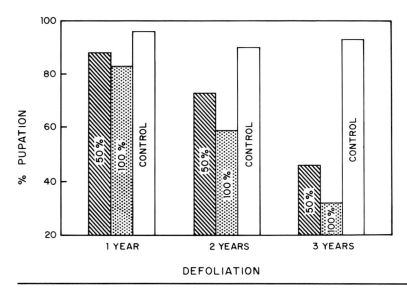

Figure 3.6. Percentage of larvae of the spear-marked black moth (*Rheumaptera hastata*) attaining the pupal stage when reared on foliage from birch trees defoliated annually for 3 yr. (From Werner, 1979.)

tortuosa) leaves whose adjacent leaves have been damaged mechanically 2 days earlier. Leaves near the damaged leaves have an average of 9% higher phenol content (compounds presumed to be defensive), and their extract inhibits trypsin approximately 40% more efficiently than those of control leaves. Larvae feeding on leaves from trees that had been defoliated 1 or 2 yr earlier experience the same type of retardation of growth and lower pupal weight (Haukioja and Hakala, 1975; Haukioja and Niemela, 1977; Niemela *et al.*, 1979).

Larval gypsy moths feeding on artificially defoliated grey birch and black oak (*Quercus velutina*) have reduced pupal weights and longer development time compared with those larvae feeding on nondefoliated trees (Wallner and Walton, 1979). Although the changes in a large number of nutrients have been evaluated, only a reduction in total free sugar concentration is correlated to reduced pupal weights. Nitrogen limitation due to seasonal patterns of decreasing concentrations of nitrogen as well as total free amino acids may exacerbate the influence of low concentration of foliar sugar (Valentine *et al.*, 1983).

Alternatively, the defense induced by gypsy moth feeding may not result from nutritional imbalances but rather from the production of defensive toxins. Leaves of defoliated red oaks have higher tanning coefficients (an index of the ability to bind proteins), total phenolics, tannins, and dry matter content, and greater toughness compared with leaves of nondefoliated trees (Schultz and Baldwin, 1982). All these characteristics describe low-quality leaves which, when consumed, might be responsible for the reduced growth reported by Wallner and Walton (1979).

Insect-induced defensive responses of trees may not necessarily be directed solely at the attacking insect. At epidemic levels the European wood wasp (*Sirex noctilio*) causes significant economic losses in Monterey pine (*Pinus radiata*) forests of New Zealand and Tasmania. Upon insertion of their ovipositor, females introduce a symbiotic fungus (*Amylostereum* sp.) into the sapwood. Polyphenols form in these trees, which recover from attack, whereas only small amounts of polyphenols are detected in unattacked trees. Qualitative differences in the polyphenols also occur. The symbiotic *Amylostereum* fungus appears to alter the metabolism of host cells so that phenolic compounds (e.g., stilbenes-like pinosylvin and its methyl ester) are formed. These compounds may play a role as fungitoxic agents. In addition, they could confine the spread of the fungus by inactivating extracellular enzymes such as cellulase or proteinase (Hillis and Inoue, 1968).

Not all changes induced by insect feeding are harmful to the feeding insect or to subsequent herbivores. In some situations feeding improves the quality of food for the feeding species by concentrating nutritional compounds near the feeding site. At low densities the silver fir woolly aphid (*Adelges* sp.) feeds on peripheral bark cracks and lenticels. Large aggregates develop at a given site, which stimulates amino acid development in the bark around the aggregates. The individuals in these aggregates increase in size and are more fecund than those at low densities (Eichhorn, 1968; Way and Cammell, 1970). Changes induced by the feeding insect (and occasionally by the ovipositing insect) often make the host tissues more attractive or suitable for subsequent dendrophagous species (see next section).

Care must be exercised in assuming that all changes in insects result from changes in host tissues. The dendrophagous species itself may undergo changes in requirements for development or in the preferences it exhibits. For example, the acceptability of a host is often dependent on the age of the dendrophagous insect. Late instar gypsy moths accept and survive well on foliage that is unacceptable (and often fatal) to early instars (Mosher, 1915; Barbosa, 1978; Barbosa *et al.*, 1986). Whereas first and second instar Swaine jack pine sawflies suffer high mortality rates when fed juvenile foliage, fifth instars are not adversely affected (All and Benjamin, 1975). Similarly, old needles of white fir are almost totally unacceptable to early instar Douglas fir tussock moths but are eaten by late instars (Mason and Baxter, 1970).

Relationship between Injury and Susceptibility

Frequently, decreases in vigor that make trees susceptible to insect attack result from the effects of an earlier injury or disruption due to a pathogen, insect, or other environmental stress (e.g., fire, lightning, storms, or compaction) (Miller and Patterson, 1927; Schimitschek and Wienke, 1966; Hodges and Pickard, 1971). Indeed outbreaks of many insects, such as the bark beetles *Dendroctonus rufipennis* and *D. ponderosae*, pine engraver (*Ips pini*) and sixspined Ips (*Ips calligraphus*), have been suggested to be natural developments following severe outbreaks of defoliators such as the spruce budworm or the lodgepole pine needle miner (Patterson, 1921; Dewey *et al.*, 1974). Defoliated trees such as *Populus tremuloides* and *P. grandidentata* are often more susceptible to attack by borers such as the bronze poplar borer (*Agrilus liragus*) (Baker, 1965).

The two-lined chestnut borer (*Agrilus bilineatus*) preferentially attacks trees weakened by drought or defoliation (Haack and Benjamin, 1982). At low densities *A. bilineatus* successfully attacks injured trees but not healthy trees (Dunbar and Stephens, 1975). The widespread defoliations in Connecticut by the gypsy moth and elm spanworm in the 1970s have provided weakened host trees and have led to increased abundance of the two-lined chestnut borer.

Even when the food source is nonliving, as with seasoned-wood insects, the condition prior to attack can determine the success of establishment and survival. The death-watch beetle (*Xestobium rufovillosum*) and the common furniture beetle exhibit enhanced larval development and survival when feeding on wood infested by white and brown rot fungi. Fungal action is believed to (a) reduce the mechanical strength of the wood by affecting the integrity of cell-wall components, (b) increase nitrogen concentrations, and (c) increase the suitability of the wood surface for egg laying and larval penetration. On the other hand, the fungal–larval association of the old-house borer (*Hylotrupes bajulus*) and the blue-stain fungus is detrimental to larval development because the fungi live on residual cell contents and reserve food materials in the sapwood without breaking down the cell walls. The two organisms are thus competitors (White, 1962).

Chemical Stresses and Host Vigor

Forest trees as well as shade trees are frequently exposed to a multitude of chemical agents including herbicides, insecticides, fungicides, fertilizers, growth regulators, and chemical and gaseous pollutants. These substances may affect a dendrophagous insect directly or indirectly by altering the vigor of its tree hosts (i.e., growth and quality). Such changes can affect a tree's suitability to herbivores or simply predispose it to attack. Black pine leaf scale (*Nuculaspis californica*) density decreases when this insect occurs on ponderosa pines damaged by weed killers (Edmunds and Allen, 1958) (Fig. 3.7). The phytotoxicity of BHC (lindane) can be sufficient

Figure 3.7.   Injury resulting from a black pineleaf scale (*Nuculaspis californica*) infestation. (Courtesy USDA Forest Service.)

to induce root malformations in Norway spruce seedlings (Simkover and Shenefelt, 1952). Absorption of elements such as phosphorous by conifer seedlings is inhibited by BHC and other insecticides (Voigt, 1954). Lindane and dimethoate induce a short-term reduction in photosynthetic activity of black poplar. Stresses such as these clearly result in major changes in tree growth, which in turn must have a significant impact on insect herbivores (Johnson, 1968).

Air pollutants of various types affect transpiration, root respiration, photosynthesis and leaf respiration, and expansion (Johnson, 1968) (Fig. 3.8). Population density and distribution in certain species can be affected by pollution levels in certain sites. Ponderosa pine is affected by a

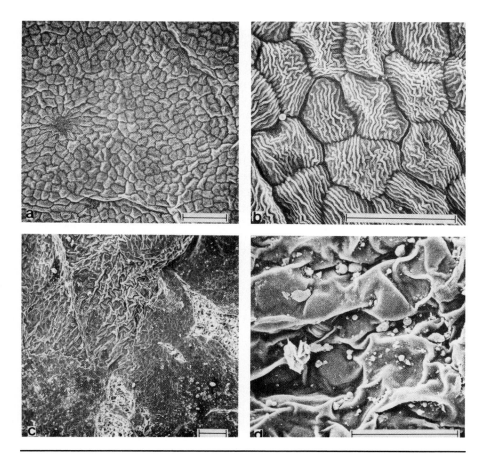

Figure 3.8.   Scanning electron micrographs of *Aesculus hippocastanum* (horse chestnut) leaves show effects of pollution: (a, bar is 100 μm) and (b, bar is 50 μm) are adaxial surfaces of leaves from nonpolluted areas; (c, bar is 100 μm) is adaxial surface and (d, bar is 50 μm) is abaxial surface of leaves from polluted areas.

disease that is associated with photochemical air pollutants. Trees exhibiting chlorotic decline symptoms are more frequently attacked by bark beetles, such as *Dendroctonus brevicomis* and *D. ponderosae*, compared with those exhibiting little or no symptoms. Although almost 62% of the insect-killed trees examined were found to have high disease ratings, less than 15% of the nearest noninfested neighbors were severely attacked (Stark *et al.*, 1968). Although no differences occurred in needle monoterpenoids of air pollution-injured and healthy trees, significant differences in percent methyl chavicol were observed (Cobb *et al.*, 1972). Pollutants, like other stresses, may simply weaken a tree host without inducing disease. For example, needles of eastern white pine subjected to air pollutants suffer from stomatal occlusions and disrupted structural integrity (Percy and Riding, 1978).

Chapter Summary

There is long standing conventional wisdom that the vigor of a tree has a significant influence on its susceptibility to forest insects. Tree vigor is not clearly defined but is related to the physiological condition of the host tree. Tree vigor can be measured by a variety of qualitative and quantitative techniques. When vigor is reduced, trees become stressed and allocate a lower portion of their carbohydrate reserves to defensive functions. Also, nutritional factors may change under stress which improves insect performance. As trees age, they tend to become more stressed, but other ontogenetic factors change independent of stress. Stressed trees may be more susceptible because of a reduced ability to recover from insect attack. In general, why stressed trees are more susceptible is not well understood, but this is an active area of current research.

CHAPTER 4

Host Selection by Forest and Shade Tree Insects

Introduction

The study of host selection by forest and shade tree insects could focus on many different insects such as parasitoids, predators, or herbivores. However, even though details of host selection differ among insect species, their overall selection behaviors are similar. The phases into which the host selection of endoparastic insects has been divided (i.e., host-habitat finding, host finding, host acceptance, host suitability, and host regulation) are usually applicable to other types of insects, including dendrophagous species, though minor modifications of the scheme may be required for specific ecological circumstances. For example, species whose hosts or food units (tissues) occur in clumps or patches may display an additional phase: patch-finding.

Conceptual models have been proposed for the generalized process of host selection (Fig. 4.1 and Table 4.1) that provide a general view of insect host selection including that of forest insects. The models described in Table 4.1, and similar models, concentrate primarily on host finding, host acceptability, and host suitability. Host-habitat finding and host regulation are perhaps the least understood aspects of host selection. An example of host regulation is the influence of an herbivore on the physiology and ecology of its hosts, such as the induction of favorable chemical changes in the attacked host.

Host selection in one form or another is an integral part of the ecology and behavior of all forest and shade tree insects. Even polyphagous species (which feed on hosts belonging to several plant families) exhibit the capacity to distinguish subtle differences among host plant species. Many species prefer upper-crown foliage (i.e., sun leaves to shade leaves) (Fig. 4.2), old foliage to young foliage, mature trees to young trees, or stressed trees to vigorous trees.

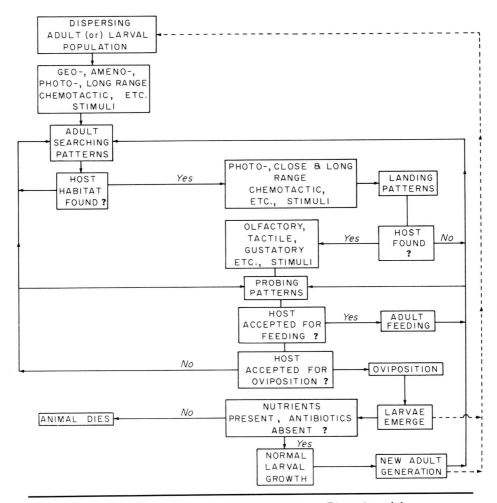

Figure 4.1.   Flowchart of generalized host-selection process. Dispersing adult or larval populations respond to physical and chemical stimuli of the environment and initiate a searching pattern that leads to the host's habitat. If the host habitat is found, the process proceeds in successive steps contingent on the presence of proper stimuli. Successive phases of the process include host-finding, acceptance, initiation, and maintenance of feeding and/or oviposition. If the larval food contains proper nutrients and does not contain antibiotics, development is completed, and a new adult generation emerges. This new adult generation may resume the host-selection process at any pattern level following the phenology of the species. If the larval food is not adequate (nutritionally deficient, or antibiotic), larvae most likely die. (Modified from Kogan, 1977.)

Table 4.1. Patterns of host selection[a]

Pattern	Characteristics
A	Nonselective oviposition, eggs scattered and/or laid on some nonplant substrate Host range of adult and larvae comparable Host-plant orientation via generalized mechanisms, e.g., taxes, appropriated physical factors, visual orientation to vertical objects Usually highly polyphagous species
B	No long-range chemical cues; nonspecific short-range cues Host-plant finding based on visual cues Oviposition and/or feeding acceptance based on generalized chemical compounds, arrestants, excitants, and so on Most host plants selected by adults acceptable to immatures Includes many polyphagous species
C	Host-finding random in adults, but utilization of appropriate host results in aggregation of others on host plant
D	Orientation for host finding directed by specific cues Acceptability of host plant determined by high degree of discrimination Larvae less discriminatory; host range limited by feeding deterrents, not excitants Usually oligophagous species
E	Highly selective host-finding behavior Larval feeding selective; biochemical basis of selection same as in adults
F	Highly selective adult host-finding behavior Host-plant range of adults and larvae are very different. Selection by adults of hosts for oviposition and for feeding, may differ from those of larvae

[a] Based on data from Kogan (1977).

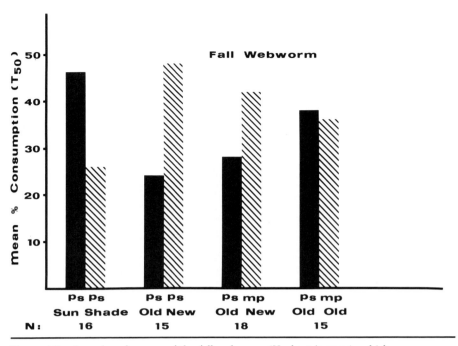

Figure 4.2.   Larval preferences of the fall webworm (*Hyphantria cunea*), which was presented leaves of *Prunus serotina* (ps) and *Malus pumila* (mp). (From Barbosa and Greenblatt, 1979b.)

Inter- and intraspecific preferences are usually manifested in feeding and/or ovipositional choices made by larvae and/or adults. Although their choices are often identical, larval and adult host-plant preferences may overlap but not coincide in some circumstances.

Movement and Host Selection

In some species, both host selection and dispersal are attributes of the larvae (Lance and Barbosa, 1981a,b). These interactions may be particularly important at the level of host-habitat finding. The role of larvae in host selection, particularly in those species whose adults are wingless or flightless (e.g., species in Lymantriidae, Geometridae, Tortricidae, Oecophoridae, and Psychidae), may be much more important in forests than in other ecosystems.

Perception of and movement toward host habitats or hosts are influenced by visual, olfactory, gustatory, and tactile stimuli. In many cases, host-habitat finding results, in large part, from a congruence of the abiotic microenvironmental requirements of both insect and host; that is, conditions favorable for the growth and survival of an insect are the same as those favorable for its host. Indeed, in some species, hosts that are acceptable and support growth are not exploited because they grow in a habitat that is not frequented by the herbivore or that is in some way suboptimal.

Orientation movements may be relatively fixed responses to physical environmental cues. For example, insects may respond to a token stimulus, such as light intensity, in such a way as to bring them to a host habitat or to an area of the host where preferred tissues are found. Larval gypsy moths travel in more or less straight lines by orienting to the plane of polarized light and to vertical objects (i.e., trees) near their line of travel (Doane and Leonard, 1975). In spring, when second instar spruce budworms emerge from their hibernacula, they are positively phototaxic and, thus, climb to the tips of branches where their preferred food source is located and from where dispersal is most likely (Wellington and Henson, 1947). Crowding, food scarcity, or changes in food quality often induce movement to new habitats or host trees. The buoyancy of first instars of insects such as scales and Lepidoptera insures effective transport by wind and the likely infestation of new habitats and hosts. In summary, directed movements, such as adult flight, locomotion by larvae; or passive transport due to biotic or abiotic factors, may allow individuals of a species to reach different hosts or parts of trees.

Visual and Physical Cues

Long-range cues that enable insects to select appropriate habitats, patches, or host trees are primarily visual and olfactory. Although it is uncertain at what distances insects can respond to hosts, it is likely that quantitative and qualitative visual cues and orientational movements (e.g., phototaxis, geotaxis, anemotaxis), modified by environmental factors, probably are responsible for selection over great distances. Short-range host finding also may be aided by visual cues.

Various behaviors of forest arthropods including attraction, landing, oviposition, and feeding may be affected by the perception of color or specific spectral qualities, such as intensity or reflectance. The mealy plum aphid (*Hyalopterus pruni*) has been shown to discriminate host-specific colors (Moericke, 1969). Larval nun moths can differentiate various colors and, when hungry, appear to prefer green foliage over brown foliage (Hundertmark, 1937a).

Perception of shape and form (i.e., pattern) is another important aspect of host finding. Orientation toward vertical objects or dominant silhouettes is a common behavior. White pine cone beetle (*Conophthorus coniperda*) adults fly above the trees in their initial attack and tend to orient toward dark silhouettes (Henson, 1962). Spruce budworm females deposit more eggs on flowering trees (which are more favorable for larval development) than on nonflowering trees, in part because flowering trees are taller and grow more in the open (Blais, 1952). Finally, the mountain pine beetle tends to orient its flight to vertical trapping devices (Billings *et al.*, 1976; Pitman and Vite, 1969); similar orientation behavior has been noted for *Ips curvidens* (Hierholzer, 1950). Visual cues also may facilitate selection of preferred parts of a host tree. The white pine weevil prefers and is most successful in the vigorous, large diameter, upright leaders of dominant or codominant trees. This selection is accomplished by preferential orientation toward vertical or near-vertical host leaders of greater than average size and length (Vander Sar and Borden, 1977b).

Some larval insects exhibit similar responses to visual cues and orient toward vertical objects. Larval gypsy moths, for example, orient toward vertical objects near their line of travel and, with their ability to cue in on polarized light, can apparently orient toward trees within 3 m (Doane and Leonard, 1975). Gypsy moth larvae show a preference for the larger of two vertical objects when given a choice (Table 4.2). Similarly, larval nun moths preferentially orient toward larger or closer vertical objects (Hundertmark, 1937b).

Close-range assessment and selection of hosts or host tissues are also influenced by visual and physical factors. The selection by *Pissodes strobi* of the large-diameter leaders of its conifer hosts involves the close-range assessment of physical characteristics of potential hosts. The volume of feeding cavities made by *P. strobi* is related to bark thickness and to the distribution of cortical resin ducts. If the weevil cannot avoid rupturing the epithelial cells surrounding a resin duct by feeding around the duct, the feeding cavity is abandoned. The likelihood that a feeding weevil will contact a resin duct in small-diameter, thin-barked host material is greater than in large leaders. Similarly, oviposition is affected by thin-barked host material. Oviposition follows the excavation of a feeding cavity that is sufficiently deep to take the female's oblong eggs. The average egg size

Table 4.2. Frequency of movement to and up tree trunk, for models of different diameters, by fifth instar gypsy moths[a]

Test	Diameter of Model (cm)		
	5	10	15
	Number of Times Models Were Chosen		
5 cm vs. 15 cm	16	—	49
10 cm vs. 15 cm	—	22	40
5 cm vs. 10 cm	11	24	—

[a] Modified from Lance and Barbosa (1981a).

provides an index of the appropriate minimum thickness of host bark (Vander Sar and Borden, 1977a,b). (For other examples see the chapter on oleoresin.) Lime aphids (*Eucallipterus tiliae*) insert their stylets and feed in the sieve tubes of leaf veins. As the aphid becomes larger it tends to select larger veins. In spring when leaves are small, aggregations result. On larger, summer leaves, wide spacing occurs due to the insects' tendency to occupy more leaf surface (Kidd, 1976a). A large portion of adults in groups are capable of maintaining aggregations despite some tendency to feed on the widely spaced minor veins. Aggregations occur as a result of a so-called social interaction that is mediated by visual cues. The wing pattern in adults apparently acts as a sign stimulus inducing other adults to approach (Kidd, 1976b).

Physical characteristics of hosts may also be important to foliage feeders in host selection. Many leaves have small hairs known as trichomes (Figs. 4.3 and 4.4). Depending on their shape, density, and type, trichomes can affect host plant choice by interfering with feeding and oviposition. Some of these fine hairs are hooked, making them effective at entrapping or impaling insects. Puncture wounds and subsequent desiccation, combined with partial starvation due to being immobilized, can cause mortality. Glandular trichomes also are common. These vary in structure and secrete a variety of allelochemicals (Levin, 1973). These and other allelochemicals, so-called secondary substances, are nonnutritional chemicals that can act as attractants, feeding stimulants, repellents, arrestants, or toxins.

As noted above, physical and chemical cues may influence host selection at the same time. Foraging by individuals of tent-making colonial species is often regulated by the chemical and mechanical stimuli of trails laid from the tent, along twigs, to the foliage. Larvae of several lepidopterous species deposit a chemical cue along silk trails that directs siblings to foliage. This phenomenon has been demonstrated in the eastern tent caterpillar (Fitzgerald and Gallagher, 1976) and in the birch tent moth (*Eriogaster lanestris*) (Weyh and Maschwitz, 1978). Some parasitoids of silk-producing forest Lepidoptera may use similar host-produced cues to find their hosts (Fig. 4.5).

Semiochemical Cues

Although visual orientation behavior enables insects to reach a host species, it may not enable the insect to differentiate among specific host plants or host tissues. Thus, although some larval Lepidoptera can visually orient preferentially to large vertical tree stems, they appear to have to climb a tree and sample foliage before they can make a decision to accept or reject it (Rafes and Ginenko, 1973). Contact, olfactory, and gustatory chemoreception and mechanoreception (e.g., for monitoring leaf geometry and texture), as well as the perception of temperature and humidity, may all be

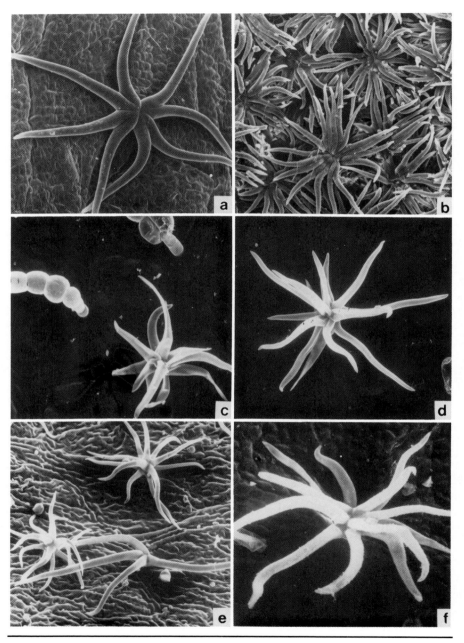

Figure 4.3. Selected trichome types in *Quercus*: (a) stellate (*Q. sinuata*) (×500), (b) fused-stellate (*Q. virginiana*) (×300), (c) bulbous and rosulate (*Q. incana*) (×300), (d) rosulate (*Q. laevis*) (×300), (e) multiradiate and fasciculate (*Q. pumila*) (×200), and (f) multiradiate (*Q. myrifolia*) (×400). (From Hardin, 1979.)

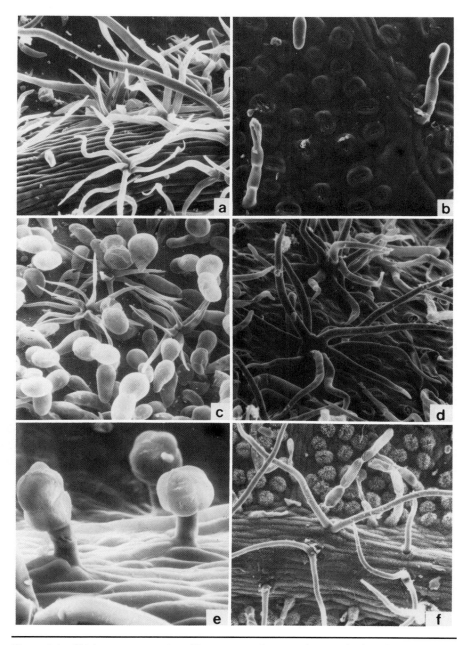

Figure 4.4. Trichome topography of immature and mature leaves of selected *Quercus* spp.: (a) *Q. phellos*, immature (×300), (b) *Q. phellos*, mature (×500), (c) *Q. falcata*, immature (×300), (d) *Q. falcata*, mature (×300), (e) *Q. michauxii*, mature (×800), and (f) *Q. michauxii*, mature (×300). (From Hardin, 1979.)

Figure 4.5. *Cotesia melanoscelus* responding to silk trails left by its host, larval gypsy moths. (From Weseloh, 1976.)

used in host selection (Schoonhoven, 1972a,b). However, the overwhelming importance of chemical (semiochemical) cues is unquestionable. A semiochemical is defined as a chemical that mediates interactions between organisms of the same or different species (i.e., pheromones or allelochemicals, respectively).

Assuming that an insect has an appropriate level of internal motivation (e.g., hunger or reproductive maturity), acceptance of the host is often regulated by allelochemicals such as terpenes or alkaloids and/or by nutritional compounds such as sugars or nitrogen. These cues, which allow fine tuning of the selection process, are particularly important when certain host tissues are more suitable for survival than others. For example, when the spruce budworm emerges in spring, larvae may mine vegetative buds or needles. However, when staminate flowers are available, many larvae never go through the needle-mining phase (Blais, 1952). In choice tests, penultimate instars prefer samples with ethanol extracts of staminate flowers to those with extracts of shoots, which in turn were more acceptable than those with extracts of mature needles (Heron, 1965). Similarly, discriminatory abilities have been demonstrated for species that choose leaves of a certain age or position in the crown. This selectivity is exhibited not only by leaf-chewing species but also by insects with piercing-sucking mouth parts (Parry, 1976).

Host Selection in Bark Beetles

There are very few forest and shade tree insects for which the details of host selection are well known. The bark beetles are an exception. Host selection in bark beetles is unusual in that it is closely linked to mating and colony formation; and although it is not typical of other forest insects, it provides an excellent example of the varied roles of semiochemicals and environmental and physical factors. In addition, bark beetles are economically important, and understanding their host selection process is important in their control.

In bark beetles, the process of host tree selection, feeding, and mating are inextricably linked, part of a sequence of events leading to maximal resource use and progeny production. Colonization of host material is usually divided into dispersal, selection, concentration, and establishment (Wood, 1972). Other schemes are quite similar but may combine selection and concentration into one phase (Borden, 1982). The dispersal phase starts as adults emerge from a brood tree and terminates with sustained feeding by pioneer beetles (initial invaders). The point at which host selection begins and dispersal ends is unclear due to our incomplete understanding of these two phases of colonization. Host selection is sometimes subdivided according to the nature of the attraction of beetles to their host. Primary attraction refers to initial orientation and responses to host trees or logs. This attraction often results from some change in the host tree. For example, trees stressed by fire, lightning, drought, disease, and other insects (see section on host quality) become attractive to bark beetles.

Secondary attraction results from orientational cues produced after the attack by pioneer beetles and often results in a mass attack on host material. Secondary attraction may be due to the increased volatility of primary attractants resulting from the boring activity of pioneer beetles. In most instances, secondary attraction is due to pheromones produced by pioneers that have fed on or been exposed to host compounds. Host discrimination may occur during this initial feeding period. For example, although *Ips paraconfusus* can produce pheromones even when forced to feed on non host material, when given a choice, sustained feeding occurs only on host species (Wood, 1972).

Primary Host Selection

Relatively little is known about primary attraction. An early theory proposed that initial attraction was due to the aldehydes or esters produced from respiratory fermentation associated with abnormal enzyme activity in subnormal host trees (Person, 1931). Similar proposals suggest that deficiency of oxygen in tissues of dying trees leads to a change from oxidation to fermentation metabolism (Graham, 1968). This activity induces the formation of attractants (see below). An alternative theory suggests that the initial selection process is random and that its success is a function of gallery initiation stimulants and/or the effectiveness of resin flow (Wood, 1972; Hynum and Berryman, 1980; Elkinton and Wood, 1980). For example, there is no evidence of primary attraction (i.e., response to host-produced volatiles) in *Dendroctonus brevicomis* or *D. ponderosae*. However, since beetles land on healthy and stressed hosts at a rate of approximately one beetle per tree, either per day or per five days, and since theoretically only one beetle boring in a suitable host can initiate attack, there is ample opportunity for host colonization (Moeck *et al.*, 1981). Although most research has emphasized chemical attractants, other factors such as visual cues may also be important in primary host selection (see introduction to host selection). For example, in *Trypodendron lineatum* and *T. domesticum*, lineatin together with visual stimuli, induces landing on tree trunks (Vite and Bakke, 1979).

Pioneer Beetles

Generally, in monogamous species (e.g., *Dendoctonus* spp.) the pioneer beetles are female whereas in polygamous species (e.g., *Ips* spp.) pioneers are usually male (Wood, 1963). Pioneers may emerge from an overwintering host tree in which they have matured. Still other beetles overwinter in nonhost hibernacula, such as leaf litter. For species with multiple generations, pioneers may simply emerge from a brood tree.

Environmental factors, such as temperature, light intensity, and rainfall, influence the quantity and quality of emergence and activity. Success of pioneers is also a function of the precision of the synchrony between emergence and the availability of appropriate host material, whether they be cones, mature trees, or windthrown trees. Finally, success in host selection ultimately depends on whether the beetles have attained a so-called suitable reactive phase (or physiological state), a condition we know little about but which might include completion of development (Perttunen *et al.*, 1970; Borden, 1982).

Host Components

Numerous volatile compounds emanating from host material can direct or orient pioneers. In a forest, attractants, repellents, and arrestants, in a multitude of combinations, represent an odor mosaic of specific host and nonhost cues that modifies beetle behaviors. Thus, sorting out how a chemical acts or how the beetles respond to each chemical has been difficult and continues to be a source of debate.

Ethanol, generated by anaerobic metabolism in host tissue, is attractive to some species of ambrosia beetles (Harring, 1978). The striped ambrosia beetle (*Trypodendron lineatum*) and other scolytids also are attracted by odors emanating from logs of various species. A greater number of entry holes are formed by the striped ambrosia beetle in the sapwood of western hemlock (*Tsuga heterophylla*) (kept in *vacuo* for 24 hr) than in aerobically treated samples (Graham, 1968). The attractiveness of anaerobically treated sapwood is detectable after 4 hr and peaks at 20–24 hr. In laboratory extractions and bioassays, ethanol, along with α-pinene, is an attractive component for both sexes of the striped ambrosia beetle (Vite and Bakke, 1979). Field tests also indicate that ethanol is a primary attractant for *Gnathotrichus sulcatus*, another ambrosia beetle (Cade *et al.*, 1970). Earlier observations might have suggested this phenomenon since some beetle species have been recorded as being attracted to beer casks (Moeck, 1970). Primary attraction to breeding sites for *Scolytus multistriatus* is apparently due to a relatively weak host attractant from decadent elms (Meyer and Norris, 1967).

Generally, scolytids have differential responses to the volatiles of different tree species (Table 4.3). Specific attractive components may vary among host species. Nevertheless, host-produced allelochemicals may induce the initial attack of pioneer beetles. *Blastophagous piniperda*, a scolytid that primarily attacks felled pines, initially responds to the odor of phloem of *Pinus sylvestris* (Perttunen *et al.*, 1970). Fractionations of the pine phloem produce an attractive extract composed of a-terpineol, *cis*-carveol, and *trans*-carveol. Some subfractions (geraniol, nerol, *cis*- and *trans*-carveol) are not attractive but are synergistic; thus, when mixed in appropriate proportions, they increase the attraction of other components such as α-terpineol (Oksanen *et al.*, 1970). The exact blend of attractive components and synergists is not known for many species.

Male and female *Dendroctonus pseudotsugae* are highly responsive to host phloem. Flight in 97% of males and 91% of females is arrested upon exposure to host volatiles (Bennett and Borden, 1971). Thus, although the roles of various specific compounds or their blends are often unknown, the role of host volatiles in primary host selection is clearly important (Vite and Gara, 1962; Chapman, 1963) (Table 4.4).

Termination of Primary Attraction: Secondary Attraction

Feeding, for most species, marks the beginning of the process referred to as secondary attraction. The factors stimulating feeding and leading to the acceptance or rejection of tree or log are not well known. Feeding may be induced or inhibited by a compound in the vast array of chemicals to which beetles are exposed. These cues may be of host origin, insect-produced, or a combination of both. Some evidence suggests regulation of feeding behavior by chemoreception. Feeding by *Scolytus multistriatus* is deterred by extracts of a non host, *Carya ovata* (Gilbert *et al.*, 1967). Stimulation of feeding in *S. multistriatus* is caused primarily by two extracts of elm bark: (±)-catechin xylopyranoside and lupeyl coerotate (Doskotch *et al.*, 1970).

Although the role of specific host monoterpenes in attraction and feeding initiation is a source of controversy (Pitman, 1969; Bedard *et al.*, 1969, 1970), the fact that most components do play some role in the

Table 4.3. Number of scolytid beetles responding to odors of various tree species[a]

Species	Douglas-fir (A)	Douglas-fir (B)	Western Hemlock	Control	Western White Pine	Western Red Cedar	Total
Trypodendron lineatum	14	9	1	0	0	0	24
Gnathotrichus retusus	14	71	72	0	7	0	164
Dendroctonus pseudotsugae	0	8	0	0	0	0	8
Xyleborus saxeseni	0	18	290	1	12	0	321
Hylastes nigrinus and *Hylurgops rugipennis*[b]	85	130	7	7	46	37	312
Other scolytids	0	8	4	1	33	2	48

[a] From Chapman (1963).
[b] Because of uncertainty of field identification and pooling of beetles on certain collection dates, these two species are treated together here.

Table 4.4. Attraction of scolytid beetles in flight to traps baited with Douglas-fir oleoresin and terpenes during a 10-h testing period[a]

Bait in Ethanol	*Dendroctonus pseudotsugae*		*Gnathotrichus sulcatus*
	♀	♂	
2.5% Df resin	342	187	48
1% α-pinene	310	157	182
1% β-pinene	10	5	62
1% limonene	161	108	14
1% camphene	403	235	88
1% geraniol	9	5	45
1% α-terpineol	0	0	11
1% myrcene (benzene solution)	0	0	0
Controls			
Douglas-fir log	103	32	36
Grand fir log	0	0	0
95% ethanol	0	0	0

[a] Modified after Rudinsky (1966b). Copyright 1966 by the AAAS.

pheromone system of bark beetles is generally accepted. For most species tested, attraction after initial attack by pioneers is substantially greater to infested hosts than to non-infested hosts (Table 4.5).

Aggregation

Aggregation in response to single compounds has been demonstrated for several species of *Ips, Pityokteines, Dendroctonus,* and others (Vite and Francke, 1976) (Table 4.6). However, for many other species, aggregating pheromones consist of insect-produced chemical and host constituents, which together attract both sexes to the host material.

Mass aggregation of the European elm bark beetle occurs due to an attractant consisting of three synergistic component: (-)-4-methyl-3 heptanol (H), α-multistriatin (M), and α-cubebene (C). The component H is produced exclusively by virgin females whereas M is produced by both mated and virgin females. Compound C is released from elms after beetles of either sex bore into the bark. After mating, boring females continue to release M, C is released, but not H. The amount of the host produced component C appears to be related to the presence of the fungal pathogen *Ceratocystis ulmi*. Although C occurs in healthy tissue, its level in tissues with Dutch elm disease (but without beetles) is approximately 15 times that in healthy tissue (Cuthbert and Peacock, 1978; Gore *et al.*, 1977).

The concentration or mass attack phase is often initiated as pioneer bark beetles come into feeding contact with the inner bark (secondary

Table 4.5. Influence of host constituents in secondary attraction of bark beetles

Insect	Source of Attraction	No. of Responding Beetles
Dendroctonus brevicomis[a]	Fresh log infested with 25 females	37
	Fresh log not infested	0
	25 females	0
Dendroctonus frontalis[b]	Frontalin	1
	Oleoresin	0
	Frontalin and oleoresin	10.7
Ips calligraphus[c]	Ipsdienol and *cis*-verbenol	19.0
	Forty male *I. calligraphus* feeding in a log	37.8
	Ipsdienol and *cis*-verbenol	15.0
	Ipsdienol, *cis*-verbenol, and uninfested logs	38.7
	Ipsdienol, *cis*-verbenol, and uninfested logs	46.0
	Forty male *I. calligraphus* feeding in a log	48.0
Ips grandicollis[d]	Myrcene	12
	Frass extract and myrcene	22
	D,L-camphene	10
	Frass extract and D,L-camphene	22
	D-limonene	3
	Frass extract and D-limonene	13
	Blank control	1

[a] Responses in 24-hr field bioassay (Vite and Gara, 1962).
[b] Average values for beetles responding within 10-min test periods (3 tests) in a field bioassay (Kinzer *et al.*, 1969).
[c] 5, 6, and 3 1-hr replicates, respectively, in a field bioassay. Note that males produce ipsdienol and *cis*-verbenol (Renwick and Vite, 1972).
[d] Number of replicates or time intervals not reported for this field bioassay (Werner, 1972a, b).

Table 4.6 Selective examples of secondary attraction

Insect Species	Compound	Produced by	Attractive to
Ips avulsus	Mixture of ipsdienol and possibly ipsenol	♂♂	♂♂ and ♀♀
Ips paraconfusus	Mixture of ipsdienol, ipsenol, *cis*-verbenol, (by both sexes), and 2-phenylethanol	♂♂ (primarily)	♂♂ and ♀♀
Ips typographus	Methylbutenol, ipsdienol, and *cis*-verbenol	♂♂	♂♂ and ♀♀
Ips duplicatus	Ipsdienol	♂♂	♂♂ and ♀♀
Ips calligraphus	Ipsdienol, *cis*-verbenol, and *trans*-verbenol	♂♂	♂♂ and ♀♀
Ips grandicollis	(−)-Ipsenol	♂♂	♂♂ and ♀♀
Ips acuminatus	Ipsenol, ipsdienol, and *cis*-verbenol	♂♂	♂♂ and ♀♀
Ips cembrae	Ipsenol, ipsdienol, and probably methylbutenol	♂♂	—
Dendroctonus ponderosae	*Trans*-verbenol, exo-brevicomin, α-pinene, myrcene, and terpinoline	♀♀	♂♂ and ♀♀
Dendroctonus jeffreyi	1-Heptanol and probably heptane	♀♀	♂♂ and ♀♀
Dendroctonus frontalis	Frontalin, *trans*-verbenol (in ♂♂ also), verbenone (in ♂♂ also), myrtenol, and α-pinene	♀♀	♂♂ and ♀♀
Dendroctonus brevicomis	Myrcene, exo-brevicomin *trans*-verbenol	♀♀	♂♂ (mainly to exo-brevicomin)
	Frontalin, verbenone	♂♂	♀♀ (mainly to frontalin)

(continued)

Table 4.6. *(continued)*

Insect Species	Compound	Produced by	Attractive to
Dendroctonus pseudotsugae	Frontalin, *trans*-verbenol, seudenol, α-pinene, *trans*-pentanol ethanol, 3,2- and 3,3-MCH (♀♀ primarily), camphene	♀♀	♂♂ and ♀♀
Pityogenes chalcographus	Chalcogran	♂♂	♂♂ and ♀♀
Platypus flavicormis	Mixture of sulcatol, 1-hexanol, and/or 3-methyl-1 butanol	♂♂	♂♂ and ♀♀
Pityokteines curvidens	(−)-Ipsenol	♂♂	♂♂ and ♀♀
Gnathotrichus sulcatus	Sulcatol (mixtures of enantiomers of 6-methyl-5-hepten-2-), ethanol, and α-pinene	♂♂	♂♂ and ♀♀
Trypodendron lineatum	4,6,6-Lineatin, α-pinene, and ethanol	♀♀	♂♂ and ♀♀
Scolytus multistriatus	Delta-multistriatin, (−)-4-methyl 3-heptanol, (−)-α-multistriatin, and (−)-α-cubebene	♀♀	♂♂ and ♀♀

phloem), at which time host monoterpenes combine with or synergize male-or female-produced compounds. There are exceptions to this sequence of events. For example, females of *Leperisinus fraxini* and *Dendroctonus brevicomis* are the pioneers of the species, but both sexes contribute to the aggregating pheromone (Borden, 1974a).

In general, the aggregating pheromones that have been identified are either bicyclic ketals or alcohols and their corresponding ketones. Terpene alcohols, which appear to be particularly effective cues, are represented by relatively simplex oxidation chains of hydrocarbon precursors in the oleoresin of host trees. Oxygenated terpenes are found in the wood of various pine species and Douglas fir; so there is a question whether they are separately biosynthesized by both tree and beetle or are passed on to the beetles from host trees, or whether both processes occur (Sumimoto *et al.*, 1974; Sakai and Sakai, 1972).

A current hypothesis suggests that the degree of host involvement, the minimal required feeding period, and thus the speed of production of aggregating pheromone are all associated with the aggressiveness of the insect species and the relative resistance of its hosts (Borden, 1974a). Thus, in some groups (e.g., *Dendroctonus spp.*) which attack living trees, a rapidly released pheromone may be required to attract the large populations necessary to overcome host resistance (Borden, 1974a). Indeed, *Dendroctonus frontalis* adults may begin pheromone production before host attack (Renwick and Vite, 1968; Coster and Vite, 1972; Renwick *et al.*, 1973) (Table 4.7). Extensive feeding inhibits pheromone release (Vite and Pitman, 1968; Coster and Vite, 1972). In other beetles (e.g., *Ips spp.*), which usually attack weakened trees or logs, pheromone release occurs after feeding or gallery formation, thus signaling host suitability (Vite and Pitman, 1968).

Table 4.7. Association of various pheromones and terpene derivatives with hindguts of *Dendroctonus frontalis* exposed only to α-pinene and β-pinene vapors[a]

	Compounds Detected after Treatment	
Treatment	Male	Female
None	Verbenone, myrtenal, and myrtenol	*cis*-Verbenol, *trans*-verbenol, and myrtenol
α-Pinene	*cis*-Verbenol, *trans*-verbenol, and 4-methyl-2-pentanol	4-Methyl-2-pentanol
β-Pinene	Pinocarvone and *trans*-pinocarveol	*trans*-Pinocarveol

[a] Modified from Renwick *et al.* (1973). Reprinted with permission from: Oxidation products of pinene in the bark beetle, *Dendroctonus frontalis. Journal of Insect Physiology* **19.** Copyright 1973 Pergamon Press, Ltd.

The importance of feeding to pheromone production has been a source of controversy. Evidence suggests both host-dependent and host-independent production of pheromones. Data vary for the same insect. For example, although some research indicates that host components are required for pheromone production in *T. lineatum* (Schneider and Rudinsky, 1969), other data indicate there is no evidence that food in the digestive tract is required for pheromone production (Borden, 1974b).

The role of feeding in pheromone production may be explained, in part, by the two mechanisms of host monoterpene utilization that apparently operate in many bark beetle species. Exposure to pine oleoresin vapors for a few hours stimulates male *Ips* to produce *cis-* and *trans-*verbenol. Similarly, other species exposed to oleoresin vapors can produce a variety of behavior-modifying pheromones, (Klimetzek and Francke, 1980; Byers, 1981) (Table 4.8). Pheromones produced in this manner have been termed contact pheromones (Vite *et al.*, 1972). Alternatively, frass pheromones (e.g., ipsenol and ipsdienol) are produced after beetles feed on host tissues (Vite *et al.*, 1972).

Beetles such as the *Dendroctonus* species can release their chemical signals when nearing host material without the delay imposed by a feeding requirement. Reliance on frass pheromones by beetles that attempt colonization of hosts that resist feeding through mechanisms such as high oleoresin exudation pressure (aggressive species) would be considerably less efficient (Vite *et al.*, 1972). Other less aggressive species (e.g., *Ips* species), which need to determine host suitability (low vigor of tree) before aggregation occurs, are best served by frass pheromones (Vite *et al.*, 1972).

In species that make frass pheromones, production appears to be under neural and hormonal control (Harring, 1978; Gerken and Hughes,

Table 4.8.   Influence of oleoresin and oleoresin components on the production of pheromones

Insect	Exposure to	Production of[a]
Dendroctonus spp.[b]	α-Pinene vapors	*trans*-Verbenol
D. brevicomis (♂♂)	Resin	Ipsdienol
D. brevicomis and *D. frontalis*	Camphene	Camphenol
D. brevicomis (♂♂, ♀♀)	Myrcene	Myrcenol
Ips pini and *I. paraconfusus*	Myrcene	Ipsenol and ipsdienol
Pityokteines curvidens (♂♂)	Myrcene	Ipsenol
P. spinidens (♂♂)	Myrcene	Ipsenol and ipsdienol

[a] From Hughes (1973a,b, 1974, 1975), Renwick *et al.* (1975, 1976), Harring (1978), Byers *et al.* (1979), and Hendry *et al.* (1980).
[b] Includes a variety of *Dendroctonus* species.

1976; Hughes and Renwick, 1977a,b; Bridges, 1982). Nevertheless, many other factors may determine the ultimate product of pheromone synthesis. For example, the chirality of single-host components may determine which pheromone is produced (Renwick *et al.*, 1976).

Termination of Secondary Attraction

The methods for stopping secondary attraction and mass attack in bark beetles range from the termination or reduction of pheromone production to a complex interaction of sonic signals and pheromone production (see Appendix). The decline in attractiveness of elms attacked by female *Scolytus multistriatus* is apparently due to the termination of pheromone production. Similarly, *Ips paraconfusus, I. calligraphus, I. pini,* and *Gnathotrichus sulcatus* reduce pheromone production when the sexes intermingle (Borden, 1974a).

Interaction of males and females can lead to the cessation of secondary attraction. This often occurs as a result of the production of masking or antiaggregation pheromones. Field studies indicate that male *T. lineatum* mask the attractiveness of logs infested with females (Nijholt, 1973). The chemical masking of the attractant produced by females is accomplished by a compound associated with the gut region of males (Borden, 1974b). Olfactory responses of flying male and female *D. frontalis* to female-produced attractants are inhibited by endo-brevicomin. Aggregating male *D. frontalis* also possess the pheromone verbenone, which at high concentration is the primary deterrent to aggregating populations. At still higher concentrations it may further inhibit the responses of both sexes (Renwick and Vite, 1969, 1970). However, at low concentrations verbenone may enhance the aggregation pheromone, at least at close range (Rudinsky, 1973). Finally, 3-methyl-2-cyclohexen-1-one (3,2 MCH) may be produced by *D. pseudotsugae* females in response to stridulation by males. The pheromone blocks the flight response of both sexes to known attractants such as frontalin and *trans*-verbenol (Rudinsky, 1968, 1969). Multifunctional activity similar to that of verbenone has been reported for 3,2 MCH. However, further investigation of this phenomenon may be required since ethanol, which is used as a pheromone solvent in many studies, can have some attractive or synergistic activity. Our current knowledge of the termination of secondary attraction demonstrates that a number of solutions have evolved for the same biological necessity.

Site of Pheromone Production

No conclusive evidence is yet available indicating the exact site of pheromone production in many bark beetle species. However, the hindgut and malpighian tubules have been implicated as the sites of elaboration of pheromones in some species (Percy and Weatherston, 1974). Indeed, extractions and positive bioassays of pheromones have been made of beetle frass and hindguts. Male *Ips paraconfusus* that have fed for several hours on ponderosa pine exhibit an increase in total malpighian tubule length. The hindguts of males removed from nuptial chambers have been found to be attractive. Posterior portions of hindgut become increasingly attractive to females over time, suggesting a concentration of pheromone (Pitman *et al.*, 1965).

Midgut, hindgut, malpighian tubules, and reproductive organs of female *Trypodendron lineatum* have been tested for attractiveness to males. Only the hindgut of unmated females evokes any response by males (Borden and Slater, 1969). Further examination indicates that pheromone production probably occurs in the posterior part of the ileum where epithelial cells enlarge in attractive females. These cells are rich in vacuoles and possess large granular nuclei, suggesting secretory activity (Schneider and Rudinsky, 1969).

Changes in the length of malpighian tubules of *Dendroctonus pseudotsugae* females correlate with the production of pheromone. Newly emerged and re-emerged females that are unmated and feeding have longer tubules than those in which pheromone is not produced (i.e., either unmated and nonfeeding or mated and feeding females). Hindguts arrest male beetle movement. Responding males tend to stay near hindgut samples (Zethner-Moller and Rudinsky, 1967). Behavior bioassays using entire guts and hindguts with malpighian tubule extracts of virgin females taken from host logs result in strong responses by males (Borden and Slater, 1969). Recent preliminary evidence indicates that components of the aggregation pheromone of *Scolytus multistriatus* may be produced by an accessory gland that opens through a so-called vaginal palpi (Gore *et al.*, 1977).

An alternate source of pheromones suggested by other researchers is based on evidence of the ability of fungi and bacteria to oxidize monoterpenes (Fonken and Johnson, 1972). Fungal symbionts of *Dendroctonus frontalis* are capable of synthesizing verbenone by oxidation of *trans*-verbenol and 3,2 MCH from its alcohol (Brand *et al.*, 1976). Similarly, a bacterium isolated from the gut of *Ips paraconfusus* is capable of oxidizing α-pinene to *cis*- and *trans*-verbenol and myrtenol (Brand *et al.*, 1975). Thus, in addition to the host tree and the invading beetle, a third component, symbiotic microflora, may have to be considered in the host-selection/pheromone production systems of bark beetles. However, it should be noted that some researchers believe that the role of microorganisms in pheromone synthesis is subject to debate (Conn *et al.*, 1984).

Pheromone Specificity

Pheromone specificity may occur at the level of production or reception. That is, specific pheromones may be produced, or beetles may respond with varying degrees of specificity. Because of insufficient understanding of pheromone systems, we can evaluate specificity in only a few genera of tree-feeding Coleoptera (e.g., *Ips* and *Dendroctonus*).

Specificity and isolation of species based on pheromones alone are difficult to document. Many pheromones are common to more than one group. Ipsdienol seems to be common to more than one species. Ipsdienol is a common pheromone in *Ips*, and frontalin occurs in many *Dendroctonus* species. The verbenols are found in both genera (Vite *et al.*, 1972).

Sympatric species of *Ips* in different groups are not cross-attractive. For example, the presence of either California or New York male *Ips pini* with male *I. paraconfusus* reduces the trap catch of *I. paraconfusus* (Birch and Light, 1977). However, those allopatric or parapatric species that are in the same group are generally cross-attractive. Sibling species with contiguous boundaries, such as *Ips paraconfusus*, *I. confusus*, and *I. montanus*, are slightly attracted to the pheromone of one another if given no choice. However, if given a choice between their own and that of their siblings, they are able to discriminate (Birch, 1978). What is known about pheromone production and responses to pheromones by species of *Dendroctonus* suggests that reproductive isolation among sympatric species is not based solely on pheromone systems (Lanier and Burkholder, 1974).

Electrophysiological techniques can be used to determine the neurophysiological responses of beetles to chemicals via antennal receptors (Dickens and Payne, 1977, 1978) (Figs. 4.6 and 4.7). However, there are

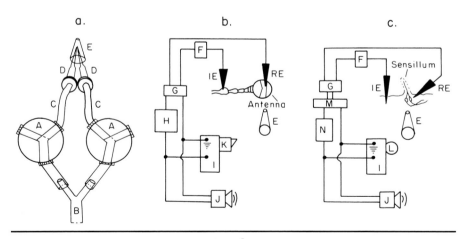

Figure 4.6. (a) Stimulus delivery system, (b) EAG recording system, and (c) single unit recording system. Note: A, 3-way solenoid air valves; B, glass Y-tube; C, teflon tubing; D, Pasteur pipette odor cartridge; E, disposable automatic pipette tips; F, calibration unit; G, high impedance preamplifier; H, low-pass filter; I, oscilloscope; J, audiomonitor; K, Polaroid camera; L, 35mm movie camera; M, AC amplifier; N, high-pass filter; RE, recording electrode; and IE, indifferent electrode. (From Dickens and Payne, 1977. Reprinted with permission of *Journal of Insect Physiology*, Pergamon Press, Ltd.)

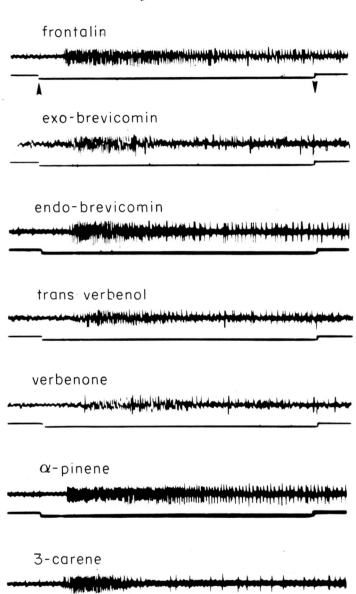

Figure 4.7.   Single-cell response from a sensillum basiconicum of a southern pine beetle (*Dendroctonus frontalis*) female to pheromones and host-tree terpenes. Displaced horizontal bars (indicated by arrows) represent 1-sec stimulations. Delay from stimulus onset to strike initiation was due to an artifact in the delivery system. (From Dickens and Payne, 1978. Reprinted with permission of *International Journal of Insect Morphology and Embryology*, Pergamon Press, Ltd.)

problems in interpreting electroantennograms (recordings of antennal olfactory responsiveness). The electroantennograms representing *D. frontalis* responses to frontalin and brevicomin are essentially the same. Brevicomin is an aggregating pheromone of a related species *Dendroctonus brevicomis*. In addition, a positive response to either pheromone can be eliminated by prior exposure to the other compound. Mixtures of insect- and host-produced synergistic compounds may be required to elicit behavioral responses from beetles. Although host terpenes alone do not usually elicit behavioral responses, both host terpenes and beetle-produced compounds can elicit positive antennal responses (Payne, 1974).

Current Status and Future Perspective of Semiochemicals

The lack of full understanding of host selection in dendrophilous species is, in great part, due to the vagaries and difficulties associated with research technology, lack of appropriate bioassays, and the inherent complexity of the interacting components of host selection. Again the problems in the study of bark beetle host selection provide an excellent illustration.

The isolation of biologically active compounds in large enough quantities for chemical identification and testing of biological activity often is, to say the least, very difficult. Many compounds produced by beetles are in nanogram quantities (10^{-9} g). The conventional approach for isolation of biologically active materials takes the material through a series of fractionations. Fractions are then tested for activity with bioassays. However, consider that just the pentane extracts of *Blastophagous piniperda* can yield 21 terpenes, naphtalin, and some unknown compounds (Francke and Heemann, 1976). Most bioassays depend on laboratory tests that may be quite comparable to field conditions. Conversely, field conditions may influence the responses of beetles or any other dendrophagous species. Further complications arise when compounds are active only as specific isomers or in combination with other particular components.

Indeed, studies of species such as *Blastophagus piniperda*, *Ips sexdentatus*, and *Trypodendron lineatum* have reported contradictory results. One reason for conflicting conclusions is the differences between laboratory and field experimentation. For example, laboratory olfactometer tests of the responses of *Ips typographus* indicate that certain components of its host's oleoresin are attractive. However, compounds that act as attractants in laboratory tests do not demonstrate any attractiveness in the field (Vasechko, 1978). Similar results have been shown for other species (Birch et al., 1977). The presence of stronger stimuli in the field is a possible reason for the discrepancy. In addition, field conditions provide the actual mosaic of stimuli to which the beetle normally responds.

Another approach to the isolation of biologically active compounds uses the principle of differential diagnosis. Inactive insect- and host-produced volatiles are eliminated from consideration by (a) using comparative gas chromatography or (b) fingerprinting compounds that are common to various biologically active materials but that may be absent in others (Fig. 4.8). Interrelations between gas-chromatographic analyses and field bioassays serve as a basis for selection of candidate compounds (Vite and Francke, 1976). Compatible techniques such as mass spectrometry and infrared, ultraviolet, and nuclear magnetic resonance (NMR) spectroscopy are also useful in determining the structures of compounds (Renwick, 1970).

Misinterpretation of data may result from the underestimation of the complexity of the behavior of the insect being studied. In general, the response of an individual of a species to a single component may result in a single movement or behavior that is only one of a large number of sequential or concommitant behaviors leading to host selection. Movement of beetles to the presence of an odor, such as that tested for *I. typographus*, does not necessarily mean that they will be stimulated to search for a host tree. Ethanol and sulcatol have been reported to be the primary attractant and aggregation pheromones, respectively, for *Gnathotrichus sulcatus*. However, detailed studies of the attack behavior of this species suggests

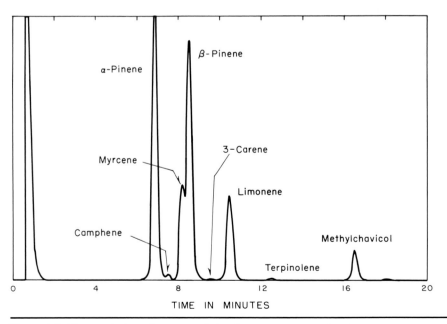

Figure 4.8 Fingerprinting of compounds in pheromones: example of a gas chromatogram of resin volatiles from *Pinus taeda*. (From Renwick and Vite, 1970.)

that ethanol is perhaps more of a boring stimulant than an attractant (McLean and Borden, 1977).

Studies of *Ips*, *Pityokteines*, and *Dendroctonus* suggest that host terpenes exert a close-range arrestant effect on beetles attracted by the multicomponent pheromone. For example, *Ips sexdentatus* and *D. frontalis* in the presence of oleoresin vapors tend to land more readily than when in the presence of pure pheromone (Vite and Francke, 1976). Frass and frass extracts attract walking smaller European elm bark beetles in laboratory olfactometer tests, but these materials are weakly attractive to in-flight beetles in the field (Peacock *et al.*, 1973).

If traps are used to assess candidate chemicals, the trap size can influence the number of each sex (the sex ratio) that is attracted to candidate materials. Southern pine beetles tend to land prior to reaching a source of attraction if the concentration of the attractant mixture is high. The extent of this behavior differs between the sexes. Males appear to be less affected by concentration than females and tend to orient closer to the source of pheromone. Thus, increasing the rate of release of frontalin or decreasing trap size appears to favor the capture of male *Dendroctonus frontalis* (Hughes, 1976). In general, distance orientation and the selection of a landing site may be two distinct behavioral events for the southern pine beetle, although they may be interrelated. Frontalin, the aggregation pheromone that directs distance orientation at low concentrations, may also stimulate landing behavior or lower the threshold for response to other stimuli (either chemical or visual) that guide landing behavior at high concentrations (Hughes, 1976).

Finally, it is becoming increasingly clear that often some apparently small differences in pheromone or kairomone chemistry may cause large differences in beetle responses. The response of the southern pine engraver (*Ips grandicollis*) to its aggregation pheromone, ipsdienol and (S)-*cis*-verbenol, drops dramatically when a ten fold higher concentration of (R)-*cis*-verbenol is part of the pheromonal complex. Recent research has made it clear that there is a strong chiral specificity in pheromone biosynthesis and olfactory receptor systems. *Dendroctonus brevicomis* produces and responds to (−)-frontalin and (+)-*exo*-brevicomin; *Ips grandicollis* and *Pityokteines curvidens* aggregate in response to (−)-ipsenol; *Ips avulsus* responds to (−)-ipsenol in the presence of ipsdienol; and *Ips typographus* responds to (S)-*cis*-verbenol (Vite *et al.*, 1976). Enantiomers of pheromones also have been found to be inhibitory. Enantiomeric specificity of an interspecific pheromonal inhibitor has been demonstrated between *Ips pini* and *I. paraconfusus* (Birch *et al.*, 1977). Thus, the use of pheromones may have to be tempered by chirality response relationships. It has been suggested that a high degree of purity may be required when synthetic pheromones are used in traps or trap trees compared with their use for inducing dispersion or for disrupting communication (Vite and Renwick, 1976).

Chapter Summary

The selection of hosts by dendrophagous insects is a complicated process that involves many specific host cues and behavioral responses by adults or larvae. Long-range selection may be mediated by relatively fixed environmental cues (e.g., anemotaxis, phototaxis, or orientation to large silhouettes) or by olfactory cues. Short-range attraction is usually related to host physical or olfactory cues. Both long- and short-range attraction can be modified by environmental factors. The process of host selection in bark beetles is better understood than that for other forest insects. Bark beetle selection includes several sequential phases: random primary attraction or attraction to specific host cues, secondary attraction via pheromones and host volatiles, aggregation, and termination of aggregation. The selection process is complicated and variable among bark beetle species. Knowledge about host selection and chemical cues (including pheromones) that affect selection behavior has great potential for future application in the management of forest insects.

CHAPTER 5

Symbiotic Associations of Forest Insects, Microorganisms, and Trees

Introduction

Symbiosis as initially discussed by deBary (1879) is a broad concept referring to the living together of two distinct organisms. The term *symbiosis* should be used synonymously with *association*. Mutualism is a type of symbiosis that refers to an association of two living organisms in which both partners benefit. Two other types of symbiosis are commensalism and parasitism. Commensalism is the living together of two organism in which one of the organisms benefits and the other is neither benefited nor harmed. Parasitism is an association in which one partner benefits at the expense of the other. The misuse of *symbiosis* to refer to the mutual benefit of organisms (mutualism) has led to much confusion among students and scientists alike.

Symbiotic relationships are often classified as endosymbiotic (living within) or ectosymbiotic (living outside). This classification is independent of the nature of the association, hence an endosymbiotic relationship can be mutualistic, commenalistic, or parasitic.

Symbiosis between insects and microorganisms, especially fungi, is a common and ecologically important association. In the forest ecosystem, this association can have a major impact on plant species composition, succession, and ultimate productivity. Perhaps the most dramatic example of symbiosis is the transmission of fungal spores by xylophilous (wood-loving) insects. Ambrosia beetles carry spores in specialized structures. The spores are inoculated into galleries where the beetles feed on the fungus reproductive structures.

Symbiotic microorganisms may also be involved in the production or modification of behavior-modifying chemicals. Gut bacteria in *Ips confusus* play a role in the production of the aggregation compound verbenol from host terpenes (Brand *et al.*, 1975). Other dendrophagous insects involved in mutualistic associations with fungi include species in the Coleoptera, Hymenoptera, Hemiptera, and Isoptera (Graham, 1967). Representative species are discussed below. The primary ectosymbiotic fungal groups associated with forest insects are represented by (a) ambrosia fungi, (b) wood-rotting fungi, (c) wood-staining fungi, and (d) species of *Septobasidium*.

The Ambrosia Beetles

The name *ambrosia beetle* refers to various species in the families Scolytidae, Platypodidae, and Lymexylonidae. They are commonly referred to as pinworms, pinhole borers, or shothole borers and usually have wide host ranges. Scolytid species occur in temperate and tropical areas and are represented by the genera *Xyleborus*, *Trypodendron*, *Gnathotrichus*, and *Anisandrus*. Platypodidae are more common to the tropics and include the genera *Platypus*, *Crossotarsus*, *Diapus*, and *Webbia*.

Ambrosia beetles bore into the sapwood and heartwood of freshly felled trees, green stumps, or highly stressed trees. Adults and larvae feed on the fungus that they introduce into the tree and cultivate in their tunnels (Fisher *et al.*, 1953). Tunnel systems consist of an entrance and several breeding tunnels (Fig. 5.1). Larvae occur in special cradles or in galleries made by the maternal beetle (Fig. 5.2). The gallery pattern is usually characteristic for each species. The breeding tunnels and the larval chambers are kept free from frass (wood dust and insect feces) and are lined with fungus.

Fungal spores are carried by the adults and inoculated into the galleries, which are generally initiated by females. The fungal spores or vegetative bodies are carried in structures called mycangia, which are specialized structures for the maintenance (via specialized secretory glands) and transport of symbiotic fungi.

Nature of the Relationship

The fungus in most mutualistic relationships benefits from its association with the insect because its spores are transmitted and introduced into suitable wood. The moisture and nutrient content of the wood allows rapid growth and penetration of hyphae. The susceptibility of the fungus to desiccation is circumvented by being transported in specialized mycangia. The fungus may also multiply in the mycangia. The insect is benefited by the ability of the fungus to weaken wood elements prior to egg hatch. This, in turn, facilitates gallery construction by larvae. Larvae derive their nutrition from the consumption of fungus (Batra, 1967). Insects generally have the capacity to use ergosterol, the most common fungal sterol. In

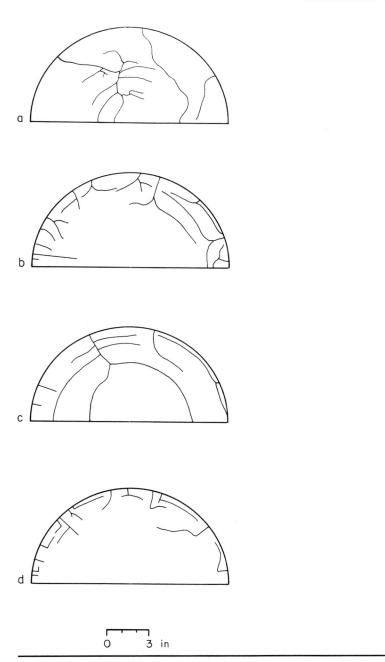

Figure 5.1. Gallery patterns of selected ambrosia beetles: (a) *Platypus wilsoni*, (b) *Trypodendron* spp., (c) *Gnathotrichus sulcatus*, and (d) *Xyleborinus tsugae*. (From Prebble and Graham, 1957.)

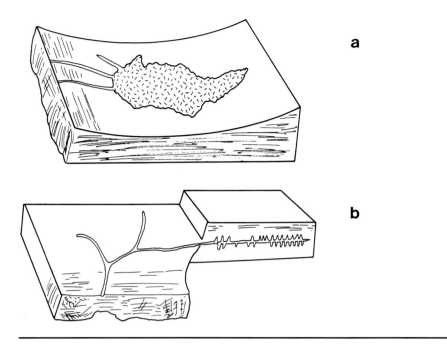

Figure 5.2.   The two general types of ambrosia beetle tunnels: (a) the communal gallery of *Xyleborus xylographus* in hickory wood; death chambers are special portions of the tunnels in which dead individuals are sealed off. The ambrosia fungus grows and sporulates over the surface of the entire communal gallery. (b) Tunnel of *Monarthrum mali,* the type characterized by individual larval galleries. The ambrosia fungus fruits on the surface of the larval galleries. Two general types of ambrosia fungi correspond to the two types of tunnels. (Redrawn from J. G. Leach, "Insect Transmission of Plant Disease," 1940. Copyright © 1940 McGraw-Hill Book Company.)

addition, fungal phenol oxidases acquired from the fungus during feeding may augment an insect's capacity to detoxify heartwood phenols encountered during boring (Martin, 1979).

Function, Storage, and Transmission of Fungi

For a long time it was assumed that the exclusive method of transmission of ambrosia fungi was by the alimentary canal or by adherence to the body surface of the beetle. It is evident now that many beetle species have specific organs for carrying and protecting fungi. The presence of these structures, mycangia or mycetangia, has been verified in over 50 species of beetles (primarily in the Scolytidae) (Francke-Grosmann, 1963). The structure and location of the mycangia vary from species to species. They usually are present in the sex that initially attacks the host tree (generally the female) (Francke-Grosmann, 1963). In the Platypodidae, the male initiates the tunnel, but the sex that carries the mycangia varies. In some species, males and females have similar mycangia, in others, male mycangia are unlike those of females; and in still others, only females have the specialized structures (Farris and Funk, 1965) (Fig. 5.3).

Some of the major types of mycangia include (a) saclike organs on either side of the prothorax (Fig. 5.4), (b) shallow cavities in the pleural

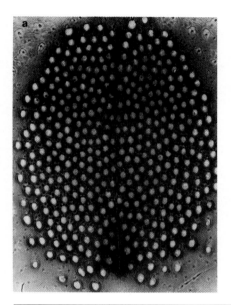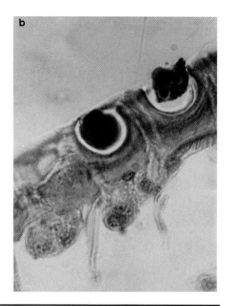

Figure 5.3.   (a) Whole mount of mycangia on the pronotum of a female *Platypus wilsoni* (×125) and (b) sagittal section of two mycangia showing the cell clusters (bottom left and center) and ducts or canals (top right and center) associated with the mycangia (×650). Males of this species do not have mycangia. (From Farris and Funk, 1965.)

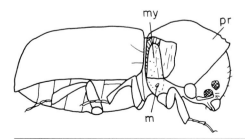

Figure 5.4.   Female of *Xyloterus lineatus* (pr, pronotum; my, mycangium; m, mouth of organ near precoxal cavity). (From Francke-Grosmann, 1967.)

region of the prothorax, (c) dorsal pouches formed by the integumental membrane between thoracic nota (Fig. 5.5), (d) cavities or sclerotized pouches in the base of the elytra (Fig. 5.6), (e) suboral cavities (Fig. 5.7), (f) oral mycangia (Fig. 5.8), and (g) single or clustered glandular pores (Francke-Grosmann, 1963; Nakashima, 1975; Livingston and Berryman, 1972; Happ *et al*, 1971). There are still other types of mycangia and a variety of modifications of various basic types (Figs. 5.9–5.13).

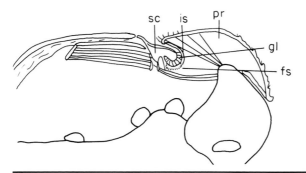

Figure 5.5.   Sagittal section of female of *Xyleborus dispar* (pr, pronotum; sc, scutellum; is, intersegmental membrane; fs, fungus spores; gl, glandular cells). (From Francke-Grosmann, 1967.)

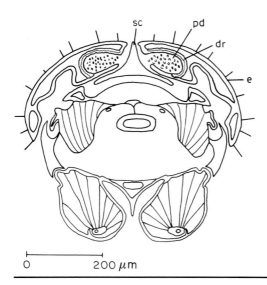

Figure 5.6.   Cross section through posterior end of female of *Xyleborus gracilis* (sc, scutellum; dr, glandular cells; e, elytra; pd, fungus stores). (After Schedl, 1962; from Francke-Grosmann, 1967.)

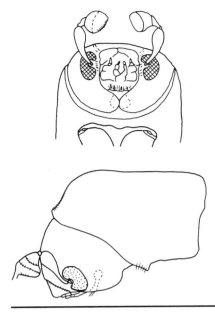

Figure 5.7. Suboral pocket (hatched) of female of *Pterocyclon bicallosum.* (After Schedl, 1962; from Francke-Grosmann, 1967.)

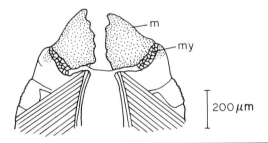

Figure 5.8. Frontal section through head of female of *Ips acuminatus* as seen from behind (m, mandibles; my, mycangium). (From Francke-Grosmann, 1967.)

Mycangia usually contain oily or mucous secretions that protect or provide nutrients to the symbiont (Fig. 5.14). Although the evidence is inconclusive, the oily secretions appear to serve in the nutrition of the fungus and the maintenance of pure cultures. In many species the ambrosia fungi in the mycangia is mixed with yeasts, blue-staining fungi [which have a direct role in the nutrition of the insect (Graham, 1967)], or other microorganisms. Yeasts may also be maintained and transported in

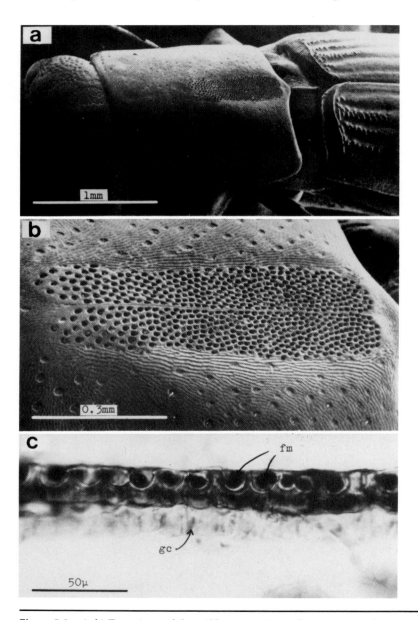

Figure 5.9. (a,b) Two views of the pitlike mycangia on the pronotum of female *Platypus severini* and (c) a longitudinal section of the pitted area (fm, funguslike material; gc, glandlike cell). (From Nakashima, 1975.)

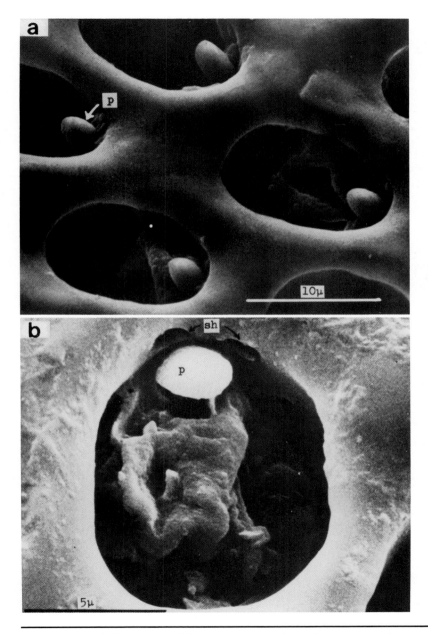

Figure 5.10. (a,b) Two views of the pit-type mycangia on the pronotum of female *Platypus severini* (p, process in the pit; sh, very small holes in the pit). (From Nakashima, 1975.)

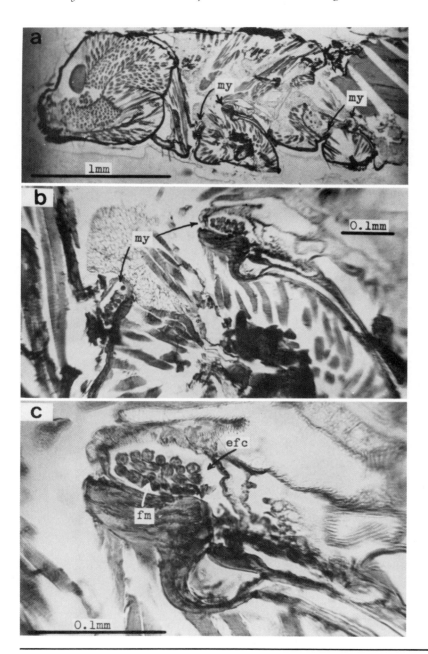

Figure 5.11.   (a) Longitudinal section and (b,c) two views of the forecoxal cavity of male *Platypus severini* (my, mycangium; fm, funguslike material; efc, enlarged forecoxal cavity). (From Nakashima, 1975.)

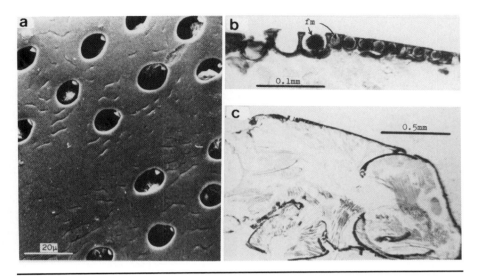

Figure 5.12.   (a) Small pit-type mycangia of female *Platypus curtus*, (b) longitudinal section of the pitted area, and (c) sagittal section of the female illustrating the position of the mycangia (fm, funguslike material). (From Nakashima, 1975.)

oily secretions on the body surface. They are believed to accelerate the growth of ambrosia fungi or to chemically render the wood more suitable for growth of the fungus.

Wood-Staining Fungi and Bark Beetles

Many species of phloem-feeding scolytid beetles are associated with fungi in their galleries. Fungal hyphae penetrate the surrounding wood and cause discoloration. However, wood decay is usually not associated with this fungal growth. The associations are more or less specific and in a very few cases are mutualistic. The association between blue-staining fungi and bark beetles may have developed from their mutual requirements for a similar host tree quality. Reduced water content in the phloem may be required for optimal fungal growth. Similarly, a reduced oleoresin exudation pressure (OEP), usually associated with water deficits, allows bark beetles to breed successfully (Francke-Grosmann, 1967).

Two groups of staining fungi can be distinguished: those disseminated by air currents and those transmitted by insects. Most staining fungi associated with bark beetles belong to the genera *Ceratocystis*. *Ceratocystis ips* is a cosmopolitan species associated with *Ips calligraphus, I. grandicollis*, and other *Ips* spp. Coniferous woods are preferred, and *Pinus* species are particularly susceptible.

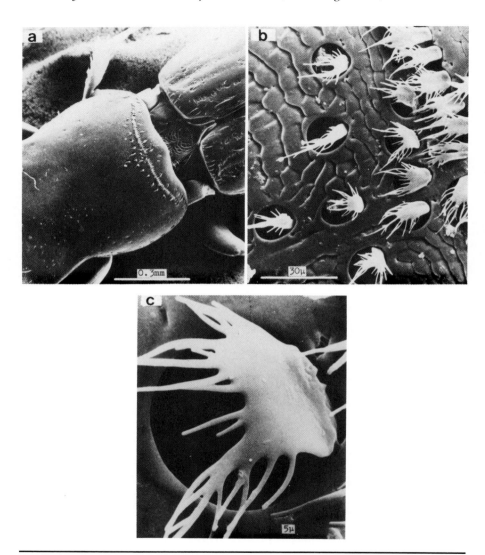

Figure 5.13.   Three views of the crevice-type and pit-type mycangia on the pronotum of *Diapus quinquespinatus*. (From Nakashima, 1975.)

Transmission of Staining Fungi

Spores of most fungal species are well adapted for transmission by their ability to adhere to beetles' integument or to pass intact through the intestinal tract. Mycangia vary from simple pits or pores in the integument (particularly on the elytra) to oral pouches or tubular structures. In beetle species with mycangia, an association exists that is similar to the symbiotic

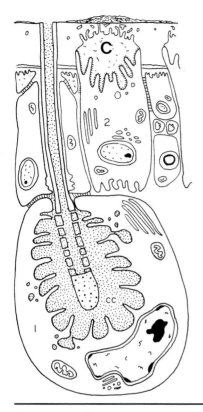

Figure 5.14. Diagrammatic representation of the secretory epithelium of the mycangium of the pine beetle (*Dendroctonus frontalis*) with its lumen at the top of the illustration. Two types of secretory cells are present. The product of type-1 cells (1) passes into a central cavity (cc). Type-2 cells (2) lie against the wall of the mycangium. Their secretions pass into a cavity C and through the overlying porous cuticle to reach the lumen. (From Happ *et al.*, 1971.)

relationship of ambrosia beetles. For example, *Scolytus ventralis* is always associated with *Trichosporium symbioticum*; *Myelophilus minor* is associated with *T. tingens* (Figs. 5.15–5.17).

Nature of the Relationship

Whitney (1982) outlined three ways in which associations of bark beetles with blue-staining fungi can provide benefit to the bark beetle: (1) overcoming tree resistance, (2) enhancing nutrition, and (3) effecting beetle aggregation. The natural resistance factors of conifers, such as the ability to mobilize resin, often present a substantial barrier to attack by bark beetles. When introduced into the tree, blue-staining fungi have the ability to

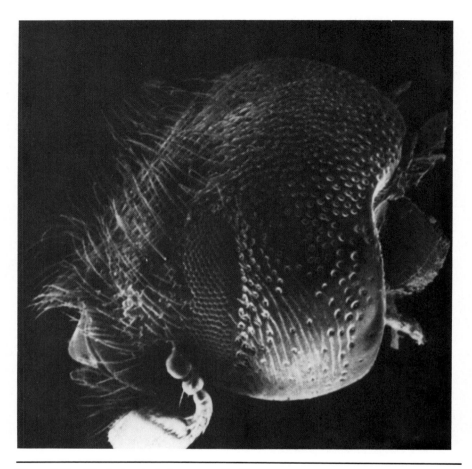

Figure 5.15.   Head of male *Scolytus ventralis*. (From Livingston and Berryman, 1972.)

interfere with translocation and subsequent resin production. Blue-staining fungi help cause the death of the host tree. Indeed, substantial differences of opinion exist among researchers about the relative role of fungi versus beetles in killing trees. Perhaps the most appropriate way to view the question of relative importance is to consider the bark beetle and its symbiotic fungi, functionally, as a single organism.

Although phloem is an adequate food source for bark beetles, staining fungi may enhance its nutritional value. When several species of bark beetle are reared without their symbionts, their general performance is reduced (Whitney, 1982). The initial quality of the host may be an important factor in determining the effect of blue-staining fungi on its subsequent nutritional value.

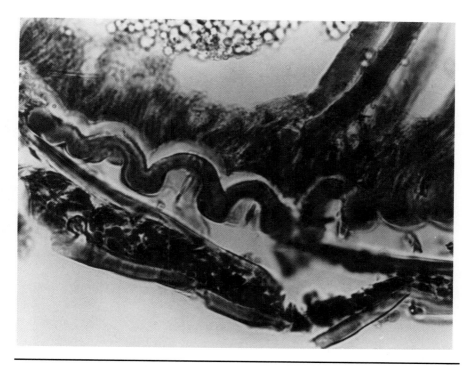

Figure 5.16.   Setae (right of center) arising from the bottom of mycangial pits
(×640) on the head of *Scolytus ventralis*. (From Livingston and Berryman,
1972.)

Wood-Rotting Fungi

Outbreaks of defoliators and bark beetles often persist for several years,
extend over a wide area, and kill many trees. In many situations the action of fungi subsequent to insect attack is of more economic importance
than the initial insect activity. Often, the trees that are attacked and killed
are overmature or less vigorous due to site conditions, competition, or
other environmental factors. In other circumstances, fungi such as root rot
(*Heterobasidion annosum*, predispose trees to insect attack (Alexander *et al.*,
1981; Livingston *et al.*, 1983). Insects also vector several root pathogens
such as *Ceratocystis wagnerii* (Witcosky *et al.*, 1986a). It is often difficult to
assess whether an apparently insect-killed tree may have died without the
aid of the insects. Regardless of the cause of death, the rate of deterioration
in relation to the ability to salvage dead trees may have major economic
implications. Many factors are involved in tree deterioration. Ponderosa
pines killed by bark beetles are not usually salvageable after two seasons.
The timber may be degraded after one season by blue-staining fungi. On
the other hand, bark bettle-killed Engelmann spruce can be salvaged for a
considerably longer period.

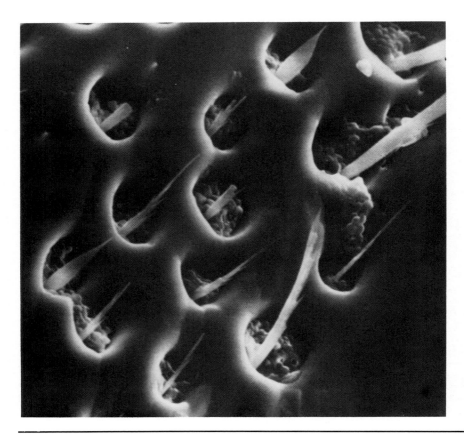

Figure 5.17.   Plumose setae arising from spore-filled mycangial pits immediately posterior to the eye in *Scolytus ventralis* (×1000). (From Livingston and Berryman, 1972.)

Perhaps the best known association of wood-decay fungi and insects is that of wood wasps (Siricidae) and their symbionts (Neumann *et al.*, 1987). Wood wasps are economically important pests that prefer weakened or newly cut trees. The association of wood wasps with wood-rotting fungi affects total yield and lumber quality. Adults are free-living, but larvae tunnel through and subsist on the wood of various tree species. Females of species in the genera *Sirex*, *Urocerus*, *Xeris*, *Tremex*, and *Xiphydria* transmit and implant fungal spores during oviposition. Species such as *Sirex juvencus* and *Urocerus gigas* (the giant wood wasps) have mucous-filled intersegmental saclike structures at the base of their ovipositors that serve as the source of oidia of the fungus. These are extruded with the eggs (Fig. 5.18). Larval wood wasps also have mycangia in the hypopleural region of the first abdominal segment.

Siricid oviposition holes in the bark are often found in association with dead or weakened trees; in turn, fungal development has been as-

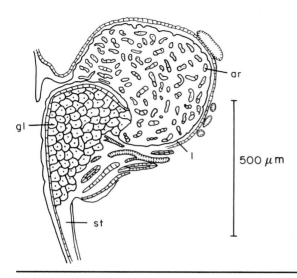

Figure 5.18.   Sagittal section through base of ovipositor of *Sirex juvencus* (i, intersegmental membrane; st, sting; gl, glandular cells; ar, arthrospores). (From Francke-Grosmann, 1967.)

sociated with these oviposition punctures. *Stereum chailletii*, a wood-destroying fungus, has been identified from fungal isolates of the wood surrounding oviposition punctures. Siricids, such as *Xeris spectrum*, the blue horntail (*Sirex cyaneus*) and the white-horned horntail (*Urocerus albicornis*) that attack trees weakened by budworm defoliation and balsam woolly aphid infestation have been implicated in the introduction of wood-decaying fungi (Stillwell, 1960). Balsam fir that have been killed or weakened by infestations of the spruce budworm and the balsam woolly aphid decay quickly. Sapwood penetration by fungi averages 0.4 in. in trees dead for less than one year. In comparable windthrown trees, relatively little decay is present in the first two years after death.

Researchers have also demonstrated a correlation between the amount and rate of fungal heartwood decay and infestation by the white-spotted sawyer beetle (*Monochamus scutellatus*) and the northern sawyer beetle (*M. notatus*). Although probably not responsible for the introduction of the fungus, their tunneling aids in the invasion of the heartwood (Table 5.1). The greatest proportion of the heartwood decay usually spreads from the larval tunnels, which traverse the heartwood (Leach, 1940). Transport of certain deteriorating fungi from dead balsam fir to dying or recently dead trees has been attributed to *Monochamus scutellatus*, the weevil *Pissodes dubius*, and the bark beetle *Pityokteines sparsus*. The spores of the fungi can gain access to the sapwood through the galleries of these beetles. In addition, trees with *Monochamus* spp. activity deteriorate at a faster rate than those with little or no activity (Basham and Belyea, 1960).

Table 5.1. Relationship between the depth of penetration of sap rot (or stain) in deteriorated wood and the density of sawyer beetle holes in killed balsam fir[a]

Years since Death	Season of Death	Study Blocks Examined (Number)	Average Depth of Radial Penetration of Sap Rot or Stain (in.)	Average Density of Sawyer Beetle Holes[d]
1–2	Summer	107	0.42	2.79
1–2	Winter[b]	34	0.47	1.55
1–2	Winter[c]	104	0.46	3.49
2–3	Summer	129	0.62	7.10
2–3	Winter[b]	21	0.47	4.15
2–3	Winter[c]	88	0.55	5.15

[a] Modified from Basham and Belyea (1960).
[b] Trees died without a previous late-summer sawyer beetle attack.
[c] Trees died following a late-summer sawyer beetle attack.
[d] The number of holes per square foot of wood surface.

Nature of the Association

The nature of the association of wood-rotting fungi and insects is highly variable. The advantages and disadvantages derived from the relationship have not been fully explored. Nevertheless, many fungi apparently use host-tree lignin, starch, cellulose, and pentosans. As they do, they proliferate and may kill all or part of a tree. For many wood wasp species, the fungal activity may be essential for survival. The siricid *Sirex juvencus* does not develop beyond the first instar in the absence of its associated fungus, *Stereum chailletii* (Stillwell, 1966). The fungal symbiont may aid in creating favorable habitat conditions by lowering wood moisture content. The fungus may also serve as food for larvae, although it is only a small part of their diet. Fungi produce several classes of enzymes that may continue to function in the gut of the consumer insect and contribute to the digestion of cellulose, hemicellulose, and pectin (Martin, 1979). Regardless of the exact interaction between fungi and insect, species such as *Tremex fuscicornis*, *T. columba* (the pigeon tremex), and *Sirex noctilio* and their associated fungi seem to have an accelerating effect on the death of weakened trees (Francke-Grosmann, 1963, 1967; Coutts and Dolezal, 1969; Parkin, 1942).

Role of Insects in Predisposing Hosts to Fungal Infections

Armillaria mellea, commonly referred to as the shoe-string fungus or armillaria root rot, causes decay in roots and root collars of many host trees. Trees are susceptible to attack when their vigor is reduced. Attack by defoliators, bark beetles, or other dendrophagous insects may be severe enough to reduce tree vigor and predispose the tree to fungal attack. The association of insects and subsequent fungal activity very often makes it difficult to determine the actual mortality agent (Boyce, 1961; Papp *et al.*, 1979).

Conversely, *Heterobasidion annosum* infection predisposes trees to bark beetle attack. This fungus is a problem, particularly following thinning in plantations because spores enter the root-systems through fresh cut stumps. Infected trees become susceptible to bark beetle attack. This phenomenon is similar to the effects of parasitic seed plants. Heavy infestation of dwarf mistletoe on a number of western conifers predisposes these trees to bark beetle attack (Stevens and Hawksworth, 1984).

Other Ectosymbionts

Septobasidium spp.

Various species in the fungus genus *Septobasidium* have a unique symbiotic relationship with scale insects. The fungus serves as a protective mat for the scale insects that live in it (Fig. 5.19). The insect transports the

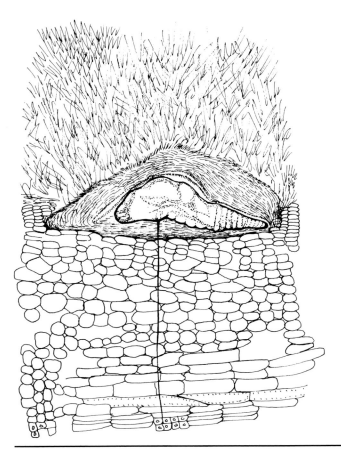

Figure 5.19.   Cross section of *Septobasidium purpureum* showing a parasitized scale insect, over which the fungus has developed a dense pad of hyphae, and the bark of the host plant through which the insect's suctorial apparatus penetrates to the medullary ray cells. (From "The Genus Septobasidium," by J. N. Couch. Copyright 1938 The University of North Carolina Press. Reprinted by permission of the publisher.)

fungus, which may serve as its primary food source. The fungal patch may be layered, consist of communicating chambers, or be a relatively simple mat, such as the patch formed by *Septobasidium pinicola*, a fungus associated with *Matsucoccus macrocicatrices* (Fig. 5.20). The scale feeds on the phloem of the tree. The effects of this symbiotic relationship on the tree are probably negligible (Graham, 1963; Watson *et al.*, 1960).

Sooty Mold

Several fungi utilize insect exudates as substrates on which to develop, and the resulting black growth is called sooty mold. The fungal growth

Figure 5.20.   Fruiting body of *Septobasidium pinicola*. (From Watson *et al.*, 1960.)

occurs on various tree surfaces, particularly leaves and other surfaces covered by honeydew. Honeydew, rich in sugars and amino acids, is excreted by many insect species. Leafhoppers, aphids, scales, mealybugs, and treehoppers produce much of the honeydew on trees; other species, such as flies, wasps, bees and coccinellids, feed on fungus and disseminate the fungal inoculum. The primary concern about severe sooty mold development is the unfavorable aesthetic qualities of leaves and the significant interference with photosynthetic activity (Carter, 1962).

Insects in Relation to Tree Disease

Tree diseases can be caused by organisms such as viruses, bacteria, fungi, or seed plants. Because insect–pathogen relationships span various scientific disciplines, progress in understanding these relationships has been slow. Nevertheless, it is clear that insects play a role in the transmission and inoculation of many causal agents of tree disease. Vector relationships are often difficult to establish. To be considered a vector an insect must occur with some regularity on an infected tree, be associated with the pathogen (internally or externally), and also occur on healthy trees. Infection of healthy trees must occur subsequent to the occurrence of the potential vector.

Many causal agents of tree disease are also transmitted by insect relatives. Walnut blight, for example, is caused by a mite-transmitted bacterium (Carter, 1962), and over 34% of mites associated with American chestnut stump sprouts in Virginia and West Virginia carry *Endothia parasitica*, the causal pathogen (Wendt *et al.*, 1983). The pine wilt disease is a nematode-caused disease of many pine species in the U.S. and Japan. The causal organism, *Bursaphelenchus xylophilus*, is vectored by a common wood borer, *Monochamus carolinensis* (Appleby and Malek, 1982). Black stain root disease fungus is vectored by root bark beetles (Witcosky *et al.*, 1986a).

Infection through Oviposition Wounds

Many causal organisms are not vectored by insects but their transmission is aided by their feeding or reproductive activities. In red pine, abnormalities of current-season foliage due to fungal invasion are associated with the ovipositional activity of a cecidomyid midge. This needle blight causes the lower parts of needles to become hooked and eventually to drop. The fungus *Pullularia pullulans* enters the bases of shoots through openings created by ovipositing midges. Approximately 90% of observed so-called bleeding cankers of red maples are associated with oviposition punctures of the narrow-winged tree cricket (*Oecanthus angustipennis*) (Taylor and Moore, 1979). Other surveys suggest that high infection levels of the pathogen *Hypoxylon mammatum* may be associated with oviposition sites of cicadas (Ostry and Anderson, 1979).

Infection through Feeding Wounds

Infestations of beech trees in the United States and Canada by the fungus *Nectria coccinea* var. *faginata* are dependent on the presence of the beech scale (*Cryptococcus fagisuga*). Both insect and fungal spores are wind disseminated (Carter, 1962). Fruiting bodies develop only in the bark of stems and branches on which the insect has occurred for at least one year. The fungus enters through lesions that result from the scales' feeding. Fungal growth extends into the bark, cambium, and sapwood. Large pieces of bark may crack and eventually fall off. Large trees succumb most readily to infection (Boyce, 1961).

The structure of many beech forests, such as those in Maine, has changed over the past 40 yr due primarily to the death (by disease) or removal (for fuelwood) of larger trees. Today few trees are found that are larger than 12 in. dbh. However, due to the ability of beech to regenerate by the release of sprouts, beech thickets occur in many stands. Interestingly, instead of being eliminated as was the American chestnut (*Castanea dentata*), in many instances there are more beech stems per acre than ever before. Areas formerly inhabited by single old trees now support dense stands of young stems. This situation, together with the cosmopolitan nature of the disease agent, suggests the possibility of epidemic levels of the disease as trees age. A second interesting development observed in the aftermath of disease development is the occurrence of another scale insect, which is much more abundant on beech than before the disease developed. This scale (*Xylococculus betulae*) causes serious erumpent wounds on trees. The importance of this scale may rival that of C. *fagisuga* (Figs. 5.21 and 5.22).

Disease Enhancement due to Insects

Various other examples of disease enhancement through insect activity have been reported. Scots pine trees over 15 years old can develop a massive crown infection that may kill the tree. The development of this condition is associated with epidemic population levels of the pine spittlebug and the fungal invasion of heavily infested trees. The Saratoga spittlebug is also associated with a fungus (*Nectria cucurbitula*) that causes burn blight on red and jack pine. The feeding punctures of the spittlebugs are the points of entry for the fungus (Carter, 1962).

The disease caused by *Endothia parasitica*, commonly known as chestnut blight, has had dramatic effects on the composition of eastern forests. The rapid spread and persistence of this fungus has created a situation in which the few sprouts that develop are soon killed. Insects appear to have played an important role in the success of the blight. Hyphae from germinating spores can enter any wound in the bark that goes through the outer green cortex of stems. Insects are probably responsible for the majority of injuries that serve as points of entry (Boyce, 1961).

Hypoxylon canker is a disease of *Populus tremuloides* induced by the fungus *Hypoxylon mammatum*. This fungus results in the loss of more than one million cords of quaking aspen per year in northern midwestern United States. The poplar gall borer (*Saperda inornata*) is a beetle that lays its eggs in small aspen branches, causing characteristic shield or U-shaped scars. This egg-laying activity and subsequent larval development induces branch galls. In observations of natural fungal infections, cankers are associated with *Saperda* galls. Studies indicate that *Hypoxylon* infection can take place in insect galls on small branches or main stems. Although there may be other means of infection, a relation between the beetle and fungus infection exists and appears to be important in the health of the tree host (Anderson *et al.*, 1976).

Figure 5.21.   Defects deep in wood tissues of American beech (*Fagus grandifolia*) due to infestations of *Xylococculus betulae* and/or infections by *Nectria* spp. that have penetrated the cambial region. (From Houston, 1975.)

Insects As Vectors of Disease Agents

Oak wilt, which is caused by the fungus *Ceratocystis fagacearum*, is responsible for the loss of highly susceptible species in the red oak and white oak groups. Whereas species in the red oak group succumb in one season, those in the white oak group may last several years (Carter, 1962).

Figure 5.22.   Heavy attack and injury of small young beech trees. The circular depressions are probably the result of *Nectria* spp. infections following *C. fagisuga* infestations, which in turn are frequently infested with *X. betulae*. (From Houston, 1975.)

Crown wilting begins as the infected leaves crinkle and become pale, eventually becoming bronzed and then brown. Premature leaf drop may also occur. The changes are primarily due to the establishment of the fungus in the outer vessels (xylem) of the sapwood (Boyce, 1961) and the formation of pluglike growths (tyloses) in the vessels. The fungus also forms a toxin that may aid in the death of the tree.

Sap-feeding beetles of the family Nitidulidae have been implicated as the agents of transmission of oak wilt. These beetles depend on the mycelial mat of the fungus on diseased trees but are also found in bark wounds of healthy trees. Since mats often do not form on wilt-killed trees in certain areas, the search continues for primary vectors. Other research indicates that various species of oak bark beetles (*Pseudopityophthorus* spp.) are probably the primary vectors (Rexrode and Jones, 1970). Although they breed in dying oak trees, the adults feed in twig crotches, at the base of leaf petioles and on twigs of healthy trees (Rexrode, 1968).

Many ring-porous tree species are highly susceptible to mortality caused by blockage of water conducting vessels. Leaf development can not proceed because any interference with the functioning of the vessels prevents water transport. Oak wilt and Dutch elm disease are probably the two best examples of diseases resulting in blockage of water-conducting vessels.

Discovered in Holland in 1919, Dutch elm disease occurs throughout northern, central, and southern Europe, the United States, and Canada. Initial signs of the disease include leaf wilt and discoloration in one or more limbs of the upper crown. Premature leaf drop may follow, and limbs may appear crooked. Cross sections of wood exhibit brown rings in the outer sapwood (Fig. 5.23). The fungus *Ceratocystis ulmi* injures the tree by clogging the water-conducting tissues (Fig. 5.24). The fungus also produces toxins that cause leaf curling and interveinal necrosis. The fungus is vectored by the smaller European elm bark beetle and the native elm bark beetle (*Hylurgopinus rufipes*) (Fig. 5.25). Various other vectors have been recorded and include *Scolytus scolytus* and *Eutetrapha tridentata*, the elm borer (Carter, 1962).

Adults of the primary vector (*S. multistriatus*) usually feed on healthy trees in the crotches of young twigs or leaf axils, thereby introducing the fungus (Fig. 5.26). The beetles then breed and establish colonies in weakened, dying or dead hosts in which the fungus is generally active (Boyce, 1961). In many places in the United States, elms are major shade trees, lining the streets of numerous urban and suburban areas. The loss of these trees is of major economic and aesthetic concern (Fig. 5.27).

Phloem necrosis of American elm is a disease caused by a mycoplasmalike organism, transmitted by the white-banded elm leafhopper (*Scaphioteous luteolus*). It affects both forest and shade trees of all ages and causes a decline of the crown. Early symptoms may include leaf drooping and curling, epinasty, and sparse and yellowish foliage. Before there are any obvious external symptoms of the disease, a large portion of the root system may be affected. Eventually the roots are killed (Braun and Sinclair, 1979). The phloem of large roots may show a yellowing and have a slight wintergreen odor (Boyce, 1961). The leafhopper usually does not occur in great numbers; although when it does, it is not evident because of the tendency of adults to concentrate in the inner parts of the crown.

Figure 5.23.   Symptoms of Dutch elm disease: (a,c) drooping and shriveled foliage, (b,d,f) brown streaks in outermost xylem of twigs, (e) flagging indicating an infected branch high in the elm crown, and (g) a dying elm. (From Sinclair and Campana, 1978; photos by H. H. Lyon, Cornell University.)

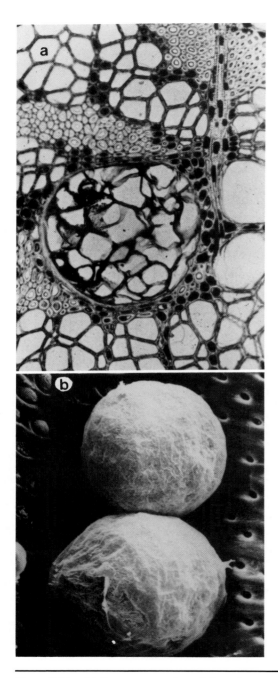

Figure 5.24. (a) Transverse section of elm xylem vessels completely plugged by tyloses and dark materials, which are responsible for the brown discolorations typical of Dutch elm disease (×200), and (b) tyloses in a vessel, which began as extrusions of the elastic walls of adjacent parenchyma cells through pits (upper left ×1600). (From Sinclair and Campana, 1978; photos by H. H. Lyon, Cornell University.)

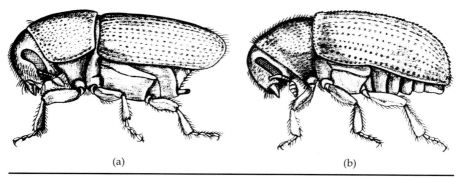

(a)   (b)

Figure 5.25. Adults of (a) the smaller European elm bark beetle (*Scolytus multisriatus*) and (b) the native elm bark beetle (*Hylurgopinus rufipes*). (Courtesy Shade Tree Laboratory, University of Massachusetts.)

Figure 5.26. Feeding by *Scolytus multistriatus* in the crotch of an elm twig. (Courtesy Shade Tree Laboratory, University of Massachusetts.)

Figure 5.27. Elm dying of Dutch elm disease. (Courtesy of Shade Tree Laboratory, University of Massachusetts.)

Infestation of foliage nearest the trunk occurs first. Eggs are laid in the bark, and any hatching nymphs wander about on the nearest foliage. The species feeds principally on the midrib of large leaf veins on the underside of leaves. Excessive feeding may cause the death of apical portions of leaves (Baker, 1949). Mycoplasmalike organisms cause several so-called yellow-type diseases of various tree species, including sandal trees (*Santalum album*) in Asia (sandal spike disease) and oil and coconut palms in tropical regions. These diseases are frequently vectored by species in the Cicadellidae (Sen-Sarma, 1982).

Chapter Summary

Numerous symbiotic associations between microorganisms and dendrophagous insects have been identified. The symbiotic associations discussed in this chapter are restricted to those that are either mutualistic or commensalistic. Ambrosia beetles are wood-boring insects that introduce a fungus into their galleries where it grows rapidly and produces fruiting structures that serve as the main food source of the beetles. This mutualistic relationship allows ambrosia beetles to develop quickly in the relatively poor resource, wood. Specialized structures called mycangia have evolved on ambrosia beetles and other wood-boring insects to support and transport fungi. These fungi benefit bark beetles by assisting in overcoming tree resistance, enhancing the nutritional value of wood, and influencing beetle behavior. Insects also vector fungi that infest wood. In some cases, there is no apparent benefit to the insect for this service. In many cases, the association between insects and symbiotic microorganisms is so intimate that an understanding of their ecological role cannot be achieved without detailed study of both organisms. Various dendrophagous insects have intimate relationships with disease-causing organisms capable of killling trees. The wounds created by feeding and oviposition also serve as points of entry for tree diseases.

CHAPTER 6

Oleoresins in Conifers: Their Nature and Function

Introduction

Attacks on and destruction of conifers by dendrophagous species cause the loss of well over five million board feet of timber per year in the United States alone. These losses would be higher if conifers did not possess defensive mechanism that provide a measure of protection against attack and colonization by numerous insects, particularly bark beetles. How these protective mechanisms work and the factors that cause them to vary partly explain why this protection of conifers is insufficient to prevent economic loss.

It is generally acknowledged that resins are important to the resistance of conifers to dendrophagous insects, particularly bark beetles and to other wood borers, needle miners, and bud borers. Resins produced in response to bark beetle attack include primary resins, which flow from existing resin canals, and resins associated only with trauma (the so-called secondary resins) (Reid *et al.*, 1967; Berryman, 1969). There are some indications that non-host-tree resins are toxic to invading individuals whereas host-tree resins merely alter insect physiology and behavior. Typically, resin permeates the tissues surrounding the point of attack or the tissues surrounding eggs and galleries. This resin permeation is referred to as resinosis.

Oleoresins are compounds composed essentially of resin acids (commercially known as rosin) dissolved in a terpene hydrocarbon oil (commercially known as turpentine). They are produced in and pass through resin ducts of living trees (Figs. 6.1–6.3). The three main groups or constituent terpenes are the monoterpenes (by far the most volatile and abundant), the sesquiterpenes, and the diterpenes (Smith, 1972).

Pines have vertical and horizontal duct systems that apparently connect and permit some degree of resin interchange. In general, resin

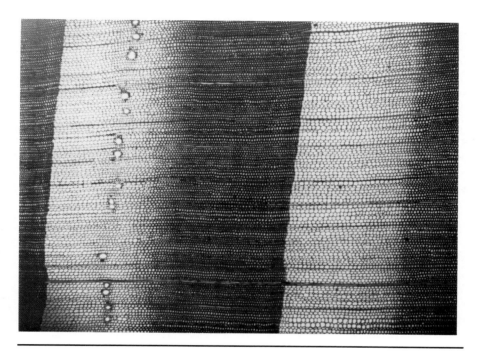

Figure 6.1. Horizontal section of Douglas-fir showing tangential arrangement of resin canals in one annual ring (left) and no canals in another ring (×25). (From Rudinsky, 1966a.)

ducts can be found in needles, shoots, and buds (cortex system) or in the trunk (xylem-resin system). These two systems differ in the size and number of ducts; however, there is a great deal of inter- and intraspecific variation in the size, distribution, and abundance of ducts within each system. The highest resin-duct densities may occur at different heights on the stem in each tree species. Characteristics of the resin system in specific tree tissues vary. In the cortical parenchyma, external to the vascular cylinder of winter buds of *Pinus*, a ring of longitudinal resin canals extends from the base to the bud apex. Also, small lateral canals may extend to the bases of bud scales (Harris, 1960).

Slash pine provides a good illustration of the nature of oleoresin variation in individual trees. The oleoresin of slash pine consists of approximately 20% monoterpenes whereas the remainder consists mainly of resin acids. Monoterpenes in the cortical tissues of branch tips are similar to those in stem xylem oleoresin except that the former has a very high content of myrcene and/or limonene in branch tips. The composition of monoterpenes in stem xylem often varies with height in the tree and depends on distance from the live crown. The cortical tissue oleoresin is

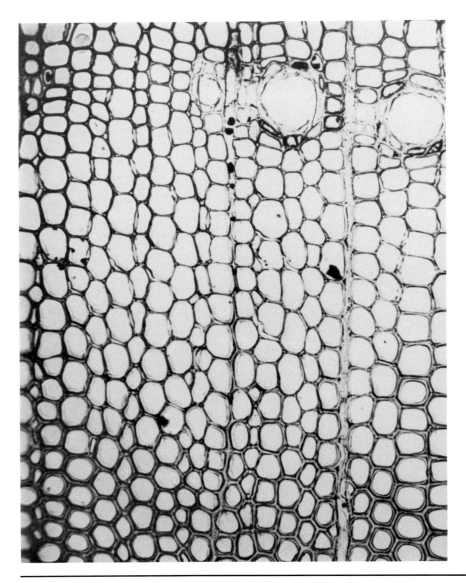

Figure 6.2.   Horizontal section of Douglas-fir showing epithelial cells enclosing resin canals (upper left) (×200). (From Rudinsky, 1966a.)

relatively constant within the crown. Monoterpene composition in needles tends to be similar to that of branch tissues (Squillace, 1977).

Very little is known about the implications of variations in size, distribution, or abundance of ducts in either system. Nevertheless, the number, position, and size of resin canals can help determine the suscepti-

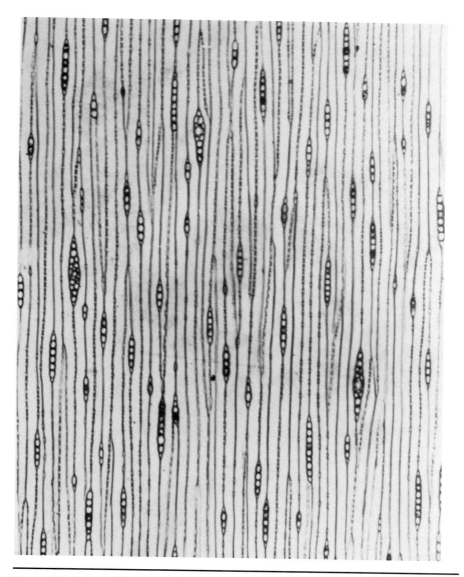

Figure 6.3.   Tangential section of Douglas-fir showing radial resin canals imbedded into the radial rays (×65). (From Rudinsky, 1966a.)

bility of tree species to severe injury. For example, there is a correlation between the internal structure of conifer needles and injury by the pine needle miner (*Exoteleia pinifoliella*). Susceptibility ranges from very low in species such as Scots pine and red pine, which have numerous and large resin ducts, to very high in species such as lodgepole pine, whose needles

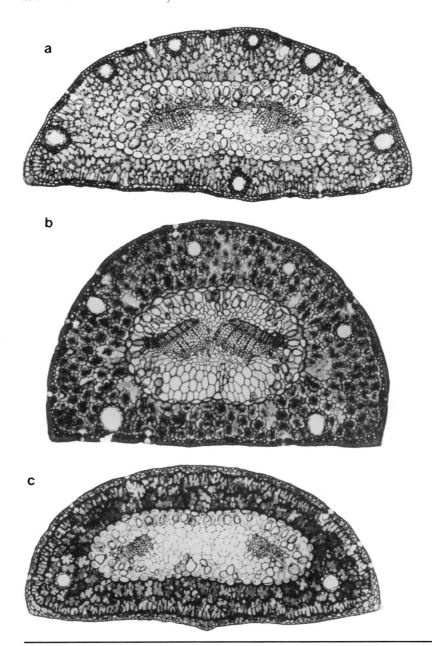

Figure 6.4.   Relative distribution of resin ducts in cross sections of the needles of (a) Scots pine (*Pinus sylvestris*), (b) red pine (*Pinus resinosa*), (c) jack pine (*Pinus banksiana*), and (d) lodgepole pine (*Pinus contorta*). (Plate 7, Plate 8, and Plate 12 reprinted with permission of the State University of New York, College of Environmental Science and Forestry, Syracuse, N. Y.; from Harlow, 1931.)

d

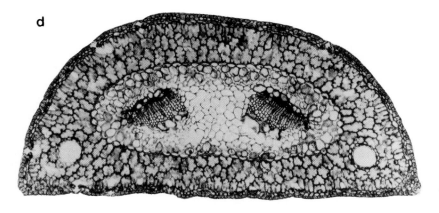

Figure 6.4. (*continued*)

have two ducts at opposite ends of the needle (Fig. 6.4). In Douglas-fir, large vertical channels (up to 90 μm diameter) and small radial channels (15–25 μm diameter) leading to the cambium comprise approximately 0.2% of the tree volume (Rudinsky, 1966a).

Variation in Oleoresin Exudation Pressure

Oleoresin exudation pressure (OEP) describes the force with which resin is expelled from damaged tissues of conifer trees. Variation in OEP occurs among species, within the geographic distribution of a single tree species, among individuals of tree species, within the tissues of a single tree and over diurnal or seasonal periods (Table 6.1 and Fig. 6.5). The daily

Table 6.1. Variation of monoterpene composition in longleaf pine[a]

		Composition of β-Pinene(%)	
	Trees Sampled	**Mean**[b]	**Range**
(a) Tissue			
Basal stem-xylem	58	25	4–47
Cortex	15	46	30–65
Needle	5	80	68–84
(b) Geographical Region			
Louisiana	20	23	8–36
Mississippi	23	26	4–47
Florida	15	27	16–37
Mean		25	

[a] Modified from Franklin and Snyder (1971).
[b] 95% confidence interval: $22 < \mu < 27$.

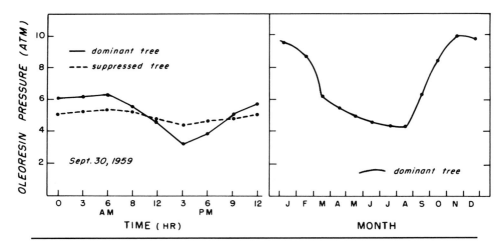

Figure 6.5.   Diurnal course of oleoresin exudation pressure in both a
dominant (———) and a suppressed (---) tree (90 yr old) on a sunny September
day; (b) seasonal course of oleoresin exudation pressure (mean of 35
codominant, 90 yr old trees). (From Rudinsky, 1966a.)

patterns in OEP in loblolly pines and other species are related to both soil
moisture levels and vapor pressure deficits of air (Lorio and Hodges, 1968)
(Table 6.2). Similarly, in species such as ponderosa pine, diurnal variation
is closely associated with changes in transpiration and those factors
affecting transpiration (i.e., temperature, light, and humidity). In general,
OEP values are highest at dawn and drop to their lowest point in early
afternoon (spring) or late afternoon (summer). In the evening and during
the night, OEP increases with declining transpiration (Vite, 1961).

Single measurements of OEP for some species are of limited value
except under extreme moisture stress (Lorio and Hodges, 1968). Variation
in OEP of ponderosa pines is generally somewhat higher in the lower bole
than at midbole or the top of the tree. Radial and circumlateral differences
can also be recorded (Vite, 1961). Seasonal variation in OEP is usually
strongly influenced by soil moisture; OEP usually decreases throughout
the season (Vite, 1961). The severity of fluctuations may be altered by site
conditions. Disease or any other debilitating factor may also cause reductions in OEP (Rudinsky, 1962).

A great deal of research has been conducted to more accurately
classify and predict the likelihood of a tree or stand being attacked by
beetles. Evaluation of trees based on their OEP has been proposed as a
method of selecting high-risk individuals. Selective removal of high-risk
trees as a silvicultural practice would reduce the incidence of outbreaks.

Resin samples taken along the stem of standing and felled trees of
ponderosa pine show only slight changes in monoterpene composition
between trees in the two conditions. Many of the trees show less than

Table 6.2. Change in resin acids and monoterpene hydrocarbons of loblolly pines induced by 4 months of moisture stress[a]

Oleoresin Component	Oleoresin Content (mg/100 mg)	
	Initial Level	Average Weekly Change
Resin acids		
Pimaric	$\underline{5.68^{b}}$ 5.29	$\underline{-0.17}$ -4.49^{c}
Sandaracopimaric + unknown	$\underline{1.57}$ 1.25	$\underline{-0.33}$ -0.025^{d}
Levopimaric + palustric	$\underline{37.48}$ 41.50	$\underline{-1.44}$ -8.68^{c}
Isopimaric	$\underline{0.67}$ 0.67	$\underline{+0.04}$ $+0.23^{c}$
Dehydroabietic + unknown	$\underline{4.95}$ 4.89	$\underline{+1.07}$ 0.31^{c}
Abietic	$\underline{7.77}$ 8.99	$\underline{+1.52}$ -0.14^{c}
Neoabietic	$\underline{7.30}$ 7.59	$\underline{+0.92}$ $+1.44^{2}$
TOTAL	$\underline{65.42}$ 70.18	$\underline{+1.61}$ -8.20^{c}
Monoterpene hydrocarbons		
α-Pinene	$\underline{11.67}$ 12.40	$\underline{-0.14^{c}}$ $+4.71$
Camphene	$\underline{0.21}$ 0.21	$\underline{-0.01^{c}}$ $+0.05$
β-Pinene	$\underline{2.18}$ 9.39	$\underline{-0.54}$ $+4.41^{c}$
Myrene	$\underline{0.80}$ 0.54	$\underline{-0.11}$ $+0.02^{c}$
Dimonene	$\underline{1.47}$ 1.58	$\underline{-0.02}$ $+0.74^{c}$
β-Phellandrene	$\underline{0.38}$ 0.55	$\underline{-0.05}$ $+0.26^{c}$
TOTAL	$\underline{26.71}$ 24.67	$\underline{-0.83}$ 10.19^{c}

[a] From Hodges and Lorio (1975).
[b] Values for control trees are underlined (upper); values for stressed trees are below underlined numbers.
[c] Difference significant at the 0.05 level.
[d] Difference not significant

a 2% change in the five major components, α-pinene, β-pinene, 3-carene, myrcene, and limonene, and only 4 of 35 trees sampled had greater than a 10% change in one component. Alternatively, monoterpene composition of living *Pinus ponderosa* xylem resin has been shown to differ in its myrcene and limonene content from that found in the pitch tubes of trees fatally attacked by *Dendroctonus brevicomis*. The content of the constituents was significantly lower in the killed trees (Smith, 1966a). Similarly, analysis of monoterpene hydrocarbons in foliage vapors of two Douglas-fir clones resistent to animal browsing indicates that more terpenes occurred in the resistant trees than in the susceptible clones sampled (Radwan and Ellis, 1975).

Resin flow is believed to be a critical determinant of the success of insect attack. The flow of resin is often directly correlated to tree vigor. The flow of resin in high-risk, susceptible ponderosa pines virtually stops within 3 days after wounding by bark beetles. In comparison, the flow in low-risk, nonsusceptible trees persists after 3 days (Smith, 1966b). Similarly, stands of ponderosa pines in northeastern California with high average oleoresin flow have less western pine beetle activity than stands that have low rates of flow (Smith, 1972). Studies of resin flow and resin quality and their relationship to ponderosa pine resistance suggest that, at comparable flow, the degree of western pine beetle success is related to resin quality (i.e., monoterpene composition). The effectiveness of limonene against the beetle is greater than 3-carene, which in turn is greater than β-pinene. Thus, quantity and quality of resin may be interacting factors that determine the degree of resistance of a tree (Smith, 1966b).

When bark beetles cause resin to flow from a living tree, the exuding resin is mixed with frass and boring particles. The mixture often forms either a globule or a mass that adheres to the bark (pitch tubes (Fig. 6.6)) or falls to the ground (pitch pellets). In the field these serve as a convenient way of locating bark beetle activity (Smith, 1966a).

Although species differences may account for some of the discrepancies observed from study to study, difficulties in establishing which factors affect the quantity and composition of resin may be due to the extensive, naturally occurring variation found within trees. Within individual trees, substantial vertical changes in the quantity of any one component can be found (Smith, 1968). Extensive sampling of the major constituents of the resin of species such as slash pine indicates that each component shows considerable within-tree and tree-to-tree variation (Squillace, 1971). In addition, the differences in the age of the trees examined in each study may also account for variations.

Regional differences in monoterpene composition also occur. Examination of the composition of cortical oleoresin samples of 16 populations of white spruce in southern Michigan has demonstrated that four out of six monoterpenes differ among geographic areas. The most dramatic differences are between eastern and western trees (Wilkinson *et al.*, 1971).

Figure 6.6. Pitch tubes: evidence of bark beetle attacks. (Courtesy of USDA Forest Service.)

Monoterpene changes for balsam fir, a common associate of white spruce, show similar east–west shifts in monoterpene composition (Lester, 1974). Differences are also present in closely related species. Major quantitative differences have been found in camphene, β-pinene, and total mono-terpenes and in the viscosity of oleoresin of eastern white pine and west-ern white pine. The two species were found to differ in the concentration of 12 of 13 diterpene resin acids (Hanover, 1975).

Mode of Action

The presence or accumulation of resin may interfere with the develop-ment and survival of dendrophagous insects. In many species of bark beetles, gallery elongation is determined by the relative speed with which the host tree responds to attack. The adult beetle abandons its gallery as soon as resin flows into it. Thus, brood mortality is often inversely related to attack density or degree of aggregation (Berryman and Ashraf, 1970). The exact mechanism by which death occurs is not always clear. Egg and larval mortality in *Scolytus ventralis, Dendroctonus ponderosae,* and various other species may result when resin acts (1) as a mechanical barrier to feeding larvae, (2) as a toxicant, or (3) as a barrier to respiration (Berryman and Ashraf, 1970). In some species, such as *Dendroctonus ponderosae*, only direct contact of resin with eggs leads to mortality; volatiles have little or no effect (Reid and Gates, 1970). For other species such as *D. brevicomis*, vapors of nonhosts, as well as resin contact, cause an increase in mortality (Table 6.3).

The fraction of oleoresin referred to as turpentine and comprised of various aliphatic hydrocarbons (terpenes) can be toxic to various insect species (Smith, 1963; Callahan, 1966). Preliminary evidence suggests that the toxicity of resins may be associated with only a few of the various resin components. However, empirical data and the obvious success of bark beetles suggest that the toxicity of oleoresin is not absolute. Polyphagous insect species are generally more tolerant of nonhost resin vapors than oligophagous species, which in turn exhibit greater tolerance than monophagous species.

Although resins may act to paralyze or kill insects, one cannot exclude the possibility that they may act by altering behavior (Smith, 1963). Experiments with *D. ponderosae* and *D. brevicomis* demonstrate that although mortality of adults exposed to resin vapors of hybrid tree hosts does not increase, their feeding activity is reduced (Smith, 1963). Studies also show that tree resins may attract or repel adult bark beetles. Douglas-fir beetles are attracted, while in flight, to vapors from resin of susceptible hosts and repelled when in proximity of nonhosts (Rudinsky, 1966a).

Successful attack by the European pine shoot moth (*Rhyacionia buoliana*) occurs only in pine species having buds with low resin activity. Although larvae have little trouble removing small beads of resin exudate from their mines in buds, large amounts of resin act as strong irritants, inducing the larvae to abandon their mines, or killing them by drowning (Harris, 1960). The vagaries of the interaction between tree resin produc-

Table 6.3. Mortality of adult *Dendroctonus brevicomis* after brief immersion in fresh resin[a,b]

| | Percent Mortality | | | |
| | Test 1 | | Test 2 | |
Source of Resin	3 Days after Treatment	5 Days after Treatment	2 Days after Treatment	3 Days after Treatment
Untreated	0	5	4	17
Ponderosa	2	17	10	26
Jeffrey × ponderosa	3	16	25	35
Jeffrey	11	21	—	—

[a] Modified from Smith (1966b).
[b] 60 beetles per treatment; non hosts are underlined.

tion, tree physiology, insect feeding activity, and site condition suggest there are a great number of complex interactions yet to be discovered.

A final example illustrates the complexity of potential interactions. Grand fir responds to attack by *Scolytus ventralis* and its associated fungus (*Trichosporium symbioticum*) with a hypersensitive reaction involving cellular phloem necrosis and traumatic resin cavity formation in the xylem. The tree's defense also involves increased repellency of its resin monoterpenes. Monoterpenes in wound resin are more repellent than those in cortical blister resin. Wound resins contain more than 46% of the more repellent compounds, myrcene and $\triangle^3$-carene, which are present only in trace amounts in blister resin. Wound resin also contains high concentrations of α-pinene, one of the most repellent compounds. In addition, α-pinene is one of the first monoterpenes produced in fungus-infected wounds (Bordasch and Berryman, 1977). Results such as these suggest a caveat for those evaluating and breeding for resistance: Evaluation of monoterpene composition of the preformed resin system or of preattack trees may not be sufficient for accurate assessment of resistance.

Oleoresin Exudation Pressure and Moisture Stress

The susceptibility of conifers to wood borers is generally believed to be associated with water balance. The expulsion of oleoresin is dependent on the turgidity of the epithelial cells lining the resin channels (Stark, 1965; Bushing and Wood, 1964). With decreasing turgidity, OEP diminishes (Vite, 1961). The OEP of some species (e.g., loblolly pine) may not decrease with increasing soil moisture stress to a point at which flow stops completely. Instead, a minimum level is maintained, apparently unaffected by moisture stress (Mason, 1971). Flow also may be directly or indirectly affected by other factors such as stand density (Fig. 6.7). OEP is therefore considered to be an indicator of water balance in a tree and provides a relative index of susceptibility. A weakened tree with decreased resin flow can be overcome by small numbers of bark beetles compared with the rapid *en masse* invasion required to overcome the resistance of healthy trees.

The changes in the composition of the resin of moisture-stressed trees remains to be discussed. The relative composition of the monoterpene hydrocarbon and resin acid fractions in stressed trees shows significant changes. For example, α-pinene concentration increases in stressed loblolly pines whereas β-pinene, myrcene, and limonene decrease (Gilmore, 1977). These and similar changes may have important effects on insect behavior, since the change in host resins may alter the intensity of attraction to host material and the concentration of pheromone precursors or cofactors. In addition, a decrease in resin acids and an increase in monoterpene hydrocarbons may lead to a decrease in viscosity and rate of crystallization, factors that reduce host resistance to insect attack.

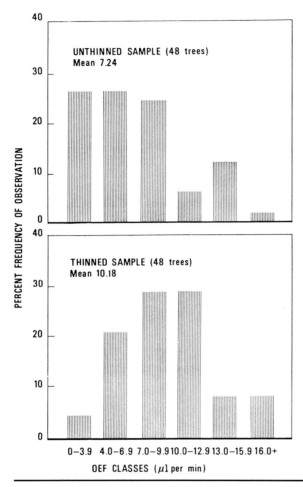

Figure 6.7. Frequency distribution of 48 loblolly pine trees with different rates of oleoresin exudation flow (OEF): (a) in unthinned plantation (mean 7.24) and (b) in thinned plantation (mean 10.18). Means are significantly different at $p < 0.005$. (From Mason, 1971.)

Viscosity and Crystallization

Differences in the physical and chemical nature of oleoresin, other than those already discussed, may be important in the defensive capacities of coniferous trees. Relatively little is known about potentially important factors such as viscosity and rate of crystallization. Like other characteristics of resin, viscosity and rate of crystallization vary widely among conifers. Viscosity is a function of the amount and chemical composition of turpentine. Typically, the crystallization of oleoresin is the result of the

Figure 6.8. Unsuccessful *Scolytus ventralis* attack (bark removed) showing extent of the resinous lesion and resin in the beetle's gallery. (From Berryman, 1969.)

precipitation of resin acids out of solution and is thus a function of the total resin content and the proportion of acids in solution. Although crystallization may occur quite readily in hard pines, wood resins of soft pines seldom exhibit rapid crystallization. In some species, rates of crystallization of shoot resin may differ markedly from those of stem resin (Santamour, 1965a,b).

  In general, rapid crystallization is beneficial to invading insects because they can dispose of semisolid resin more efficiently than viscous resin (Harris, 1960). The ability of an insect to survive potential immobilization by viscous resin is also enhanced by rapid crystallization (van Buijtenen and Santamour, 1972). However, rapid volatization associated with rapid crystallization may dissipate vapors that are toxic to the insect (Stark, 1965). The ability to induce rapid crystallization has been demonstrated in a few insects. Small drops of vomitus from larvae of the pine shoot moth mixed with resin induce crystal formation within a few hours whereas untreated resin remained viscous for several days (Harris, 1960).

A limited capacity to induce crystallization was also demonstrated in the white pine weevil (Santamour, 1965a). White pine cone beetles also are capable of inducing crystallization (Santamour, 1965b). However, some tree species are resistant to insect-induced resin crystallization. The physical properties of resin are important because total flow, flow rate, viscosity, and time to crystallization are some of the best ways to classify southern pine-beetle resistant and susceptible loblolly, shortleaf, longleaf, and slash pines (Hodges *et al.*, 1979).

Traumatic Resin Ducts

As a wood borer tunnels through the bark to the cambium-sapwood interface and constructs a small nuptial chamber and egg galleries, resin ducts are incised and primary resin is released. This resin flows out of the entry hole and often forms a mass or globule on the outer surface of the stem (pitch tube). However, some conifers, such as grand fir, lack resin ducts in the wood and only have resin-containing cavities in the outer bark. Thus, primary resin flow is rarely observed. Alternatively, if secondary resin production in response to attack by a bark beetle, such as

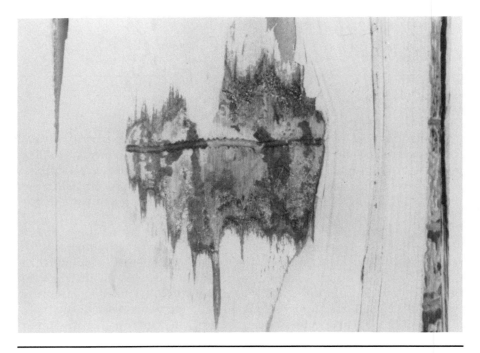

Figure 6.9.   Resinous lesion in 21-yr-old attack by *Scolytus ventralis* in grand fir stem. (From Berryman, 1969.)

S. ventralis, occurs within 12 hr, the insect will be prevented from reaching the vascular cambium (Berryman, 1969; Berryman and Ashraf, 1970) (Figs. 6.8–6.10).

Secondary resin originates from the parenchyma cells of the bark, which are not normally associated with resin production (Figs. 6.11 and 6.12). The tissues next to the wound become soaked with resin. Cells in the

Figure 6.10. Resinous lesion in a 4-month-old wound (area around gallery) and vertical extension of nonresinous dead inner bark. (From Berryman, 1969.)

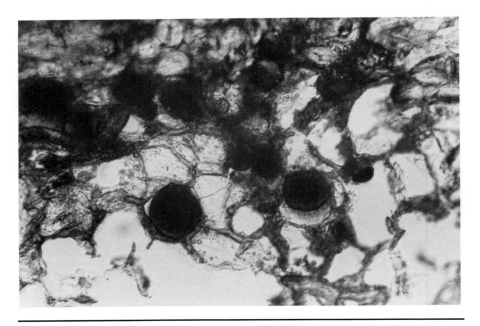

Figure 6.11.   Transverse section of the inner bark surrounding an attack by *Scolytus ventralis* showing resin globules (dark staining) in the distended cells of the phloem parenchyma. (From Berryman, 1969.)

inner bark break and separate to form cavities that fill with secondary resin and are often referred to as traumatic resin cavities. Secondary resin may also flow out the entry hole or flow vertically along the stem in what are sometimes referred to as traumatic resin ducts (Reid *et al.*, 1967).

Other Defensive Reactions

The defensive mechanism that has received the most attention and has been researched most frequently is the production of oleoresin. This type of resistance exists in the tree before attack. Recently, attention has been centered on mechanisms of resistance or defense that are directly induced or initiated by attack. When the inner bark is infected by insects and/or fungi, a tissue reaction characterized by rapid cellular desiccation and necrosis is initiated around the wound site. In addition, tree tissue response may include chemical synthesis and release of various compounds including terpenes and polyphenols from dying parenchyma cells. Wound periderm or callus tissue may form at the wound site. Chemicals toxic to insect and/or fungi may be produced in the wound periderm. Lesion formation represents a containment of invading organisms. Previously unsuccessful attacks may be evident as darkened reaction zones in

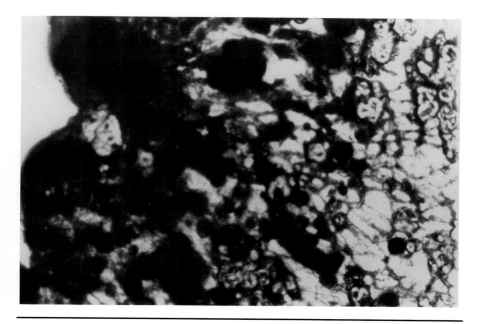

Figure 6.12. Transverse section of the reactive tissues surrounding a *Scolytus ventralis* attack showing resin flow toward the insect's tunnel (left of photograph). (From Berryman, 1969.)

the wood. The proliferation of necrotic lesion callus tissue may result in the formation of traumatic resin canals (Berryman, 1972).

Chapter Summary

Conifers have a vast array of defenses with which they enhance their survival. One of the most important defenses is an elaborate system of resin production. Resins protect trees by mechanically blocking entry, killing insects outright, or modifying behavior. The force with which resin is expelled from damaged tissues is referred to as oleoresin exudation pressure (OEP). Many environmental factors influence OEP, and OEP is an indicator of the overall survivability of a tree after attack. The composition of the resin is also an important aspect of the defensive capability of the tree. Basic understanding of the resin production processes in conifers is necessary to understand the various roles resins play in defense against insects.

Forest and Shade Tree Insects: Their Impact on Trees

CHAPTER 7

Defoliating Insects

Introduction

One of the major feeding guilds in the forest ecosystem is insect defoliators. Most of these forest species are characterized by small populations that rarely exhibit major density fluctuations. However, some species are characterized by periodic fluctuations and high numbers; this often represents a severe disruption of typical pattern of plant–defoliator interactions and the stable state of the ecosystem. These species are the more renowned forest defoliators such as the gypsy moth, spruce budworms, tussock moth, pine butterfly, and others. The impact of forest defoliators on their tree hosts ranges from imperceptible changes to the death of trees (Fig. 7.1).

How Defoliation Affects Trees

Feeding injury to leaves (the primary photosynthetic organs of trees) by defoliators disrupts the production, utilization, and transfer of substances required for growth, regulation of growth, and survival. Growth perturbations can be exhibited as short- or long-term changes in increment (e.g., diameter, height, or shoot growth), bud survival, leaf size and number, seed and cone production, vigor (susceptibility to secondary attack by insects and pathogens), or overall survival. Dramatic changes may occur in the forest stand as repeated defoliation affects species composition and nutrient cycling in the forest (Fig. 7.2).

Various distinct changes in insect-defoliated trees have been documented in the scientific literature. Selected studies illustrate a variety of these effects (Tables 7.1 and 7.2). The issue of the impact of insect defoliation on trees may be approached by understanding the associated variables that influence changes in trees. For example, there are major differences in the impact of defoliators on angiosperms and on gymnosperms.

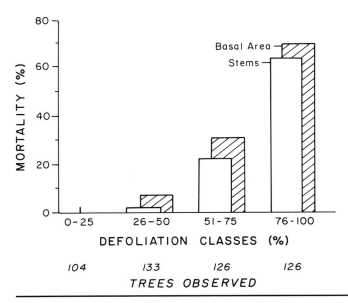

Figure 7.1. Relationship between defoliation by the hemlock looper (*Lambdina fiscellaria lugubrosa*) and mortality of western hemlock (*Tsuga heterophylla*). Shaded bars refer to basal area and white bars to stems. (From Johnson *et al.*, 1970.)

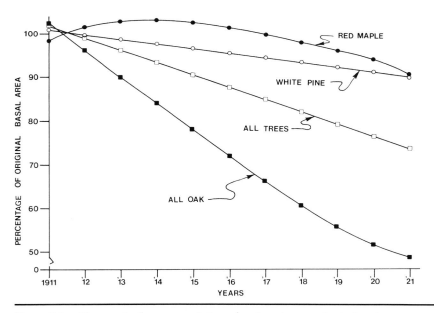

Figure 7.2. Changes in the representation of various tree species in forest stands subject to defoliation by the gypsy moth (*Lymantria dispar*) (1911 represents the 100% base). All oak, ■; all trees, □; white pine, ○; red maple, ●. (From Campbell and Sloan, 1977b.)

172

Table 7.1. Effects of insect defoliation on forest trees

Cause of Defoliation and Parameter Affected	Tree Species	Year of Evaluation	Reduction per Degree of Defoliation	Photosynthetic Organ Consumed or Removed	Method of Evaluation	Reference
LEAF SIZE Artificial defoliation	Larch (*Larix larcina*)	1[a] 2 3	Degree of Defoliation 0 · 25 · 50 · 70[b] 24.8 · 18.7 · 20.4 · 15.5 22.3 · 16.3 · 15.0 · 12.9 26.6 · 16.9 · 18.0 · 12.0	Needles	Dry foliage weight (mg)	Ives and Nairn (1966)
Larch sawfly (*Pristophora erichsonii*)	Larch (*Larix larcina*)		Degree of Defoliation 0 · 25 · 50 · 75[b] 25 · 19 · 19 · 17	Needles	Dry weight changes due to larvae feeding	Graham (1931)
Artificial defoliation	Red oak (*Quercus rubra*) and red maple (*Acer rubrum*)	Year after defoliation	50% · 75% · 100% Red oak[c] 85 · 67 · 56 Red maple[c] 63 · 76 · 57	Leaves	Measurement of regrowth foliage	Heichel and Turner (1976)
HEIGHT GROWTH Artificial defoliation	Eastern white pine (*Pinus strobus*)	0[e] 1	Degree of Defoliation I · II · III · IV · V · VI · VIII[d] 3.25 · 8.0 · 8.05 · 1.9 · 3.5 · 5.05 · 17.25 10.19 · 7.38 · 6.35 · 5.24 · 7.93 · 4.18 · 4.56	Various ages of needles	Increase in height (in.)	Linzon (1958)
European pine sawfly (*Neodiprion sertifer*)	Scots pine (*Pinus sylvestris*)		Colonies/Tree 10 · 25 · 50 · Control · Hand Defoliation 14 · 23 · 37 · 0 · 63	All ages of needles (younger needles preferred)	Terminal shoot elongation (average growth over 3 yr in cm)	Stark and Cook (1957)
Red pine sawfly (*Neodiprion nanulus nanulus*)	Red pine (*Pinus resinosa*)	1[f] 2 3	21% 64% 70%	Old needles	% reduction in height growth compared with previous base year with no defoliation	Kapler and Benjamin (1960)

(continued)

Table 7.1. (continued)

Cause of Defoliation and Parameter Affected	Tree Species	Year of Evaluation	Reduction per Degree of Defoliation	Photosynthetic Organ Consumed or Removed	Method of Evaluation	Reference
Lodgepole needle miner (*Coleotechnites milleri*)	Lodgepole pine		**Degree of Defoliation** 0 — 10 — 40 41.8 — 38.1 — 29.6	All ages of needles (younger needles preferred)	Terminal shoot elongation (average growth over 3 yr in cm)	Stark and Cook (1957)
RADIAL INCREMENT European sawfly (*Neodiprion sertifer*)	Scots pine (*Pinus sylvestris*)		**Colonies/Trees** Control — Hand Defoliation 10 — 25 — 50 · 71 18 — 47 — 53 · 0 — 71	Old growth	% loss per increment core boring (mean annual ring width in mm)	Wilson (1966)
White fir sawfly (*Neodiprion abietis*)	White fir (*Abies concolor*)	$n + 4$[g] $n + 5$	**Degree of Defoliation** 65% 58% less than n-1 33% less than average of period between n-10 and n	One-year-old or older foliage	Radial growth reduction as determined by increment cores	Strubble (1957)
Artificial defoliation	Larch (*Larix larcina*)		**Degree of Defoliation** 0 — 25 — 50 — 75 — 100 1873 — 1318 — 1124 — 665 — 226	Needles	Actual average total volume (cm²)	Graham (1931)
Jack pine budworm (*Choristoneura pinus*)	Jack pine (*Pinus banksiana*)	0[h] 1 1 2 2	**Degree of Defoliation** Light — Medium — Heavy — Very Heavy 32(sm) — 60(sm) — 83(sm) — 99(sm) 27(sp) — 44(sm) — 73(sm) — 91(sm) 54(sp) — 76(sp) — 99(sp) — 99(sp) — — — — 20(sm) — 86(sm) — — — — 52(sp) — 86(sp)	New needles	% reductions in average ring width in cross sections at 1, 4, 5, 9.5, 14.5, and 19.5 ft above the ground	Kulman *et al.* (1963)

Tip Damage Class [i,k]

				I	II	III	IV			
Western spruce budworm (*Choristoneura occidentalis*)	Grand fir (*Abies grandis*)	Preoutbreak period		3.96	2.56	2.89	2.85	Needles	Average radial growth determined in cross sections at various heights on the tree	Williams (1967)
		Outbreak period		2.55	2.12	1.77	1.74			

				I	II	III			
Western spruce budworm (*Choristoneura occidentalis*)	Engleman spruce (*Picea engelmanni*)	Preoutbreak period		1.52	1.47	1.29	Needles	Average radial growth determined in cross sections at various heights on the tree	Williams (1967)
		Outbreak period		1.53	1.35	1.06			

ROOT MORTALITY

Degree of Defoliation

			Root Type	0	25	50	70[l]			
Artificial defoliation	Larch (*Larix laricinia*)	3 yr after initial year of defoliation	Living	20	15	11	8	Needles	Total length (m) of larger roots	Ives and Nairn (1966)
			Dead	2	1	1	4			

Degree of Defoliation

			Normal (15% dead)	Medium (31–50%)	Heavy (75%)	Complete (100%)			
Spruce budworm (*Choristoneura fumiferana*)	Balsam fir (*Abies balsamea*)	Type I[m]	6–7	2–3	2	—	Current-year foliage	Loss is average number of trees with specific rootlet mortality[n]	Redmond (1959)
		Type II	—	1	14	—			
		Type III	—	8	8–9	10			

CONE AND SEED PRODUCTION

Degree of Defoliation

			0–Lt	Lt	Med	Hvy	V.hvy	V.hvy[a]			
Jack pine budworm (*Choristoneura pinus*)	Jack pine (*Pinus banksiana*)	3-yr after initial year of defoliation	32	60	40	12	0	0	1st and 2nd year cones and in cone bearing portion of the pine twig	% of 100 tips producing male cones	Kulman *et al.* (1963)

(continued)

Table 7.1. (continued)

Cause of Defoliation and Parameter Affected	Tree Species	Year of Evaluation	Reduction per Degree of Defoliation — Degree of Defoliation				Photosynthetic Organ Consumed or Removed	Method of Evaluation	Reference
			0–25%	26–50%	51–75%	76–100%			
TREE MORTALITY									
Red-headed pine sawfly (*Neodiprion lecontei*)	Loblolly pine (*Pinus taeda*) (1–5 ft tall)						Late summer defoliation of trees (%)	Mortality of trees (%)	Beal (1942)
			Shaded 0	33.3	55.0	96.5			
			Unshaded 0	0	0	84.2			
			Crown Class[p]	Medium	Heavy	Very Heavy	Needles	Mortality of trees (%) over 3-yr outbreak period	Kulman *et al.* (1963)
Jack pine budworm (*Choristoneura pinus*)	Jack pine (*Pinus banksiana*)		Progressive	4	2	29			
			Provisional	4	6	40			
			Regressive	9	13	4			
			Suppressed	71	50	87			

[a] 0,1,2,3 denote year of initial defoliation and 1, 2, and 3 yr after defoliation, respectively.

[b] Values denote average foliage weight (mg) per fascicle. Differences among defoliation classes were significant.

[c] Average percent of the area of the undefoliated controls.

[d] I through VII denote, respectively, removal of current-year foliage, removal of 1-yr-old foliage, removal of 2-yr-old foliage, removal of current-year and 1-yr-old foliage, removal of current-year and 2-yr-old foliage, removal of 1-yr-old and 2-yr-old foliage, and no foliage removed on check trees.

[e] 0 and 1 denote measurements in the fall of the year of defoliation (spring defoliation) and the year following defoliation.

[f] 1,2,3 denote evaluations after 1 yr of severe defoliation, 2 yr of severe defoliation, and 2 yr of severe defoliation plus 10% defoliation on the third year, respectively.

[g] n = first year of outbreak (defoliation); $n − x$ or $n + x$ denote x number before or after that year.

[h] Year 0 = year of initial defoliation, followed by years 1,2,3, etc.

[i] sm is summerwood reduction, sp is springwood reduction.

[j] Classes I, II, III, and IV indicate, respectively, <30% of tips dead, no top killing; about 50% of tips dead, tops stunted or killed in the past, new leaders developing; more than 50% of tips dead, top dead, 50:50 chance of tree survival; and >50% of branches dead, little crown foliage, tree dying.

[k] For *Abies grandis*, adjusted values are statistically significant. For *Picea engelmannii*, adjusted values for damage classes I and II are not significantly different, but those for damage classes I and III and II and III are significantly different.

[l] Data manifest high variability. Although a trend is indicated, differences are not statistically significant.

[m] Types I, II, and III denote sequences of previous defoliation (1950–1956) in which there were, respectively, early average low levels of defoliation (<25%) and average moderate-to-heavy levels (26–80%) the last 2 yr; early low levels (1950–1952), moderate levels (1953–1954) followed by heavy levels (1955–1956); and generally heavy and very heavy levels throughout the period of previous defoliation.

[n] Categories of mortality denote the percent mortality of 20 random rootlets from each root examined.

[o] Defoliation concentrated in the top of the tree.

[p] Progressive, provisional, regressive, and suppressed trees denote, respectively, lead dominants and strong codominants; conditional dominants and codominants; weak

Table 7.2. Progressive impact of spruce budworm defoliation on balsam fir[a]

Years of Severe Defoliation[b]	Impact
1	Flowers and cone crops die. Radial growth loss occurs in the upper crown.
2–3	Small roots begin to die. Radial growth loss occurs over the entire stem. Height growth ceases. Some treetops die.
4–6	Suppressed trees in the understory and mature and overmature trees in the overstory begin to die. Tree growth and wood production nearly cease.
7–15	Budworm populations begin to collapse. More trees die, particularly balsam fir. Some seedlings and saplings die. Dead trees begin to deteriorate as a result of disease, secondary insect attack, and wind breakage. Protective cover in deer yards is diminished.

[a] From Schmitt *et al.* (1984).
[b] Seventy-five percent or more of current-year growth.

In general, angiosperms (broadleaf trees) have greater carbohydrate reserves, are able to refoliate, and are thus better able to withstand defoliation. Many deciduous trees may survive several years of defoliation with minimal changes in growth. Differences in defoliation tolerance can be observed among conifers. Pines, for example, have greater reserves (and thus tolerance to defoliation) than spruces (Sukachev and Dylis, 1964). Deciduous conifers such as larch (tamarack) may also be relatively resistant to injury (Graham, 1931). Conversely, evergreen conifers such as spruce, hemlock, and balsam fir are more susceptible to injury or death. Interspecific tree variation and the vigor of trees also may be important considerations (Figs. 7.3 and 7.4).

Importance of Timing of Defoliation

Another important variable influencing the impact of insect defoliation on trees is the timing of defoliation in relation to the age of the lost foliage (Larson, 1964; O'Neil, 1963). Leaf consumption toward the end of the season's growing period often has much less impact than early season defoliation. Spring defoliation of a tree species in temperate North America may cause great reductions in growth. Defoliation of the same tree species by another generation of the same insect species in August may have little or no effect. This differential impact may be related to seasonal differences

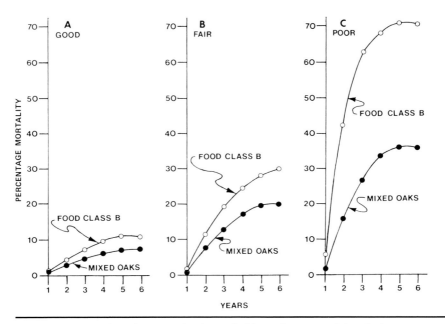

Figure 7.3. Mortality of tree species in good, fair, and poor condition during the years following one heavy defoliation by the gypsy moth (*Lymantria dispar*): (○) tree species that are acceptable to the gypsy moth as food but are not particularly favored, and (●) a composite of red, black, scarlet, and white oak. (From Campbell and Sloan, 1977b.)

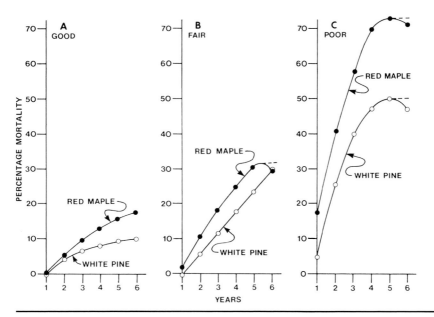

Figure 7.4. Mortality of white pine (○) and red maple (●) trees in good, fair, or poor condition following one heavy defoliation by the gypsy moth (*Lymantria dispar*). (From Campbell and Sloan, 1977b.)

in photosynthate production or the differential impact of the loss of old versus new needles. In many species, such as white pine, removal of new foliage has more of an effect than removal of old needles (Table 7.3). Although defoliation of old and/or new needles causes growth reductions in red pine, only when new needles are removed is the growth reduction evident 2–3 yr after defoliation. In stands defoliated by the spruce budworm, host trees that are subjected to the loss of some old foliage in addition to the current-year foliage die sooner than those losing only the current-year growth (Blais, 1958b). The preferences exhibited by some insect defoliators for either old or new foliage can thus serve as guides to their potential to injure specific hosts (O'Neil, 1963; Kulman, 1965a).

The timing of the life cycle of the insect in relation to the occurrence of new foliage can influence the impact of its defoliation. The first generation of the sawfly *Neodiprion rugifrons* consumes only old foliage of jack pine and causes little injury. When a second generation occurs, new foliage is also consumed, and 75% of the completely defoliated trees die (Wilkinson *et al.*, 1966).

Another aspect of timing is the relationship between the life history of defoliators and the timing of budbreak. Onset of budbreak can be altered from year to year by the duration and severity of defoliation. A 4-yr analysis of the timing of defoliation and budbreak in red maple and red oaks has led to a hypothesis which proposes that defoliation might play a role in the regulation of dormancy of buds (Heichel and Turner, 1976). The hypothesis suggests that growth-promoting substances, such as gibberellins, may be more abundant in defoliated trees, thereby mediating the release of dormant buds in the spring.

The degree of synchronization between egg hatch and budburst varies but can have a significant influence on insect survival as well as host tree survival. Budburst of white fir and egg hatch of the Douglas fir tussock moth are closely related to accumulated degree-days. Peak egg hatch occurs when 77–97% of the buds burst (Wickman, 1976a,b).

The late opening of the buds on black spruce (*Picea mariana*) apparently is an important factor in its immunity to the spruce budworm. Budburst does not take place until 10–14 days after it occurs in balsam fir and white spruce. Consequently, the larvae are forced to feed for a longer period on old foliage or unopened buds, which do not provide larvae with adequate nutrition. Thus, larval starvation or abandonment of trees results in the reduction of the spruce budworm population on this host species (Blais, 1958a).

In England, winter moth larvae generally hatch 2 weeks before most of their host's leaves have flushed. This results in high larval mortality (Varley *et al.*, 1973). Similarly, the survival of hatching winter moth larvae in Nova Scotia is closely correlated to the degree of synchronization between egg hatch and host budburst (Embree, 1965, 1967). Winter moth attack is heavier on trees that flush early. The basal area increment of the

Table 7.3. Shoot growth (in cm) of eastern white pine before and after artificial defoliation. Note that annual growth in white pine is completed by July[a]

| | Shoot Growth (cm) | | | | | |
| Treatment | July 1960 Defoliation[b,c] | | September 1960 Defoliation | | April 1961 Defoliation | |
	Before	After	Before	After	Before	After
Complete defoliation	15.8	0.0[d]	12.2	5.5[e]	16.2	0.0[d]
Removal of current needles only (1960)	14.1	6.4	14.4	7.8	16.2	4.8[f]
Removal of all but current needles	11.6	19.6	14.8	21.6	16.4	21.1
No defoliation	15.9	22.1	15.1	22.8	17.2	23.5

[a] Modified from Dochinger (1963).
[b] July 1960 was within growing season, September 1960 was during dormant season, and April 1961 was before candle elongation.
[c] Before and after refer to growth in 1960 and 1961, respectively.
[d] No surviving pines.
[e] Two surviving pines.
[f] One surviving pine.

trees that flush late usually continues to increase during an outbreak, whereas that of trees that flush early drops sharply (Embree, 1965, 1967). In oak forests in Russia, 73% fewer oak defoliators occur on trees that flush 20–30 days late (Sukachev and Dylis, 1964). The synchrony of hatch and budburst may be so important that crucial behavior patterns may evolve to insure synchrony. *Pristiphora abietina* females lay eggs only on buds of their hosts that are beginning to open. If the buds are not at the correct stage, many of these sawfly females do not lay eggs (Sukachev and Dylis, 1964).

Finally, the pattern of feeding in relation to the occurrence of budbreak, leaf flush, and leaf expansion can also be important. Radial growth of aspen, the primary host of the forest tent caterpillar (*Malacosoma disstria*) begins approximately 1 week after budburst and larval emergence and quickly reaches a high rate. The final instar of the tent caterpillar occurs when at least one-fourth of the radial growth has been added and leaf production and linear growth of vigorous shoots are approximately one-half completed. Thus, when larval feeding and, consequently, major foliage destruction are completed, approximately one-half of the annual radial growth has been laid down, and linear growth and leaf production are still 2–3 weeks from completion (Rose, 1958). A clear understanding of the chronological sequence of feeding in relation to specific physiological or growth changes in a tree allows a more precise evaluation of the actual impact of a defoliator.

Regardless of the modifying influences of these factors, defoliation often reduces shoot growth, diameter increment, and survival, although the extent of the injury is a function of the severity of defoliation (Fig. 7.5). Reduction in leaf or needle size may be very localized and occur only in those parts of the tree in which injury actually occurs. Individual branches may be affected by disbudding or localized defoliation (Kulman, 1965b).

Effect of Location on Defoliation

Attack by defoliators may be concentrated in specific crown zones. Feeding by the spruce budworm and other defoliators and subsequent mortality may be concentrated in the tops of trees (Batzer, 1973; Kulman and Hodson, 1961). Changes in tree shape may result from top-killing or heavy feeding and the associated leaf size changes that may occur in the upper parts of trees (Graham, 1931; Duncan and Hodson, 1958; Kulman and Hodson, 1961; Kulman et al., 1963; Kulman, 1965a).

The concentration of feeding in specific areas of the crown may be influenced by intracrown variation in foliage characteristics. Upper-crown leaves of red oaks in dense stands are approximately 43% heavier per unit area than leaves in the lower crown (reflecting differences in food reserves). In open stands, upper-crown leaves are approximately 18% heavier than those in the lower crown. However, although needles from the top of the crown of white pine in dense stands are only approximately

a

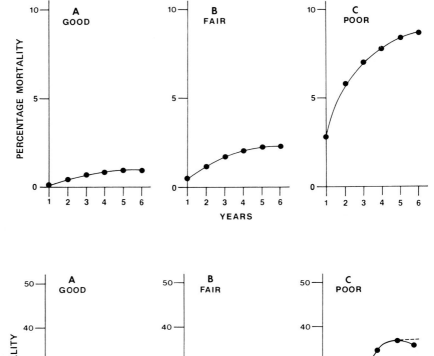

b

Figure 7.5.   Relationship between severity of defoliation by the gypsy moth (*Lymantria dispar*), and mortality among mixed oaks rated in good, fair, or poor condition during the 5-yr period following (a) low defoliation levels, (b) one heavy defoliation, and (c) two heavy defoliations. (From Campbell and Sloan, 1977b.)

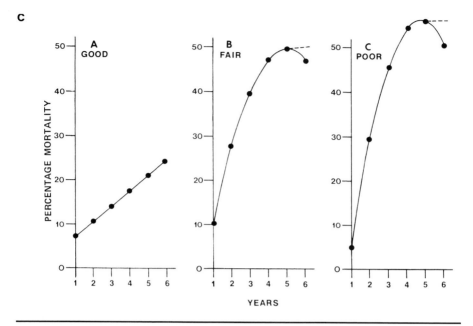

C

Figure 7.5. (*continued*)

12% heavier per given length than those in the lower crown, the upper-crown needles of trees in open stands are 33% heavier than lower-crown needles. In deciduous species such as red, white, and scarlet oak, American elm, Norway spruce, and beech, the area of leaves from the upper crown is usually smaller than the area of leaves from the lower crown (Potts, 1938).

The so-called sun leaves in the upper crowns of forest trees or the upper and outer crowns of shade trees are different from the shade leaves in the lower crowns or interior areas of the trees. Sun leaves are generally smaller, thicker, hairier, and more deeply lobed than shade leaves (Vogel, 1968). These and other associated differences can affect the development and survival of defoliators. Larvae of *Lymantria dispar* and *Lambdina athasaria*, for example, attain greater weight when fed foliage from the top of the tree crown compared with larvae consuming lower-crown foliage (Potts, 1938). Supporting evidence has been provided by experiments in which gypsy moth larvae exhibit a feeding preference for sun leaves when given a choice between leaf types (Schoonhoven, 1977). This preference, as well as a preference for full-grown young leaves, is exhibited in other species (see Fig. 4.2 in the chapter on host selection).

Finally, the degree of shading of the whole tree may influence its survival after defoliation. Pines in open areas are less subject to mortality due to red-headed pine sawfly defoliation than are shaded pines

Table 7.4.   Mortality of shaded and unshaded pines defoliated by the redheaded pine sawfly (*Neodiprion lecontei*)[a]

	Mortality							
	0–25% Defoliation		26–50% Defoliation		51–75% Defoliation		76–100% Defoliation	
	No. Trees	Percent Dead	No. Trees	Percent Dead	No. Trees	Percent Dead	No. Trees	Percent Dead
Shaded trees	18	0	15	33.3	20	55.0	57	96.5
Unshaded trees	16	0	19	0	17	0	38	84.2
Totals	34	0	34	14.7	37	29.7	95	93.7

[a] Modified from Beal (1942).

(Table 7.4). Other factors may enhance this differential susceptibility since shading of pines due to over-topping hardwoods may lead to root competition and other stresses (Beal, 1942). (For further detail see Recovery and Compensatory Responses of Defoliated Trees.)

Overall Impact of Defoliation

In general, reduced growth of trees in the years after defoliation is a result of reduction in needle (or leaf) number and total leaf area. Although tree mortality is not always to be expected, defoliation can cause extensive mortality (Fig. 7.1). Often, tree mortality is not observed for a number of years after the initial outbreak. Balsam fir mortality in areas defoliated by the spruce budworm can be the same on control and defoliated plots until the fourth year of infestation (Batzer, 1973; Hatcher, 1964). The defoliation history may also be critical (Table 7.5).

When defoliation does have an impact, it can be substantial. Above ground woody production in quaking aspen ranged from 2 to 6 kg annually before a period of severe defoliation (1959–1968) by the forest tent caterpillar. By 1963, all trees were producing less than 1 kg, a reduction of approximately 90%. The severity of the growth reduction was, in part, enhanced by lower-than-normal rainfall (Pollard, 1972). Similarly, defoliation by spruce budworm reduces volume of balsam fir by one-fourth at the end of a 7-yr outbreak and by two-thirds, 5 yr later (Batzer, 1973). Even relatively short-term defoliation by the gypsy moth and the elm spanworm results in differences in bole shrinkage in defoliated trees of approximately 13 μm compared with a 157-μm increment in undefoliated trees (Stephens et al., 1972) (see also Table 7.1). Finally, defoliation can affect specific types of cambial activity. Defoliation of alpine ash (*Eucalyptus delegatensis*) by a phasmatid (*Didymuria violescens*) reduces diameter growth and affects the

Table 7.5. Influence of history of defoliation, site, and age on the mortality of trees killed between 1955 and 1961[a]

Category	Number of Live Trees in 1955	% Mortality from Various Causes							Total Killed (%) (1955–61)
		Fomes	Hypoxylon	Nectria	Insect	Wind	Mechanical	Unknown	
All trees	4458	0.4	5.4	0.6	1.1	0.3	1.7	20.6	30.1
History[b]									
L	465	0	4.9	0	0.2	0.2	2.4	19.1	26.8
H-L	377	1.3	4.5	0	0	0.5	1.1	15.6	23.0
L-H-L	353	0.3	6.8	0.3	0.6	0.3	1.4	22.4	32.1
H-H-L	405	0	7.2	0	0.5	0	1.5	15.1	24.3
H-H-H	309	0.3	7.4	2.3	2.9	0	1.6	34.6	49.1
Site									
Good	1267	0.3	5.4	0	2.2	0.3	1.9	18.9	28.7
Medium	2710	0.4	5.9	0.9	0.7	0.3	1.4	20.2	29.8
Poor	437	0	4.4	0.7	0.4	0.2	1.6	26.7	33.7
Age									
<43 years	2644	0.3	5.9	0.5	0.8	0.2	2.1	22.8	32.6
≥43 years	1770	0.4	4.9	0.7	1.6	0.3	0.8	17.4	25.8

[a] Based on: Long-term effects of defoliation of aspen by the forest tent caterpillar. G. B. Churchill, H. H. John, O. P. Duncan, and A. C. Hodson, (1964). *Ecology*, **45**: 630–633. Copyright © 1964 by Duke University Press. Reprinted by permission.
[b] L or H denotes 1 yr of light or heavy defoliation.

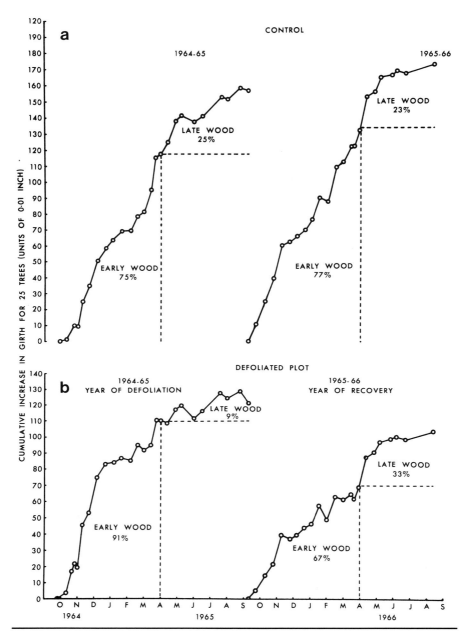

Figure 7.6.   Diameter growth at breast height (dbh) and estimated proportion of late and early wood formation in alpine ash (*Eucalyptus delegatensis*) trees: (a) the undefoliated, control trees and (b) trees defoliated (1964-1965) by a phasmatid (*Didymuria violescens*). (From Mazanec, 1968. Reprinted with permission of Australian Forestry.)

Table 7.6.   Average annual cone production in 1975
and 1976 for balsam fir trees for three spruce
budworm defoliation classes[a]

Year	Average Number of Cones per Tree[b]		
	Lightly Defoliated in 1976	Severely Defoliated in 1976	Severely Defoliated in 1975 and 1976
1976	402	1	0
1975	0	0	2

[a] Modified from Schooley (1978).
[b] The decreased 1975 shoot development probably reduced tree
vigor and limited the availability of sites upon which reproductive
buds could have developed.

production of late wood during the year of defoliation and that of early
wood during the following recovery year (Mazanec, 1968) (Fig. 7.6).

Recovery often ensues in the years after defoliation. Depletion of
reserves due to refoliation and/or growth reductions may have indirect
and long-term effects. Defoliation can reduce cone and seed production
(Table 7.6). In spruce stands defoliated by the nun moth, only nondefoliated trees produce seeds (Kulman, 1971). Flowering, fruiting, and seed
production can be a considerable drain on reserves and thus further
vegetative growth reduction may ensue (Morris, 1951a; Tappeiner, 1969;
Kozlowski and Keller, 1966; Stark and Cook, 1957). Conversely, fruit size
and seed yield may decrease greatly in the years following the reduction in
the number of leaves below some species-related critical value.

Compensatory Growth

Recently, several authors have suggested that the rate of growth reduction
in plants may not be a linear function of defoliation (Schowalter *et al.*,
1986a; Wickman, 1980; McNaughton, 1983). This ability of heavily im-
pacted trees to compensate for loss of foliage is generally referred to as
compensatory growth. Only recently have researchers begun to consider
the possibility that trees can compensate for or recover some of the growth
impact sustained after attack by insect herbivores. This compensation can
occur through a thinning effect in which the herbivore causes mortality of
weaker individuals in the stand. When the weaker individuals are killed,
nutrients and water become more available to the remaining trees (Mattson
and Addy, 1975).

Several other mechanisms of compensatory growth have been suggested (Chew, 1974). One proposed method is an increase in the nutrient
cycling rate due to insect feeding. The hypothesis in this case is that, by

feeding on old foliage that would soon drop from the tree, insects accelerate the decomposition of foliage, which effectively increases the available
nutrients. Grace (1986) did not find evidence to support this in eastern
forests defoliated by gypsy moth. Another possible mechanism is the
so-called photosynthetic efficiency hypothesis. In this case the hypothesis
states that heavy defoliation followed by refoliation with young, high-
efficiency foliage allows the tree to more efficiently utilize the available
resources. Reducing the amount of foliage on trees also increases the total
exposure of the remaining foliage to sunlight. In this way, a lower total
amount of foliage in the young age class could result in equal or greater
biomass accumulation. Clearly, considerably more research is needed to
determine the extent and importance of compensatory growth in forest
ecosystems.

Variation Associated with Growth Patterns

Although there are various patterns of shoot growth in temperate-zone
trees, a brief discussion of two important patterns illustrates its influence on the impact of defoliation. In various tree species (i.e., maple, pine,
and beech) shoots are preformed in the winter bud. Differentiation within
the bud occurs during the first year, and extension of the preformed
parts into a shoot occurs during the second year. Leaves in this preformed
type of shoot have similar morphology. In a second (heterophyllous) type
of growth pattern, some shoots (short shoots) are preformed in the winter
bud, but other (long) shoots are not fully preformed inside the winter bud.
Early leaves of long shoots emerge at budbreak; late leaves (of the second
flush of growth) appear later in the growing season after the early leaves
have expanded. There is considerable dimorphism in the venation, size,
toothing, thickness, and stomatal development in the two leaf types.
Examples of this latter type of tree include white birch (*Betula papyrifera*),
yellow birch, red gum (*Eucalyptus rostrata*), several species of maple, larch,
and some tropical pines. Differences in shoot growth patterns are correlated with patterns of photosynthate use, which in turn can determine the
impact of defoliation (Kozlowski and Clausen, 1966).

   Shoot growth of tree species with predetermined shoots generally
occurs relatively quickly, within a short period after the frost-free season.
In heterophyllous species, shoots usually grow for a long time and rely
heavily on current-year photosynthate. Although some current photosynthate may be used, predetermined species generally use stored reserves for shoot elongation and respiration. This is accomplished early in
the season before any extensive production of new photosynthate occurs
(Kozlowski and Clausen, 1966). Similarly, in many evergreen species the
old needles supply the carbohydrates needed for expansion of the new
shoots (Kozlowski and Keller, 1966). Thus, defoliation of old leaves of
certain species by early defoliators can be devastating to the growth that
occurs during the remainder of the growing season. Deleterious effects

due to defoliation by late-season defoliators are evident in the tree species that use current photosynthate (e.g., temperate species with heterophyllous shoot development, recurrently flushing southern pines, and some tropical tree species).

Defoliation and Hormonal Physiology

The growth of buds or shoots in trees may be influenced by the distal buds of certain shoots. Apical dominance is a relationship that determines whether or not a bud grows. If there is strong apical dominance, the bud does not grow into a shoot. Once the bud is growing, apical control influences how much it will grow. Thus, in a given year the lower shoots of a tree may be shorter than the leader due to the apical control the leader exerts. In defoliated red pine trees, many secondary axes elongate more than the terminal leader, and secondary axes in lower whorls frequently grow more than those in upper ones. This suggests that reserves from foliage may be necessary to maintain greater elongation of the terminal leader (Kozlowski and Winget, 1964).

The physiological events associated with hormonal controls can be disrupted by the activities of defoliators. Plant hormones such as auxins occur in high concentrations in stem tips, young leaves, and flowers and are very important in cell elongation in shoots and in stimulation of cambial activity and root primordia. Destruction of buds and lack of photosynthetic activity in defoliated species can result in decreased availability of carbohydrates and the hormones that are translocated with them.

Species with intact shoots produce 20 times as many roots than debudded plants (Kozlowski and Keller, 1966). A mortality of rootlets of more than 30%, and more than 75% results from defoliation of new shoots of balsam fir of 70% and 100%, respectively. Although young trees immediately produce new rootlets as defoliation pressures are reduced, mature trees often fail to recover (Redmond, 1959). In artificially defoliated tamarack trees, complete defoliation results in the death and decay of all fine roots. However, in some partially defoliated trees (even up to 75% defoliated), the root system may appear perfectly normal (Graham, 1931). Severely defoliated red maples refoliate the same season and exhibit greatly reduced starch content in their roots (Wargo *et al.*, 1972) (Fig. 7.7). Thus, it is very clear that any feeding that impairs or destroys buds or young leaves can have far-reaching effects on neighboring or distant parts of a tree.

Site Characteristics

The quality and structure of the site in which trees grow can have a tremendous impact on the development of epidemic populations of defoliators, the extent and severity of defoliation and the recovery of

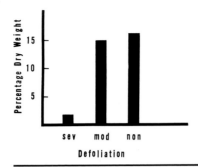

Figure 7.7. Autumn starch content in roots (% dry weight) from severely defoliated (sev), moderately defoliated (mod), and nondefoliated (non) sugar maple (*Acer saccharum*) after one defoliation by the saddled prominent (*Heterocampa guttivitta*). (From Wargo *et al.*, 1972.)

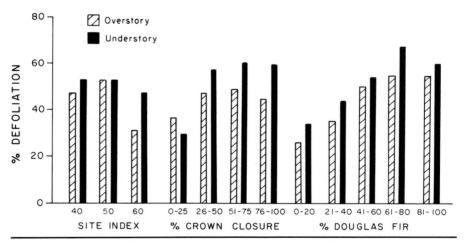

Figure 7.8. Influence of site characteristics on the amount of defoliation of overstory and understory trees by the spruce budworm (*Choristoneura fumiferana*) in a western Montana forest. (From Fauss and Pierce, 1969.)

defoliated trees (Fig. 7.8 and Tables 7.4 and 7.7). These factors are of such importance to all dendrophagous insects that they are dealt with elsewhere in separate sections.

Species differences, crown (dominance) position, and tree age actively interact with site variables in determining host susceptibility to defoliation. Most tree species show a trend toward greater susceptibility to defoliation when individuals are very young and overmature (Fig. 7.9).

Table 7.7.   Soil moisture content during July of various sites defoliated by the gypsy moth (*Lymantria dispar*) and the elm spanworm (*Ennomos subsignarius*) and of undefoliated sites.[a]

Drainage Class	Slope Position	Aspect	Soil Moisture Content (%)[b,c]	
			Undefoliated Sites	Defoliated Sites
Moderately well drained	Lower	North	39.5	50.4
Well drained	Upper	South	32.3	31.4
Well drained	Ridgetop		20.5	29.5

[a] From Stephens *et al.* (1972).
[b] At 0–15 cm level.
[c] Relatively high moisture content of sod, in part, reflects deposition of 46 mm of rain during storm 2 weeks prior to sampling.

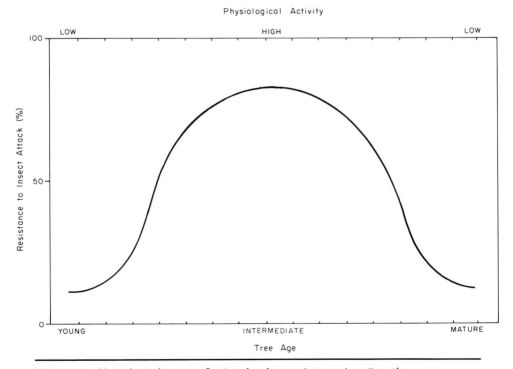

Figure 7.9.   Hypothetical curve reflecting the changes in tree vigor (i.e., the ability to resist or recover from injury) as a function of growth rate (metabolic activity) and age.

Recovery and Compensatory Responses of Defoliated Trees

Different tree species may not respond similarly to the same degree of foliage loss. Complete defoliation of yellow birch, tough buckthorn (*Bumelia tenax*), eastern hop hornbeam (*Ostrya virginiana*), and American basswood (*Tilia americana*) saplings results in 30%, 20%, and 0% tree mortality and 78%, 76%, and 8% bud mortality, respectively (Kulman, 1971). Even within the same species, differences among individuals in their physiological condition (vigor), as well as in their dominance in the crown, may cause differences in mortality due to defoliation (Fig. 7.10).

The effects of defoliation on budbreak in red maple and red oak provide a good illustration of differential species responses to defoliation. In maples that were 100% defoliated for 2 yr, budbreak occurs approximately 2 days earlier than in the year prior to defoliation. On the other hand, in red oaks that were completely defoliated for 2 yr, budbreak occurs 4 days later than in the year prior to defoliation. Both red maple and red oak produce equivalent proportions of primary foliage in the years following 100% defoliation, however, oaks average 67% of the area of the primary foliage of control trees in years following 75% defoliation, and maples produce 76% of the area of foliage in controls (Heichel and Turner, 1976).

The severity of defoliation-related injury or the ability of a tree to recover from that injury may be a function of the growth rate of trees. This physiological activity level, in turn, may be related to tree age and crown

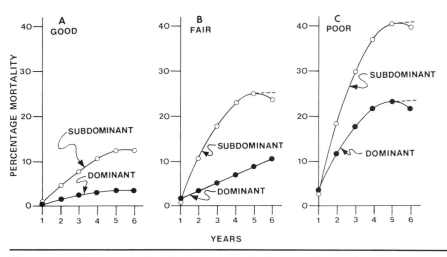

Figure 7.10.   Percent mortality due to gypsy moth (*Lymantria dispar*) of mixed oaks (a composite of red, black, scarlet, and white oak) rated in good, fair, or poor condition for dominant (●) and subdominant (○) crown classes during the 5-yr period following one heavy defoliation. (From Campbell and Sloan, 1977b.)

position (dominant versus suppressed trees). In areas defoliated by the nun moth, defoliation is often concentrated among dominant trees. However, severity of injury is generally greatest in suppressed trees. Medium to heavy defoliation of progressive and provisional trees by jack pine budworm (*Choristoneura pinus*) can cause 2–6% mortality, whereas regressive trees suffer 9–13% mortality within 2 yr after defoliation (Kulman, 1971).

In pines, losses due to defoliation can be recouped by the formation of secondary short (epicormic) shoots between the nodes (often at intermediate nodes of multinodal stems) from dominant shoots, and by rosette shoots, which bear abnormal needles formed from the last resources of a dying tree (Sukachev and Dylis, 1964). The formation of adventitious shoots may be observed in various species undergoing defoliation (Miller, 1966). In some species of pine undergoing early spring defoliation, trees show a tendency to retain older needles rather than shed them as they normally would. Scarious bracts, which are also normally shed, may elongate and develop chlorophyll (Craighead, 1940).

Under appropriate conditions, and in tree species in which it is possible, refoliation represents a response to defoliation occurring in the same season of foliage loss. Refoliation is usually at the expense of stored reserves, and the recovery is frequently incomplete. Tree species on sites where there has been defoliation by insects such as the gypsy moth and the elm spanworm may ultimately suffer at least a 40% reduction in leaf area in spite of refoliation (Stephens *et al.*, 1972). Canopy regeneration (leaf area) is usually proportional to the severity of defoliation (Heichel and Turner, 1976; Sukachev and Dylis, 1964; Kulman, 1971) (Table 7.8).

The time before initiation of refoliation may vary with differences in site, species, and other factors (Table 7.5). In addition, the photosynthetic capacity of the new foliage may be limited by its predisposition to pathogenic infestation. Injury or infection of secondary leaves may result from their predisposition (due to defoliation stress and microclimatic

Table 7.8. Influence of defoliation severity on the timing of refoliation in two deciduous species[a]

Tree Species	Defoliation %	Days before Refoliation
Red oak (*Quercus rubra*)	100	8–22
	75	22–34
	50[b]	>30
Red maple (*Acer rubrum*)	100	11–22
	75	22–27
	50[b]	>30

[a] Based on data from Heichel and Turner (1976).
[b] Very few cases where any refoliation occurred.

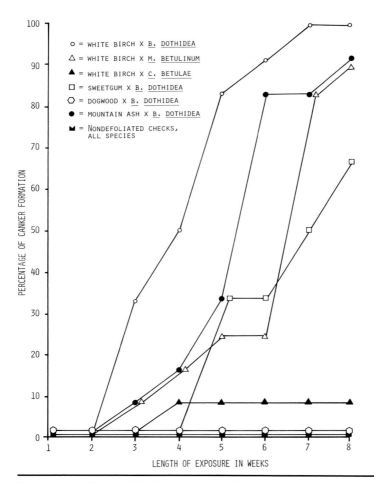

Figure 7.11.   Effect of defoliation stress on disease susceptibility in woody stems: relation between length of exposure to defoliation stress and percentage of canker formation at inoculation points in 2-yr-old, container-grown seedlings of European white birch (*Betula alba*) inoculated with *Botryosphaeria dothidea* (○), *Melanconium betulinum* (△), and *Cryptospora betulae* (▲); sweetgum (*Liquidambar styraciflua*) (□), red osier dogwood (*Cornus stolonifera*) (○), and European mountain ash (*Sorbus aucuparia*) (●) inoculated with *B. dothidea*; and nondefoliated checks of all species (◣). Plants were completely defoliated artificially at weekly intervals for 7 weeks. (From Schoeneweiss, 1975. Reproduced with permission from *Annual Review of Phytophathology* **13**. Copyright © 1975 by Annual Reviews, Inc.)

conditions during the period of refoliation) to the proliferation of pathogens (Crist and Schoeneweiss, 1975; Heichel *et al.*, 1972) (Fig. 7.11). Although defoliation predisposes species such as European white birch (*Betula alba*), sweetgum, and European mountain ash (*Sorbus aucuparia*) to attack by *Botrosphaeria dothidea*, other species such as red osier dogwood (*Cornus stolonifera*) may not be affected.

Evaluation of Growth Reduction due to Defoliation

The effects of defoliation are so closely associated with the growth patterns of trees that any measurement or evaluation of those effects without considering tree growth would result in a superficial or inaccurate evaluation. Cambial activity (wood increment) can be viewed as adding a sheath of wood to the previous year's growth. These sheaths are seen as growth rings in a transverse (cross) section. Cambial activity and the effects of stresses, such as defoliation, on that activity can be assessed by sheath analysis or examination of growth rings. Analyses of radial growth can be made using the Duff-Nolan growth sequences (i.e., horizontal, vertical, and oblique sequences) (Duff and Nolan, 1953; Mott *et al.*, 1957) (Fig. 7.12).

The horizontal sequence is the simplest sequence involving the evaluation of growth rings in transverse sections. It is useful only for dramatic changes in growth (Fig. 7.13). Growth rings vary in width at different heights on a tree. Under conditions of environmental stress the rate of radial growth may also vary at different points on the stem. The

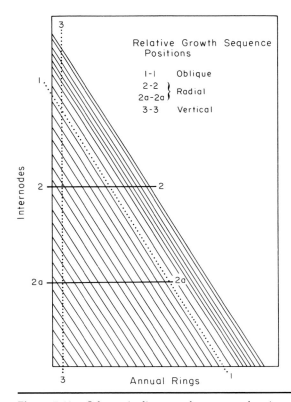

Figure 7.12. Schematic diagram of tree stem showing relative growth sequence positions: 1–1, oblique; 2–2 and 2a–2a, radial; and 3–3, vertical. (From Stark and Cook, 1957.)

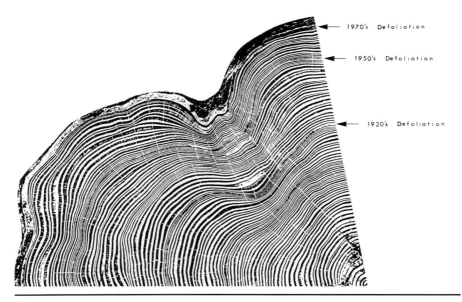

Figure 7.13. Western hemlock stem section showing reduced ring widths (cambial activity) as a result of defoliation by the hemlock sawfly (*Neodiprion tsugae*). (From Hard and Torgersen, 1975.)

Figure 7.14. Relative differences in crown shape and position of forest trees: (a) dominant, (b) subdominants, and (c) shade trees.

vertical sequence is useful because it eliminates variability due to position. With this type of sheath analysis, cambial activity is compared at a single point through time; that is, one can follow the growth of a particular ring during each of the successive years of growth (Fig. 7.12). Thus, if one followed the first ring from the pith, one could tell how well the tip of the tree (i.e., first ring) grew every year. The most realistic description of cambial activity, but the most difficult to generate, is the oblique sequence. It follows the cambial growth of a particular year at all internodes where it occurs (Fig. 7.12).

Each procedure or sequence provides some information but may have limitations that influence the assessment of the effects of defoliation. For example, the greatest ring width often occurs at the base of the crown. This point of greatest cambial productivity may differ among forest trees (dominant versus suppressed) or between forest and shade trees (Larson, 1963) (Fig. 7.14). Ring width in dominant conifers may be greatest at a point in the living crown and decrease above and below that point, except for an increase at the base. In suppressed trees, the point of maximal ring width may be higher, with very little growth occurring at the base (Kozlowski, 1963). Additional variability may result from density, age, and species (Kozlowski, 1963; Larson, 1963). Several studies have demonstrated that defoliation affects wood increment differently at various positions on a stem (Craighead, 1940; Larson, 1963). The effects of defoliation seem to be manifested first in the living crown.

Most studies evaluate the effects of defoliation using horizontal sequences or diameter increment at breast height, as was noted. Although this method can lead to a distorted understanding, recent studies suggest it gives a reasonable estimate of activity throughout the tree (Alfaro *et al.*, 1985). The oblique sequence is usually the most useful in analyzing the influence of defoliation on radial increment. The expense of obtaining oblique sequence data on large numbers of trees, when compared with collecting horizontal sequence data, often precludes its use in all but research applications. The changes in the distribution of radial growth may show up in all three growth sequences. Studies of balsam fir trees defoliated by spruce budworm (Mott *et al.*, 1957) provide an example. In these trees the oblique sequence demonstrates progressive reduction throughout the stem. During early severe defoliation the maximal ring width, typical of the upper internodes, is suppressed. The ring in the upper stem is eliminated for progressively greater numbers of internodes. The horizontal sequences at middle and lower crown points exhibit significant reductions of recent rings, but the influence of defoliation may be complicated by the effects of flowering. The vertical sequences show reduction of all rings and early elimination of radial growth in the youngest regions. Similar analyses have been performed in larch defoliated by the larch sawfly and in lodgepole pine defoliated by the lodgepole pine needle miner (*Coleotechnites milleri*) and various other insect defoliators (Stark and Cook, 1957).

Another important consideration in the evaluation of growth reduction is the presence of missing rings. For some forest species, particularly those growing on very poor sites, defoliation may result in no detectable radial growth. For example, K. K. Miller and M. R. Wagner (unpublished observations) found that 80% of the ponderosa pines defoliated by the pandora moth added no radial growth in the year of defoliation. Because no radial growth had occurred, the annual rings are missing. This usually causes the previous year's growth to be counted as growth for the year of defoliation and results in an underestimation of the growth reduction.

The only accurate means for accounting for missing rings is to use dendrochronology (Swetnam et al., 1985), a well-developed method for relating patterns of true ring growth to specific dates. Patterns are usually the result of year-to-year variation in climate. The methods of dendrochronology are specialized but should be applied whenever possible to accurately estimate the impact of defoliators. In addition to the methods of growth analysis, the way in which the methodology is applied may influence the validity of estimations of the effects of defoliation. A number of comparative procedures can be used; for instance, measurements can be made of defoliated and nondefoliated trees at a given time, or comparisons can be made of pre- and postdefoliation growth rates. The limitations of these procedures must be considered as well, especially at the critical points of experimental designs and interpretation of data.

Long-Range Changes in Forest Stands

Changes in canopy regeneration rates and canopy structure due to defoliation may lead to significant changes in site conditions, understory growth, and subsequent stand species composition. For example, defoliation may increase the amount of light available to understory trees and may decrease transpirational depletion of soil moisture. Changes in temperature and other environmental conditions may lead to alterations of physiological events such as budbreak and also affect the growth of understory tree species (McGee, 1975). Defoliation of aspen and some brush species by the forest tent caterpillars favorably affects understory balsam fir and leads to an increase in radial growth of this species (Duncan and Hodson, 1958).

In woodland sites where upper-crown host species are defoliated by the gypsy moth, the shoot elongation period of understory red maple continues beyond its normal duration. No red maple mortality occurs in the defoliated plots, and a 6.1% decrease in the number of stems can be observed in undefoliated plots (Collins, 1961). Such differential mortality can have a significant impact on forest species composition (Fig. 7.2). In addition, an analysis of the relationship between insect defoliation and tree mortality in Connecticut forests for a period of 53 yr indicates that defoliation, particularly repeated defoliation, increases mortality of oak but has little effect on maple (Stephens, 1971). Even when some mortality of

Table 7.9.   Changes in tree species composition
resulting from defoliation by the spruce sawfly
(*Diprion hercyniae*)[a]

	Total Stems	Percentages by Stems		
		White Spruce	Black Spruce	Balsam Fir
Original stand in 1938	331	21.3	8.8	69.9
Stand in 1946	253	12.0	7.6	80.3

	Total Volume (cords)	Percentages by Volume		
		White Spruce	Black Spruce	Balsam Fir
Original stand in 1938	32.6	35.0	7.3	57.7
Stand in 1946	21.1	14.6	6.0	79.4

[a] From Reeks and Barter (1951).

relatively nonpreferred species occurs, mortality of preferred species of the gypsy moth can be substantially greater (Kegg, 1971, 1973). Similarly, spruce budworm defoliation of young stands of black spruce, spruce and fir, or spruce, fir and hardwoods leads to an extensive reduction of fir from its being a major stand component (45% by volume) to a relatively minor component (11% by volume) (Hatcher, 1964). Conversely, in plots containing white spruce, black spruce, and balsam fir that have been heavily defoliated by the European spruce sawfly (*Diprion hercyniae*) there is a considerable increase in the proportion of balsam fir and a decrease in the proportion of white spruce (Reeks and Barter, 1951) (Table 7.9).

Chapter Summary

The effect of defoliation depends on the herbivore species, the amount of tissue removed, the number of defoliations, and the season or timing of defoliation. Concurrently, the impact of defoliation also depends on the species, age, size, stage of development, location of foliage in the crown, and vigor of the affected tree, as well as the condition of nearby vegetation. To further complicate the interpretation of defoliation impact, trees may have the ability to compensate for lost foliage and "catch up" with nearby undefoliated trees. In addition to the effects of defoliation on the amount of biomass accumulated, defoliation can also predispose trees to other agents such as diseases. Methods for evaluating growth reduction due to defoliation are well defined but, depending on the purpose of the evaluation, may involve multiple samples per tree along the bole and/or dendrochronology techniques.

CHAPTER 8

Insects with Piercing–Sucking Mouthparts

Introduction

A great number of insects and insect relatives with piercing–sucking mouthparts can be found feedings on almost any tissues of trees. The injury resulting from mechanical disruption of vascular tissues or the deposition of secretions during feeding varies in its severity and impact on a tree. In some cases, feeding by very small insects (or other arthropods such as mites) that are present at very low densities can cause significant changes in the growth, vigor, and form of trees. In other cases, even though the insect's stylet penetrates cells and disrupts or kills them, the resulting changes can be minor. In addition to changes induced by feeding, many arthropods with piercing–sucking mouthparts injure trees when they oviposit (Figs. 8.1 and 8.2). A third type of disfunction is associated with fluid loss from tissues being fed upon. Associated with this fluid drain is a disruption of photosynthesis and photosynthate production. Finally, the feeding and/or oviposition activity of arthropods with piercing–sucking mouthparts is often associated with disease transmission.

The impact of a population can be best understood in relation to the impact of an individuals. In one day a willow aphid can ingest the carbohydrate equivalent of the photosynthetic product of 5–20cm^2 of leaf (Mittler, 1958). Similarly, a Saratoga spittlebug adult can withdraw an average of 0.42 ml of plant fluid from a twig, which is equivalent to the total moisture content of 3–6cm^2 of inner bark. Thus, if fluids are not replaced, the phloem in an average twig (15 cm $\times$ 0.6 cm) will become dry as a result of four to nine adult feeding periods (i.e., 4–10 hr/day/adult). Indeed, if one considers population density, the impact of such species is quite apparent. A 20-m sycamore tree with approximately 116,000 leaves may, at any one time, have a population of 2.25×10^6 aphids (Dixon, 1971a). Similarly, a 14-m lime tree (*Tilia vulgaris*) with approximately

Figure 8.1. (a) Epinasty of branches of Kousa dogwood (*Cornus kousa*) due to differential growth of tissues opposite cicada oviposition sites; note stunted growth of laterals on affected branches compared with growth of unaffected branches. (b) Typical flagging of branches due to severe mechanical damage caused by oviposition of female periodical cicadas. Dieback of terminal shoots of (c) white flowering dogwood (*Cornus florida*) and (d) European white birch (*Betula alba*) due to cicada damage; note the stunted growth on laterals on the dogwood and the strongly developing laterals on the birch originating below the damaged terminal. (From Smith and Linderman, 1974.)

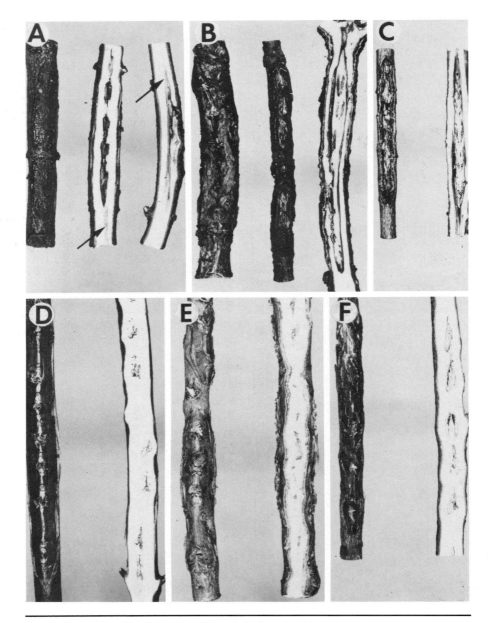

Figure 8.2.   External and internal views of cicada oviposition sites on
(a) Amur maple (*Acer ginnala*), (b) leatherleaf viburnum (*Viburnum
rhytidophyllum*), and (c) European smokebush (*Cotinus coggyria*), all showing
progressive internal tissue decay (see arrows). In contrast, note the
well-healed oviposition sites and relatively little internal tissue decay on
(d) Japanese maple (*Acer palmatum*), (e) American holly (*Ilex opaca*), and
(f) *Rhododendron mucronulatum*. (From Smith and Linderman, 1974.)

58,000 leaves may support 1.07 million aphids at any given time (Dixon, 1971b). The magnitude of the drain imposed by such a population is illustrated by estimates of the number of aphids that are required to use up the net annual primary production of leaves. If no compensating growth occurs, it would require 30 lime aphids (*Eucallipterus tiliae*) per leaf (or 776 aphids per 10 cm^2 of leaf) during a season to drain the annual production of a host tree, i.e., approximately 0.95 g/week/10 cm^2 (Dixon, 1971b). Schowalter *et al.* (1981) uses consumption rates and endemic biomass in eastern deciduous forests to calculate that sucking insects annually remove 30% of the potassium in the foliage standing crop at biomass levels <1% of the foliage biomass.

The stylet bundle of Hemiptera consists of dovetailing paired mandibles and maxillae. The inner structures (maxillae) contain two grooves that join to form a double canal system through which saliva is pumped or through which fluids are brought into the body (Fig. 8.3). Salivary enzymes aid the passage of the stylets through tree tissues. Stylet paths through plant tissue may branch along the same plane or at 90° angles

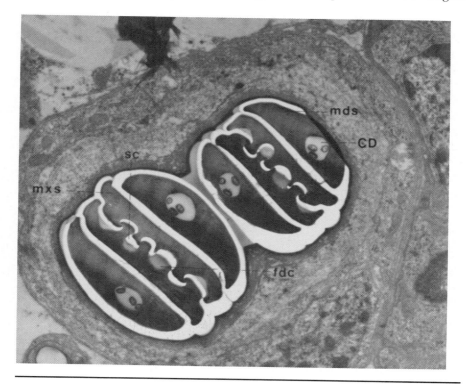

Figure 8.3. Cross section of mouthparts of the balsam woolly aphid (*Adelges piceae*) midway along the looped stylet bundle; CD (central duct), fdc (food canal), mds (mandibular stylet), mxs (maxillary stylet), and sc (salivary canal). (From Forbes and Mullick, 1970.)

to each other. Salivation appears to occur continuously in parenchyma feeders. However, in some Hemiptera, salivation ceases when stylets penetrate a phloem vessel (Miles, 1968).

Two major methods of feeding are observed in insects with piercing-sucking mouthparts. In some species, the movements of inserted stylets disrupt cells, there by releasing cell contents, which are taken up by the stylets. The second method of feeding is evident in tree tissues when the proteinaceous sheath forms around the stylets; the proteinaceous sheath remains behind, tracing the routes taken by the probing stylets. The movement of stylets may be inter- or intracellular. Depending on the species, feeding occurs in phloem, xylem, or tracheids. Insects using the latter method are among the most important forest and shade tree pests and include species of aphids, scales, leafhoppers, and treehoppers. Feeding in nonvascular tissues also has been observed (Carter, 1962).

A number of changes are induced when stylets penetrate plant tissues. For example, cells around the stylet sheath may show increased permeability, loss of starch, degeneration of chloroplasts, and other changes (Miles, 1968). Various substances that account for these changes are found in salivary secretions. Between 4 and 15 free amino acids, glucose, and glycerol are transferred unchanged from the hemolymph to the saliva. Salivary enzymes aid the passage of stylets by softening the pectate layers of the middle-lamella in plant tissues. Pectin polygalacturonase and perhaps cellulase are components associated with stylet penetration, particularly in those species whose stylets follow an intercellular path. In addition, phloem and xylem feeders may have proteinases and lipases in their saliva (Carter, 1962). Esterases, proteinases, amylases, phosphatases, phosphorylases, and other enzymes that hydrolyze carbohydrates have been found in the saliva of mesophyll and seed feeders (Miles, 1968). A few enzymes that hydrolyze various sugars have been found in phloem and xylem feeders.

Regulators in Saliva

Research on a variety of Hemiptera suggests that feeding may affect the activity of plant hormones or the response of plant tissues to hormones. The similarity between growth changes in plants caused by insects and those caused by hormone treatment, the occurrence of enzyme activity associated with feeding, and the ability to suppress insect damage with hormonal treatments suggest a relationship between feeding and hormonal changes. Although much more research is required to determine the nature of the relationship, enough data are available to consider it a viable and general relationship.

The stimulation of plant growth by gall insects can be viewed as paralleling the responses of tissues to plant growth-promoting substances. Development of adventitious buds and reductions in internode spacing can be induced by the application of hormones and also by insect feeding (Allen, 1947; see similar information in Allen, 1951). Plant growth-inhibiting substances have been found in several hemipterans such as the aphid *Cinara piceae* (Carter, 1962). The saliva of many aphids also contains amino acids, indole acidic acid, and various phenolic compounds that could affect growth (Dixon, 1971a,b).

Many species of the economically important genus *Cinara* cause growth reductions in their conifer hosts. The white pine aphid (*Cinara strobi*) is capable of causing serious injury to eastern white pine. *Cinara winokae* (which feeds on both stems and roots) and *Cinara todocola* seriously reduce the growth of young *Thuja orientalis* (oriental arborvitae) and *Abies mayriana*, respectively. Similarly, other species reduce the growth of Douglas-fir seedlings. Among the species commonly infesting Douglas-fir are *Cinara pseudotsugae*, *C. pseudotaxifoliae*, and *C. taxifolae*. In a 2-yr aphid infestation of seedlings, the average growth in height of infested trees was 2.8 in. and 8.9 in., respectively, compared with 11.1 in. and 21.8 in. for uninfested trees (Johnson, 1965).

Alterations in the growth of trees is often due, at least in part, to the influence of feeding on the production, mobilization, and storage of photosynthate (Dixon, 1985). Leaves of *Picea sitchensis* that have been fed on by the aphid *Liosomaphis abietinum* have a permanently lowered rate of photosynthesis and are thus less efficient than unattacked leaves (Dixon, 1971a). *Acer pseudoplatanus* infested with sycamore aphids exhibit significant changes in growth. Heavy aphid infestation of trees in the spring may result in the production of leaves that are 40% smaller than noninfested trees. This reduction in leaf size is greater than would be expected merely from the depletion of available nutrients by the aphids. The calculated drain imposed by the aphids accounts for only a small proportion of the observed diminution in leaf area. However, the leaf size reduction is correlated with the number of aphids feeding on the leaves. This suggests that the aphids influence leaf growth in some other fashion, perhaps as a result of the injection of a component in the saliva. As a result of reduced leaf size, annual ring width can be reduced by up to 62% (Dixon, 1971a). Dixon (1971a) estimates that in the absence of an aphid infestation a sycamore tree could produce as much as 280% more stem wood than when infested.

An interesting contrast is provided by the infestation of *Tilia vulgaris* by the aphid *Eucallipterus tiliae*. Although the growth in diameter, height, leaf number, and size of infested saplings is normal, the roots do not grow. The leaves of infested saplings are heavier per unit area and contain more nitrogen than uninfested individuals; however, they are shed earlier. A 1-yr infestation results in smaller leaves in the following year. However, these leaves have a net production that is 1.6 times greater than that of uninfested saplings (Dixon, 1971b).

Depending on the specific circumstances, feeding by insects with piercing–sucking mouthparts may cause partial or total mortality of the host. The golden oak scale, a pit-making scale, can cause major growth reduction and mortality of chestnut and white oaks. The scale's stylets move intracellularly; and although they do not extend into the phloem or cambium, they induce significant changes in the cortical tissues. The tissues immediately surrounding the stylets are killed and become lignified. Beyond the cone-shaped group of lignified cells the tissues proliferate radially. This collenchyma activity appears to be both hypertrophic and hyperplastic (i.e., cells are formed and elongated). This abnormal activity, probably induced by a component in salivary secretions, forms a gall-like growth characteristic of this species. As time passes, thin-walled cells of the proliferated areas begin to disintegrate. Wound tissue may also develop, even when there are only a few insects on a twig (Parr, 1937).

Phytotoxemia

Feeding by insects with piercing–sucking mouthparts often results in clearly defined symptoms of disease. This phenomenon is referred to as phytotoxemia. The insect whose feeding induces the symptoms is referred to as the toxicogenic agent (Carter, 1962). Although specific components of the saliva may be directly responsible for certain symptoms, the possibility exists that some toxic substances may originate from the host tree or that salivary secretions induce the production of substances that alter the physiological state of the tree (Carter, 1962).

Phytotoxemia can be classified into broad arbitrary categories: leaf spotting, tissue malformation, and systemic translocation. Leaf spotting, or stippling, often appears as localized lesions (cellular disruptions) at the point of feeding. These lesions may develop secondary symptoms including cankers, cork formation, leaf or needle drop, and premature seed drop. These symptoms result from the formation of numerous, tiny chlorotic (discolored) spots whose tint, shape, and distribution are often characteristic of the species involved. *Lecanium coryli*, which feeds on ash trees, causes spotting by destroying walls along the stylet track and by altering the chloroplasts. Leaf stippling is particularly evident among mesophyll feeders such as leafhoppers (Carter, 1962). As indicated above, localized lesions may develop secondary symptoms. For example, *Matsucoccus feytaudi*, a coccid pest of *Pinus pinaster* var. *mesogeenis*, causes local cellular disruptions. In addition, multiple punctures induce resinosis (Carter, 1962).

Another category of phytotoxemia includes various primary tissue malformations such as curling of leaves (Fig. 8.4), rosetting of leaves, shortening of internodes and petioles, and formation of adventitious buds. Primary malformations such as leaf curling or leaf rolling are frequently induced by aphid species. Another primary malformation is the development of witches brooms, which are characterized by either dense or loose clusters of branches, often originating from an enlarged axis. Branches may develop numerous lateral shoots from one point, forming a compact

Figure 8.4.   Leaf curling caused by the feeding of the woolly elm aphid (*Eriosoma americanum*). (Courtesy Shade Tree Laboratory, University of Masssachusetts.)

so-called witches broom. The leaves or needles may be reduced, the buds may be small and close together, and the internodes may be shorter than normal. Although witches brooms on both hardwoods and conifers can result from insect feeding, many other agents including fungi, viruses, and mistletoes may have the same effect (Boyce, 1961).

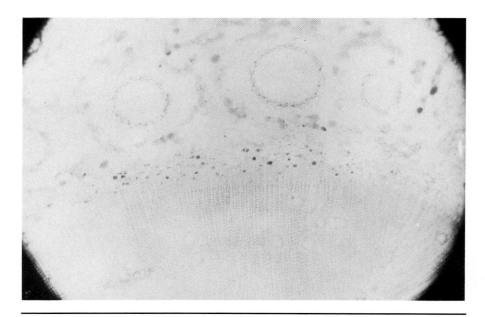

Figure 8.5.   Cross section of uninjured tissues of a red pine twig. (Courtesy USDA Forest Service.)

A final category consists of injuries resulting from the translocation of toxic substances and includes limited chlorosis (i.e., leaf-clearing and leaf-banding) or systemic injuries (i.e., death of roots, wilting, phloem necrosis, and vigor reduction). The stylets of the Saratoga spittlebug move directly through the cortex of twigs and stop at the cambium (Figs. 8.5 and 8.6). Feeding scars consist of necrotic resin-filled pockets in the phloem and varying amounts of injured cambium. As the season progresses, pitchy areas develop in the xylem (Fig. 8.7). When the cambium does not heal over for some time, pitch-filled necrotic streaks may extend through two or more growth rings (Fig. 8.8). These residual scars occur only after several seasons of heavy attack. Discoloration of the cambial–xylem interface can be detected as early as 17 hr after feeding. The effect of exposure of the cambial–xylem interface to excised spittlebug salivary glands demonstrates the necrosis of cortical tissues is probably due to some component in the spittlebug's saliva (Ewan, 1961).

The treehopper *Stictocephala festina*, when feeding on black locust seedlings, makes a ring of punctures 1–2 in. above the ground. When the girdling is incomplete or is on older stems, the tree survives, but a gall-like swelling (or callus) forms above the girdle. Eventually, many affected trees die (Boyce, 1961).

A systemic phytotoxemia is characterized (a) by toxic effects some distance from the feeding site (generally some type of chlorosis), (b) by the

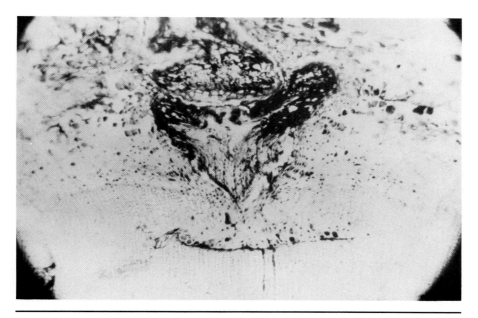

Figure 8.6. Cross section of the center of a feeding scar in red pine twigs caused by the Saratoga spittlebug (*Aphrophora saratogensis*). (Courtesy USDA Forest Service.)

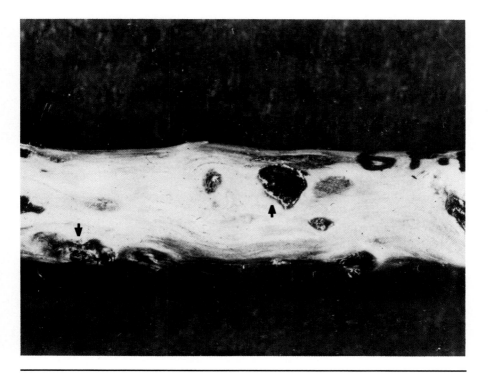

Figure 8.7. Surface of the xylem of a 3-yr-old red pine twig showing residual scars (arrows) and characteristic lumpy appearance resulting from injury by the Saratoga spittlebug. (Courtesy USDA Forest Service.)

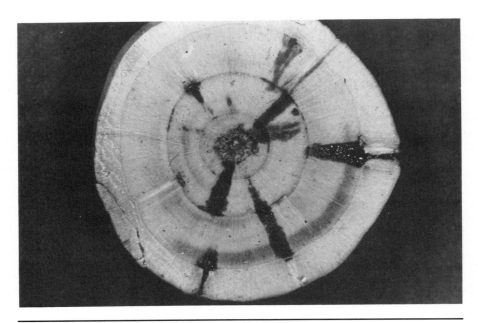

Figure 8.8.   Cross section of a 4-yr-old red pine twig showing residual feeding scars of the Saratoga spittlebug. (Courtesy USDA Forest Service.)

expression of symptoms on new growth in the absence of the toxicogenic insect, or (c) by demonstrable symptoms in the whole plant. Limited systemic effects are well illustrated by the aphid *Hamamelistes spinosus*, which is associated with birches and induces vein clearing and necrotic corrugations of the leaf (Varty, 1963). The process of leaf mortality in American beech infested with oystershell scale (*Lepidosaphes ulmi*) (Fig. 8.9) begins with a marginal necrosis that progresses inward toward the midrib. Scales may be found feeding on the twigs, petioles, or midribs of leaves (De Groat, 1967).

Not all systemic phytotoxemia result in short-term changes; some changes in the tree extend over the entire growing season. The leaves of sycamore are sometimes shed in autumn while still green. The nitrogen content is very high in the leaves (at leaf fall) of trees that had a high spring density of the sycamore aphid. However, when aphids are scarce in the spring, the leaves shed in the autumn contain very little nitrogen. No significant association exists between leaf nitrogen content and the aphid population throughout the season or the population in autumn when nitrogen is being translocated out of leaves. Thus, in this insect–tree interaction, aphids affect the nitrogen metabolism of their host at the end of the growing season if a high density of aphids is attained at the beginning of the same year (Dixon, 1971a). The ponderosa pine twig scale (*Matsucoccus bisetosus*) is suspected of being associated with the appearance

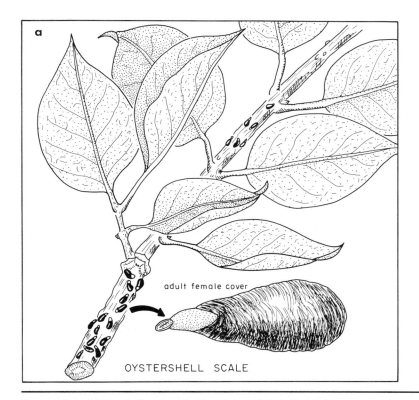

a

adult female cover

OYSTERSHELL SCALE

Figure 8.9a.   Oystershell scales (*Lepidosaphes ulmi*) infesting twig.
(Illustration courtesy J. A. Davidson, University of Maryland.)

of a needle blight of pines, such as jeffrey and ponderosa pines (McKenzie, 1941; McKenzie *et al.*, 1948).

A number of injuries caused by insects with piercing-sucking mouth-parts are of particular importance in urban forests and in the protection of shade trees because of the unaesthetic alteration of leaves and distortion of tree crowns. Flagging (death of a twig or branch beyond a girdling canker or injury, usually with distinct discoloration; or the final stage of wilting) is another aesthetically unacceptable deformation that can result from the feeding or oviposition of insects with piercing-sucking mouthparts. Extensive flagging can cause substantial stem deformation. A severe infestation of the Prescott pine scale (*Matsucoccus vexillorum*), can result in flagging and subsequent deformation of ponderosa pines.

Damage by the periodical cicada (*Magicicada septendecim*) illustrates some typical twig distortions (see Figs. 8.1 and 8.2). The mechanical injury resulting from ovipositional activity causes many changes in host trees. Epinasty of branches is due to differential growth of tissues, which in turn is due to cicada oviposition injury. In the years following the oviposition

Figure 8.9b.   Oystershell scales *(Lepidosaphes ulmi)* infesting twig. (Photograph courtesy Shade Tree Laboratory, University of Massachusetts.)

period of the periodical cicada, terminals and laterals may exhibit various stages of progressive dieback, which is generally associated with incomplete healing of oviposition wounds (Smith and Linderman, 1974).

Many of the interactions associated with phytotoxemia remain unclear and require detailed investigation. Many of the symptoms of phytotoxemia and virus diseases are very similar (Van der Plank, 1975). Often, virus diseases of trees are suspected when, in fact, the symptoms are due to phytotoxemia. The witches brooms of the mimosa tree (*Albizia julibrissin*), the yellow-line pattern of yellow birch foliage, and the chlorotic islands bordered by small veins of hackberry (*Celtis occidentalis*) foliage were all at one time believed to be caused by virus infections. These symptoms are now suspected of being due to damage by leafhopper feeding.

The Balsam Woolly Aphid: A Case Study

The numerous ways in which an insect with piercing-sucking mouthparts can affect the growth and survival of its host tree can be illustrated by a detailed consideration of one species. The interactions between the balsam woolly aphid and its hosts, which include balsam fir, grand fir and other *Abies* species (Fig. 8.10), provide such an example. The aphid's life cycle includes an alternation of hosts. Spruce species are always primary hosts whereas the secondary host is another conifer species. The insect was probably introduced into eastern North America from Europe (Balch, 1952). Timber loss due to this insect is important in some provinces of Canada, for example, where balsam fir constitutes almost half of the current merchantable growing stock. Balsam fir is an important resource for the pulp and paper industry in many areas of the world.

Feeding is accomplished by the insertion of stylets intercellularly past the epidermis and into the cortex. Although, in young shoots, the phloem is occasionally penetrated, feeding generally takes place in the parenchyma. Probing is a trial-and-error process of repeated insertion, partial withdrawal, and reinsertion of the thread like mouthparts (Fig. 8.11). Feeding by the balsam woolly aphid is the cause of an abnormal condition of balsam fir that is commonly referred to as gout disease (Fig. 8.12). The external symptoms consists of swelling and distortion of twigs and small branches, particularly in the upper crown of trees. Immature aphids tend to move toward the new growth and settle at the base of buds and new shoots (Balch, 1952). Stems of small trees may swell at lenticels where the aphid has been feeding. In general, the swellings are most noticeable at the nodes and around the buds (Figs. 8.13 and 8.14). Internally, small, purple pockets of tissues are found inside the outer bark. Around the point of stylet insertion the growth of the bark and wood is stimulated, giving the characteristic swollen appearance. This may occur even if the insect dies before actually feeding. The swollen appearance is enhanced by the fact

Figure 8.10. Heavy population of the balsam woolly aphid on the main stem of subalpine fir (*Abies lasiocarpa*). (From Doerksen and Mitchell, 1965.)

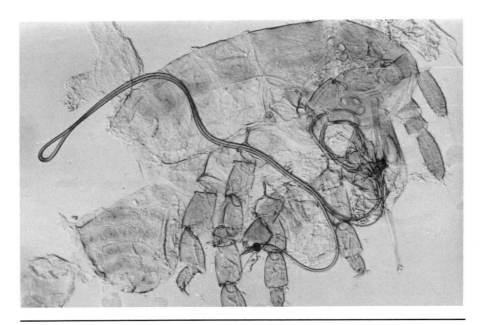

Figure 8.11. Balsam woolly aphid with long, threadlike mouthparts. (From Forbes and Mullick, 1970.)

Figure 8.12.   Effects of gout disease produced by the balsam woolly aphid; repeated gouting of the main terminal produces a stunted leader. (From Carrow, 1974.)

Figure 8.13.   Gouting of twig nodes caused by balsam woolly aphid feeding. (From Carrow, 1974.)

that infested buds either fail to open or produce short shoots. Elongation of the branch stops, and diameter growth increases. The branch may also have a characteristic taper toward its tip. The most dramatic symptoms are manifested in the uppermost, vigorously growing branches. The degree of swelling is directly correlated to the degree of infestation but may appear even at a density averaging 10 immature aphids per shoot over several years. Branches in which all bud growth has ceased will start to die in approximately 2 yr (Balch, 1952).

When a large part of the stem surface is heavily infested, the bark dies down to the cambium, and the foliage turns yellow and eventually

Figure 8.14. Balsam woolly aphid-infested node where no gout is evident. (From Carrow, 1974.)

reddish. Individual trees show progressively more severe levels of damage with increasing height, crown class, diameter, and age. In general, taller (older) stands that are under moisture stress are most seriously affected by aphid attack (Page, 1975).

Infestations of balsam woolly aphid also cause major reductions in the carbohydrate reserves of the needles and twigs of grand fir. This can lead to reduced growth and vigor due to a reduction in the reserves upon which the growth of new shoots depends. Before budbreak, aphid infestation reduces starch content by 28% in foliage and reduces sugar concentration in twigs. After budbreak, the major carbohydrate (starch) accounts for 65% of the total carbohydrate content in noninfested foliage and 53% in infested foliage (Puritch and Talmon-De L'Armee, 1971).

Balsam fir wood is occasionally found to be hard and brittle and exhibits a tendency to warp and split. These changes in wood quality are due to balsam woolly aphid infestation. Even at low aphid density, some changes are noticeable; for example, the surface of the wood exhibits reddish-brown blotches. At the points of stylet insertion, a component of saliva or a substance produced in the cortical tissue in the presence of saliva stimulates cell division in the cambium. This activity causes the production of tracheids having thickened cell walls and the formation of dark, hard, and brittle wood (Fig. 8.15). This heavily lignified wood is called rotholz (Figs. 8.16 and 8.17). Rotholz looks much like compression wood (wood formed because of compression stress in leaning branches or trees) but occurs on all sides of the stem or branch. Moderately infested, fast-growing stems produce the most abnormal wood.

Even in species such as *Abies grandis*, where there is no indication of rotholz formation, aphid feeding interferes with the water conduction system. Although aphid feeding does not cause any obvious structural change in *Abies grandis*, xylem function is nevertheless affected (Table 8.1). Low permeability (restricted water flow) results from a reduced number of openings or pores in the bordered-pit membranes (Puritch and Petty, 1971). In aphid-infested trees all the pit membranes from the sapwood are incrusted and resemble those from the heartwood of aphid-free trees. Incrustations are absent in the sapwood of noninfested trees (Puritch and Johnson, 1971). Various studies suggest that the aphids accelerate the normal aging of the sapwood and cause heartwood to form prematurely (Puritch and Johnson, 1971).

Identical phenomena have been observed in Pacific silver fir (*Abies amabilis*) and subalpine fir in the Pacific Northwest. In balsam woolly aphid-infested trees, abnormal wood reveals secondary walls that are frequently marked with checks, an increased proportion of thick-walled latewood-like tissue, and atypical proliferations of traumatic resin canals. In addition, cell walls of the earlywood are approximately 50% thicker than normal, and tracheids are approximately 40% shorter. Calculations of the

Figure 8.15.   Cross-sectional disk from balsam woolly aphid-infested grand fir (*Abies grandis*). The irregular outer rings were formed during a period of infestation by the aphid. (From Doerksen and Mitchell, 1965.)

average amount of cross-sectional cell lumen area for two 10-mm wide annual rings indicate that the rotholz ring of an infested tree may have 66% less lumen area than a normal ring. If latewood is accepted as non-functional tissue, this would mean an 82% loss of lumen space. Such an inhibition of flow represents a considerable stress on the tree (Doerksen and Mitchell, 1965; Mitchell, 1967). Interestingly, although there are great differences in the susceptibilities of grand, subalpine, and Pacific silver firs, there are no significant differences in the anatomical changes in infested trees. Similar, but less severe, interference with conduction may be caused by other species. The Saratoga spittlebug causes feeding scars

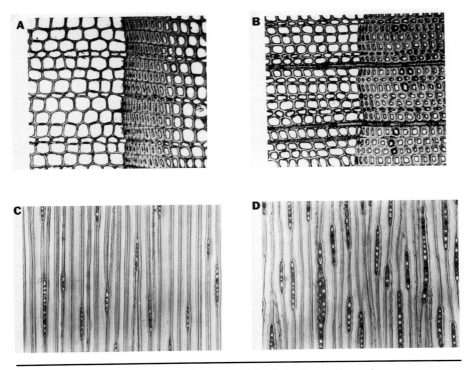

Figure 8.16.   Comparison of normal wood and the dark, abnormal wood (rotholz) associated with balsam woolly aphid infestations on grand fir (*Abies grandis*): (a) cross section of normal xylem, (b) cross section of abnormal xylem (rotholz) produced in the stem (note thickened cell walls in springwood and rounded cells and intercellular spaces in summerwood), (c) tangential section of normal xylem (springwood), and (d) tangential section of abnormal xylem (rotholz springwood). (From Doerksen and Mitchell, 1965.)

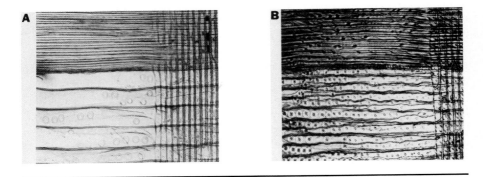

Figure 8.17.   (a) Radial section of normal xylem in grand fir (*Abies grandis*) and (b) radial section of abnormal xylem (rotholz) produced in the stem of grand fir. Note serpentine cell walls in springwood and diagonal checks in summerwood. (From Doerksen and Mitchell, 1965.)

Table 8.1.   The effect of *Adelges piceae* infestation on
the permeability of *Abies grandis* (expressed as the flow
rate)[a]

| Condition | Tree | Permeability (Q) (ml water hr^{-1}) | |
		Sapwood	Heartwood
Noninfested	A	128.9	—
	B	139.0	0.82
	C	147.1	0.98
	D	172.6	1.66
Infested	A	17.8	—
	B	6.1	—
	C	1.8	—
	D	6.9	—
Lightly infested	B	97.3	—
	D	101.5	—

[a] From Puritch (1971). Water permeability of the wood of grand fir (*Abies grandis* (Doug.) (Lindl.) in relation to infestation by the balsam woolly aphid, *Adelges piceae* (Ratz.). Oxford University Press.

consisting of resin-filled pockets in the phloem and xylem. Although the cambium may be gradually repaired, the pitchy defects in the xylem remain as a small, permanent blockage to fluid conduction (Ewan, 1961).

Recovery

The nature and extent of recovery from injury vary widely. Again, the balsam woolly aphid provides some useful examples. Although infestation of balsam fir by the balsam woolly aphid may lead to swelling distortions, bud inhibition, and crown dieback, the aphid populations may not persist on individual trees, and recovery may occur (Fig. 8.18). Studies have shown that the original leader of 60 out of 100 trees can resume normal growth. This growth resumption occurs in spite of moderate or moderate to severe damage sufficient to cause reduction in main stem growth or complete inhibition of leader development from the bud stage. Original leaders can also be replaced by new leaders, which are formed by reorientation of main stem nodal or internodal or secondary branches (Fig. 8.19). Relatively few trees form severe crooks or forked stems. Thus, one period of aphid infestation of relatively short duration may have little impact on the trees in a young stand.

Recovery may be observed following heavy infestation and severe injury by the oystershell scale. New twig growth arises from formerly dormant lateral buds in branches proximal to killed terminals. This results

Figure 8.18.   Balls of new foliage in tops of two Pacific silver firs (*Abies amabilis*) indicating the beginning of host recovery from balsam woolly aphid infestation. (From Johnson *et al.*, 1963.)

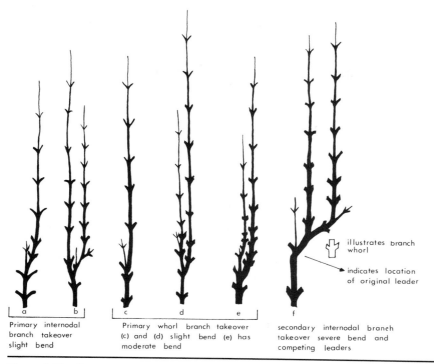

Figure 8.19. Recovery patterns in young balsam fir trees damaged by the balsam woolly aphid. (a,b) Primary internodal branch takeover, with a slight bend. (c,d) Primary whorl branch takeover, with a slight bend; (e) same, with a moderate bend. (f) Secondary internodal branch takeover, with a severe bend and competing leaders. (From Schooley, 1976.)

in an aggregation of dense foliage at the tree crown periphery. These changes are very similar to symptoms in beech and other hardwoods attacked by defoliators (De Groat, 1967). Although these are but a few examples, the variation in the extent of recovery is almost as great as the variation in types and extent of injury caused by piercing-sucking arthropods.

Death and Injury: A Summary

In summary, death of an entire tree may result from feeding by piercing-sucking insects because of a series of changes occurring throughout the tree: inhibition of bud development, depletion of food resources, production of abnormal wood, and interference with conduction by xylem tissues. The production of secondary periderm, which induces resinosis or the death of the cambium, may combine with the above changes to cause the death of trees during heavy infestations.

Mortality is often localized, affecting only certain branches or crown levels. Twig and branch mortality of American beech can result from heavy infestations of oystershell scale. Extensive foliage death may occur as a result of scale feeding on the current season's twigs, petioles, or leaf midribs. Injury usually commences with a marginal necrosis and may ultimately lead to twig mortality. Even the gall-forming activities of insects such as *Pineus similis* may result in the death of small twigs. Occasionally, the presence of numerous wingless females may cause stem distortions where no galls develop (Cumming, 1962).

Although tree mortality due to the activity of piercing–sucking insects is relatively rare, trees may occasionally succumb, particularly when predisposed by other environmental factors. Poor site conditions, low host vigor, high tree density, and other factors influencing the degree of tree competition enhance the possibility of host mortality. When populations of the Saratoga spittlebug are low, little or no twig mortality results from adult feeding on saplings. However, when populations are dense, stunting, twig mortality (flagging), and whole tree mortality can be observed in as little as two to three seasons (Benjamin *et al.*, 1953; Ewan, 1961).

Chapter Summary

The damage caused by insects with piercing–sucking mouthparts varies widely and is often underestimated. Only a few individuals on a twig or leaf can cause significant damage. Insects with sucking mouthparts cause injury by mechanical disruption of vascular tissue, removal of plant fluids, injection of toxic compounds into plants, and transmission of disease. Feeding affects the activity of hormones, which can cause a variety of plant deformations. The balsam woolly aphid (actually an adelgid) is an example of an insect with piercing–sucking mouthparts that can cause significant damage to several important coniferous species.

CHAPTER 9

Phloem and Wood-Boring Insects

Introduction

Many economically important rural and urban forest pests are insect borers. Perhaps the most important group of insect borers, especially in the western and southern United States, is the bark beetles. The overriding effect of insect borers on trees is the destruction or disruption of the structural and functional integrity of primary and secondary growth. Insects that affect primary meristems (shoots, roots, and buds) are discussed in Chapter 11; this chapter concentrates on those insects that affect secondary meristematic tissue.

Secondary growth arising from the vascular cambium accounts for all radial expansion of trees, including the formation of xylem and phloem (Fig. 9.1). Like all living organisms and biological communities, tree organs and tissues are interacting, interlocking systems that are separated only as a pedagogical convenience. The survival of primary growth is essential for the occurrence of secondary growth. In temperate regions the resumption of cambial growth in the spring is correlated with the renewed activity of bud and leaf development.

Mechanical support and long-distance transport of water and minerals from roots to aerial shoots are achieved in trees by the xylem system. As xylem vessels age they gradually lose their ability to conduct water. Embolized (air-filled) vessels fill with tyloses (balloonlike outgrowths of ray or axial parenchyma cells through pit membrane into the vessel lumen), resulting in the formation of heartwood. This central core of heartwood provides the mechanical support of the stem.

Phloem transport in trees is responsible for the translocation of photosynthetic products from leaves to the stem through conducting elements (i.e., sieve tubes). Carbohydrates are the most important constituents of photosynthate and thus the prime component of phloem in the

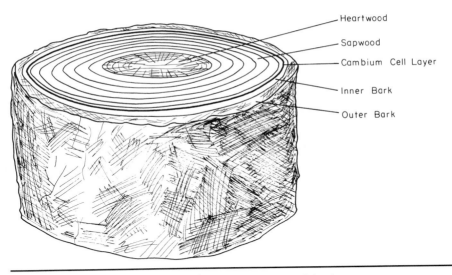

Heartwood

Sapwood

Cambium Cell Layer

Inner Bark

Outer Bark

Figure 9.1.   Cross section of a tree stem illustrating some of the tissues involved in secondary growth.

form of sugars. Sucrose is the most common and, quantitatively, the most important sugar. Other translocated materials include oligosaccharides of the raffinose group such as raffinose, stachyose, and verbascose. Sugar alcohols such as D-mannitol and sorbitol are other translocation carbohydrates. Other substances have been reported in the phloem and are often associated with particular events such as incisions, leaf senescence, and cellular growth in the cambium. Some of these substances include enzymes, amino acids, amides, hormones such as auxins, inorganic ions such as sodium and potassium, and vitamins. The destruction of the living cells of phloem tissue renders the transport of vital carbohydrates impossible. Subsequent implications to growth and survival are obvious since phloem provides the basic building blocks of new tissues. *Agrilus liragus*, for example, causes the deterioration and often the death of hosts (i.e., poplar) by infesting the stem and branches and injuring the phloem. Injury to the phloem and cambium disrupts normal translocation of nutrients. In the year of attack, premature leaf chlorosis and abscission can be observed (Barter, 1965).

Finally, feeding may occur, beyond the phloem, in the functional xylem. The impact of such feeding depends, to an extent, on the functional physiology of the tree species. For example, the severity of the disruption can depend on the size and distribution of vessels, that is, whether the species is ring porous or diffuse porous (Fig. 9.2). Nevertheless, xylem sap transport is crucial to the tree, not only to supply its water requirements but also because xylem carries other important constituents. Amino acids

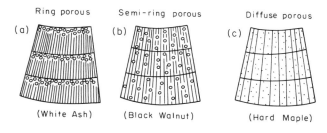

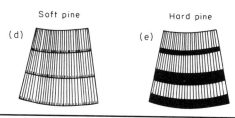

Figure 9.2. Selected examples of xylem system development in various trees. Hardwoods (deciduous trees): (a) white ash, ring porous; (b) black walnut, semi-ring-porous; (c) hard maple, diffuse porous. Softwoods (coniferous trees): (d) soft pine and (e) hard pine.

and enzymes, as well as essential hormones, are believed to be transported from the root system through the xylem.

Phloem Feeders

The effects of phloem feeding vary with insect species, abundance, the nature of the injury, and the ability of the host tree to resist and recover from injury. Host resistance and recovery are covered extensively in Chapter 3. One important aspect of phloem-feeding insect injury is the location of the attack on a tree. For example, ponderosa pine may come under simultaneous attack from several species of bark beetles that distribute themselves along the bole. *Dendroctonus valens* normally attacks the lower 1–3 ft of the bole and occasionally the upper root crown. *Dendroctonus brevicomis* attacks the main bole up to a diameter of approximately 6–8 in. The Arizona Ips (*Ips lecontei*) attacks only very small branches (3–6 in. in diameter). Those borers that attack the main bole are usually the most important. It is interesting to note that most of the serious pests in this group are in the genus *Dendroctonus*. The word *Dendroctonus* literally means *tree killer*. Tree death is often the ultimate outcome of borer attacks, but more subtle effects also occur.

Phloem–Borer Interactions

Feeding by borers such as bark beetles and others may result in direct or subtle, indirect changes in their food, the phloem. These changes, in turn, may affect the survival of the trees as well as that of the dendrophagous species. The changes resulting from insect attack may not be due merely to the feeding of the insect but may result from the association between the insect and its microorganisms (see Chapter 5). The southern pine beetle and its fungal symbionts commonly infest loblolly pine in the southern United States. The composition of the host phloem changes dramatically in infested trees compared with those free from infestation. Analyses of phloem sucrose, glucose, and fructose indicate that total sugar content of stored bolts (logs) decreases by 31% from that found in fresh bolts, whereas no effect in the level of reducing sugars is observed. In bolts infested by a beetle–fungal complex there was a 74% decrease in reducing sugars. Total sugar content also decreases in infested bolts. Fungi such as *Ceratocystis minor*, which are associated with bark beetles, utilize some phloem sugars. Decreases in sugar levels also occur in ponderosa pine logs infested with *D. brevicomis* and its fungal flora (Barras and Hodges, 1969). Infestation of loblolly pines with a beetle–microorganism complex also leads to a decrease in most free amino acids and soluble nitrogen and an increase in insoluble nitrogen, total nitrogen, and most protein-bound amino acids (Hodges *et al.*, 1968).

The quality of the cambium and phloem as well as its quantity can influence the survival of an insect borer. The bronze birch borer (*Agrilus anxius*) preferentially attacks mature or overmature stands. The young larvae need living cambial tissue, but development to pupation is dependent on the death of the tissues attacked. Thus, larvae in attacked trees only partially complete their development. If the level of attack is sufficient to severely weaken or kill the branch or tree, development may continue to the adult stage. If not, galleries are healed over, and tree growth resumes (Balch and Prebble, 1940).

Moisture Loss and Bark Beetle Activity

The bark beetle entrance holes and extensive tunneling formed in the phloem exposes the wood surface to the atmosphere, and water loss occurs. If a pine bole infested with *Dendroctonus frontalis* is coated with wax after the bark is made smooth, and ventilation holes made by beetles are filled, the moisture content of the tree remains approximately equal to that of uninfested trees. The beetle brood will not complete development, and the leaves will remain green. In unwaxed boles the moisture content is consistently lower than in uninfested or waxed trees, and the differences increase dramatically with tree height. Dehydration of the wood may lead to air accumulation in the rings. This, in turn, creates resistance to the vertical and lateral transport of water, and the outer growth rings lose the capacity for transport. Bark beetles often introduce symbiotic fungi. Some of these fungi are believed to accelerate drying by their growth in tree tissue. The failure of wood to conduct water and nutrients properly is often closely associated with the penetration of the wood by fungi (Caird, 1935).

Girdling

The impact of phloem feeders can be best understood by evaluating the results of severe injury, that is, complete girdling (Fig. 9.3). Although most forest insects only partially girdle their hosts, an understanding of the effects of complete girdling provides a basis for comparison. Complete girdling is defined as the complete severing of the phloem and removal (e.g., by consumption) of all tissues external to the secondary xylem.

The effects of girdling on foliage vary. In some tree species leaf fall may follow girdling, whereas in others full foliage may remain 3 months after girdling. Budbreak may follow girdling, but subsequent leaves are

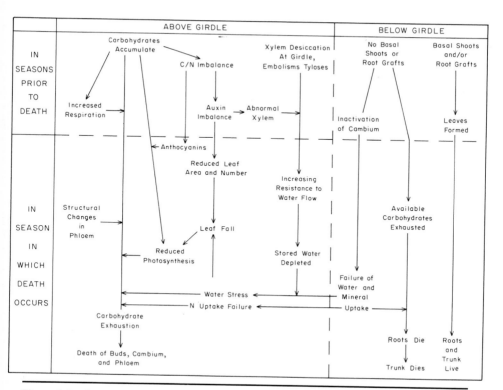

Figure 9.3. Interaction of the factors leading to the death of girdled trees. (From Noel, 1970.)

smaller and fewer in number. Girdling may induce flowering and some-times increase seed production; this may result from carbohydrate accu-mulation above the girdle. In Douglas fir, even partial girdling stimulates cone production (Ebell, 1971). Basal sprout development is stimulated by girdling, except in many pines and other conifers. This is due, in part, to the removal of apical inhibitions. These and other changes are external manifestations of internal alterations resulting from girdling.

Physical changes in the tree occur in addition to biochemical changes. Girdling leads to drying out of the trunk, although this process rarely occurs uniformly throughout the tree. In northern temperate regions, the water content of the wood above a girdle can be higher than that below the girdle. These changes affect the susceptibility of the tree or parts of the tree to further attack by bark beetles, wood borers, and pathogens.

Cambial activity above the girdle is often only temporarily inter-rupted, if at all. The accumulation of carbohydrates or auxins can lead to increased wood formation. Other organic nitrogenous substances such as amino acids also accumulate. Cambial activity below the girdle (in the absence of basal shoots) is generally more severely and permanently cur-tailed, resulting in poorly differentiated seasonal growth rings. The exact effects depend to a great extent on the timing of the girdle relative to the growth cycle, the extent of localized auxin synthesis and auxin transport, and various other tree characteristics such as whether it is a angiosperm or gymnosperm, ring porous of diffuse porous. Finally, girdling may cause leaf vein swelling, excessive development of secondary phloem, and callusing of primary phloem. Below the girdle, phloem degneration is marked by callusing, collapse of sieve tubes, and hypertrophy of phloem parenchyma. In spite of the changes delineated above, many trees may survive for a considerable period after girdling, even in the absence of girdle healing. Various species of North American, European, and African trees may remain alive for 1–5 yr after girdling. In general, it may take 1 or 2 yr to kill a tree by girdling. Some of the factors that may be involved in this survival (i.e., insuring the transport of nutrients to roots) are basal shoot development, depletion of stored root reserves, mycorrhizal asso-ciations, and natural grafts with roots of adjacent trees (Noel, 1970).

Sapwood/Heartwood Borers

The designation of phloem and xylem feeding among wood and bark borers are generalizations and are rarely consistent. Although many species are restricted to the phloem throughout their lives, other species, such as the bronze birch borer and the two-lined chestnut borer, spend most of their larval life in the phloem, with periodic diversions into the xylem. Other insects, such as species of the bark beetle genus *Pseudothy-sanoes* and the hickory bark beetle (*Scolytus quadrispinosus*), remain in the

Figure 9.4. Pupal cells of the northern pine weevil (*Pissodes approximatus*) in the sapwood–cambial interface. (Courtesy Shade Tree Laboratory, University of Massachusetts.)

phloem and pupate in the inner–outer bark interface or in the wood–cambial interface (Fig. 9.4). Many insects complete their development in the inner bark (phloem) although their tunnels etch the xylem (wood) surface (Fig. 9.5). Xylem etching is often species-specific and is an important identification tool. Alternatively, although small larvae of the pine sawyer beetles (*Monochamus* spp.), the carpenterworm (*Prionoxystus robiniae*), the locust borer (*Tetropium parvulum*) (Fig. 9.6), and others initiate wide galleries along the cambial–wood interface (Fig. 9.7), development is completed within the extensive galleries they make in the xylem.

Some species live exclusively in the sapwood and heartwood and thus either degrade the quality of wood or disrupt the physiological functions of the tissue (Fig. 9.8). The sugar maple borer (*Glycobius speciosus*) attacks the trunk and large branches of sugar maple. The attacks are marked in later years by transverse, spiral, or longitudinal swellings and ridges on the bark (Figure 9.9a). Symptoms of current damage may include leaf color changes, premature leaf drop, or smaller than normal leaves in infested trees. Borer feeding cuts into the conductive tissues, disrupting the translocation of food and water (Fig. 9.9b). In some cases, comparisons of average growth in diameter and height show no differences between attacked and nonattacked trees (Talerico, 1962). However, the trees examined were those that had recovered from an attack, perhaps indicating the recovery capacity of survivors. Nevertheless, most research

Figure 9.5. Galleries of a pine bark beetle (*Orthotomicus caelatus*) in the sapwood–cambial interface. (Courtesy Shade Tree Laboratory, University of Massachusetts.)

shows that borers inflict considerable damage to sapwood and heartwood and thereby reduce wood quality, weaken standing trees, and often hasten wood decay by providing entry to pathogens such as fungi.

Perhaps a primary reason for the importance of borers is the effect they can have in reducing the economic value of newly cut trees or timber.

Figure 9.6. Adults of the long-horned beetles (a) *Monochamus scutellatus oregonensis* and (b) *Tetropium paravulum*. (From Raske, 1972.)

Figure 9.7.   Initial gallery excavations of early instars of pine sawyer beetle (*Monochamus* sp.) (top of log and below knife blade). Note frass (center) produced during excavations. (Courtesy Shade Tree Laboratory, University of Massachusetts.)

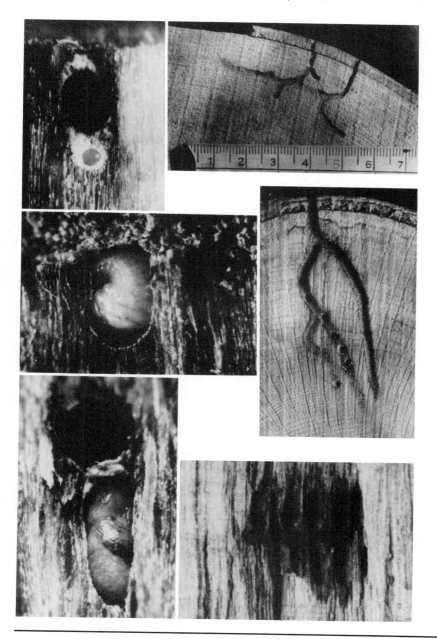

Figure 9.8. Ambrosia beetle damage to sapwood: Left, top to bottom, egg (gray circle) surrounded by symbiotic fungus below a common tunnel; larva in its solitary chamber; pupa nearing emergence. Right, three examples of damaged timber. (From Nakashima, 1978.)

Figure 9.9a.   Transverse scar left as evidence of sugar maple borer (*Glycobius speciosus*) activity. (Photograph courtesy of Johann Bruhn.)

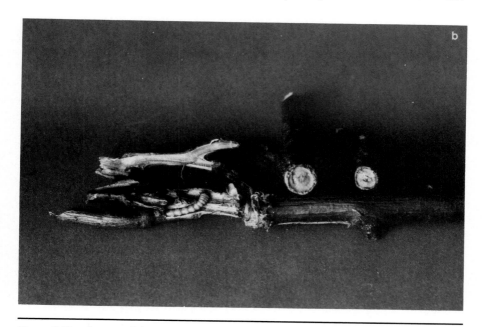

Figure 9.9b.   Larvae of the oak twig pruner (*Elaphidionoides villosus*) and the typical damage it causes. (Courtesy USDA Forest Service.)

Trees that are killed by any of the common mortality agents, such as competition, fire, insects, disease, or lumbering, are often subsequently attacked by various insect species whose larvae make deep and extensive burrows into the wood. The larvae of Cerambycidae and Buprestidae species, such as *Monochamus scutellatus*, the pine sawyer beetle (*M. titilator*), the heartwood pine borer (*Chalcophora virginiensis*), and *C. oregonensis*, can riddle a log and make it essentially worthless (Fig. 9.10). The severity of attack and the extent of damage is determined by the cause of tree mortality, the factor responsible for the weakening of the tree, the density of ovipositing females, the rate of wood drying as affected by site and time elapsed, and other related variables (Butterick, 1913; Richmond and Lejeune, 1945; Ross, 1960). Wood damage and accompanying value losses generally increase with time (Gardiner, 1975). Degradation of wood quality may result not only from gallery formation but also from staining of wood, which results from the growth of fungi associated with bark beetles. This type of effect is dealt with in detail in Chapter 5.

Stems or shoots are often structurally weakened by the action of boring insects (see Fig. 9.9b). Parts of trees that are weakened in this fashion are susceptible to breakage. Aside from the loss of the infested tree part, the point of breakage may serve as a point of entry for pathogenic fungi or other microorganisms. Twigs are structurally and functionally injured by the irregular galleries of the mottled willow borer (*Cryptorhyncus*

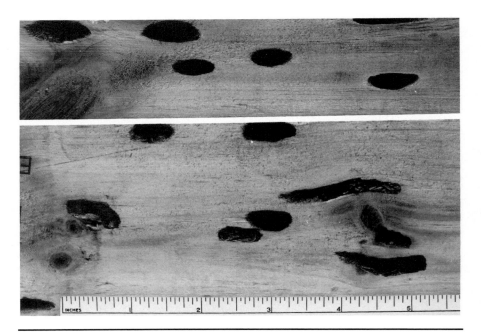

Figure 9.10.   Pine log riddled with galleries formed by the pine sawyer beetle (*Monochamus* sp.). (Courtesy Shade Tree Laboratory, University of Massachusetts.)

lapathi). These twigs are subject to breakage or may develop a cankerlike structure around the injured sites (Fig. 9.11).

Many other species, such as *Styloxus, Elaphidionoides*, and *Xylotrechus*, girdle small twigs and branches, both as larvae and as adults (Linsley 1959). Often their attacks are not accompanied by fungal entry. The Douglas fir hylesinus (*Pseudohylesinus nebulosus*) feeds in the phloem of thin-barked branches and stems of Douglas fir. These attacks can kill small trees and the tops of old trees. Maturation feeding (required for sexual maturation) by teneral adults occurs primarily in the xylem of twigs. Twigs are mechanically or structurally injured by the formation of irregular galleries and are subject to breakage or development of cankerlike structures around the injured sites (see Fig. 9.11). Larvae of *Saperda* living in large branches or twigs also are associated with abnormal tissue growth and gall formation.

Wood Degenerators

Whereas most of the previous groups of bark- and wood-boring insects require at least some consumption of cambial tissues, other groups include species that feed exclusively on wood (sapwood and heartwood) and

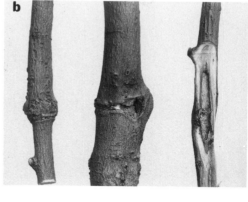

Figure 9.11. (a) Internal damage, pupal cells, and adults of the mottled willow borer (*Cryptorhynchus lapathi*) and (b) damage by the mottled willow borer. (Courtesy Shade Tree Laboratory, University of Massachusetts.)

species that inhabit and destroy wood but do not depend on the wood for their nutrition. The type of wood attacked by wood borers ranges from standing green timber to seasoned wood. Included in this group of borers are horntails, woodwasps, carpenter ants, termites, powder-post beetles, and weevils. Although most species attack felled trees, a few infest standing timber. For example, the western cedar borer (*Trachykele blondeli*) bores through undamaged bark and directly into the sapwood and heartwood. Still other species attack living trees and gain entry at scars left in the aftermath of fire, lightning, or logging. Major economic injury, however, is caused by species that attack green lumber and seasoned wood.

Degenerator species generally require moist wood, that is, at least year-old wood in contact with the soil or moist heartwood of living trees. Examples include horntails (e.g., *Tremex* spp., *Sirex* spp., and *Urocerus* spp.), carpenter bees (*Xylocopa* sp.), carpenter ants (*Camponotus* sp.) (Fig. 9.12), and termites (*Reticulitermes* sp.) (Fig. 9.13). The degeneration of wood that contains galleries is closely linked to the activity of soil microflora (protozoa, fungi, and bacteria).

Insects that attack sound, dry (seasoned) wood are also capable of almost complete destruction of wood (Fig. 9.13). The requirements necessary for the successful utilization of dry, relatively nonnutritious wood probably include the evolution of internal symbiosis with a microorganism. Thus, relatively few species fill this seasoned-wood feeding niche. How-

Figure 9.12.   Typical galleries formed by carpenter ants (*Camponotus* sp.).
(Courtesy Shade Tree Laboratory, University of Massachusetts.)

ever, many household wood products, tools, furniture, and structural
timbers are susceptible to attack by those insects capable of surviving on
wood (Fig. 9.14). Unlike many other wood borers, insects such as the
powder-post beetle can reinfest host material and render it a powdery
residue.

General Ecological Effects of Wood Borers

Major climatic disturbances may create conditions that enhance the development of epidemic levels of wood-boring insects. In the past, storms
have leveled vast areas of forests and have uprooted, weakened, or killed

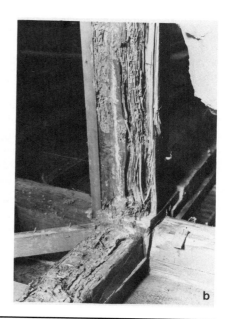

Figure 9.13. Structural damage to a building resulting from termite feeding. (Courtesy Shade Tree Laboratory, University of Massachusetts.)

Figure 9.14. Emergence holes and damage by (a) anobiid powder-post beetles on seasoned wood and by (b) lyctid powder-post beetles on a wooden handle. (Courtesy Shade Tree Laboratory, University of Massachusetts.)

numerous trees. In western areas of the United States the destruction caused by high winds has led to bark beetle epidemics that have covered hundreds of square miles of watershed and killed up to 80% of the forest trees (Bethlahmy, 1975). Such an outbreak can have a significant influence on the composition of the forest (Fig. 9.15). For example, before a bark beetle epidemic, the dominant species in certain areas of the western United States were Engelmann spruce and subalpine fir, which occurred in a ratio of 4:1 with a 34 m²/ha basal area and a 343 m³/ha volume. Twenty-five years after the epidemic, the dominant species were spruce and fir, but in a ratio of 1:4, a basal area of 10 m²/ha, and a volume of 60 m³/ha (Bethlahmy, 1975).

The death of trees also has major implications for the hydrology of an area. Loss of trees (vegetative cover) may lead to a greater accumulation of snow. Because the water from spring snow melt is not taken up from the soil by the vegetation, areas that have been seriously affected by bark beetle epidemics experience increased annual streamflow. The severity of changes resulting from injury depends on several factors including the degree of tree loss, tree species composition as it relates to the attack, specificity of the insect borers, and speed of regeneration of tree species (Bethlahmy, 1975).

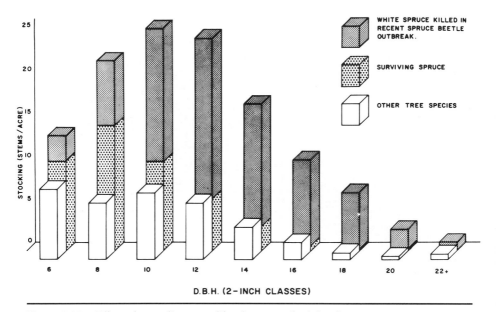

Figure 9.15.   Effect of mortality caused by the spruce bark beetle (*Dendroctonus rufipennis*) on stand composition. Note: "other tree species" represents 22% paper birch (*Betula papyrifera*), 1.2% balsam poplar (*Populus balsamifera*), and mountain hemlock (*Tsuga mertensiana*). (From Werner *et al.*, 1977.)

Perhaps one of the most interesting ecological roles of bark beetles is the regulation of forest succession. The mountain pine beetle kills large acreages of lodgepole pine and thus creates high fuel loads that result in widespread fire. Fire releases the seeds from the serotinous cones and results in maintenance of the lodgepole pine forest. This seral tree species would otherwise be succeeded by species that are more shade tolerant. In this way the mountain pine beetle serves to enhance the species survival of lodgepole pine (Peterman, 1978). Other examples of the role insects play in forest succession have been discussed by Schowalter *et al.* (1981) (southern pine beetle) and Flieger (1970) (spruce budworm). Thus, the short- and long-term influences of various phloem and wood-boring insects may be quite different. Similarly, the benefits and disadvantages of their activities may differ depending on whether one has an economic or ecological perspective.

Chapter Summary

Phloem- and wood-boring insects constitute a major group of economically important forest and shade tree pests. Bark beetles in particular are well known for their potential to kill vast acreages of coniferous forests. Feeding by borers such as bark beetles results in direct girdling and subtle changes in phloem that affect the survival of the tree. The response to girdling is dependent on the extent of the girdle and the tree species. Many species can survive several years of girdling. Wood-boring insects have specialized to feed on selected woody tissues. Often the pattern of feeding is related to the nutritional quality of the tissue and ecological association with microorganisms. Many of these wood borers are important problems of fresh-cut logs and wood in service. Wood borers play an important ecological role in nutrient recycling and forest succession. Some forest communities are sustained because of the activities of wood-boring insects.

CHAPTER 10

Gall-Forming Insects

Introduction

A number of organisms are able to stimulate plant tissues to form a variety of abnormal swellings on leaves, twigs, stems, and roots (Fig. 10.1). The occurrence of these structures, known as galls, on shade trees can have a significant socioeconomic impact because of people's emotional response to the reduced aesthetic quality of gall-infested ornamental trees. In addition, heavy gall infestations can result in physiological disfunctions in trees.

Galls are abnormal growths consisting of pathologically developed cells, tissues, or organs of plants, resulting mostly from hypertrophy (overgrowth) and hyperplasy (cell proliferation). These are induced by organisms such as bacteria, fungi, nematodes, viruses, mites, and insects. Although some tree species are more frequently utilized as hosts, host preferences vary throughout the world (Table 10.1). This section reviews some examples of mite and insect gall-makers with emphasis on the major groups (Table 10.2).

In the strict sense, a gall is any growth response of a plant to the attack of some foreign organism. In the interactions between an insect or mite gall-maker and its tree host, the gall-maker appears to be the only organism to benefit from the association. The gall serves as food and shelter for the insect or mite. The tree, on the other hand, has the growth of its tissues redirected from its own uses (e.g., photosynthesis, transpiration, water absorption, and shoot development). In addition, infestation by gall-forming insects can result in the relocation or loss of chemical elements needed for growth. Trees may also suffer from disturbances of xylem and phloem flow and premature tissue decay caused by these gall-forming insects.

A number of theories have been proposed to explain why plant tissues react to insect attack by forming galls. One hypothesis suggests that a plant produces a gall to limit or isolate the plant parasite (i.e., the insect

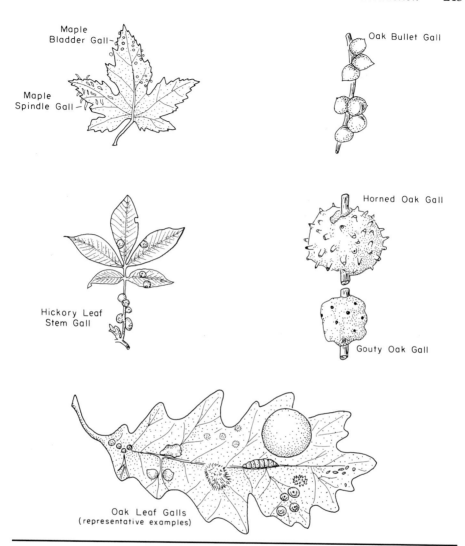

Figure 10.1. Some of the variations in galls of trees: Maple bladder and spindle galls, oak bullet gall, hickory leaf-stem gall, horned oak and gouty galls, and oak leaf galls (representative examples). (Courtesy of J. A. Davidson.)

or mite) in time and space and "force" it into extreme specialization. A second hypothesis proposes that gall formation enables the plant to neutralize toxins produced by the attacking insect. The abundance of gall-inducing insects with piercing-sucking mouthparts and the ample evidence of translocated and systemic phytotoxemia (see section on insects with piercing-sucking mouthparts) give some credence to this hypothesis. However, this second hypothesis provides only a partial explanation since,

Table 10.1. Gall-forming insect preferences (indicated by number of species) for certain plant genera from different sections of the world[a]

Plant Families	Number of Species Forming Galls				
	Asia	Dutch East Indies[b]	South Europe	North America	South & Central America
Pine	17	—	84	47	—
Willow	71	—	569	115	2
Walnut	4	3	7	65	—
Birch	9	—	110	34	—
Beech	304	9	993	419	16
Elm	207	125	65	37	7
Maple	17	—	80	48	—

[a] Modified from Felt (1940).
[b] Currently, Republic of Indonesia.

Table 10.2. Important groups of gall-producers, showing the number of species known in different sections of the world as of 1940[a]

Area	Gall Mites	Psyllids	Aphids	Gall Midges	Gall Wasps	Thrips
Asia	130	42	47	176	102	53
Dutch East Indies[b]	355	145	62	535[c]	27[d]	124
South Europe	263	38	172	398	262	—
North America	162	11	47	682	444	—
Middle Europe	196	10	96	325	125	1
South & Central America	19	20	10	127	6	1

[a] From Felt (1940).
[b] Currently, Republic of Indonesia.
[c] Figures for Diptera only available.
[d] Figures for Hymenoptera only available.

in a few cases, galls induced by insects and mites are not localized but involve a distant organ (Mani, 1964). The potential for such distant gall induction exists for various gall-forming species. Saliva or its components are translocated through the phloem to all parts of a plant (Dixon, 1973).

Gall-Related Tree Injury

In general, there are insufficient data to develop generalizations describing the relationship between tree physiology and susceptibility to gall infestations. Some data suggest that an increase in gall formation may occur on slowly growing trees (Wilford, 1937; Rosenthal and Koehler, 1971) (Fig. 10.2).

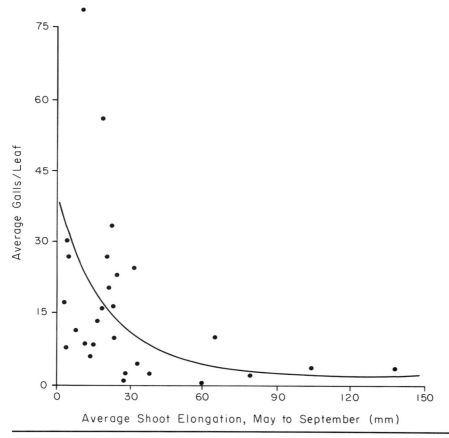

Figure 10.2. Relationship between growth of shoots of the valley oak (*Quercus lobata*) and the density of galls of the agamic generation of *Neuroterus saltatorius*. (Redrawn from Rosenthal and Koehler, 1971.)

By far the greatest impact of gall-forming insects is on the aesthetic value of ornamental trees. Unpleasant-looking trees occur as a result of an abundance of galls, discoloration of the tree and stem, and twig malformations. Physiological injury may also occur in individual trees. Discoloration of leaves or needles and premature needle (leaf) drop may occur in heavily infested trees (Baker, 1972). For example, in balsam fir that is infested with the balsam gall midge (*Dasyneura balsamicola*) an abscission layer forms in infested needles by the end of the first growing season. By the following spring, 90% of the needles have dropped. Balsam fir normally retains its needles for four or more years (Giese and Benjamin, 1959).

Heavy infestations by species such as *Callirhytis gemmaria* (the ribbed bud gall wasp), *Rhabdophaga* sp., *Pineus pinifoliae* (the pine leaf adelgid), *Mayetiola rigidae* (the willow-beaked gall midge), and others can lead to twig mortality and the death of young trees (Balch and Underwood, 1950; Wilson, 1968a,b). Seed crop reduction due to infestation by insects such as *Callirhytis operator* has also been observed (Baker, 1972). Injury may be manifested in a series of changes in the growth of the tree and in subsequent injurious interactions with tree pathogens. For example, the spruce gall midge (*Mayetiola piceae*) attacks the current year shoots of white spruce and causes them to curl, twist, and swell to twice their normal diameter. The majority of the heavily infested shoots die the following year. Subsequently, fungal infestations such as that caused by *Ascochyta piniperda* may become established in the tree (Smith, 1952).

Mechanisms of Gall Induction

Although the initial stimulus for gall induction may come from the mechanical stimulus of oviposition (Fig. 10.3), this cannot account for all gall formation since a continuing stimulus may be necessary. In addition, many insect species pierce plant material without inducing galls, and some gall-inducing species deposit their eggs on leaves.

Experiments have demonstrated that neither punctures, incisions, nor the physical stimuli of foreign bodies are capable of producing lasting plant outgrowths. Research performed as early as the late 1800s established the chemical nature of many of the stimuli leading to gall formation. The role of chemical factors has been demonstrated by showing that injected extracts of the eastern spruce gall aphid (*Adelges abietis*) cause the induction of galls in buds of Norway spruce. Similarly, injected extracts of the salivary glands of two gall-making coccids (*Matsucoccus gallicola* and *Asterolecanium variolosum*) induce symptoms typical of the damage caused by the insects themselves (Carter, 1962). Chemical stimuli often originate in the secretions associated with feeding of adults and/or larvae, oviposition, or in some species both activities. The production of chemical

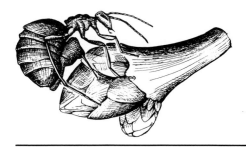

Figure 10.3.   Oviposition by *Biorrhiza pallida* inside the bud of *Quercus* sp. (Redrawn from Mani, 1964, by A. Bartlett. Reproduced by permission of W. Junk B. V.)

mediators of gall formation may be reflected in the morphology of important organs. The salivary glands of the gall-forming generation of the aphid *Chermes* are bettter developed than in the non-gall-forming generation (Mani, 1964). The initial stimulus for leaf gall formation in white willow (*Salix alba*) is due to the fluid formed by accessory glands of the female *Euura pacifica* sawfly and injected into the plant at oviposition. However, growth-stimulating substances are associated with both adults and larvae. If the larva is removed, the gall stops growing. But if extracts of female accessory glands are injected into an empty gall, growth resumes (Osborne, 1972).

Galls are generally categorized as either kataplasmic or prosoplasmic. Kataplasmic galls are primitive galls characterized by little tissue differentiation and lack of constant external shape. Prosoplasmic galls have tissues that are differentiated into well-defined zones and are characterized by a definite size and shape. The layers of cells bordering the gall chamber, with which the insect comes in direct contact are called nutritive cells. Beyond the nutritive layer is a zone of hard tissues composed of thick-walled lignified cells, followed by cortical-parenchyma cells. Soluble sugars and amino-soluble products are abundant in nutritive tissue (Rohfritsch and Shorthouse, 1982).

Nature of Stimulus

The growth-stimulating effect of insect secretions suggests that the active components of these secretions are plant-hormonelike substances. Some researchers have proposed that gall-makers such as aphids produce their own plant-growth modifiers or, perhaps nonspecific plant hormones (Miles, 1968). The presence of IAA (indole acetic acid) in aphid saliva and the induction of galls by injected mixtures containing IAA support this hypothesis. In addition, experiments have demonstrated that a non-gall-forming insect can produce galls when it is injected with excessive amounts of metabolic precursors of IAA. Other hypotheses have suggested that amino acids (also present in salivary secretions) may provide the stimulus for gall formation (Carter, 1962).

Hypotheses to date do not appear to provide adequate explanations for the full range of observations on gall formation. Although substantial concentrations of auxins (plant hormones), gibberellinlike substances, and amino acids have been found in salivary secretions, this evidence is insufficient because these substances could be of plant origin. Auxins have been fed to aphids but were not subsequently detected in the saliva (Osborne, 1972). High levels of IAA in aphid saliva may only reflect high levels in the aphid's food. It is possible, however, that aphid saliva could inhibit plants' IAA oxidase activity, which would cause high IAA levels (Dixon, 1973). Another problem with the idea that substances such as IAA act as universal inducers of galls is that this mechanism does not account for the specificity of galls (Carter, 1962). A counter proposal to this dilemma is that different hormone concentrations and different times of inducement may cause morphologically different galls (Byers *et al.*, 1976; Brewer and Johnson, 1977).

However, the possibility of specific multicomponent gall-inducers still exists. Gall initiation may also result from the ingestion of plant substances, which are then concentrated or modified in the insect and ultimately reintroduced into the plant. This process may require feeding on specific plant parts by specific insects at a particular time in the season before the substances induce the appropriate response (Osborne, 1972).

Trees and Tree Parts Affected

In general, although few galls occur in cryptogams and gymnosperms, they are remarkably abundant on dicots. Over 93% of galls are formed on dicotyledons (Mani, 1964). The variety of plants that are galled and the differences in susceptibility among closely related species are remarkable. A single species such as *Celtis occidentalis* may have 17 distinct types of galls (Carter, 1962). Similarly, 750 of approximately 805 galls caused by wasps are restricted primarily to oaks.

Whatever the host tree, almost all the parts of the tree can produce galls, from the root tip to the growing shoot and on the stems, buds, petioles, leaf blades, flowers, and fruits or seeds. Although the occurrence of galls on the tree is widespread, some plant parts may be the site of many more galls than others. These differences are demonstrated by cynipid (gall-wasp) galls on *Quercus* (Fig. 10.4). In general, the largest number of galls are leaf galls, followed by twig or branch galls, bud galls, and various other less common types (Fig. 10.4). Many types of organisms are capable of inducing galls, including nematodes, fungi, bacteria, viruses, mites, and insects. The latter two will be emphasized here.

The state of the host plant at the time of oviposition may also determine the form of the gall that results. The button top gall midge (*Rhabdophaga heterobia*) makes several types of galls (lateral bud, button, and catkin galls) depending on the stage of plant growth. Similarly, the bat willow gall midge (*R. terminalis*) causes two types of galls depending on the state of willow growth (Barnes, 1935).

Gall-Forming Groups

Mites

Two mite families contain the major gall-forming species, Tarsonemidae and Eriophyidae. Eriophyidae contains the species of most importance to trees (Figs. 10.5–10.7). The mouthparts of mites are adapted for piercing

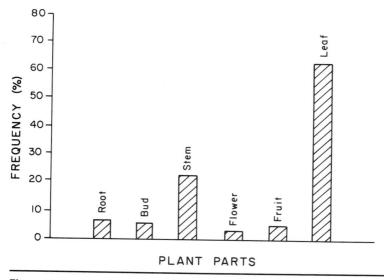

Figure 10.4. Frequency with which galls are formed by cynipids on various tissues of *Quercus* species worldwide. (Redrawn from Mani, 1964. Reproduced by permission of W. Junk B. V.)

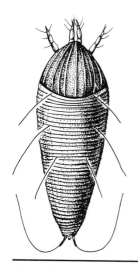

Figure 10.5. Dorsal view of the adult *Eriophyes*, a typical gall-forming mite. (Redrawn from Mani, 1964, by A. Bartlett. Reproduced by permission of W. Junk. B. V.)

plant parts and sucking out their contents. An interesting diagnostic characteristic of eriophyids is their large salivary glands. Over 500 species of gall-forming mites have been described, primarily in the genus *Eriophyes* but also in the genera *Phyllocoptes, Anthocoptes, Oxypleurites,* and *Epitremerus.*

Figure 10.6.   Linden mite gall (*Eriophyes abnormis*).

Figure 10.7.   Maple bladder gall (*Vasates quadripedes*). (Courtesy F.E. Wood.)

Insects

The principal insect gall-formers belong to the orders Thysanoptera, Hemiptera (Heteroptera and Homoptera), Coleoptera, Hymenoptera, Lepidoptera, and Diptera. The most important gall-formers on trees belong to the orders Hemiptera, Hymenoptera, and Diptera.

Homoptera

Gall-forming species are found in several families including various types of aphids (Eriosomatidae, Chermidae, and Phylloxeridae), spittlebugs (Cercopidae), and some Coccoidea. As in many taxonomic schemes, the nomenclature used to classify related forms of Homoptera varies from expert to expert (an unfortunate complication in an introductory survey of a subject). Aphidlike gall-formers, for example, can be grouped into the Eriosomatidae (woolly and gall-making aphids), Chermidae (pine and spruce aphids), or Phylloxeridae (phylloxerans). Other schemes use the family Aphididae (aphids or plant lice) and include gall-formers (i.e., the gall-former Eriosomatidae) and non-gall-forming free-feeding aphids. In addition, the family Phylloxeridae can include the former Chermidae and Phylloxeridae (Baker, 1972). Regardless of the taxonomic categories, the life histories and host preferences of gall-formers establish more reasonable groupings.

Chermidae.   The members of this group feed exclusively on conifer host species. Most species alternate between two different conifer species but form galls only on the so-called primary host. Adults exhibit antennal polymorphism such that antennae of winged forms have five segments, sexual forms have four segments, and wingless forms have three segments. All females are oviparous (egg layers). Not all species in this group form galls.

The eastern spruce gall aphid is a common northeastern representative of this group (Fig. 10.8). Both of the two generations, which consist only of females, occur on spruce. Overwintering nymphs complete their development in the spring by feeding on needles of new shoots. Eventually, adults lay their eggs; new nymphs settle on swollen needles; and, as twig swelling continues, nymphs are enclosed and develop in the galls (Borrer *et al.*, 1964). The Cooley's spruce gall aphid (*Adelges cooleyi*) (Fig. 10.8), which occurs across the northern and western United States, alternates hosts. Its primary hosts include white, Colorado blue, Sitka, and other spruces; its alternate host is Douglas fir (Fig. 10.9). In this species, galls begin to form and enclose the nymphs immediately after nymphs settle down to feed at the bases of young needles. Other species that are injurious to forest and ornamental trees include the balsam woolly aphid and the pine leaf adelgid. The latter forms terminal compact galls on red and black spruce and alternates on white pine. Infestations may be severe on white pine, and even though galls are not formed, diameter and shoot growth can be reduced. The normal angle of growth of shoots may be disrupted, and a mild crooking at nodes often results.

Phylloxeridae.   The few important species in this group feed and form galls on deciduous trees. Although they do not produce waxy threads, they are often covered with waxy, powderlike material. Several

Figure 10.8. (Top) Gall of eastern spruce gall aphid *(Adelges abietis)* and (bottom) gall of Cooley's spruce gall aphid *(adelges cooleyi)*. (Courtesy Shade Tree Laboratory, University of Massachusetts.)

gall-forming species occur on hickories (*Carya*) (e.g., the hickory gall aphid, *Phylloxera caryaecaulis*) and on pecan (*Carya illinoensis*) (e.g., the pecan and pecan leaf phylloxera, *P. devastatrix* and *P. notabilis*, respectively) (Baker, 1972). These phylloxeran galls are characterized by an opening, usually surrounded by hairy lobes, on the upper or lower leaf surface (Felt, 1940).

Eriosomatidae. The cornicles of these aphid are generally reduced or absent whereas wax glands are abundant. Most species in the group alternate between hosts, the primary host usually being a tree on which

Figure 10.9. Cooley's spruce gall aphid (*Adelges cooleyi*) on needles of its alternate host, Douglas-fir. (Courtesy Shade Tree Laboratory, University of Massachusetts.)

overwintering eggs are deposited. The secondary host is usually a herbaceous plant, on which galls are not usually produced. Other species such as *Eriosoma lanigerum* form galls on roots of apple trees and cause leaf rosetting on elm. *E. lanigerum* generally feeds on roots, leaves, and bark, overwinters on elm, spends its first generation on elm, and then the remainder of the growing season is spent on apple, hawthorn, and related tree species. Other common elm-feeding species include the cockscomb gall aphid (*Colopha ulmicola*), which forms a characteristic gall on leaves, and the woolly elm bark aphid (*E. rileyi*), which attacks American and slippery elm (*Ulmus rubra*), causing a knotty growth at sites of injury. Other gall-forming aphids of poplar in this group are *Pemphigus papulitransversus*, *P. bursarius*, *P. spirothecae*, and *P. filaginis* (Fig. 10.10), and also the poplar vagabond aphid (*Mordwilkoja vagabunda*), which forms unusual convoluted galls on leaves and twigs (Baker, 1972) (Fig. 10.11).

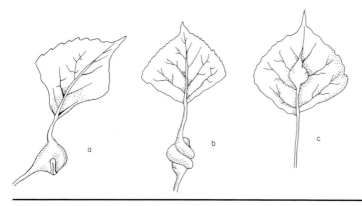

Figure 10.10.   Galls induced on the leaves of poplar (*Populus nigra*) by (a) *Pemphigus bursarius*, (b) *P. spirothecae*, and (c) *P. filaginis*. (Redrawn from Dixon, 1973.)

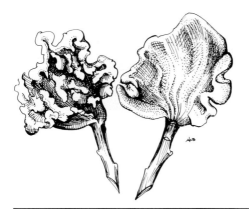

Figure 10.11.   Vagabond poplar gall formed by *Mordwilkoja vagabunda*. (Drawing by A. Bartlett-Wright.)

Diptera

Most gall-forming flies are mosquito-like insects and are among the most important gall-inducing insects. The Cecidomyiidae (gall midges) is a large group of approximately 5000 species, many of which are gall-formers (Fig. 10.12). Gall midges, along with gall wasps, comprise approximately 1500 to 2000 species of gall-formers in the United States. The representatives of this group are found on a great variety of hosts (Mani, 1964). Most adult midges never feed, and the larvae (which are characterized by a sclerotized structure at the front end called a spatula) form the galls. The irregular potato-shaped gall, common on small twigs of willow, is formed by the midge *Rhabdophaga batatus* and may contain several larvae, each in an individual cell (Felt, 1940). Other galls such as the maple leaf spot gall (formed by *Cecidomyia ocellaris*) support only one larva in each gall (Borrer *et al.*, 1964). The pine needle gall midge (*Cecidomyia pinirigidae*) causes swelling at the base of needle clusters. The activity of some dipteran gall-formers may result in tree disfunctions.

Hymenoptera

Gall-forming species occur in the Tenthredinoidea (sawflies), Cynipoidea (gall wasps), and Chalcidoidea (chalcids) (Fig. 10.13). Many of the gall-forming tenthredinidae (e.g., *Pontania desmodioides*, *P. californica*, and *P. pisum*) form rather large, fleshy, hollow galls on leaves of several species, particularly willows. Species in the genus *Pontania* are different from most other hymenopteran gall-formers because plant cell proliferation is initiated by substances formed in the adult female's accessory gland and injected when eggs are deposited. In other Hymenoptera there is no gall induction until after larval eclosion.

Cynipoidea are among the best-known gall insects in Europe and North America. They occur on various plants and give rise to a great variety of galls. These insects are particularly common on oaks. Species may be parthenogenetic or may alternate between parthenogenetic and bisexual generations. Major differences in adult and gall forms may be associated with this sexual alternation. Gall-forming chalcids are minute

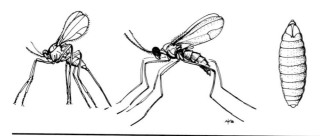

Figure 10.12. Typical cecidomyid adult gall-makers. (Drawing by A. Bartlett-Wright.)

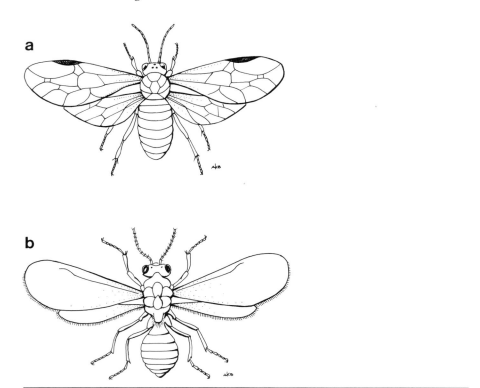

Figure 10.13.   Typical hymenopterous gall-forming insects similar to those in the genera (a) *Pontania* and (b) *Isthmosoma*. (Drawing by A. Bartlett-Wright.)

insects which represent only a small portion of the chalcid group and are mostly entomophagous parasitoids (Mani, 1964).

Many other insect groups have species that induce galls. However, compared with the major groups these other species are of minor importance. Included among the minor gall-formers are the Cerambycidae, Buprestidae, and Curculionidae in the Coleoptera and the Psyllidae (Mani, 1964). Clearly, a number of issues remain to be evaluated. The induction of gall growth, the development of gall tissues, and the impact of gall growth on trees are among the most important areas of future research.

Chapter Summary

Insects from several different taxonomic groups are capable of inducing plant tissue to form galls. Galls are abnormal growths of plant tissue resulting from hypertrophy and hyperplasy that disrupt normal physiological processes and decrease the aesthetic value of trees. This results in significant impact on some natural forests and especially on trees in the urban environment. The mechanisms by which galls are formed are still not well understood, but their induction by chemical stimuli is a likely mechanism. The nature of galls suggests that the chemical stimuli are plant hormonelike compounds. Galls are very common in some tree groups such as oaks, but relatively less common in most conifers. Mites and insects in the orders Homoptera, Diptera, and Hymenoptera are the most notorious inducers of galls in trees.

CHAPTER 11

Insects That Attack Buds, Shoots, and Roots

Introduction

Insects that attack the buds, shoots, and roots of trees are a specialized and economically important group. Such insects attack the actively growing tissues of host trees and can have a major influence on meristematic tissues. Apical meristems at the tips of shoots and roots are responsible for all primary growth, which is accomplished by the addition of new cells and the control of the pattern of tissue differentiation. Buds are essentially embryonic shoots, possessing apical meristems and leaf primordia enclosed by bud scales. All leaves arise from apical meristems; and in seed plants, branches arise in close association with leaves. The apical meristem of the root apex produces not only all primary tissue but also the root cap. Lateral roots form some distance behind the apical meristems, arising from deep within the root tissues.

Thus, the effects of apical meristem damage by insects are profound. Destruction of embryonic buds and elongating shoots and roots causes a drastic reduction in shoot growth and nutrient uptake. Some types of damage can reduce or completely eliminate the growth of the tree. Another important aspect of such damage is its influence on tree form. Changes in tree form due to insect attack are characteristic of some insect species and can be as important as loss in growth.

Many shoot and root insects are problems in reforestation. Because nurseries and plantations represent substantial financial investments due to high establishment costs, losses in growth or poor form can have a substantial economic impact. As forestry becomes more and more intensive, the importance of this group of insects will increase.

Bud and Shoot Insects

Species in this category injure or destroy apical meristems (new growth centers) of trees. The typical outcome of this type of injury is the development of adventitious buds. Depending on the frequency and extent of bud destruction and the degree of apical dominance expressed in the host, deformities may occur. These deformities may include leader distortion (Fig. 11.1), posthorn formation (Fig. 11.2), bushiness (Fig. 11.3), and forking (Fig. 11.4). Repeated attack by insects that damage terminal shoots may result in a substantial change in overall tree form (Fig. 11.5). Widespread damage from repeated attack, a common occurrence in southwestern forests, has a major impact on reforestation efforts.

Figure 11.1.   Leader distortion of Scots pine (*Pinus sylvestris*) caused by feeding of European pine shoot moth (*Rhyacionia buoliana*). (From Carolin and Determan, 1974. Reproduced by permission of the Society of American Foresters.)

Figure 11.2.   Posthorn: A severe crook resulting from larvae feeding on only one side of the terminal shoot. This illustrates a deformity on ponderosa pine (*Pinus ponderosa*) due to feeding by the southwestern pine tip moth (*Rhyacionia neomexicana*). (From Lessard and Jennings, 1976. Courtesy USDA Forest Service.)

One species affecting reforestation and plantation programs, *Pissodes strobi*, has been the subject of a great deal of research. This insect is considered one of the most important factors limiting the successful cultivation of eastern white pine (Belyea and Sullivan, 1956). Adults become active at approximately the time of budbreak of pines and feed primarily on the buds and bark of terminals. Feeding is usually marked by evidence of pitch flow. The eggs are laid in chambers in the inner bark, and subsequently larvae bore downward and feed on the inner bark of the shoots. As larvae become fully grown, they move into the pith and form pupal cells.

Although eastern white pine is the favored food of white pine weevil, other conifers such as Norway spruce, Scots pine, and jack pine are also attacked. This insect was described as a forest pest as early as 1817; however, injury by the weevil was not considered serious until the turn of the 20th century (Belyea and Sullivan, 1956). The occurrence of the weevil and its damage is associated with pure stands of white pine growing in open

Figure 11.3. Bushiness: An abnormal increase in the number of branches at a whorl. Adventitious buds produce multiple shoots after the destruction of terminal and lateral shoots. This illustrates a deformity on ponderosa pine (*Pinus ponderosa*) due to feeding by the southwestern pine tip moth (*Rhyacionia neomexicana*). (From Lessard and Jennings, 1976. Courtesy USDA Forest Service.)

conditions. In increasingly shaded areas, such as that due to dominant deciduous species, injury decreases. Weeviled leaders usually have a characteristic crooked appearance (Fig. 11.6). Laterals of the topmost unweeviled whorl compete for the dominant position; if one succeeds, the tree may develop a bayonet-shaped trunk. The crook may be dramatic and persist for the life of the tree (Fig. 11.7). When two laterals are codominant, the tree takes on a forked appearance. Often growth of all the laterals causes the so-called bushy or cabbage pines.

Similar injury can be observed in trees damaged by several other species of bud and shoot borers (Butcher and Hodson, 1949; Friend 1931). Indeed, after-the-fact determinations of the cause of damage to pines based on tree shape must be done with care. Other causes of leader damage may induce very similar damage. The pine grosbeak (*Pinicola enucleator*) feeds in the winter on buds and seeds of trees. The severity of the injury it inflicts it varies, but often the central bud in a cluster is removed by the bird, which

Figure 11.4.   Forking: Stem deformation resulting when two or more laterals assume dominance. One ultimate effect of forking may be a crook. This illustrates a deformity on ponderosa pine (*Pinus ponderosa*) due to feeding by the southwestern pine tip moth (*Rhyacionia neomexicana*). (From Lessard and Jennings, 1976. Courtesy USDA Forest Service.)

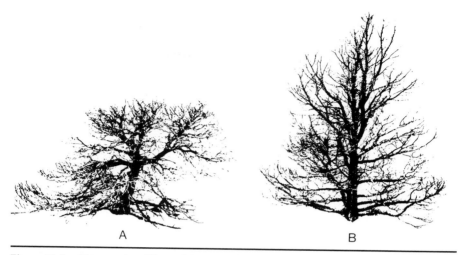

Figure 11.5. Pinyon pine (*Pinus edulis*): (a) a tree that was repeatedly attacked by *Dioryctria* spp. and (b) an unattacked tree. (From Whitham and Mopper, 1985. Copyright 1985 by the AAAS.)

may result in multiple leaders. Red squirrel (*Tamiasciurus hudsonicus*) injury to the leaders also may cause a crooked stem. To the untrained observer, injury by bud freezing is perhaps most similar to white pine weevil damage. Growth patterns such as prolepsis may also be mistaken for white pine weevil damage.

Other insects that produce damage similar to that of white pine weevil include the pine shoot moth (*Eucosma gloriola*), the aphid (*Dilachnus strobi*, and the Sitka spruce weevil (*Pissodes sitchensis*), a serious pest of young Sitka spruce throughout the coastal northwestern United States and Canada. Injury similar to that of white pine weevil can be observed in various species of western U.S. trees as a result of the feeding of the lodgepole terminal weevil (*Pissodes terminalis*). Similarities also occur in adult behavior and larval feeding and in their preferences for open-grown hosts (Amman, 1975).

Weeviled trees that reach merchantable size may be of poor quality. Even straightened, weeviled white pine stems show cross-grain defects, large knots, and compression wood (Spurr and Friend, 1941). Height loss resulting from a single weeviling represents loss of both the previous year's growth and what would have been the current growth (terminal buds). Height growth reductions and growth defects result in a reduction in merchantable lumber (Morrow, 1965; Waters *et al.*, 1955) (Table 11.1).

Larval mining by the western pine shoot borer (*Eucosma sonomana*) in current shoots inhibits shoot and needle elongation as well as the formation and subsequent development of new buds (Fig. 11.8). Newly hatched

Figure 11.6.   Typical damage caused by the white pine weevil (*Pissodes strobi*). (Courtesy USDA Forest Service.)

Figure 11.7.   White pine tree with large distinct crook, the result of white pine weevil damage several years earlier.

larvae bore into expanding buds, develop in the pith of the shoot, emerge, drop to the ground, and pupate (Stoszek, 1973). Larval boring does not kill the shoot or bud but significantly reduces shoot growth. The needles on damaged shoots are usually shorter and stouter than those on undamaged shoots.

Disbudding can be extremely injurious to trees because auxins, though absent prior to the expansion of buds, usually become abundant as the buds start to expand. The impact of disbudding varies with the tree species, type of vessel development, and timing of the injury. Little shoot growth occurs in trees with diffuse-porous wood if the buds are removed before auxins have accumulated to functional concentrations. In ring-porous trees, dormant and adventitious buds and the cambium may have reserves of auxin-precursors that were formed the previous season.

Table 11.1.   Board-foot volume on nonweeviled and weeviled saw-timber white pine trees[a]

dbh (in.)	Volume (b.f.)		Loss in Volume of Weeviled Trees		
	Nonweeviled Trees	Weeviled Trees	Due to Loss in Height (b.f.)	Due to Weevil Culls (b.f.)	Total (%)
9	30	20	10	2	40
10	43	29	14	3	40
11	56	39	17	5	39
12	74	50	24	6	41
13	97	64	33	7	41
14	116	79	37	9	40
15	141	94	47	11	41
16	166	109	57	13	42
17	196	127	69	15	43
18	224	144	80	17	43
19	254	160	94	19	44
20	291	179	112	21	46
21	327	196	131	23	47
22	365	209	156	24	49
23	405	223	182	26	51
24	442	232	210	27	54
				Average	46

[a] From Waters *et al.* (1955).

Disbudding trees with diffuse-porous wood may result in a reduction of growth because in these trees little or no reserve of auxin precursor is formed in the previous season. Variations of the basic diffuse-porous and ring-porous patterns can produce exceptions to the above general trends (Kulman, 1965a).

In addition to growth reductions, bud damage may ultimately result in stem distortions. The spruce bud midge (*Rhabdophaga swainei*) commonly attacks the terminal buds of 6–12 ft open-grown white and black spruce trees. The larvae, which feed and overwinter in the terminal bud, have a considerable impact on height gain and stem shape of their hosts (Cerezke, 1972) (Table 11.2 and Fig. 11.9).

Root Insects

Insects that feed on roots are abundant in both natural and managed stands. Many species, however, become more abundant and injurious in the managed plantation or nursery than in the natural forest. Thus, forest insects that were once considered innocuous or species that were not tree feeding species have become important pests. The strawberry root weevil (*Otiorhynchus ovatus*), the black vine weevil (*O. sulcatus*), and the Japanese

Figure 11.8. Ponderosa pine shoot: Left, damaged by *Eucosma sonomana;* note the shorter needles compared with an undamaged shoot (right).

beetle (*Popillia japonica*) are examples of this phenomenon. Feeding by these and other species can severely injure coniferous seedlings, saplings, or transplants.

White Grubs

Adults of the genus *Phyllophaga*, commonly referred to as June or May beetles, as well as species in the genera *Serica, Geotrupes, Melontha,* and *Anomala,* are important plantation pests. Their larvae, called white grubs, cause serious damage to trees in nurseries (Fig. 11.10). The larvae are similar in appearance and have a characteristic body which is strongly curved, shiny, and milky white. Larvae also have three pairs of prominent legs and highly sclerotized dark heads and mouth-parts (Fig. 11.11).

Table 11.2.   Number of top whorl branches and percentages of terminal buds of top whorl branches and leaders on black (Plm) and white (Plb) spruces infested by *Rhabdophaga swainei* in central Alberta, Canada[a]

| Height (ft) | Class (cm) | Average No. of Top Branches Per Tree | | Terminal Buds Destroyed (%) | | | |
| | | | | Leader Terminals | | Top Whorl Terminals | |
		Plm	Plb	Plm	Plb	Plm	Plb
1	30	2.1	3.5			0	4.8
2	60	2.1	4.6			7.3	0.6
3	90	2.8	4.3	15.7[b]	38.3[b]	14.5	13.4
4	120	2.7	4.7	($n = 83$)	($n = 47$)	28.3	24.3
5	150	3.9	4.3			23.4	42.3
6	185	4.0	4.4			28.7	45.7

[a] From Cerezke (1972).
[b] Percentages based on numbers of buds pooled from all height classes; values in brackets designate sample size.

Various deciduous and coniferous species are attacked; however, severe damage is concentrated on pines. Severe mortality among species of other conifer genera such as larch and spruce is rare.

Injury to plantation trees by white grubs is, at least in part, an outcome of land use patterns or the nature of a habitat before it is used for the establishment of a plantation. Open land that has been under sod for two or more years tends to have a high grub population density. Even managed grass-covered land can have large populations (e.g., 132 grubs/m^2) compared with forb-covered fields (approximately 5–6 grubs/m^2) (Fig. 11.12). Establishment of a plantation in an area with a high density of these insects can result in extensive root consumption, which in turn can quickly result in mortality or impose considerable physiological stress on young trees. Environmental conditions may add more stress; in a dry year damage increases.

Land Management and Pest Problems

Several dendrophagous insects breed and develop in tree stumps, root systems, or wood refuse, and thus are not generally a threat to living trees. However, large insect populations may become injurious to very young trees in managed plantations. The adults of species breeding in dead wood may feed on the bark of the stems and branches of seedlings or transplants.

Still other previously innocuous species have become so-called pests with the increasing use of land for plantations and reforestation projects.

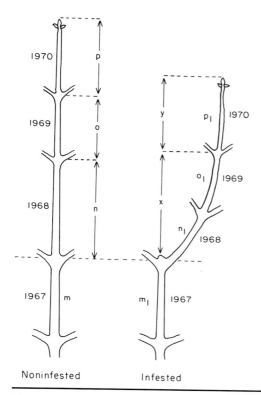

Figure 11.9. (a) Pattern of development of terminal branch on young white spruce (*Picea glauca*) after removal of the 1967 leader terminal bud, compared with (b) normal development. Note: m and m_1, n and n_1, o and o_1, and p and p_1 represent comparable areas on uninfested and infested branches, respectively. (From Cerezke, 1972.)

Steremnius carinatus was known as a scavenger that fed on ground vege-tation and detritus. In the last 15 yr or so, it has gained the status of a pest species of coniferous plantations in areas of natural regeneration in coastal British Columbia. This pest is now of concern because the adults feed on newly planted seedlings. This provides an excellent example of what has happened with many other species. The general behavior and biology of *Steremnius carinatus* is similar to that of related species in Europe and North America, such as *Hylobius abietis*, *H. pales* (the pales weevil), and *Pissodes approximatus*. The key to its success (i.e., greater abundance) appears to be due to an increase in the availability of suitable breeding material. Large-scale continuous logging, right-of-way construction, and similar activities have provided the densities of breeding material that can lead to massive population increases. Add this to the long-lived nature of insects such as adult weevils and the high economic value of nursery-grown seedlings used for reforestation, and small losses over an extended

Figure 11.10.   Range of severity of root system damage in red pine (*Pinus resinosa*) due to white grub feeding. (From Sutton and Stone, 1974. Courtesy of Canada Department of Agriculture, Research Branch, Bio-Graphic Unit, Ottawa, Ontario.)

time period can represent a major disruption of any restocking program (Condrashoff, 1968; Carter, 1916; Speers, 1958).

Larvae of other species are of little economic importance; but adults feeding on bark, buds, or foliage of pine seedlings can kill them. An example is the pitch-eating weevil (*Pachylobius picivorus*). In areas where overstory pines are cut within 3 months of planting pine seedlings, this weevil kills 35–60% of the loblolly, shortleaf, and slash pine seedlings (Thatcher, 1960).

Clearing of forests to establish plantations may also provide a greater than normal attraction to dendrophagous insects, which in turn results in aggregations and subsequent injury in the cleared area. The pales weevil is a stout-beaked, reddish-brown to black beetle that overwinters in the forest litter. In the spring it is often attracted to the pine odor in cut-over areas where it feeds on the tender bark of sapling or at the base of seedlings. Eggs and larvae occur in the inner bark of newly cut logs, pine stumps, or large pieces of slash.

The adults of the closely related pine root-collar weevil feed on tender bark, but the most serious injury is caused by larval feeding in the inner bark and cambium of the root-collar area (Fig. 11.13). *Hylobius warreni, Hypomolyx piceus,* and *Hylobius abietis* are other important species with similar habits. Larvae of some of these species (e.g., *H. warreni*) may

Figure 11.11.   Typical white grub (*Phyllophaga* sp.) larva. (From Sutton and Stone, 1974. Courtesy of Canada Department of Agriculture, Research Branch, Bio-Graphic Unit, Ottawa, Ontario.)

partially or completely girdle the main lateral roots or the stem of older trees, not just saplings. Pine stems that were 50% girdled to simulate *H. warreni* feeding, exhibit leader height reductions of approximately 6–16% during the second and third years after girdling. Stand site and age also affect the degree of growth reduction. Reductions due to simulated girdling are believed to underestimate *H. warreni* damage. Girdling by the related *Hylobius radicis* of *Pinus sylvestris* and of *P. resinosa* (12–30 in. tall) also causes significant leader growth reductions. The effects on radial increment are usually delayed. Root starvation and possibly partial depletion of food reserves in the roots for 3 yr lead to a decline in the radial increment of *H. warreni*-girdled lodgepole pine (Cerezke, 1974).

The best preventative measure is to eliminate the material that attracts these insects to cleared areas. An alternative would be to delay replanting in an area for two to four seasons. However, in managed stands every procedure has consequences that must be dealt with. Thus, in certain areas of the United States rapid encroachment by competing hardwoods makes it critical to establish seedlings within the first year after

Figure 11.12.   Population of white grubs exposed by rolling back grass sod. The population density is approximately 85 grubs/m². (From Sutton and Stone, 1974. Courtesy of Canada Department of Agriculture, Research Branch, Bio-Graphic Unit, Ottawa, Canada.)

the old stand is removed (Thatcher, 1960). The previously mentioned problems with white grub population buildup cannot be ignored.

In many insect species the larvae feed on root tissue, and the adults feed on foliage. Species exhibiting this feeding dichotomy include the curculionid *Scythropus californicus*, which feeds on Monterey pine (Jensen and Koehler, 1969). Larvae of several species of *Hylobius* (such as *H. rhizophagus* and *H. aliradicis*) make meandering tunnels in the upper tap root, hollow out smaller roots, and pupate in the root system. A 40–50% infestation rate of young pines by *H. aliradicis* has been observed in areas of intensive site preparation, making the insect a potential threat to plantation trees (Ebel and Merkel, 1967).

Microflora–Pest Interactions

Finally, the overall impact of insect injury may be enhanced by the relationship between insect activity and the invasion of pathogens. The scolytid genus *Hylastes* contains a variety of species that are associated with coniferous forests, the majority preferring hosts in the genus *Pinus*. Most *Hylastes*, such as *H. cunicularius* and *H. ater*, breed in the roots and bases of

Figure 11.13.   Damage caused by *Hylobius radicis* to Scots pine (*Pinus sylvestris*): (a) cross section of root collar showing fluted edge following recovery from old attack; (b) resin-impregnated soil around root collar with feeding wounds; and (c) completely girdled root collar resulting from weevil feeding. (From Finnegan, 1962.)

dead or dying trees or in stumps. Occasionally, slash and logs in contact with soil are also infested.

The larvae of root-feeding insects participate in the breakdown process of dead roots. Fragmentation, as a result of feeding, increases the surface area for microfloral decomposition. The adult, on the other hand, may cause severe economic damage to newly planted conifers due to the requirement of a so-called maturation feeding on roots, root collars, or stems. Some adults apparently require this type of feeding before oocytes can differentiate and reach a state of maturity. In the northwestern United States, *Hylastes nigrinus* maturation (Douglas-fir root bark beetle) feeding appears to take place on thin roots of old, dead, or dying Douglas-fir and roots of weakened Douglas-fir seedlings. Maturation feeding by various *Hylastes* species may kill young conifers or lead to their death by allowing invasion of fungi such as *Armillaria mellea* or *Heterobasidion annosum*.

Chapter Summary

Bud, shoot, and root insects are a group of forest and urban pests that specialize in attacking apical meristems of trees. Once apical meristems are killed, secondary buds initiate, changing the overall form of individual branches and entire trees. Widespread damage by these insects can have significant effects on reforestation efforts. Modifications of tree form reduce the value of the tree for lumber and is aesthetically displeasing. Root insects are common species that are generally most important in managed plantations and nurseries. On occasion, these insects become serious problems in reforestation. Most recently, root insects have been implicated in the vectoring of causal agents of some important root rot diseases.

CHAPTER 12

Insects That Attack Seeds and Cones

Introduction

Seeds and cones are rich food sources for forest and shade tree insects. Seeds and cones can be abundant, especially in good years when species such as white spruce may average from 2000 to 8000 cones per tree (Tripp and Hedlin, 1956). Insects that feed on seeds, cones, and cone-bearing shoots have an adverse impact on natural and artificial reforestation, often causing seed crop failures over large areas. Seed destruction is increasingly troublesome due to our increasing reliance on seed sources for the production of seedlings or for trees of known genetic characteristics. Seeds from trees with high growth rates, desirable aesthetic qualities, or resistance to disease, insects, or pollution can be valuable. Seed and cone insects can greatly reduce the value of seeds produced (Table 12.1). Trees are cultivated as seed sources over relatively large aeas, about 10,000 acres in the southern United States alone (Goyer and Nachod, 1976). To a great extent, the success of reforestation and reclamation programs is dependent on the activities of seed and cone feeders (Lyons, 1956). In recent years there has been an increasing demand for a higher quality and quantity of seed with known heritable traits.

Economic Impact

Seed and cone insects can destroy a large portion of the annual seed production (Table 12.2). Unlike many other dendrophagous species, seed insects such as the red pine cone beetle consume a high percentage or often nearly all of their available food each year (Mattson, 1971). A study of insects affecting the seeds and cones of ponderosa pines in New Mexico indicates that 5 of approximately 122 species found in the survey cause important economic losses. Together these species destroy an average of

Table 12.1. Comparison of the value of sound seed produced per tree for cone crops without insect damage and with insect damage[a]

	No Insect Damage	Insect Damage 1982[b]	Insect Damage 1984
Initial number of cones/tree	200	200	200
Cone mortality (%)	0	28	60
Surviving cones	200	144	80
Average number of sound seed/cone	47	21	0.9
Sound seed/tree	9,400[c]	3,024	72
Seeds/pound	11,400[d]	11,400	11,400
Pounds of sound seed/tree	0.82	0.27	0.01
Value/pound of seed[e]	$38.00	$38.00	$38.00
Value of seed/tree[e]	$31.60	$10.26	$ 0.38

[a] From Blake *et al.* (1986).
[b] Data from Schmid *et al.* (1984).
[c] Based on U.S. Forest Service estimate of 75% sound seed per tree.
[d] Estimate from Schopmeyer (1974).
[e] Estimate from U.S. Forest Service.

Table 12.2. Loblolly pine conelet and cone losses to insects and other factors[a]

Causal Agent	Conelet Damage, Mar–Sept 1973		Cone Damage, Oct 1973–Sept 1974		Total Damage, Mar 1973–Oct 1974	
	Number	Percent	Number	Percent	Number	Percent
Insect (subtotal)	1127	65.8	160	34.7	1287	75.1
Dioryctria	820	47.9	151	32.7	971	56.7
Pityophthorus sp.	9	0.5	—	—	9	0.5
Thrips ?	82	4.8	—	—	82	4.8
Rhyacionia frustrana	1	0.1	2	0.4	3	0.2
Unknown	215	12.6	7	1.5	222	13.0
Noninsect (subtotal)	125	7.3	19	4.1	144	8.3
Abortion	20	1.2	—	—	20	1.2
Missing	94	5.5	13	2.8	107	6.2
Equipment	7	0.4	2	0.4	9	0.5
Unknown	4	0.2	4	0.9	8	0.5
Total (insect and noninsect)	1252	73.1	179	38.8	1431	83.5

[a] From Goyer and Nachod (1976).

Table 12.3. Damage to ponderosa pine cones from attack by *Conophthorus ponderosae* in New Mexico, 1964 to 1967[a]

National Forest	Year	Number of Samples	Average Percent Cones Attacked
Lincoln	1964	10	60
Gila	1964	13	44
Lincoln	1965	19	4
Gila	1965	11	76
Lincoln	1966	3	47
Gila	1966	3	7
Apache	1966	1	0
Carson	1966	4	5
Santa Fe	1966	2	0
Lincoln	1967	13	13
Gila	1967	19	5
Apache	1967	2	0
Carson	1967	7	26
Cibola	1967	9	24
Santa Fe	1967	6	17

[a]From Kinzer (1976). Reprinted with permission from *Nature* (London) **221**:477–478. Copyright © 1976 Macmillan Journals, Ltd.

82% of the usable seed crop (Kinzer, 1976). In addition, surviving injured cones may have a low proportion of viable seeds. However, the fluctuations in cone production by trees and the relative abundance of pest insects with differing damage potential result in variable amounts of insect damage from place to place and year to year (Table 12.3).

Seed Crop Yield and Insect Abundance

Several factors influence the variability of insect populations and subsequent damage from seed and cone insects. Insect damage to cones of white spruce (*Picea glauca*) may vary from 15 to 89% over a 5-yr period (Werner, 1964). Cone-feeding insects survive fluctuations in cone abundance primarily by the common occurrence of extended diapause (Tripp and Hedlin, 1956; Williamson *et al.*, 1966). In other species, such as the cone midge *Mayetiola thujae*, a portion of the population remains in diapause each year (Hedlin, 1964a,b). In still other insect genera such as *Conophthorus*, there is no extended dormancy. Species such as *C. resinosae* compensate for the absence of extended dormancy by their ability to breed in other parts of the tree (i.e., current-year shoots). In other species, variation in cone abundance may be countered by having a wide host range (Lyons, 1956).

White spruce is a sporadic cone producer, with occasional heavy crop years followed by scarce years. Heavy cone crops are rarely produced for two consecutive years. An unusually heavy crop leads to extensive damage by the cone insect *Dioryctria reniculella*. The insect's population reaches high levels during heavy crop years and declines to low levels in intervening years. *D. reniculella* is probably able to survive because larvae behave as defoliators in the absence of cones, mining in needles, boring into buds, and feeding on foliage. Occasionally, if a cone is completely consumed before larval development is completed, adjacent foliage is consumed (McLeod and Daviault, 1963).

In seed production areas dominated by red pine, cone insects annually use a high percentage or often most of the crop. Annual cone damage may be limited in areas where small crops occur in alternate years. However, cone damage can be substantial in areas where there are no small crops. Thus, cone damage tends to increase annually unless or until cone abundance is limiting. Approximately 66% of the variation in damage is due to variations in cone abundance. The remaining variation in damage results from biotic and physical factors that affect the survival and efficiency of the insect species (Mattson, 1978). The implication of this research on cone insect dynamics is that management of red pine seed orchards for increasing cone yield will most likely lead to expanding insect populations.

Evidence of Seed and Cone Damage

The damage caused by seed and cone insects is relatively easy to assess because any of the insects that attack the cone cause obvious external damage. For example, *Conophthorus ponderosae* produces an obvious pitch tube (Fig. 12.1) and kills the cones early in the season. Damaged cones can easily be sorted from green cones at harvest time. On the other hand, damage by some of the seed-infesting insects is very difficult to assess. For example, *Megastigmus albifrons* feeds entirely within the seeds of ponderosa pine. The only external evidence of infestation is the adult emergence hole. Because adults emerge in spring, fall-harvested seeds have no external signs of insect damage. The only method for detecting the damage is with X-ray (Fig. 12.2). In the absence of a means of systematic detection of this insect, infested seeds may be planted in the nursery, which of course will not germinate. Considerable expense may be incurred before the damage of this insect can be detected.

Natural Regeneration and Seed and Cone Insects

The role of seed and cone insects in commercial seed production has received the most emphasis in research because this area is where they cause the greatest economic loss. However, damage from seed and cone insects can also be important to natural regeneration. In many forested areas of the United States, mechanical site preparation is used to increase

Figure 12.1. Pitch tube caused by *Conophthorus ponderosae* girdling a ponderosa pine cone. (Courtesy T. Koerber, USFS.)

regeneration success. But heavy damage from seed and cone insects can result in very poor regeneration and a 1-yr delay in regeneration. Subsequently, it may be necessary to repeat the site preparation, at considerable expense. This is especially a problem for those species of trees that are sporadic seed producers. This emphasizes the importance of estimating the available seed yield (after seed and cone insect losses) prior to initiation of site preparation.

Taxonomic Association

Four orders of insects are associated with most of the destruction of cones and seeds: Coleoptera, particularly the cone beetles and also cone weevils; Lepidoptera, including the families Pyralidae (cone worms) and Olethreutidae (seed worms and cone moths); Hymenoptera, primarily

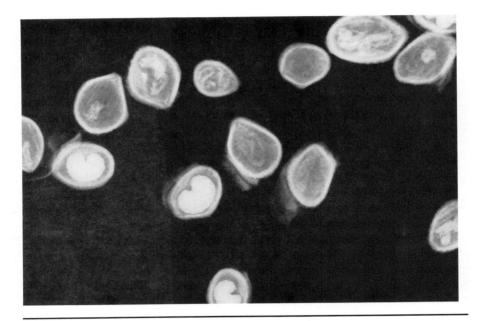

Figure 12.2.   X-ray of ponderosa pine seeds infested by *Megistigmus albifrons*. (Courtesy T. Koerber, USFS.)

the seed chalcids (Torymidae); and Diptera, for example, the cone or seed midge (Cecidomyiidae). Many species of forest trees are host for all or nearly all of the major groups. As an example, almost every species of pine is attacked by at least one species of cone beetle. In several cases each tree has its own species of *Conophthorus*.

Coleoptera: Scolytidae (Cone Beetles) and Curculionidae (Cone Weevils)

Cone beetles are in the family Scolytidae and the genus *Conophthorus*. The beetle genus *Conophthorus* includes a large number of injurious species that attack a variety of hosts. The beetle *Conophthorus resinosae* is particularly important on red pine, as are *D. edulis* on pinyon pine, *C. coniperda* on eastern white pine, *C. ponderosae* on ponderosa pine, and *C. lambertianae* on sugar pine.

In most species of the genus *Conophthorus*, adult females, after leaving their overwintering hibernacula, go through a sequence of behaviors leading to brood establishment. These include the initial penetration, construction, and filling of egg pockets and the plugging of the axial tunnel. The entrance tunnel into the cone is constructed by the female. When she reaches the center of the axis, she turns and bores distally, forming blind egg pockets in the cone as she goes along. In some species, such as *C. ponderosae*, the female usually girdles the axis near the base, thereby severing the conductive tissues and causing early cone death.

Girdling appears to be a prerequisite for brood development (Kinzer, 1976). In other species (e.g., *C. flexilis* on south western white pine, *Pinus strobiformis*), construction of the egg gallery by the female kills the cone.

The adult male *Conophthorus* beetle is rarely active in any of the phases of gallery formation. A so-called nuptial chamber may be formed in which the male is often located and where copulation takes place. In most cases the method or point of entry into the cone appears to be a function of cone structure. If cones have a long petiole, the adult (e.g., of *C. lambertianae*) enters through a tunnel initiated in the petiole. Beetle entry into cones with very short petioles occurs directly through the cone base (as in the case of the mountain pine cone beetle, *C. monticolae*).

Tunnel or gallery construction in the cone varies with each species. Larvae, boring in the cone, consume much of the cone contents (Fig. 12.3). The method by which the parent adult exits is variable. Some species, such as *C. resinosae*, exit through the entrance tunnel, whereas *C. coniperda* and *C. lambertianae* exit through the distal end of the cone. New adults exit through the plug in the axial tunnel or out through the top or sides of the cone. Overwintering may occur in cones or shoots which are attacked after adult emergence and then break off and fall to the ground.

Cone weevil damage to coniferous tree species is not usually of great economic importance; however, many of these insects do cause substantial damage to deciduous tree species. Various species in the genera *Curculio* and *Conotrachelus* breed in the acorns of a large number of oaks, butternuts,

Figure 12.3. Internal damage to western white pine cones caused by larval boring by *Conophthorus ponderosae*. (Courtesy T. Koerber, USFS.)

walnuts, hickories, and filberts. *Amblycerus robiniae*, a member of the family Bruchidae, is a pest of honey locust pods *(Gleditsia triacanthos)*. The larvae of the moth *Melissopus latiferreanus* completely consume many nuts of oaks. Finally, from 20 to 50% of the seeds in injured cones of yellow birch (*Betula alleghaniensis*) are destroyed by the weevil *Apion walshii* (Fig. 12.4).

Figure 12.4.   Yellow birch seeds (top) injured by the weevil *Apion walshii* (center), compared with uninjured seeds (bottom). (From Shigo and Yelenosky, 1963. Courtesy USDA Forest Service.)

Lepidoptera: Pyralidae (Cone Worms) and Olethreutidae (Cone Moths, Seed Worms)

Pine cone worms are widely distributed throughout the United States and the world and constitute an economically important insect group. Most members of this group are in the genus *Dioryctria*. *Dioryctria amatella*, the southern pine cone worm, is a major cone pest in the southern United States, much as *D. auranticella* is a major pest in the west. Hedlin *et al.* (1980) list 21 species of *Dioryctria* in North America.

There is considerable variation in the life histories of cone worms, making it difficult to generalize about their life cycle. Many of the species begin feeding in the cone only when they are mature larvae. Consequently, a distinct larval entrance hole, often surrounded by silk, can be seen on attacked cones (Fig. 12.5). The larvae create a pupation chamber and pupate in the cone. When adults emerge, they leave the pupal cavity in the cone (Fig. 12.6). Boring usually results in cone mortality and complete seed loss.

Seed and cone insects in the family Olethreutidae include the cone moths and seed worms. Cone moths occur in the genus *Barbara*. Although this is a taxonomically small group, some members are economically important. *Barbara colfaxiana* is a particularly important pest of Douglas-fir (Fig. 12.7).

Figure 12.5.   External evidence of attack by *Dioryctria auranticella* on ponderosa pine. Note the large entrance hole and silk webbing.

Figure 12.6. Pupal case of *Dioryctria auranticella* within a pupal chamber in a ponderosa pine cone. (Courtesy T. Koerber, USFS.)

Figure 12.7. Damage to Douglas-fir cones (left) by the larvae of the Douglas-fir cone moth (*Barbara colfaxiana*) (right) . (Courtesy of A.F. Hedlin and the Canadian Forestry Service.)

Seed worms, *Cydia* (*Laspeyresia*), as their name implies feed almost exclusively on seeds. Adults lay eggs on the exterior of the cone. Larvae bore into the cone and move from seed to seed, feeding as they go along. Entrance holes are obvious when cones are dissected (Fig. 12.8). Damaged seeds tend to adhere to the scale when the cone opens.

Hymenoptera: Torymidae (Seed Chalcids)

Seed chalcids are small wasps with long ovipositors that are adapted for inserting eggs through pine scales and directly into the seeds. Seed chalcids are represented by the single genus *Megastigmus*. Single eggs are usually laid into seeds by the adult. Larvae feed entirely within the confines of the seed. The larvae pupate, and adults emerge the following season. Because almost the entire life cycle is spent within the seed, the

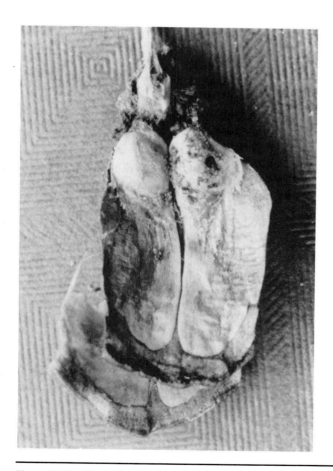

Figure 12.8.   Seeds damaged by *Cydia piperania*.

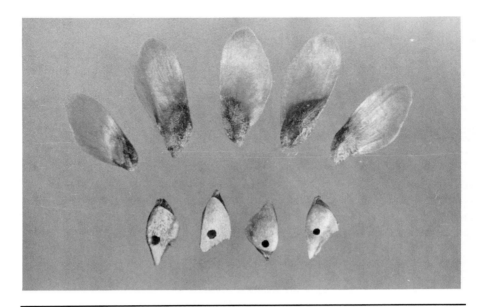

Figure 12.9.   Normal seeds of Douglas-fir (top) and seeds damaged by
Megastigmus spp., the Douglas-fir seed chalcid (bottom). (Courtesy of A.F.
Hedlin and the Canadian Forestry Service.)

only external evidence of attack is the adult emergence hole (Fig. 12.9). As
previously mentioned, the only feasible method of assessing damage from
this insect is by X-raying the seeds (Fig. 12.2).

Diptera: Cecidomyiidae (Seed Midges)

Seed midges are a large group of specialized insects that feed on seeds.
Many seed midges cause galls to form around seeds. For example, the
Douglas-fir cone midge (*Contarinia oregonensis*) induces the formation of a
gall that surrounds the seed or fuses to the seed coat (Fig. 12.10). The seed
is destroyed or fails to detach during seed fall or extraction. As many as
99% of galled seeds remain in the cone during processing, and an average
14% of the seeds are damaged. In addition, premature necrosis of cone
scales follows midge infestation. It is not clear whether this is due to the
insect itself or perhaps due to fungal activity following injury (Johnson and
Heikkenen, 1958; Johnson, 1963).

Other Seed and Cone Pests

Although not very conspicuous, thrips have a great potential for injury to
seed crops. Many factors are responsible for the lack of attention given to
these insects, including their small size (less than 1/16 in.) and the typical

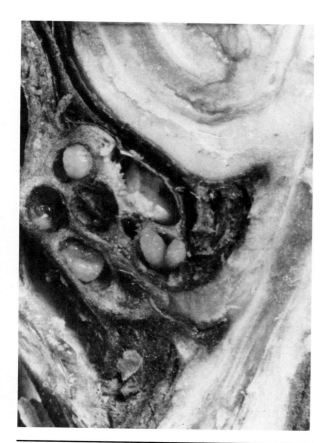

Figure 12.10.   Cross section of Douglas-fir cone showing mature cone midge (*Contarinia oregonensis*) larvae (top left and bottom chambers) in gall around seeds (top center chamber). (From Johnson and Hedlin, 1967.)

adult behavior of hiding under bud scales or in crevices. In addition, damage usually occurs in the early stages of flower development in the upper crown. Those flowers that are killed by thrips quickly dry and fall. Male catkins, female flower buds, developing female flowers, and young conelets may all be infested with thrips. Almost half of the observed flowers on heavily infested slash pines may be killed by thrips (De Barr, 1969). Other studies have indicated that thrip damage ranges from negligible to as high as 20% mortality (Ebel, 1961). However, their size and cryptic habits, may have some effect on mortality estimates.

Species in the family Coreidae (particularly in the genus *Leptoglossus*) often destroy seeds in developing cones as a result of their feeding. Losses due to species such as the pine seedbug (*Leptoglossus corculus*) occur when feeding by second instar nymphs induces abortion and abscission of

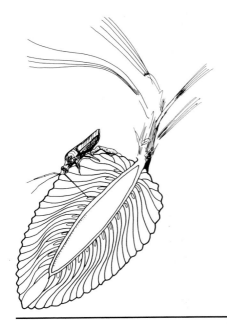

Figure 12.11.   Penetration of cone scales and feeding on seeds by the
leaf-footed pine seedbug, *Leptoglossus* sp.

conelets of their hosts (i.e., loblolly and shortleaf pine) (Fig. 12.11).
Conelets exposed to late summer feeding may also be aborted. Feeding
over a 24-hr period can cause 60% of the conelets to abort. Conelet abortion
and abscission may occur even when most ovules are undamaged (De Barr
and Kormanik, 1975).

Chapter Summary

Seed and cone insects are a group of dendrophagous insects that can have
a major impact on forest regeneration. They represent a broad taxonomic
range of insects and occur in every major forest species. Their impact can
result in the total loss of seed crops in some years. Managing seed and cone
insects is an important part of any seed production operation. The
practicing forester needs to be aware of their potential for damage to
ensure an adequate seed supply for forest regeneration efforts.

Ecology of Forest and Shade Tree Insects

CHAPTER 13

Forest Insect Communities

Introduction

The forests of the world are often categorized by the dominant tree species in the ecosystem. The animal communities associated with each forest type, including insect communities, are both influenced by and have a major influence on tree species composition. The insect communities of coniferous and deciduous forests often have different taxonomic compositions; only 4–5% of the species are common to both. The functional interrelationships between insects and other forest organisms and the biological structure of the forest may be more important than the insect species composition of a given forest type. Thus, a cicadellid species in a coniferous forest may be represented by another species, in a deciduous forest, that has a similar life history.

Each forest and its associated insect communities are also shaped by a variety of environmental constraints. However, even the influence of physical factors may be modified by the nature of the forest. At the same latitude, a deciduous forest may have a mean temperature approximately 4°C higher and a relative humidity 5–6% lower than a coniferous forest (Blake, 1926). Differences in the activities of associated insect communities often parallel differences in physical factors.

The functional structure or pattern of insect communities in the forest may result from the environmental gradients of physical factors. A study of the critical aspects of the Great Smoky Mountains indicates that moisture determines both the quality and quantity of vegetation and insect associates. In this ecosystem the size of the insect communities decreases from the moist so-called cove-forest to dry oak and pine forests. Concurrently, fungivores and herbivores are important in mesic forests, whereas herbivores of vascular plants and pollen feeders numerically predominate in xeric or open forests. Finally, the more taxonomically primitive insect groups are centered in mesic environments, and the more advanced groups (presumably better able to cope with stress) are associated with xeric environments (Whittaker, 1952).

Stratification of Arthropod Fauna

Within a forest there are vertical layers or levels that are biologically distinct. The number and type of strata differ, for example, according to the age of the forest or the latitude at which the forest is located. Tropical forests typically have more strata than temperate forests. Arthropods (primarily insects) are stratified in deciduous and coniferous forests of temperate and tropical areas (Weese, 1924; Allee, 1926; Blake, 1926, 1931; Smith, 1928, 1930; Fichter, 1939; Adams, 1941; Dowdy, 1947, 1951; Elton, 1973).

The occurrence of strata and the associated stratal distribution of arthropods in forests can be the result of a number of biotic and abiotic factors. Differences in physical factors at various levels of a forest provide a basis for faunal stratification. Critical factors include temperature, rate of evaporation (moisture), and light. These affect the type, quantity, diversity, and structural complexity of vegetation at any given level and, in turn, the nature of insect associates. The availability of prey (or hosts) or other critical resources, as well as competition or predation pressures, may also be important in faunal stratification. The amount of stratification or the insect species composition of each stratum can be dynamic, that is, affected by temporal changes such as seasonal migrations or diurnal shifts in physical factors.

Arboreal Insect Communities

Insect communities in forest canopies are composed of many insect species that interact with each other, their host plants, and the physical environment. There has long been debate as to which of these factors is most significant in structuring these communities (see Strong, 1974 a, b for detailed discussion). Until recently little effort was expended to understand these complex associations. Crossley *et al.* (1987) report on the insect biomass and nutrient dynamics in several forest watersheds in the Southeast. Clearly there are many separate guilds of insects that occur in forest canopies. Forest management practices and natural disturbances can have a significant effect on these communities (Schowalter and Crossley, 1987; Schowalter, 1985) and should be considered in a variety of forest practices (Schowalter, 1986a).

There are many ways that insects can influence the nature of forests, but there are two in particular which are important and often overlooked: their effect on succession, and nutrient cycling (Schowalter, 1981). Insects can preferentially feed on some individuals in the stand and push successional processes to favor those species (Schowalter *et al.*, 1981a,b). Catastrophic insect outbreaks, like those caused by bark beetles, can force whole forest communities back to early stages of succession. Insect activity can affect nutrient cycling by stimulating primary productivity, increasing nutrient content of litterfall, and altering nutrient cycling pathways, to name a few (Schowalter, 1981). The relative importance of these factors is yet to be determined, but clearly there is likely to be wide variation in response depending on forest species, stand age, and level of insect feeding (Ohmart *et al.*, 1983).

Structure and Variation of Soil Communities

The diversity in forest insect soil communities parallels the variation in vegetational and physical characteristics of forests (Fig. 13.1). A large number of biotic interactions in the soil are influenced by its physical structure. Soil is a complex medium that is directly and indirectly influenced by climate, parent mineral substrate, vegetation and fauna, and the action of water. The forest soil system is delineated on the top by the external surface of the litter and on the bottom by the deepest penetration of tree root systems. Within those limits insects and related arthropods are provided with their required microclimate, nutrients, and shelter. As in the above ground portion of the forest, there are important biotic and abiotic stratifications in the soil, often called soil horizons (layers). Together a number of these layers form the soil profile.

In typical mineral soils the profile commonly exhibits three primary horizons that can be distinguished by color, texture, porosity, amount and type of organic matter, root growth, and type of organisms present. Rapid decomposition, mixing of the mineral fraction and incorporation of organic matter characterize the surface layer, the A horizon. The B horizon is generally light in color due to the build up of chemically weathered mineral complexes and is usually devoid of organic matter, although some humus may be washed down from the A horizon. The C horizon is a relatively unaltered zone of rock, sand, and clay (Jackson and Raw, 1966; Schaller, 1968).

Further subdivision of the A horizon into L, F, and H layers is possible based on the degree of decomposition and other characteristics. In the upper portion of the A horizon (L layer), various types of vegetative matter and the remains of soil fauna and their feces are still recognizable although, leaching and discoloration of vegetation can be observed. Fungal hyphae permeate the vegetation. In the lower areas (F layer) of this horizon, animal and vegetative remains are not easily distinguishable without magnification. Finally, the H layer of the A horizon is the humus layer, consisting of well-decomposed organic matter. The A horizon is the most inhabited portion of the soil profile.

Three soil profiles are important in determining faunal composition. The *mor* profile is similar to the generalized profile described above; it has a clear-cut H layer that is distinctly delineated from the mineral soil. In the *mull* profile there is no clear-cut or true H layer; the humus material is intimately mixed with the mineral soil and is sometimes called the A_1 horizon. Finally, the *duff mull* (or *moder*) profile is intermediate between the previous two types and consists of L, F, H, and A_1 layers (Leaf *et al.*, 1973).

Figure 13.1.   Forest floor: cross section of microhabitats and selected life forms in the forest soil.

Certain animals are usually associated and/or linked with the formation of mor or mull soils. Earthworms and various arthropods are believed to be the principal agents in mixing organic matter with mineral soil. Earthworms are absent in the typically acidic mor soils. In intermediate formations, the activity of ants and termites may function as substitutes for missing earthworms (Murphy, 1955).

A significant difference between mor and mull soil formation is the compression of the biotic community. In a mor-soil forest area in Britain, 96% of the arthropod population is concentrated in the upper 2 1/4 in. The remaining 4% is spread through the other layers. The maximum density occurs in the first 1/4 to 1/2 in. of the upper horizon. In the mull-soil forest areas, there is a more even vertical distribution. In general there are larger organisms, greater biomass (a large portion of which is earthworms), and greater diversity and depth of substrate habitation in mull soils. Conversely, in mor soils one finds smaller biomass and smaller organisms with relatively little capacity to mechanically influence the medium in their narrow life zone (Murphy, 1955). The depth of soil horizons can have a significant effect on the response of soil arthropods to forest management activities (Seastedt and Crossley, 1981).

Certain soil types are characteristic of major forest ecosystems. Coniferous forests, characteristic of cold temperate regions, often are associated with a podzol soil formation. Although this acidic soil profile has a typical L and F layer, the other layers are typical of the mor or moder formations. In deciduous forest zones, the soil is a characteristic brown soil more typical of the mull association.

The texture (the proportions of different sized particles and the way the particles are aggregated) can have an effect on the distribution of invertebrates and on the ability of many insects to maneuver through the soil. Although some soil insects, such as mole-crickets, cicada nymphs, and beetles, are well adapted for movements through soil, other small species, such as many mites and Collembola, depend on preformed spaces. Thus, porosity can be a biologically important aspect of soil. Depending on the soil type and horizon, soil particles occupy only 40–70% of total soil volume. The pore spaces are filled with air or water and provide critical living space or appropriate moisture levels for soil arthropods and thus can determine the distribution of a species. Oribatoid mites orient themselves vertically according to the moisture of the soil (Jacot, 1936). The least drought-resistant arthropod species in the soil, which are also the smallest, frequently occur deep in the soil in the minute crevices where high saturation deficits are unlikely (Kevan, 1962).

Pore space, soil aggregate formation, root channels, and fissures provide a diversity of microhabitats that vary in temperature, humidity, and size. This variety in living conditions provides opportunity for a variety of species to inhabit the soil and may restrict distribution for some species. *Isotoma viridis*, a large collembolan, is restricted to the leaf litter, whereas the smaller *Tullbergia crausbaueri* or *T. quadrispina* inhabit deep layers of the soil (Birch and Clark, 1953). Neither of these two species occur in soil with a high clay content, which generally restricts pore spaces.

Soil Arthropods

Soil insect species can be categorized by their trophic preferences. Major groups include (a) phytophagous species (feeding on living plants such as leaves, phloem, and roots), (b) mycophagous species (feeding on fungi), (c) zoophagous (carnivorous) species such as predators and parasitoids, and (d) saprophagous species [feeding on dead plant matter, carrion (cadavericoles or necrophages), or dung (coprophages)]. Some species may be omnivorous or may change feeding habits as they age. All categories include some of the smallest and some of the largest species inhabiting the soil (Kevan, 1962; Jacot, 1936).

The dependence of a species on the soil for its survival may characterize its relation to this forest stratum. Most forest soil arthropods can be categorized as residents, their life cycles closely linked to the soil. A second group, the nonresidents or temporal residents, utilize the soil habitat for only a portion of their life cycle or may prey on soil organisms while inhabiting other strata. Temporal residents may include so-called nidifactious species, which deposit their eggs in the soil but feed elsewhere; species that live in the above ground strata but pupate in the soil (Lepidoptera and sawfly larvae, parasitoids, etc.); species that undergo dormancy within the soil; arthropods that migrate from less protected habitats (e.g. meadows and clearings); or parasitoids and predators that stalk and feed on soil-inhabiting prey but that may not inhabit the soil themselves.

Finally, soil inhabitants can be viewed on the basis of size (micro- and macroarthropods). In addition to size, these groups can be characterized by other traits. Generally, microarthropods of all stages live in the soil. These relatively primitive species tend to be minute, wingless, poorly sclerotized (hardened), not highly pigmented, and lacking in organs for defense or offense. These species are usually saprophagous, coprophagous, or predaceous, and some occur in abundance. Microarthropods are the most common arthropods in mineral soil (Jacot, 1936). Obligatory soil microarthropods, particularly those that occur in the deeper soil strata, are characterized by low mobility, low resistance to physical factors, and low susceptibility to natural enemies. Macroarthropods include the temporal or nonresident species discussed above. These species are generally more specialized and varied and exhibit all categories of feeding behavior.

Microarthropods

By far the most important microarthropods and the most abundant of all soil arthropods are the Collembola (springtails) and the mites (Tables 13.1

Table 13.1. Total mesostigmatid and trombidiform mites collected from 18 envelopes of three leaf species exposed (for 42 months) on the forest floor of a pine and a hardwood stand near Raleigh, North Carolina[a]

Taxon	Number in Pine Stand[b]	Number in Hardwood Stand[b]	Total Adults + Nymphs
Mesostigmata			
Ascidae	25	31 (6)	62
Phytoseiidae	12 (4)	24 (1)	41
Veigaiidae	2 (4)	14 (21)	41
Parholaspidae	2 (2)	22 (3)	29
Zerconidae	—	26	26
Laelaptidae	3	3	6
Uropodidae	—	3 (3)	6
Aceosejidae	1	2	3
Digamasellidae	1	—	1
Summary		171 adults plus 44 nymphs = 215	
Trombidiformes			
Bdellidae	6	1	7
Erythraeidae	2	2	4
Rhagidiidae	2	2	4
Stigmaeidae	3	—	3
Cryptognathidae	3	—	3
Cunaxidae	—	2	2
Pachygnathidae	1	—	1
Trombiculidae	—	—(1)	1
Summary		24 adults plus 1 nymph = 25	

[a] From Metz and Farrier (1969).
[b] Numbers in parentheses represent nymphs.

and 13.2). Other groups included in the microarthropods are the Diplura, Thysanura (so-called bristletails), and Protura.

Many microarthropods such as mites and springtails feed on fungi. However, much of that mycophagy involves the consumption of fungi integrally associated with plant debris. Thus, fungal feeding in these species is an integral part of litter breakdown. Most microarthropods cannot ingest freshly fallen leaves, particularly if the leaves are dry, but prefer moist, partially decomposed litter. Different microarthropods exhibit specific preferences among available leaves from various hosts and among the various leaf tissues. For example, microphytophagous mites in the genera *Galumna* and *Ceratoppia* occur on newly fallen *Quercus* litter, whereas some of the most abundant oribatid mites occur primarily among fine detritus. Similarly, gut content analysis of Collembola shows that over 90% of the particles in the gut of a species such as *Orchesella flavescens* are of higher plant origin, suggesting that they feed on relatively undecomposed surface

Table 13.2. Oribatids recovered from litter envelopes exposed (for 42 months) on the forest floor of a pine and a hardwood stand near Raleigh, North Carolina[a]

Genus	Number	Genus	Number
Oppia	1035	Allodamaeus	6
Caleremaeus	573	Hoplophthiracarus[c]	6
Tectocepheus	323	Phthiracarus	6
Suctobelba	185	Oripoda	5
Belba	117	Quadroppia[c]	5
Anachipteria[b]	58	Malaconothrus[b]	4
Eremaeus[c]	54	Trimalaconothrus	3
Peloribates	52	Microzetes[c]	2
Parachipteria[c]	38	Oribatella[b]	2
Hoplophorella	33	Trhypochthonius[c]	2
Rhysotritia	32	Trichoribates[b]	2
Eupelops	27	Xenillus[c]	2
Carabodes	25	Arcozetes[c]	1
Ceratoppia	20	Ceratozetes[c]	1
Cepheus[b]	15	Cultroribula[b]	1
Galumna	14	Gymnodamaeus[b]	1
Scheloribates[c]	14	Liacarus[b]	1
Acaronychus	13	Nanhermannia[c]	1
Hermaniella[c]	13	Oribatula[b]	1
Rostrozetes	12	Scapheremaeus[b]	1
Dolicheremaeus	10		
		Oribatid adults	2716
		Oribatid nymphs	429
		Total Oribatids	3145

[a]From Metz and Farrier (1969).
[b]Recovered from pine stand only.
[c]Recovered from hardwood stand only.

litter. On the other hand, approximately 40% of the gut material of *Tomocerus* sp., another collembolan, is of fungal origin, which suggests the partially humified leaf-matter diet of a deep-soil inhabitant. Microarthropods such as mites and collembolans also play a part in the decomposition of wood, particularly heartwood. Coprophagy among microarthropods appears to be rare, although some predatory microarthropods may colonize vertebrate dung.

Diplurans and proturans are minute (0.5–0.6 mm) insects that belong almost exclusively to the soil and litter fauna and are commonly found in moist leaf litter and moss, under bark and rotting logs, and in soil. Proturans are whitish insects that lack eyes, wings and antennae. Little is known of their food. Diplurans typically have two cerci (stylets) attached at the posterior end of the abdomen and are considered primitive. A number of other minor microarthropods complete this portion of the soil fauna. These groups of minute and sometimes scarce arthropod species are very difficult to study; therefore, little is known about their ecology or role in the soil community (Harding and Stuttard, 1974).

Macroarthropods

The critical role of macroarthropods in the breakdown of plant litter and animal remains makes these arthropod species among the most important in the forest. The larvae of Diptera are among the most common insects in leaf litter and decaying wood. Among the inhabitants of deciduous woodland litter are the larvae of tipulids (craneflies), chironomids (midges), platypezids (flatfooted flies), xylophagids, stratiomyids (soldier flies), and bibionidids (March flies).

Many dipterous larvae (larval Platypezidae, for example) live primarily on fungi. Other muscoid fly larvae feed in vegetation or litter that is in an advanced state of decay. Species of certain genera such as *Calliphora* and *Lucillia* specialize in feeding on carrion. Still others, such as the larvae of the mydas flies (Mydidae), stiletto flies (Therevidae), and the window flies (Scenopinidae), occur in decaying wood, fungi, or soil and are predaceous or parasitic. Their prey include other soil insects, annelid worms, and molluscs. Although the two-winged adults generally do not remain in the soil after completion of development, a few species do inhabit the soil and litter throughout their life cycles.

Most Hymenoptera can be considered transient or temporal soil fauna. Most species such as bees, wasps, sawflies, and parasitoids are not intimately associated with the soil. Many ants, however, represent the major exception. Some of the plant-feeding species of the Myrmicinae, Dolichoderinae, and Formicinae break down a great deal of leaf litter. Although ants are active soil-surface foragers, the extensive subsurface excavations of ants are important in soil translocation and mixing. Fungus or leaf-cutting species carry off large quantities of leaf and fecal material which are fragmented and used as a substrate for fungal growth. Their omnivorous foraging often takes them to other forest strata.

The Coleoptera probably have the greatest diversity of all soil-inhabiting species. Although many of the common forms such as the carabids (ground beetles) and staphylinids (rove beetles) are predaceous, many species including some in the latter two groups are plant feeders. As in the Diptera, there are carrion feeders (e.g., Silphidae and Histeridae), fungivores (e.g., Mycetophagidae, Endomychidae, and Erotylidae), coprophages (e.g., Scarabaeidae), and many other feeding types in the Coleoptera. The larvae of Scarabaeidae in the genera *Melolontha, Amphimallon*, and *Phyllophaga* are of particular importance as pests of nurseries and plantations. Their larvae (white grubs) often destroy the roots of seedlings, and adults eat the foliage of trees. The white, curved, legless larvae of various weevil (Curculionidae) species are also important soil inhabitants and feed on seedling roots and root collars. As adults, they feed on the tender bark of seedlings.

The natural microsuccession that occurs in a fallen log or other parts of a tree and eventually results in its incorporation into the soil involves a large variety of Coleoptera. A few of these groups include the Scolytidae (bark beetles), Cerambycidae (long-horned beetles), Buprestidae (metallic wood-boring beetles), Elateridae (click beetles), and Pselaphidae (short-winged mold beetles). These insects generally use the wood at particular points in the decomposition process. The activity of certain species often modifies the wood and determines its suitability for subsequent species.

Although most termite (Isoptera) species are tropical, they are also one of the most important macroarthropods in temperate forests. Their colonial nest building and feeding use decaying vegetation and wood almost exclusively. Termite activity results in both the fragmentation of plant matter and the mixing of organic material into the soil. Termites are one of the few groups capable of chemically breaking down cellulose and lignin. However, more cellulose than lignin is broken down. Species in the family Termitidae forage for and carry leaf litter to their nests and are thus important in its breakdown. Species of harvesting termite groups, such as the Hodotermitinae and Nasutitermitinae, prefer to use pieces of living plants. Species in the Kalotermitidae, Rhinotermitidae, Hodotermitidae, and Termitidae use and feed on dry or decaying wood. It is probable that the fungi responsible for some of the decomposition of plant material also comprise part of the diet of the termites. Termites that feed on leaf litter or wood exhibit specific host preferences.

Number and Diversity of Forest Soil Arthropods

One of the most dramatic aspects of forest soil systems is the large number of individuals and species that occur in this habitat. Comparisons of abundance and diversity among forest types or even forest soils are difficult. Because soil organisms are often vertically stratified, sampling and methodology are rarely comparable within or between sites. Available studies often give incomplete details of methodology. For example, three-dimensional sample sizes are rarely given. A final difficulty is that researchers sometimes fail to recognize that many soil organisms are not randomly distributed but exhibit a distribution pattern that reflects differences in soil moisture, food availability, soil structure, and soil pH. Nevertheless, available data demonstrate the high diversity and abundance of species in any given forest.

Mites and Collembola are the dominant arthropods in soil and humus. Estimates of the sizes of these two groups range as high as 6/7 of the total number of animals in the soil. A great abundance of mites and Collembola is often characteristic of raw humus. Mites are generally the most abundant of the two groups. Published estimates of mean density for oribatid mites range from 6.7×10^3 m^{-2} to 134×10^3 m^{-2} (Harding and Stuttard, 1974). Other soil organisms also occur in great abundance (Table 13.3).

Table 13.3. Total population of micro- and macroarthropods in selected forest soils[a]

Habitat	Total Sample Depth (in.)	Number per Sample	Number per m² (thousands)	Number per cm³	Volume (cm²)
Beech mull	4 (?)	619	6.2	0.06 (?)	10,000 (?)
Beech mor	6 1/2(?)	1,943	19.4	0.11 (?)	17,000 (?)
Spruce mor	5	1,194	11.9	0.09	13,000
Beech–oak mull	—	1,938	—	1.94	1,000
Spruce mor	—	2,260	—	2.26	1,000
Spruce–pine–birch mor	4	1,184	1,089.6	10.90	400
		1,430			100
Beech mor	1 1/2	1,033	333.5	8.79	40
		214			4,000
Forest soil[b] *Pinus sylvestris* forest in Scotland		10,000 mites and collembola per m² 38,250 collembola per m²			

[a] Based on Murphy (1955), Birch and Clark (1953), and Millar (1974).
[b] Underestimation due to imperfect extraction.

The species diversity of forest soils can also be relatively high. A study of the soil in beech woodlands in Austria found 110 species of beetles and 229 species of mites (Birch and Clark, 1953). Similarly, 22 species or genera of Mesostigmata (a major group of mites) were sampled from the F layer of a *Pinus sylvestris* stand in Scotland. Twenty-four species of Collembola occur in the F layer of a Scottish *Pinus sylvestris* forest (Millar, 1974). In a pine and a hardwood forest of North Carolina, nine families of Mesostigmata, eight families of Trombidiformes, and 41 other genera of mites were collected from envelopes filled with leaf litter that were placed on the forest floor for 42 months (Metz and Farrier, 1969) (Tables 13.1 and 13.2).

Adaptation to Soil Existence

As in any ecosystem, successful utilization of a habitat is related to the ability of organisms to adapt to biotic and abiotic stresses. Forests are horizontally and vertically diverse enough so that one cannot point out generalized adaptations that would apply to all soil arthropods. For species that live in the litter or the humus layer appropriate adaptations differ.

Nevertheless, as in other major habitats, certain stimuli predominate and others are less common; thus, certain receptor organs typically are enlarged or highly differentiated. For example, vision plays a minor role in orientation and movement within the soil, whereas tactile senses and thermoreception are very important. Highly differentiated tactile hairs of various sizes and shapes clothe the integument of mature and immature soil arthropods. Similarly, antennae of soil-dwelling Collembola are short; their prominent organs are believed to be associated with chemoreception (detecting chemical signals). In spiders and other animals, various mouthparts have been developed as specific tactile and olfactory receptors (organs of smell).

Soil arthropods also have well-developed organs for the detection of humidity and temperature. This capacity is demonstrated by the strong orientation behavior of these organisms when placed in an appropriate habitat. When placed in a moisture or temperature gradient, most soil arthropods invariably avoid the drier and warmer areas of the gradient. Such avoidance places them in conditions in which they are less subject to water loss. Finally, although eyes and light receptors of most soil animals are poorly developed, behavioral responses to light exist and are important. Even some blind forms have light-sensitive receptors in the cuticle. Soil dwellers often exhibit light avoidance, which maintains them at an appropriate level in their habitat (Schaller, 1968).

Some of the most dramatic and obvious adaptations to soil habitation are external morphological adaptations. Leg modifications evident in many species include flattened spade like legs, enlarged claws or spines, and streamlined bodies. The legs of many litter dwelling adult scarabaeoid and carabid beetles are obviously adapted for digging, whereas their larvae, which spend most of their existence in the soil in a more or less motionless state, are short legged. If one looks at a mole-cricket, the nymphs of cicadas, and other soil insects, the function of their adaptations is obvious. An alternative method of burrowing is found in other insects (e.g., dipterous larvae, certain lepidopterous larvae, and certain coleopterous larvae that are legless and rigid with shiny integuments); these have narrow wedgelike heads to cleave through litter and some soils (Kevan, 1962).

Function of Micro- and Macroarthropods

Forest insects live in an ordered and complex ecosystem. The underlying source of the cohesiveness of this ecosystem is the availability of energy at all trophic levels and the basic building blocks for growth. Insects require approximately 30–40 elements for growth and development. Biogeochemical cycles or transfers represent the main mechanism by which these essential elements, as well as nutritional compounds, are maintained and made available to living organisms. Soil micro- and macroarthropods play important roles in the decomposition process, which is a major part of many biogeochemical cycles. However, available studies indicate that although the microflora is of major importance, insects are also important as biological facilitators.

Two of the most important cycles, the nitrogen and carbon cycles, can be used as examples of the impact of insects. A major portion of the nitrogen cycle is the conversion of protoplasm to ammonia. This conversion is generally accomplished by decay (decomposition) and excretion. Similarly, the degradation of plant and animal material is an important phase in the sequence leading to the recycling of carbon. This process is simply the breakdown of complex tissues and compounds such as carbohydrates and fats into simpler elemental substances. An extensive variety of organisms require decomposition. Plant material, ranging from fungal hyphae to hard woody substances, and animals, ranging from unicellular protozoans to large vertebrates, and their excrement require the activities of many decomposers. In general, insects facilitate the breakdown of organisms and are one of the major groups involved in the initial breakdown of larger plants and animals.

Fragmentation, Soil Mixing, and Aeration

The functions served by the numerous arthropods inhabiting the soil can be characterized as mechanical or chemical manipulations. Fragmentation or comminution (physical breakup) of leaf or animal material is a major function of soil animals. Fragmentation is accomplished in temperate areas primarily by earthworms, enchytraeid worms, isopods, diplopods, larval dipterans, mites, and collembolans. This process increases the surface area susceptible to invasion by microorganisms. Conversion of plant material to fecal matter is also important in the successful invasion by microorganisms. Microarthropods produce minute fragments of litter and fungal hyphae ranging from a few micrometers to more than a hundred (Harding and Stuttard, 1974).

The importance of soil animals can be clearly demonstrated with exclusion experiments. In experiments in which leaves have been buried in bags capable of excluding invertebrates, the leaves have remained essentially intact after as many as 3 yr. Similar experiments indicate that in bags containing oak litter, which partially exclude invertebrates, less than 50% of the litter is decomposed (Edwards *et al.*, 1970). In litter without arthropods, decomposition in a 1-yr period was only 75% of the decomposition that occurs in the presence of arthropods (Witkamp and Crossley, 1966). Finally, the rate of red maple litter decomposition in a central Canadian woodlot is approximately 1.48–2.19 $g/m^2 \cdot day$. If treated with an insecticide lethal to arthropods, decomposition is reduced to between 0.99 and 1.26 $g/m^2 \cdot day$ (Weary and Merriam, 1978). Not only is the degree of arthropod exclusion important, but so is the plant species being decomposed (Fig. 13.2).

A second mechanical manipulation is achieved by detritus feeders which form soil aggregates composed of mineral and organic matter. Soil arthropods

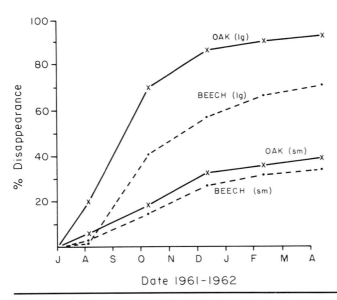

Figure 13.2. Decomposition of leaf discs by soil animals. Oak (large) denotes oak leaf discs in 7-mm mesh bags; beech (large), beech leaf discs in 7-mm mesh bags; oak (small), oak leaf discs in 0.5-mm mesh bags; and beech (small), beech leaf discs in 0.5-mm mesh bags. (From Edwards and Heath, 1963.)

are also important in the mixing and translocation of soil. This churning function can be accomplished simply by insects moving through the soil or by nesting behavior such as the colony formation of ants and termites. Mixing can also result from the foraging and cleaning behavior of ants and termites. Leaf and litter material are brought into the colony, and plant and colony debris are taken to the surface. This results in considerable mixing and soil enrichment, and sometimes in the concentration of elements in parts of the soil. Ants and termites become particularly important in forests where the nature of the soil excludes habitation by earthworms.

The interactions of the flora and fauna of the soil system are intricate. Some soil collembolans may be surface dwellers that feed on plant material; other species may live in the soil and feed on microorganisms. Many insects that feed on plants may actually be feeding on the microorganisms living on plant debris. Even in the early steps of decomposition, the falling leaves harbor a complex of microflora that may serve as food or as decomposers. The contributions of insects to decomposition and soil development often occur as a result of a relatively trivial activity. For example, in the advanced stages of humification, soil arthropods such as dipteran larvae mechanically mix the soil by their vertical movements, which may be responses to temperature or humidity extremes, food availability, or predation. After consumption and fragmentation, plant debris may chemically altered by passage through the gut. Some cellulose can also be broken down in those soil arthropods that have gut symbionts.

Decomposition of Leaf Litter

The rate of decomposition of plant litter varies with plant species, soil type, seasonal changes, and other site conditions. Leaves decompose within 8 to 15 months in temperate mull soils but within weeks in tropical soils. Although deciduous litter may be acted upon by some fungi, it is not eaten by litter animals for several weeks. The initial changes in the litter consist primarily of the action of microflora and the leaching out of water-soluble compounds such as acids, polyphenols, some carbohydrates, and pectins. These changes are accompanied by a characteristic darkening of the leaf litter. This weathering process makes the litter more palatable to insects and other animals, which then begin to fragment the litter (Edwards *et al.*, 1970).

The character of the soil can change dramatically depending upon the sequence of the above mentioned events and the changes in the types and number of soil dwellers. Acidic humus is usually soil in which plant tissues are relatively intact, fungi are numerous, and few animals can be found. As the soil becomes less acidic (mor soil), more springtails and mites are found. When most of the plant fragments, fecal matter, or similar residues become incorporated, a moder type soil forms which consists of loosely mixed organic and mineral matter. When transformation to mull humus soil occurs, an increase in larger arthropods leads to formation of soil aggregates. These are formed by the ingestion by arthropods of food and mineral residues and lead to a more complete mixture in which clay and humus components become inseparable (Edwards *et al.*, 1970).

Rates of decomposition differ in various forest types. For example, coniferous litter is more resistant to decomposition than is that of broad-leaved forests. Earthworms are not common in coniferous forest soils, which generally have a pH of less than 4.5. Thus, arthropods play a relatively greater role in decomposition in coniferous forests than in deciduous forests. Mites are one of the main agents of coniferous needle litter decomposition. They are primarily fungal feeders and eat fungi growing in the needles or graze needle-surface-dwelling fungi. The promotion of fragmentation by collembolans feeding on fungi, humus, and feces is also an important aspect of coniferous needle litter decomposition (Jensen, 1974; Millar, 1974).

Finally, environmental factors such as moisture and temperature affect the rate of litter breakdown. The rate of decomposition is often correlated to seasonal increases in temperature. Canopy structure can also play an important role in determining soil temperatures. Thinning in a white spruce stand in Alaska, for example, stimulates organic decomposition of litter (Table 13.4). The excessive weight loss (decomposition) in the most thinned plots compared with unthinned plots is believed to be due to higher average seasonal temperatures (Table 13.5). Soil moisture is also important. In the same study discussed above, decomposition of litter was greatest when samples were placed between the fermentation–humus and humus–mineral soil layers, compared with those samples placed on the surface (Table 13.5).

Decomposition of Wood

The bulk of plant materials to be decomposed in forest ecosystems consists of wood. Groups of insects closely associated with the decomposition of wood are members of the orders Isoptera, Diptera, and Coleoptera and a few blattids. Although cellulose components may be digested with the aid of internal symbionts, the breakdown of wood is generally accomplished through the aid of a succession of attacks by different insects. Dead, dying, or newly cut trees attract flat- and round-headed borers (Buprestidae and Cerambycidae), bark beetles, wood-boring weevils, and a f w other insects that feed and tunnel in the wood and thereby introduce fungi. As the wood undergoes changes, it becomes increasingly suitable for other organisms, such as Tipulidae, Sciaridae, Mycetophilidae, and Cedidomy-idae. The action of these and other species results in the mechanical destruction of wood tissues. As the moisture content of the wood in-

Table 13.4.   Effect of thinning on average percent weight loss of cellulose bags placed in different forest floor layers of a white spruce (*Pinus glauca*) forest in Alaska[a]

Sampling Time	Days since Start of Experiment	Plot[b]	L[d]	F–H[d]	H–M[d]
				Percent Weight Loss[c]	
t_1 (Sept. 3, 1971)	55	Heavy	0.07 (0.35)	18.86 (3.96)	14.99 (4.83)
		Light	1.94 (1.09)	20.62 (6.07)	26.54 (8.41)
		Control	0.74 (2.05)	57.53 (13.53)	46.11 (15.42)
t_2 (June 2, 1972)	328	Heavy	15.80 (3.80)	48.18 (4.40)	43.89 (4.08)
		Light	6.80 (0.95)	63.70 (7.53)	63.60 (10.12)
		Control	5.58 (1.14)	53.03 (7.96)	57.20 (10.22)
t_3 (July 10, 1972)	366	Heavy	20.17 (4.32)	73.12 (3.71)	72.54 (5.79)
		Light	9.50 (3.05)	73.28 (3.92)	74.33 (4.81)
		Control	8.00 (2.32)	53.42 (5.27)	57.36 (7.35)
t_4 (Sept. 10, 1972)	397	Heavy	32.13 (9.21)	79.98 (4.68)	85.84 (3.11)
		Light	23.81 (4.12)	80.70 (7.37)	74.73 (6.00)
		Control	27.70 (7.52)	74.68 (5.44)	75.00 (6.50)

[a] From Piene and van Cleve (1978).
[b] Thinned to 1480 trees/ha (light); thinned to 740 trees/ha (heavy).
[c] Standard error in parentheses.
[d] L denotes forest floor litter; F–H, fermentation to humus layers; H–M, humus to mineral layers.

Table 13.5   Effect of heavy and light thinning on average seasonal temperature and moisture content in the forest floor L and H–M layers[a]

Layer[b]	Temperature (°C)			Moisture (% dry weight)		
	Heavy	Light	Control	Heavy	Light	Control
L	22.8 ± 0.2[c]	20.7 ± 0.1	18.9 ± 0.1	30.9 ± 1.3	38.8 ± 1.5	41.0 ± 1.4
H–M	11.2 ± 0.2	8.9 ± 0.2	10.2 ± 0.2	70.1 ± 2.8	71.4 ± 2.8	53.8 ± 1.8

[a] From Piene and van Cleve (1978).
[b] L denotes forest floor litter; H–M, humus to mineral layers.
[c] 95% confidence interval.

creases, the degree of invasion by microorganisms and colonization by various animals also increases. Although the process may be slow, eventually the material becomes incorporated into the soil. In a study of 86 dead but standing Virginia pines (*Pinus virginiana*), approximately 184 species of insects and related forms were found attacking and colonizing the trees or attacking colonizing insects (Figs. 13.3 and 13.4).

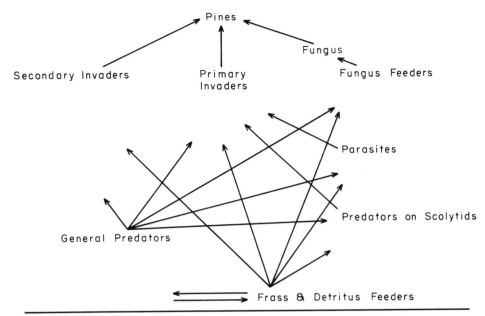

Figure 13.3.   Generalized food web of some of the insects inhabiting the early stages of dead standing pine. (Based on Howden and Vogt, 1951.)

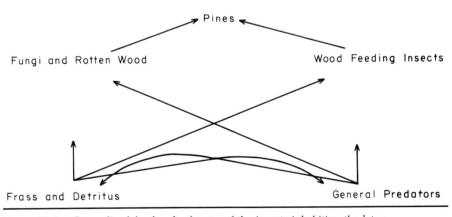

Figure 13.4.   Generalized food web of some of the insects inhabiting the later stages of dead standing pine. (From Howden and Vogt, 1951.)

Decomposition of Carrion

The activity of insects in the degradation of animals and the eventual transfer of elemental substances is similar to their role in the decomposition of plants. Again insects, by feeding and moving through tissues, inoculating dead tissues with microorganisms or modifying the substrate enough to allow for subsequent invasion, act as biological facilitators.

Several studies have dramatically shown the importance of insects in this process. Although important in all habitats, the effective decomposition of animal material may be even more essential in certain areas, for instance, in those habitats with rapid leaching of minerals, high acidity, or low organic content in the soil. Certain sandy habitats exhibit these conditions.

A great variety of insect types may be involved in the decomposition of animal tissues. Although representatives from various orders can act as decomposers, the dominant groups are the Coleoptera and Diptera. Dipterans, such as blowflies and fleshflies, and coleopterans, such as carrion beetles, dung beetles, staphylinids, and hister beetles, are common. Experiments have been conducted in which the role of insects has been well illustrated. In one such study, dead baby pigs were placed in a mixed mesophytic hardwood-pine community of South Carolina (Payne, 1965). The type and succession of insects were observed. In an attempt to delineate the specific role of insects, carrion was placed in very fine mesh screen cages to exclude insects. Control cages allowed the insects access to the carrion. Graphs of the resultant decay clearly show how much more rapid the disintegration of insect-attacked carrion was compared with the non-insect-attacked carrion (Figs. 13.5 and 13.6). A total of 522 species of animals were identified during the experiment. Of this number, 422 were insect species representing 11 orders and 107 families. Although some of these species may be opportunists using various available resources (i.e. predators, parasites, and "accidental" inhabitants), many were species characteristic of this microhabitat (Payne, 1965).

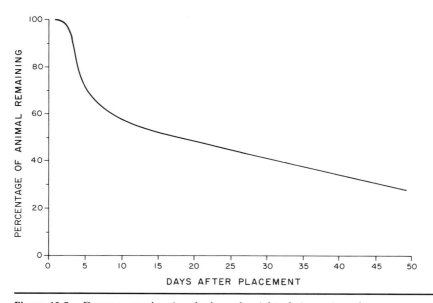

Figure 13.5. Decay curve showing the loss of weight of pig carrion when free from insects; summary of three experiments. (From Payne, 1965. Copyright 1965 the Ecological Society of America.)

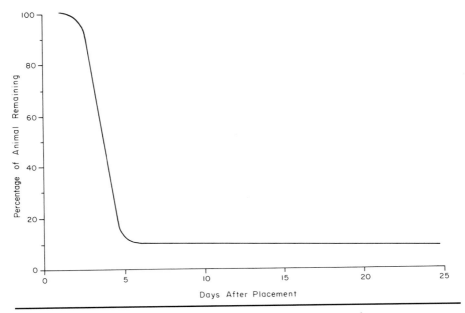

Figure 13.6.   Decay curve showing the loss of weight of pig carrion when exposed to insects; summary of three experiments. (From Payne, 1965. Copyright 1965 the Ecological Society of America.)

Decomposing carrion can dramatically modify the soil in its immediate vicinity. Decomposition products pass into the soil during putrefaction and fermentation stages. In addition, excretory products of various carrion-feeding insects (e.g., ammonia of fly larvae) and the acids associated with fermentation alter soil chemistry. Finally, soil structure is often modified by the burrowing activities of flies, other arthropods, and their predators associated with carrion (Bornemissza, 1957).

Chapter Summary

The forest arboreal and soil communities are only two of several distinct communities that exist in forest ecosystems. Factors such as inter- and intraspecific competition, host plant traits, and the physical environment play important roles in the structuring of forest communities. Forest communities consist of vertical layers or strata that have distinct components but interact with other strata. The canopy and soil strata are discussed as examples of the kinds of relationships that exist. Though research on the nature of forest insect communities is in its infancy, knowledge about these communities can be important aspects to consider in making forest management decisions.

CHAPTER 14

Physical Factors and Forest Environments

Introduction

The ability to regulate or manage insect populations that have the potential to alter the health and growth of a forest is ultimately based on an understanding of forest ecology and particularly the ecology of forest and shade tree insects. A number of central issues in insect ecology are of particular importance to forest and shade tree entomology. Among these are the questions: What factors maintain endemic populations at relatively low densities? Why do populations build up to epidemic levels? What regulates survival? Although a number of the factors that we have discussed play important roles in these processes, physical factors are also important in determining the abundance, rate of change, and overall survival of insects. These factors can modify endocrine activity, mortality, development, and reproduction. In addition, subtle changes in behavior, activity, and distribution are often related to changes in physical factors. Climatic variation from area to area and microclimatic differences within forests can affect the population dynamics of insects and help determine which insects become pestiferous and what their impact will be on the forest ecosystem.

A Problem of Scale

Physical factors of the environment operate on a broad scale. For purposes of discussion we often speak of micro-, meso-, and macroenvironmental levels. As Wellington and Trimble (1984) define these levels, microscale is where the things that concern us live; mesoscale is where we can see events taking shape as far away as the horizon; and macroscale is where regional, subcontinental, and hemispheric events occur. Though it is reasonable to subdivide these levels of physical environment for purposes of discussion, it is important to realize that they are related and that all scales can be very important to forest insect population dynamics.

Microclimate

Microclimate is often difficult to define because it is a relative concept. It encompasses the array of physical factors that affect organisms in a restricted space. It is easy to describe the average climate of forests of northeastern United States; however, the microclimate of a small woodlot, a single tree, or an individual leaf can also affect an insect. It is important to recognize that physical factors in a forest can vary both horizontally and vertically.

Stratification of light, temperature, humidity, and wind is pronounced in the forest. The foliage canopy of most forests acts as a semipermeable umbrella that modulates various weather factors. The canopy cover is the primary element that controls the ambient conditions in a forest during the growing season (Christy, 1952).

Temperature

Although there is some variation from area to area, the highest temperatures in a deciduous forest generally occur at the point of initial interception of solar radiation, that is, the upper canopy. Temperatures decrease along a gradient from the canopy to the forest floor and continue to decrease below the soil surface. In many coniferous forests the temperature gradient is reversed, so that the coolest temperatures occur in the upper canopy and temperatures increase down to the forest floor (Smith, 1966).

As with other physical factors contributing to the microclimate, temperature gradients differ diurnally and seasonally. Temperature measurements at eight different levels in a central Ohio beech forest give an illustration of a vertical temperature gradient and how the patterns change over a season (Table 14.1). In summer, the greatest difference between minimum and maximum temperatures occurs in the canopy (i.e., a 43°F (6°C) difference). Temperatures below the canopy and above the forest floor show little or no defined stratification; temperatures beneath the soil (down to 4 ft) decrease relative to the soil surface. In winter, the absence of the canopy produces more uniform temperature profiles at all levels, with minimum temperatures occurring near the forest floor. Temperatures in the upper soil remain well above air temperature for most of the winter. In the spring, the leaf-litter layer becomes the most important stratum. When

Table 14.1. Maximum and minimum temperatures for a summer at eight different levels in a beech forest, compared with macroclimate extremes[a]

	Forest Level								
	-4 ft	-6 in.	Under Leaf Litter	Leaf Litter Surface	5 ft	20 ft	62 ft	82 ft	Weather Bureau
Max. temp. (°F)	61	66	72	83	84	86	86	89	94
Dates	7/17 9/6	7/20 8/18	7/17 8/17	7/17 —	8/17 —	8/17 —	8/17 —	8/17 —	6/26 8/17
Min. temp. (°F)	51°	54	54	49	48	48	48	46	50
Dates	5/23	5/23	5/23	6/18	6/18	6/18	6/18	6/18	6/4
Range (°F)	10	12	18	34	36	38	38	43	44

[a] From Christy (1952).

minimum and maximum air temperatures are measured at the lower levels, the greatest extremes in temperature occur at the forest floor. Above the floor there is no defined stratification (Christy, 1952). The differences found in the Ohio forest represent changes that do not differ significantly from what might be termed the general pattern of forest microclimate (Sukachev and Dylis, 1964). Differences in air temperature between stands of different age within a forest type often surpass those between different forest types.

A variety of macroclimatic factors can also influence microclimatic conditions. The type of cloud formation or the rainfall over a given area can change the foliage–air-temperature relationship. Rain can cause small but sometimes biologically important differences between air and tree temperatures. After a rainfall, evaporation from raindrops on foliage can lower foliage temperature.

A number of biotic factors, including tree species, the foliage color, structure, or density, and the developmental stage or species of insect affected, can affect these relationships. It is important to note that although insects are poikilotherms, internal body temperatures are not always identical to ambient temperatures (Table 14.2). Internal body temperatures can be altered by physiological activity, the nature of the microhabitat in which the insect finds itself, and the insect's physical orientation to solar radiation. The impact of physical factors such as temperature can be

Table 14.2. Internal temperatures of two sixth instars of *Choristoneura fumiferana* during a 24-hr period in June[a]

| Time (EST) | Air Temp. (°C) | Larval Temp. − Air Temp. (°C) | |
		$\Delta L_1 T - AT$	$\Delta L_2 T - AT$
0000	3.4	−0.1	−0.5
0200	1.7	−0.1	−0.5
0400	1.7	0.0	0.0
0600	6.9	0.2	0.0
0800	16.5	1.5	2.7
1000	21.6	2.9	1.8
1200	28.2	0.5	−0.2
1400	27.5	−0.1	0.2
1600	26.2	1.1	0.8
1800	25.7	0.5	0.9
2000	16.0	−0.1	−0.3
2200	10.0	−0.1	−0.4
2400	9.0	−0.5	−0.2

[a] Modified from Wellington (1950).
[b] L_1 and L_2 represent each of two larvae, and AT represents air temperature.

determined by the shape and structure of an insect's microhabitat. Similarly, air circulation, surface-to-volume ratio, color or absorptivity, and orientation of the microhabitat to solar radiation can also influence microclimate (Table 14.3). Thus, the gall, the leaf roll, the mine, or the twig gallery may have significantly different temperatures and humidities even when located on the same tree or tissue (e.g., foliage) (Henson, 1958) (Fig. 14.1).

Ultimately, the microclimate in an insect's immediate surroundings (e.g., leaf surface, leaf mine, flower bud, silken mat, or bark flap) is biologically very important. Major differences between these insect microhabitats and ambient conditions have been demonstrated. For example, under the influence of direct sunlight, leaf temperatures rise well above those of ambient air. In spring, the surface temperature of the previous year's spruce foliage rises 2.8–5.6°C above the ambient temperatures of 12.8 to 15°C. Under clear skies, foliage surface temperatures in the shade range from 0.8°C above to 2.2°C below the air temperature. In needles of lodgepole pine, a difference of 6.4°C between ambient and the internal temperature of mines of the lodgepole needle miner (*Coleotechnites milleri*) is due entirely to needle angle (Henson and Shepherd, 1952).

Tree species may also have an influence on foliage surface temperatures. At air temperatures between 15 and 26°C, the surface temperature of aspen leaves ranges from 0.5 to 1.6°C above ambient air, whereas coniferous foliage is usually 5°C and as much as 8°C above air temperatures. The difference in surface temperature between foliage types may be a result of the angle of the foliage (i.e., whether foliage surfaces lie perpendicular to incoming radiation) (Wellington, 1950; Henson and Shepherd, 1952; Henson, 1958) (Figs. 14.2a and 14.2b). Indeed, aspen leaf rolls formed by tortricid larvae project outward at approximately 40° instead of vertically as do normal leaves. The leaf roll temperatures more closely approximately those of coniferous needle surfaces than that of normal aspen leaves (Wellington, 1950). Finally, stresses on trees may also alter needle temperatures (Table 14.4).

Moisture

Transpiration and attenuation of air circulation by tree foliage produces a forest interior with high humidity relative to areas outside the forest. Transpiration can be a substantial source of moisture. One hectare (2 1/2 acres) of mature oak trees can transpire approximately 25,000 liters of water per day. If transpiration continues at this rate for 10 days, a 2.5-cm rainfall would be required to replenish the water (Clarke, 1954). Thus, humidity is generally higher in the forest than in the open, particularly at night.

As with all other terrestrial organisms, water loss is a limiting force in forest survival. Although relative humidity is a commonly used parameter

Table 14.3. Internal temperatures of sixth instars of *Choristoneura fumiferana* concealed in feeding tunnels, together with temperatures and relative humidities (R.H.) of air inside and outside the tunnels[a]

Time (EST)	Air Temp. (°C)	Tunnel Temp.– Air Temp. (°C)	Larval[b] Temp.– Air Temp. (°C)	Air R.H.	Tunnel R.H.
1145	30.6	−1.6	−0.7	49.0	89.0
1156	27.0	2.0	3.8	58.0	77.0
1207	31.7	−4.7	0.4	43.5	96.0
1223	28.1	1.4	1.6	59.0	54.0
1233	29.0	0.3	0.7	49.0	53.5
1244	28.0	0.1	2.3	53.0	63.5
1302	29.5	0.2	3.4	49.5	89.0
1313	27.5	2.2	4.4	56.0	77.0
1323	28.1	4.3	7.2	73.0	76.0
1335	28.1	4.6	8.9	60.0	70.0
1347	28.8	2.3	0.7	52.0	65.0
1355	27.0	1.1	1.8	57.0	67.0
1407	25.9	2.2	2.9	67.0	61.0
1415	26.4	0.0	0.6	50.5	66.0
1445	27.0	−0.1	1.0	73.0	89.0
1457	29.5	−1.5	−3.6	57.0	75.0
1506	28.1	0.9	−0.1	76.0	77.5
1517	29.5	0.0	−0.7	66.5	65.0
1527	32.2	−1.7	−2.3	53.0	73.0

[a] Modified from Wellington (1950).
[b] Recordings made using several larvae.

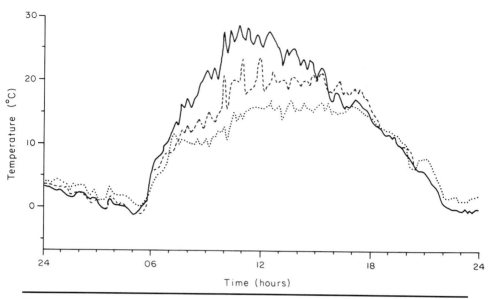

Figure 14.1. Comparison of the internal temperatures of staminate flowers and vegetative buds of white spruce with the surrounding air temperature within a typical spring polar air mass. Air temperature (...); flower temperature (——); and vegetative bud temperature (---). (From Wellington, 1950.)

in research, measurements of the drying power of air have greater biological significance. Most available research data are reported in terms of relative humidity, although a more appropriate index of water loss is the measurement of saturation deficit. The general effects of humidity on developmental parameters are illustrated in Figure 14.3.

The uneven vertical distribution of foliage also produces a stratification of increasing and decreasing humidity. The capacity of air to hold water is a function of air temperature (i.e., the warmer the air the greater its capacity to hold water). Thus, diurnal variations in humidity in a forest reflect, to some degree, the pattern of temperature changes (Smith, 1966). Environmental humidity gradients represent the prime climatic variable affecting the activity and distribution of anopheline mosquitoes (*Anopheles bellator* and *A. homunculus*) in the forests of Trinidad. Not only is rainfall level correlated with mosquito density, but vertical humidity gradients determine the relative distribution of each species. With the decrease of humidity in the mornings, the mosquitoes shift their vertical distribution downward into the more humid levels of the vertical gradient. *A. homunculus* occurs at lower levels than *A. bellator* in the midday humidity gradient. In the evenings they move into the highest canopy levels. Thus, interspecific differences in the humidity requirements of the two species are

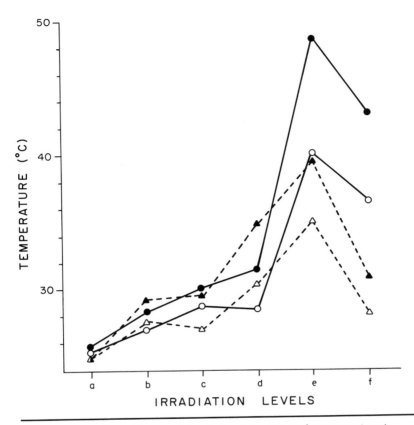

Figure 14.2a. Effect of increased irradiation intensity on the temperature t_L of aspen leaves and on the temperature t_A of air near the leaves; leaf position is denoted by subscripts 1 and 5 (see Fig. 14.2b). The curve for t_{L1} is denoted by •; curve for t_{A1} ○; curve for t_{L5} by ▲; curve for t_{A5} by △. Irradiation levels (in J m^{-2} s^{-1}) are denoted by *a* through *f*; for leaf position 1, *a* = 283, *b* = 583, *c* = 783, *d* = 1350, *e* = 1667, and *f* = 2000; for leaf position 5, *a* = 283, *b* = 500, *c* = 800, *d* = 1367, *e* = 1750, and *f* = 2167. (From Thofelt, 1975.)

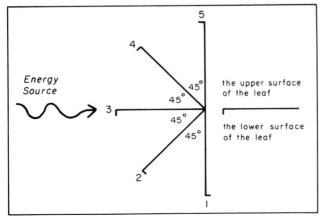

Figure 14.2b. Leaf positions used for observations represented in Figure 14.2a. (From Thofelt, 1975.)

Table 14.4. Mean needle temperature of three Engelmann spruces infested by spruce beetles (*Dendroctonus rufipennis*) and three noninfested Engelmann spruces on a typical August day when significant differences developed[a]

Local Sun Time[b] (LST)	Mean Needle Temperature (°C)			
	Infested	Noninfested	Difference[c]	*t*
0850	11.6	14.0	−2.4	3.73[d]
0900	12.3	13.2	−0.9	2.00
0930	13.2	15.2	−2.0	2.41[d]
0950	15.4	14.7	+0.7	0.83
1000	15.8	14.6	+1.2	1.41
1030	14.9	13.2	+1.7	2.46[d]
1040	15.1	15.8	−0.7	0.92
1100	14.6	15.4	−0.8	1.62
1110	15.4	14.6	+0.8	1.79
1130	15.8	15.9	−0.1	0.17
1210	19.6	15.0	+4.6	10.25[d]
1220	20.3	15.4	+4.9	6.93[d]
1230	17.3	16.0	+1.3	2.45[d]
1248	19.6	16.1	+3.5	5.43[d]
1256	21.0	16.8	+4.2	9.74[d]
1332	20.7	16.6	+4.1	4.31[d]
1352	19.1	16.7	+2.4	2.38[d]
1412	19.0	14.7	+4.3	4.14[d]
1422	14.1	13.9	+0.2	0.50
1512	19.0	14.6	+4.4	6.17[d]

[a] From Schmid (1976).
[b] Temperatures after 1512 LST were not presented because the sky became cloudy.
[c] Plus signs in the difference column indicate the infested trees were warmer than the noninfested trees.
[d] Value exceeds the 0.05 level value of *t* of 2.23.

maintained by shifting positions during the day as the vertical humidity gradient changes (Pittendrigh, 1950).

The penetration of rain can be interrupted by the canopy in the same way that the forest canopy intercepts light. Generally, conifers intercept more rain than deciduous tree species. The degree of interception varies and depends on such factors as the intensity of rainfall, tree species, tree age, and tree density. In a white and red pine forest, approximately 60% of a season's rain penetrates through the canopy to the forest floor. In comparison, 80% of a season's rainfall penetrates through the canopy of a hardwood stand that is dominated by American beech, sugar maple, and yellow birch (Beall, 1934). A number of insect behaviors and life-history parameters are affected by rainfall. In addition to mortality, emergence from overwintering hibernaculae, general locomotion, dispersal, and several other behaviors requiring movement can be affected by rainfall.

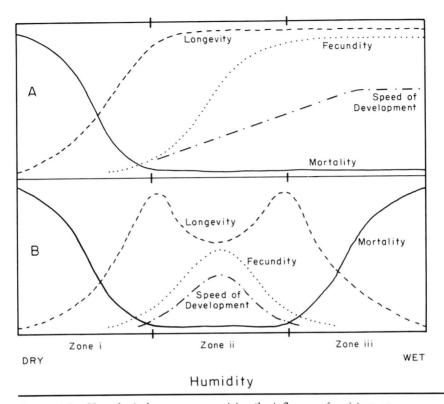

Figure 14.3. Hypothetical curves summarizing the influence of moisture on the longevity, fecundity, and speed of development of an animal; (a) relates to those kinds of animals that do not appear to be harmed by exposure to high humidity, and (b) relates to those kinds of animals that are harmed by exposure to high humidity. In zone (i) for (a) and (b) and in zone (iii) for (b) only, interest centers on the lethal influence of dryness and wetness, respectively. In zone (ii) interest centers on speed of development and fecundity. In (b) both these curves have well-defined optima; in (a) the curves approach a maximum of 100% relative humidity. Fecundity in this case is represented by an asymptotic curve; speed of development is represented by a straight line. (Reprinted from "The Distribution and Abundance of Animals," by H.G. Andrewartha and L.C. Birch, by permission of the University of Chicago Press.)

A number of important life-history events are affected by the moisture content of food. The invasion and colonization of host trees by insects such as bark beetles are often correlated with the moisture content of trees, which varies seasonally and diurnally (Figs. 14.4 and 14.5). Success of the mountain pine beetle is associated with changes in lodgepole pine moisture content. Trees infested by the beetle for 1 yr have a sapwood moisture content of only 16% of oven dry weight. Trees that do not succumb to the beetle retain a high moisture content (Reid, 1961) (Fig. 14.6). Although the

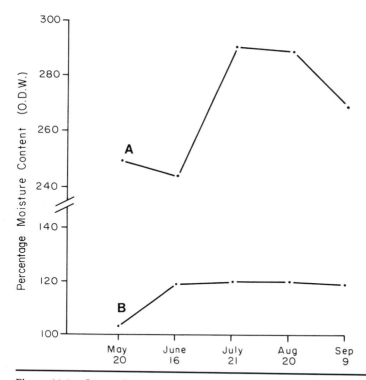

Figure 14.4. Seasonal moisture in the inner bark (curve A) and outer sapwood (curve B) of lodgepole pine (*Pinus contorta*) at the 3 1/2 foot level. (From Reid, 1961.)

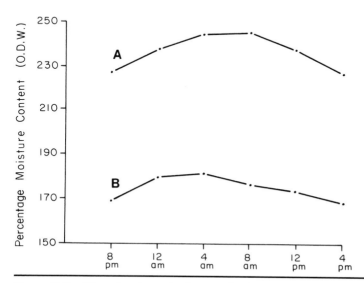

Figure 14.5. Diurnal moisture in the inner bark (curve A) and outer sapwood (curve B) of lodgepole pine (*Pinus contorta*) at the 3 1/2 foot level. (From Reid, 1961.)

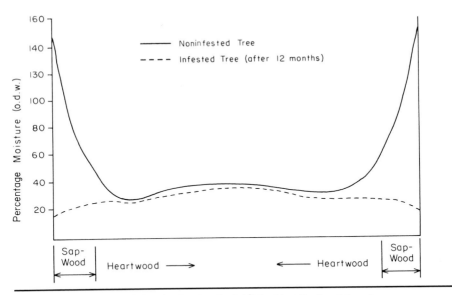

Figure 14.6.   Horizontal distribution (at the 3 1/2 foot level) of moisture in a noninfested (———) lodgepole pine (*Pinus contorta*) and in a lodgepole pine infested (----) for 12 months with the mountain pine beetle (*Dendroctonus ponderosae*) (From Reid, 1961.)

drop in moisture content appears to be associated with colonization of the wood by a blue-stain fungus, which is introduced by the beetles, the drop in moisture content also appears to be important to the successful development of the bark beetle brood (Figs. 14.7 and 14.8). In general, low moisture content (and low nitrogen content) is believed to be responsible for the significantly lower larval performance (i.e., the rate of consumption, assimilation, and conversion of plant biomass to insect tissue) and larval growth rates of leaf feeders compared with forb feeders (Scriber, 1979).

Finally, the indirect effects of moisture include the detrimental effects of excessive moisture and the subsequent increase in the incidence of insect diseases, particularly those of fungal origin. In habitats where water accumulates, excess water may be an important mortality factor. Cocoons of the larch sawfly remain in the moss or duff for 9–10 months of the year. Although the overwintering diapause stage is resistant to flooding, in the spring, postdiapause stages may be easily killed by flooding (e.g., 50 days of flooding results in 100% mortality) (Lejeune and Filuk, 1947; Lejeune et al., 1955).

Light

The upper surface of the forest canopy receives the highest light intensity. Below the tree tops the light intensity is dramatically reduced so that

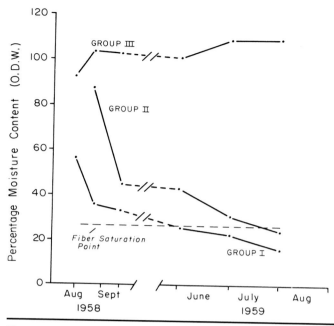

Figure 14.7. Seasonal moisture content in the outer sapwood of lodgepole pine (*Pinus contorta*) successfully and unsuccessfully infested by the mountain pine beetle (*Dendroctonus ponderosa*) at the 3 1/2 foot level. Group I were successfully infested in July, Group II were infested in August, and Group III were unsuccessfully infested. Fiber saturation is given by the horizontal dashed line. The Group I trees had lost considerable moisture before measurements were started. (From Reid, 1961.)

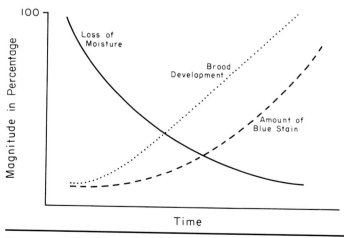

Figure 14.8. Diagrammatic illustration showing blue-stain penetration (---), drying out of the sapwood (——), and degree of brood development (. . .) in infested lodgepole pine (*Pinus contorta*). (From Reid, 1961.)

on the forest floor one perceives only a small fraction of the sunlight incident on the canopy. The forest floor is often a mosaic pattern of spots of light: a design of contrasting light and dark areas caused by light interception by foliage. Light intensity under the canopy ranges from approximately 0.1 to 20% of full sunlight. Even in the so-called sun flecks on the forest floor the intensity seldom surpasses 20% of daylight (Shirley, 1929).

The intensity of the light reaching the litter of an oak forest floor is approximately 0.4% of the intensity of light in the upper canopy. At midday, the light reaching the forest floor is approximately 6% of the total sunlight above the canopy (Smith, 1960). In tropical rain forests the upper canopy may receive 400 times more light than the forest floor (Papageorgis, 1975). The penetration of light is dependent on a number of factors, including stand maturity (tree age), stand density, and tree species composition. Most pines, for example, form a closed canopy that severely restricts light penetration and is a major factor influencing the type of understory vegetative growth. This in turn has an impact on the arthropod fauna living in the forest. Spruce, fir, larch, and deciduous tree species influence light penetration in ways that are characteristic of their forest types but also are dynamic in space and time. There are seasonal differences in light penetration, the most obvious of which occur in deciduous forests in early spring, when the floor of the leafless forest receives maximum illumination. A lower second peak in light penetration occurs in the fall (Smith, 1966).

Changes in the quality and quantity of light at each level of the forest are indirectly linked to biologically significant changes in temperature, moisture (e.g., humidity, soil moisture, and water stress on trees), and photosynthetic rates. Direct effects on forest insects result from changes in light intensity, wavelength, direction, and other components of light. These factors may vary within a forest and between forests and other habitats. For example, although spectral energy curves in a Michigan maple forest peak at 560 nm, simultaneous measurements of the spectral irradiance of open sky show a spectrum peak below 500 nm. These differences have been described as a quantification of what is generally perceived as dimness or greenness in a forest. These qualitative differences, plus the differences in spectral qualities of light reflected from branches, trunks, and so on, provide a visual mosaic that affects numerous life-history parameters including host selection, orientation, dispersal, and mate-finding (McFarland and Munz, 1975).

Light intensity appears to affect the number of larvae of the psychid moth (*Luffia ferchaultella*) found on certain areas of the trunk of its host tree. Larvae are found only on areas with a light intensity of approximately 2.5% of that in the open. However, it is still unclear whether associated factors such as temperature and humidity are of equal or more importance (McDonogh, 1939).

The quantity and quality of light have a major effect on orientation by forest insects. Practically all adult insects have image-forming eyes. Visual acuity can be extremely well developed in some species, depending on their niche. Predators, particularly flying species, generally have well-developed vision. Many immature insects, however, do not perceive form well but can perceive differences in light intensity and other quantitative characteristics of light.

Movement by some adults and many larvae of forest insects is conditioned by their photic responses. Fifth instar larch sawflies are strongly photosensitive and characteristically move toward outer branches while feeding. Defoliation in a larch stand often exhibits stripped upper crowns of trees because extremities of individual branches and of the crown are always defoliated first (Heron, 1951).

After eclosion, larvae of the sawflies (*Neodiprion pratti banksianae* and *N. lecontei*) begin to move about on the foliage. Upon contact with other larvae of their own species they form feeding groups. Although the wandering behavior of larvae appears random, it is a photopositive orientation. Starved early instars of *N. banksianae* and first instar *N. lecontei* change their behavior from relative indifference to a direct light source to movement that is strongly photopositive. Starvation of late instars results in increasing indifference to a light source.

A compassing reaction also may be exhibited by insects. This reaction is a directed movement in which an animal need not travel in line with a light but instead orients itself at an angle to the light. Observations of larval dispersion in the field illustrate how the photic responses of starved larvae can prove advantageous. A strong photopositive response by young larvae can lead to rapid formation of feeding colonies as larvae move directly to the ends of branches. As larvae age and foliage is consumed, a greater degree of indifference to light results in the dispersal of starved larvae over a wider area than would occur if they were still photopositive (Green, 1954a).

Direction reversal toward areas of lower light intensity has a definite survival value because it moves larvae away from high temperatures associated with insolation (sunlight). In a forest habitat, the pattern of temperatures may be a mosaic (i.e., not following predictable gradients), and orientation toward the upper or lower limits of a preferendum may be difficult if the insect orients only toward temperature. Thus, light acts as a token stimulus. Orientation to areas of higher or lower light intensity can be an effective method of reaching areas with temperatures in the appropriate range. Thus, it is clear that the quantity and quality of light can have a major impact on the behavior, physiology, and ecological interactions of forest and shade tree insects.

Although light intensity may be used as a token stimulus, insects still respond independently to temperatures, particularly at the extremes of their tolerated range. For example, many insects exposed to high temperature exhibit an adaptive reversal of photopositive orientation. Such changes in behavior are not always the same in all species, but rather are linked to the bionomics of the insect. Although high temperatures cause a reversal in *Malacosoma disstria* larvae from photopositive to photonegative, only the first three instars of *M. americanum* and *M. pluviale* exhibit similar reversals. Older instars of the latter species change from weakly photopositive to strongly photopositive. The difference is due to the fact that *M. disstria* does not form a tent and must move toward shade when exposed to high temperatures. The other two species form tents, which significantly modulate ambient conditions (Table 14.5). When overheated, these species need to be photopositive to exit the tent (Wellington, 1948; Wellington *et al.*, 1951; Sullivan and Wellington, 1953).

The responses of larvae are often more complex than described simply because a number of stimuli can be affecting larvae at any time. One of many examples is the role of pattern and shape in the orientation of insects. Larval movements often involve orientation to specific objects or objects at specific angles. Spruce budworms respond preferentially to striped rather than solid objects. This type of orientation may be a response to images closely approximating the needles of balsam fir (Wellington, 1948). Similarly, gypsy moth larvae, crawling across a woodland clearing, orient to vertical objects within a 3-m distance (Doane and Leonard, 1975). Nun moth larvae also orient to large close objects (Saxena and Khattar, 1977).

Table 14.5. Temperatures on and in a colonial tent of the western tent caterpillar (*Malacosoma pluviale*)[a]

Time (EST)	Air Temp. (°C)	Tent Surface[b] (°C)	Tent Center[b] (°C)
0800	12.6	13.6 (+1)	17.5 (+4.9)
0900	15.2	15.7 (+0.5)	23.6 (+8.4)
1000	18.8	23.5 (+4.7)	35.6 (+16.8)
1100	24.2	24.7 (+0.5)	35.5 (+11.3)
1200	23.7	24.7 (+1)	36.0 (+12.3)
1300	22.0	26.0 (+4)	32.7 (+10.7)
1400	23.3	23.2 (−0.1)	34.0 (+10.7)
1500	25.6	30.5 (+4.9)	39.7 (+14.1)
1600	22.5	24.3 (+1.8)	33.5 (+11)

[a] Modified from Wellington (1950).
[b] Parenthetic values represent the temperature differences between the air and the tent (either surface or center). A + or − represents a higher or lower temperature, respectively.

Mesoclimate

Wind

The attenuation of wind, primarily by the canopy, is among the most pronounced aspects of the forest mesoclimate and results in a vertical stratification of wind speeds in any given stand (Fig. 14.9). Wind velocity within the stand may be reduced by 90%; velocity above the forest floor may be only 1–2% of the velocity outside the forest. Movement of air within the forest can modify other important physical factors. The rate of cooling and evaporation, overall humidity, and humidity and temperature gradients can all be significantly affected by air movement. Air is the medium into which secretory and excretory products are released by forest insects. The ability of an insect to perceive and locate these chemical messages is affected by the force, magnitude, and physical conformation of wind patterns in the forest (Fig. 14.10).

Major movements of air outside the forest stand can be as important as those inside. Many insects are carried by wind over long distances. This passive dispersal can be important in colonizing new habitats, spreading

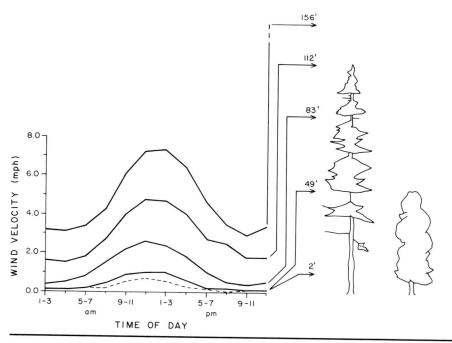

Figure 14.9.   Average wind velocity attenuation at increasing heights (in ft) in an 80- to 100-year-old conifer stand. (Redrawn from Barry *et al.*, 1975.)

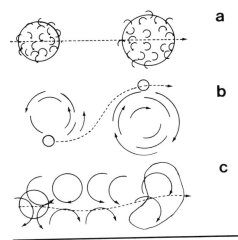

Figure 14.10. Hypothetical model showing how a cloud (puff) of pheromone might be dispersed by eddies of various sizes. (a) In eddies smaller than the puff, the puff grows slowly, and concentration decreases slowly. (b) Eddies larger than the puff transport it intact, and its concentration is only slightly reduced. (c) Eddies approximately the size of the puff tear the puff apart and rapidly diminish concentration. In the atmosphere, all eddy sizes act simultaneously. (From Aylor, 1976.)

outbreak populations, and reducing the competitive pressures of high population density. Transport can also result from turbulent wind action in which insects are swept along in surface winds (Greenbank, 1957).

Although much of the windborne dispersal is passive, certain components of this type of dispersal require the active participation of individuals or evolved adaptations. Many dispersal stages are morphologically adapted to wind transport. First instars of many Lepidoptera, such as the gypsy moth, enhance wind dispersal with dorsal aerostatic and lateral aculminate hairs which increase surface area and buoyancy. Other species, such as the tussock moths, have similar adaptations. The production of silk strands by larvae as they become airborne dramatically reduces their terminal velocity (rate of fall per second) (Table 14.6). The wind speed required to break the silk decreases as its length increases. Thus, 21-mph winds are required to break a 6-in. thread formed by spruce budworm larvae, and 4.4-mph winds are needed to break a 12-in. thread (Wellington, 1954b).

Generally, adult insects are released into the atmosphere at night, whereas most immature insects appear to be released during the day. Airborne insects may be found from as low as 6 m to as high as 5000 m and above, although their concentrations vary. Insects in the upper atmosphere may be carried hundreds of miles; however, relatively diminished air movements high above the canopy and reduced wind

Table 14.6. Terminal velocity of Douglas-fir tussock moth larvae as functions of length, weight, and silk length[a]

Larval Length (cm)	Silk Length (cm)	Larval Weight (mg)	Terminal Velocity (m/s)
0.3[b]	90	—	0.252
	60	—	0.328
	30	—	0.467
	20	—	0.544
	10	—	0.651
	0	—	0.812
0.318[c]	0	0.65	1.906
0.635[c]	0	1.00	2.103
0.794[c]	0	1.10	2.790
1.100[c]	0	7.10	2.900

[a] From Edmonds (1976).
[b] First instar.
[c] Other instars.

speeds beneath the canopy (often less than 1 m/s) increase the chances of the settling of the dispersants. Although wind speed can affect dispersal distance (Fig. 14.11), most insects do not appear to travel very far (Table 14.7). Maximum distances of wind transport become an important limiting factor in the population dynamics of a species, particularly if their dispersal is essentially wind dependent (e.g., the crawler stage of most scales and first instars of flightless Lepidoptera). Figure 14.12 illustrates the

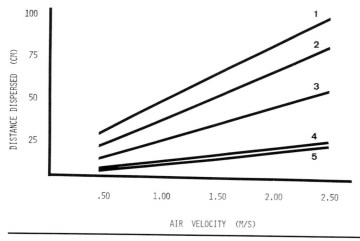

Figure 14.11. Dispersal distances of first instar gypsy moths at several air velocities, 1 to 5 days after hatch. (From Capinera and Barbosa, 1976.)

Table 14.7. Predicted downwind concentration of all tussock moth larvae (% concentration at 10 m from source) in typical morning and afternoon conditions[a]

Distance from Source (m)	Larvae Concentration at 10 m (%)	
	Morning[b]	Afternoon[c]
10	100	100
100	5.2	6.4
1,000	0.24	0.27
10,000	0.006	0.003

[a] From Edmonds (1976).
[b] Wind speed = 0.5 m/s; terminal velocity = 0.252 cm/s.
[c] Wind speed = 3.0 m/s; terminal velocity = 0.252 cm/s.

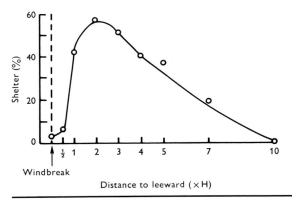

Figure 14.12.   The mean shelter profile to leeward of a windbreak in direct ($\pm 40°$) winds. Tree height $H = 20$ m. Shelter (%) is a function of wind direction relative to the windbreak (tree). A zone of relatively calm air occurs between 1H and 4H. (From Lewis, 1970.)

generalized form of wind turbulence beyond a windbreak (trees) composed primarily of 20- to 22-m-high *Pinus sylvestris*. Within the approximate height H of the trees of the windbreak, little shelter is provided by trees, and the wind (carrying insects) blows through. This pattern varies depending on the nature of the trees, the speed and direction of the wind, and the source of insects (Lewis, 1970). However, research on windbreaks suggests that insects originating in the windbreak should be concentrated immediately leeward and within 2H of the trees (H being the height of the tree). The actual pattern of deposition of crawlers of the hemlock scale

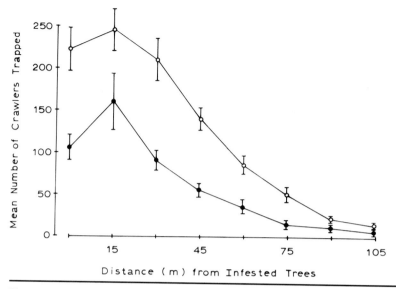

Figure 14.13.   Mean number (±SE) of *Fiorinia externa* crawlers that fell during September (closed circles) and October (open circles) over 14-day trapping periods, on sticky slides located at eight distances from an infested hemlock stand. Means are based on crawlers that settled on an area of 125 cm² at each distance for each of four transects. The greatest number of crawlers were collected along the NE transect, leeward of the infested stand. Prevailing winds were from the S and SW. (From McClure, 1977.)

(*Fiorinia externa*) (Fig. 14.13) shows the highest concentration occurring between approximately 8 and 23 m from the stand. The height H of the trees was approximately 15 m, and thus maximal concentrations occur within $2H$ of the trees (McClure, 1977).

Behavior and Physical Factors

The behavior of forest and shade tree insects is a series of directed movements mediated by biotic factors and a complex of physical factors acting either in conjunction or in sequence. Behavioral responses to physical factors generally appear to maintain the insect within an optimal range for survival. Mature larvae of the larch sawfly drop to the ground and spin cocoons in the moss or duff of tamarack bogs. The uneven ground is covered with moss that varies in moisture content. Since newly formed cocoons are susceptible to mortality from submersion in water, the avoidance of moisture extremes is adaptive. Indeed, in laboratory experiments on larval orientation, a primary response of larvae is avoidance of very wet or very dry areas of a moisture gradient (Ives, 1955a). On the other hand, the pine defoliating diprionid sawflies (*Neodiprion americanus banksianae* and *N. lecontei*) are commonly found in exposed sites on their host and on hosts growing in open or fringe (border) habitats. This preference for dry sites is confirmed by the behavior of the larvae of both species. Both respond to gradients of evaporation by aggregating in areas with a high evaporation (i.e., dry sites). There is also an important interaction with temperature. *N. lecontei* larvae appear to be able to control water loss from their bodies to some degree. However, as the temperature approaches the point at which a sudden increase in the rate of water loss occurs (approximately 40°C), there is a reversal of their photopositive orientation. Thus, the sawfly's behavioral response enhances its survival under variable conditions (Green, 1954a,b).

The combined influence of temperature and humidity (i.e., rate of evaporation) is exhibited in the responses of the spruce budworm, even though larvae of this species do not exhibit a temperature preference. Instead, temperature limits provide a threshold beyond which larvae are inactive. These limits vary with age but range from approximately 28°C for first instars to approximately 38°C for final instars. Below certain temperature thresholds all larvae respond to rate of evaporation. In a given gradient, larvae aggregate in a preferred combination of temperature and humidity required to achieve that rate (Wellington, 1949a,b).

A review of the spruce budworm's life cycle can place its responses to physical factors in proper perspective. Larvae overwinter in silk hibernaculae (spun in empty staminate cups), in bark crevices, or under lichen or other debris. Larvae emerge in the spring when temperature and humidity rise. Hibernaculae that remain under dry conditions do not yield larvae. Since larvae are mobile at temperatures of approximately 5–6°C, low temperatures are not generally inhibitory, particularly if there is direct insolation on larvae and foliage. Low rates of evaporation (i.e., saturated air), as might occur on cloudy days or during foggy weather, cause larvae to be sluggish.

After larvae are dispersed and become established on host foliage their feeding is strongly affected by temperature and moisture. The relatively low spring temperatures of the budworm's northern forest habitats can shorten or eliminate daily feeding periods. However, insolation can insure that bud or webbed feeding-shelter temperatures rise well above air temperatures. Mature larvae, in addition to becoming immobilized in saturated air, aggregate and move slowly in a specific preferred zone of evaporation rate. This preferred rate, which results in less movement and steady feeding, coincides with the evaporation rate inside the small tunnellike shelters of older instars. This rate of evaporation can usually be maintained in these shelters even when outside air is dry. Because high ambient humidity causes the air space within the shelter to reach saturation, the budworm prefers dry, sunny weather (Wellington, 1954a,b).

Effects of Polarized Light on Behavior

In addition to being able to discriminate light intensity and wavelength, insects can detect polarized light. Polarized light is made up of light waves that vibrate in the same plane. Light can be polarized by passing it through a filter or by reflecting it from a dull surface. Insects' perception of polarized light appears to be critical to the orientation and dispersal behavior of larvae and adults. Although light-compassing reactions are exhibited in certain specific circumstances in the field and in the laboratory, orientation based on patterns of polarization appears to be more prevalent. Because patterns of polarization have a fixed relationship to the sun, an insect can maintain any one orientation angle with respect to the sun by reference to the plane of polarization during its movement, even when the sun is not actually visible. In some species, orientation by polarization patterns occurs only when the polarized light is in certain wavelengths (e.g., short wavelengths or UV) (Menzel, 1975).

Many variables affect an insect's use of polarized light. Cloud cover, smoke, and ice crystals disturb the plane of polarization, which can disorient the insect (i.e., cause frequent turning, circling, or even cessation of movements until the sky clears). The amount of plane-polarized light varies diurnally and is significantly reduced at midday when the sun is highest. At sunset and sunrise, zenith polarization is most intense and thus available as a directional cue. Many insects engage in feeding and oviposition flights during these crepuscular periods. Although other physical conditions may be required for these activities, the intensity of polarization patterns may be an important variable.

As mentioned, no physical factor acts in isolation. The response to one factor is tempered or changed by other factors. If the plane of polarization remains unchanged or changes infrequently, insects orient to it for long periods of time, shifting direction as the pattern shifts. However, in the absence of disruptive influences such as cloud cover, insects are capable of disregarding one polarization pattern and adopting a pattern derived from another plane. This may also occur in response to a temperature change. The hotter the insect becomes, the less apt it is to maintain its orientation. Indeed, temperature is likely to be the trigger for movement, whereas light probably has a greater influence in maintaining the orientation angle after movement begins (Wellington, 1955, 1974).

Ecological Roles of Physical Factors

Directly or indirectly, temperature changes affect almost every ecological interaction among forest and shade tree insects. These effects may be manifested in changes in insect physiology (the magnitudes or rates of physiological processes) or behavior, or in the intensity of group interactions. In general, very few insects can grow when temperatures are continually lower than a few degrees below 0°C or above approximately 45°C (Clarke, 1954) (Table 14.8). Often the magnitude and rate of temperature change over a day may be more critical to survival or to activities such as feeding than any specific temperature. Insect species must also cope with seasonal variations in temperature. Finally, subtle but critical temperature differences may occur in different parts of an insect's microhabitat. These major differences in the microhabitat result from the direction of solar radiation (Fig. 14.14).

Table 14.8.   General effect of high and low temperatures on brood stages of the western pine beetle (*Dendroctonus brevicomis*)[a]

Temperature (°F)	Effect
−20	Fatal to all brood stages.
−15, −10, −8	Fatal to all brood stages; a low percentage of eggs and larvae that are adapted to severe winter climates may survive these temperatures.
−8, −5, 0, 5	Critical temperatures for larvae and pupae; mortality at −5° exceeds 50%.
5, 10	Critical temperatures for adults; high mortality of adults occurs at 10°; low mortality of larvae and pupae may occur at 5°.
10, 15	Larvae freeze, but recover if warmed to temperatures at which activity takes place.
15, 20, 35, 40	Dormancy occurs in all brood stages.
40, 45	Larvae sluggish.
45, 55	Larvae and adults active.
55, 95	Normal activity of larvae and adults.
95, 100	Larvae less active.
100, 105	Larvae sluggish.
105, 110	Paralysis of larvae after 1–2 hr of exposure; recovery possible if returned to lower temperatures.
110, 115	Paralysis of larvae after brief exposure; recovery possible if exposure is not prolonged.
115, 118	Paralysis of larvae after brief exposure (20 min), from which they do not recover.
120	Fatal to all brood stages.

[a] From Miller (1931).

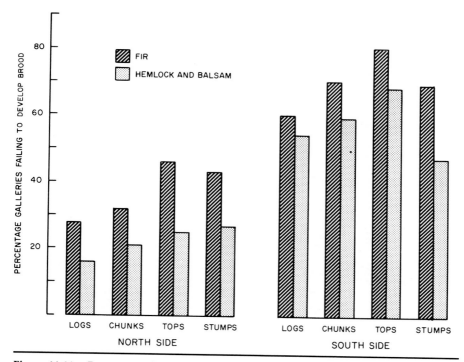

Figure 14.14. Percentage of galleries that failed to develop brood on the north and south sides of various types of slash. *Logs* represent pieces longer than 6 ft and 12 in. in diameter at midpoint. *Chunks* are shorter than 6 ft and greater than 6 in. in diameter. *Tops* are longer than 6 ft, between 6 and 12 in. in diameter, and with more than 3/4 of their surface undamaged. *Stumps* are more than 6 in. in diameter. (From Dyer, 1963.)

Temperature and Insect Development

The relationship between temperature and developmental rate is relatively straightforward; that is, high temperatures within the temperature preferendum of an insect produce rapid development. The relationship between temperature and developmental rate is generally not linear within the favorable range. Nevertheless, the relationship between temperature and development is exact enough to predict when insects will reach a given stage in their development. Such predictions are based on the calculation of so-called degree-days, a value that accounts for all the days in which temperatures exceed an insect's developmental threshold. Thus, the number of degree-days required to reach a developmental stage can be calculated for a given insect species (Fig. 14.15 and Table 14.9).

An alternative approach is provided by Hopkin's bioclimatic law (Fig. 14.16), which states that variation in the timing of life-history events that occur during spring and early summer is generally 4 days later for each

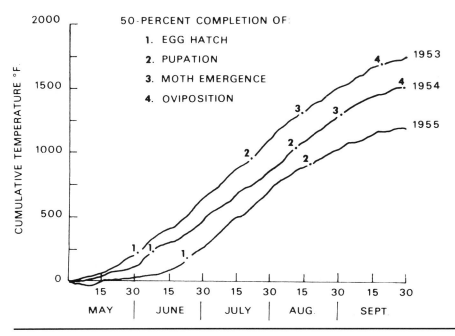

Figure 14.15.   Relationship between major events in the life cycle of the blackheaded budworm (*Acleris gloverana*) and cumulative daily temperatures above 42°F. Numerals denote 50% completion of egg hatch (1), pupation (2), moth emergence (3), and oviposition (4). (From Hard, 1974.)

degree of latitude northward, each 5° of longitude eastward, and each 400 ft rise in altitude in most of temperate North America. The reverse is true in late summer and fall. The changes described by both of these approaches are inevitably linked with other important environmental changes such as food quality, moisture stress, or interactions with other organisms. Thus, the relationship they describe never perfectly fits reality.

Temperature and Reproduction

The range of favorable temperatures for egg production and oviposition is very similar, but not always identical, to that for development. Even though reproduction is influenced by many factors, exposure to certain temperatures results in optimal reproduction (Table 14.10). Usually, fecundity is maximized at a moderately high temperature within the favorable range and drops off dramatically at the upper and lower limits of that range. The temperature at which immatures are reared may indirectly affect egg production because rearing temperatures often affect the size of pupae or adults (Table 14.11). Finally, temperature changes may indirectly affect insects as a result of their effects on activity level (Figs. 14.17 and 14.18) or on key behaviors affecting reproductive success.

Table 14.9. Number of days (±S.E.) and accumulated degree-days above a threshold of 7.2°C from pupation to adult emergence for male and female spruce budworm pupae under field conditions[a]

Date of Pupation	Male			Female		
	No.	Days to Emergence	Accumulated Degree-Days above 7.2°C	No.	Days to Emergence	Accumulated Degree-Days above 7.2°C
June 7	68	21.7 ± 0.18	127	48	21.7 ± 0.08	127
8	106	21.6 ± 0.13	119	53	21.6 ± 0.10	131
9	125	21.1 ± 0.07	123	50	21.1 ± 0.11	123
10	114	20.1 ± 0.12	122	86	19.9 ± 0.07	122
11	46	19.7 ± 0.16	133	89	19.1 ± 0.09	122
12	56	18.9 ± 0.07	133	104	18.3 ± 0.07	121
13	9	18.1 ± 0.59	128	79	17.4 ± 0.09	116
14	49	17.4 ± 0.14	124	61	17.0 ± 0.13	124
15	34	17.8 ± 0.25	127	32	16.6 ± 0.25	123
16	38	16.7 ± 0.25	123	34	16.2 ± 0.26	119

[a] From Sanders (1975).

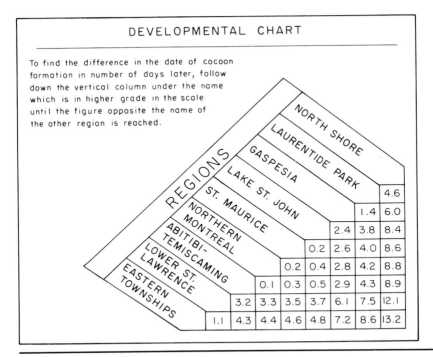

Figure 14.16. Example of Hopkin's bioclimate law: development chart showing the differences in days between the dates of cocoon formation by the European spruce sawfly in various regions. (From Paquet, 1943.)

Weather and Insect Survival

The most obvious and dramatic effects of weather (temperature, moisture, light, wind, etc.) on insects are changes in survival. Both sub- and supraoptimal levels of one or more physical factors can lead to mortality. Knowledge of high and low temperature or humidity thresholds, however, is often not sufficient to predict the lethal impact of physical factors. The type and availability of microhabitats (refugia) in a forest, the insulative capacity of these microhabitats, or the physiological state of an insect determine the impact of any given temperature, humidity, or other aspect of weather (Figs. 14.19 and 14.20).

The overwintering period for most insects is a time of high mortality (Table 14.12). The winter survival rate is partly dependent upon the proper choice of a hibernacula (overwintering microhabitat) and the insect's physiological state. For example, temperatures that are tolerated by diapausing (dormant) individuals during the nongrowing season would kill individuals of the same species during the growing season. Survival, even below freezing temperatures, is due in part to the production of

Table 14.10.   Attack and oviposition success of mountain pine beetles (*Dendroctonus ponderosae*) at low laboratory temperatures[a]

Test Temperature and Duration[b]	Beetle Attacks No.	Beetle Attacks %	Total Brood[c] (No.)	Egg Galley Length (cm) Mean	Egg Galley Length (cm) Range
4.4°C					
2 weeks	1		0	1.2	1.2
4 weeks	4		0	1.7	1.0–2.5
6 weeks	1		0	1.2	1.2
Total	6	20			
7.2°C					
2 weeks	5		0	2.1	1.1–3.3
4 weeks	5		0	2.2	0.7–4.0
6 weeks	8		8	6.3	0.9–12.0
Total	18	60			
10.0°C					
1 week	7		0	1.7	0.1–3.0
2 weeks	6		2	4.2	0.6–7.0
3 weeks	7		42	7.4	3.3–11.7
Total	20	56			
12.8°C					
1 week	10		3	2.7	0.0–5.7
2 weeks	9		36	6.5	0.8–9.3
3 weeks	9		61	9.7	3.2–15.7
Total	28	93			
20.6°C					
1 week	9		120	9.4	2.2–14.5
2 weeks	10		309[d]	22.3	1.9–35.3
3 weeks	9		103[d]	37.0	20.0–49.5
4 weeks	10		255[e]	35.0	0.0–59.5
6 weeks	8		112[e]	30.1	1.1–66.0
Total	46	92			

[a] From McCambridge (1974).
[b] Ten pairs of beetles tested at each temperature and time period indicated.
[c] All eggs unless otherwise indicated.
[d] Eggs and larvae.
[e] Larvae and pupae.

cryoprotectants, which act like automobile antifreeze. This biochemical protection allows some insects to cool below freezing to a so-called supercooling point. A great many factors determine the ability of a species to supercool, including its physiological state, the rate of cooling, and the duration of exposure.

When mortality due to extreme cold does occur, it results from the slowing and cessation of biochemical reactions, rupture of cells due to ice

Table 14.11.   Mean pupal weight by sex of the noctuid *Polia grandis* at two rearing temperatures[a]

		Pupal Weight (mg)	
Rearing Temp. (°C)	Sex	Mean[b,c]	Standard Deviation
25	Female	319	17
	Male	295	45
30	Female	257	59
	Male	223	35

[a] From Miller (1977).
[b] All possible comparisons except female–male at 25°C were statistically significant at the 0.05 level based on student *t*-tests.
[c] Means based on 4–7 pupae.

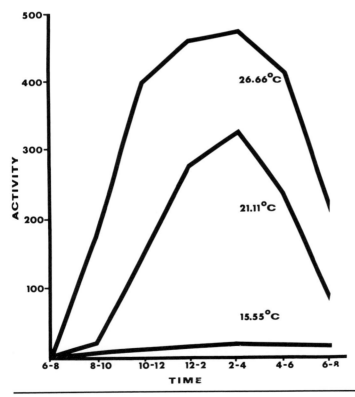

Figure 14.17.   Flight activity of *Brachymeria intermedia*, a chalcid parasitoid of the gypsy moth, exposed to 15.5°C, 21.1°C, and 26.6°C. (From Barbosa and Frongillo, 1977.)

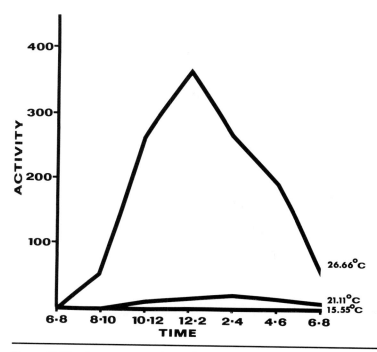

Figure 14.18.   Locomotory activity of *Brachymeria intermedia*, a chalcid parasitoid of the gypsy moth, exposed to 15.5°C, 21.1°C and 26.5°C. (From Barbosa and Frongillo, 1977.)

crystal formation, or cessation of circulation. Conversely, at maximum temperatures, insects suffer from desiccation and excessively high rates of chemical activity. Different stages of the life cycle of an insect have different limits (or preferenda) and may respond differently within these limits.

The killing effect of a component of weather, such as temperature, can be modified by another component, such as the amount of snow (Figs. 14.19 and 14.20). Although some gypsy moth eggs are killed by temperatures of −15°F (−26°C), all eggs in a mass are generally killed at temperatures between −20 and −25°F (−29 and −32°C) (Summers, 1922). Relative survival of gypsy moth eggs is affected by the height of deposition and the protection of snow cover (Table 14.13). Snow cover can dramatically increase egg survival in areas where temperatures are well within the lethal range (Leonard, 1972). In Fredericton, Canada, the maximum air temperature (at 1.3 m) for a 21-yr period was 36.1°C and the minimum air temperature was −34°C. The maximum and minimum ground temperatures (at 2.5 m below the surface) were 22.2°C and −13.9°C, respectively. For many of those years, snow cover ranged from 5 to 35 cm (Salonius *et al.*, 1977).

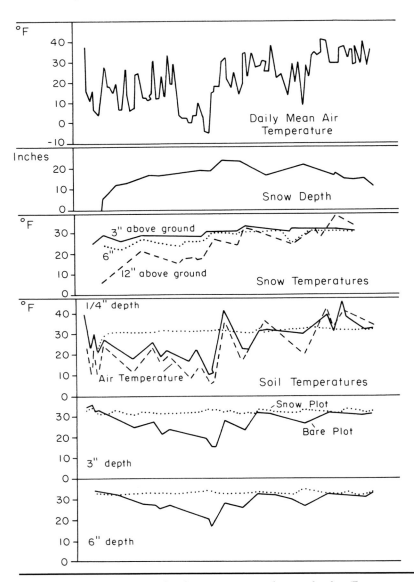

Figure 14.19.   Air, snow, and soil temperatures and snow depths. (From Hart and Lull, 1963.)

The timing of weather patterns may be more important than the nature of the weather; for instance, a sudden cold snap in spring or an early fall frost can have a substantial impact on the size of an insect population. Most of the year-to-year variation in the survival of the European pine shoot moth is associated with the timing of the onset of favorable weather, third-instar survival rate, and time of year when host

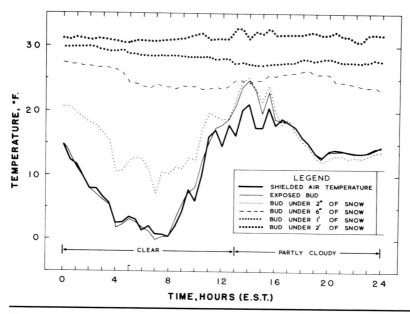

Figure 14.20. Red pine bud temperatures under various depths of snow compared with shielded air and exposed bud temperatures at Sault Ste. Marie, Ontario, Canada. (From Green, 1962.)

Table 14.12. Mortality of the European pine shoot moth (*Rhyacionia buoliana*) coincident with cold period (December)[a]

Area	Number Tips Examined	Number Live Larvae	Number Live per Tip	Number Dead	Number before Cold Period	Mortality Coincident with Cold (%)
1	30	4	0.13	8	12	66
2	30	4	0.13	21	25	84
3	50	7	0.14	12	19	63
4	60	10	0.17	27	37	73
5	70	9	0.13	103	112	92
6	50	5	0.10	24	29	83

[a] From West (1936).

buds are attacked. Large differences in population size at the end of the growing season can occur in any two successive years. This differential is often due not to the difference in the initial size of populations but to the strong influence of temperature on seasonal development. Thus, in any year a warm summer results in early adult emergence and oviposition, egg hatch, and larval development, whereas the following year a cool summer

Table 14.13.   Vertical distribution and hatch of gypsy moth egg masses in Maine[a]

	Feet above Ground								Total
	0–1	1–2	2–3	3–4	4–5	5–8	8–12	12+	
Number of masses	118	99	93	94	95	95	14	9	617
(% total)	(19.1)	(16)	(15.1)	(15.2)	(15.4)	(15.4)	(2.3)	(1.5)	
Percent of egg masses hatching	66.1	15.2	6.5	3.2	2.1	1.1	0	0	

[a] Modified from Leonard (1972).

will produce a slower rate of larval emergence and development. During the warm year, over half of the larvae may have mined buds by the end of July, whereas in the cool year, most bud mining occurs in late August and September. By September, the production of resin in buds is prolific. In mature stands, the resting buds develop earlier than in young stands, and presumably so do resin defenses. Thus, in years with cool summers, retardation of the development of young larvae may result in greater exposure to resin defenses, resulting in a population reduction (Harris, 1960).

Macroclimate

Climate and Population Dynamics

Changes in major patterns of weather (i.e., macroclimate) can also be important to survival, numerical increase, and outbreak development of dendrophagous insects. The influence of weather on the population dynamics of forest insects can be direct or indirect, depending upon whether the insects themselves, their environmental requisites, or their controls are affected. Microhabitats, parasitoids, predators, diseases, competitors, and the quantity and quality of food are examples of such requisites and controls.

One illustration is the relationship between climate and parasitoid effectiveness. An analysis of the relative effects of temperature, humidity, and light on natural enemies (e.g., parasitoids) and their hosts reveals extensive variability in the degree of developmental synchronization between parasitoids and their hosts. Although some parasitoids emerge from overwintering hibernacula at the exact time when the appropriate host stage is available and abundant, other parasitoids may emerge several weeks before or after this optimum period. One reason for this type of asynchrony is the differential effect of physical factors (or weather patterns) on hosts and natural enemies. These same weather factors may also act differentially on habitat selection, reproduction, general activity, and longevity.

A second example of the potential importance of macroclimatic effects is that of the western spruce budworm. Outbreak frequencies of the western spruce budworm were correlated with several macroclimate-influenced variables (Kemp *et al.*, 1985). Three broad categories of outbreak frequencies were identified (Fig. 14.21). These broad climate-based outbreak frequency maps can be important management tools.

The principle of climatic release conceptualizes the relationship between climate and population growth (Wellington, 1954b). The concept suggests that an area where a population persists at an endemic level and biotic conditions favor population growth, no initial growth is likely to occur until seasonal climatic restraints are relaxed. Favorable weather may have to recur over several years before a major increase in a population can

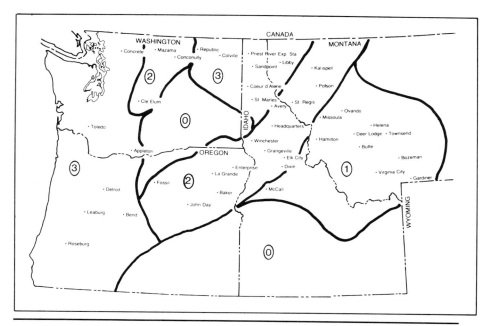

Figure 14.21. Classes of outbreak frequency (1947–78) developed for forested areas of Idaho, Montana, Oregon, and Washington (high = 1, medium = 2, low = 3). Zero indicates no outbreak. (From Kemp *et al.*, 1985.)

take place. Although intuitively appealing, this concept has been the subject of debate.

Major climatic events significantly influence weather patterns. The movement of major air currents controls the nature of the weather of terrestrial habitats. These movements form air masses that when they remain over a surface for an extended period of time, develop a vertical equilibrium and a uniform horizontal distribution of temperature and humidity. The regions where they are produced are known as air-mass source regions (e.g., arctic, polar, and tropical sources). These air masses exhibit characteristic qualities when they leave source regions. Specific major air-mass types may dominate a given region of the country. Transition zones between air-mass types are generally distinct, with areas of low pressure (cyclones) along zone borders. These develop into active frontal systems (e.g., warm or cold fronts) that exhibit markedly different weather from adjacent air-mass weather (Wellington, 1954a). Other air masses move out from their source regions with areas of high pressure (anticyclones) (Wellington, 1952, 1954a).

In warm fronts, warm air replaces cold air at the surface; the reverse occurs in cold fronts. Passage of fronts and their associated air movements are often responsible for the transport of insects over vast distances. Convective transport of both sexes of the spruce budworm has been observed in association with the passage of cold fronts (Wellington, 1954a; Henson, 1951).

The relationship between distinct meteorological events and the development of outbreaks in insect populations can be best illustrated by the events preceding development of spruce budworm and forest tent caterpillar outbreaks. The type and total number of cyclonic centers passing over a given area (expressed, for example, as a percentage of the annual total) preceding outbreaks can be calculated or determined from meteorological records. A determination of this kind in Canada indicates that an above-average number of passages of cyclonic centers of a southern or southwestern origin during spring and summer results in an increase in the forest tent caterpillar population. The development is thus preceded by several years of recurring humid, warm, and partly cloudy summers (Wellington, 1952). In addition, increasing populations are associated with cooler overwintering periods and warmer early feeding periods than are decreasing populations (Ives, 1973). These general conditions associated with epidemics are also in concert with specific biological requirements. The forest tent caterpillar is exposed to the atmosphere during feeding because it does not form the protective tent that is characteristic of other species in the genus. Thus, this insect species requires moist air to remain active (Wellington, 1952).

Conversely, spruce budworm populations move toward outbreak levels when the annual number of cyclonic passages is below average in the late spring and summer and the source of most air masses is of polar-continental or polar-maritime origin. The weather associated with this series of events includes dry air and clear skies. As in the previous example these circumstances provide optimal physical conditions for the larvae. Movement of young larvae from hibernating to feeding sites is important to survival. Movement is optimal under dry and warm conditions because larvae are sluggish at low rates of evaporation (Wellington, 1954b).

The influence of meteorologic or climatic events on insect populations may be direct or indirect. Perhaps the most direct effect is the relationship between numerical increase and mortality due to physical extremes. The balsam woolly aphid in New Brunswick and Nova Scotia, Canada, occurs in at least three climatically distinct areas. Survival and population increases in each area are determined, in great part, by prevailing climate. In maritime areas the population buildup and intensity of attack are great enough to kill or severely injure the majority of host trees. Survival of individual insects in the upper crown enhances the probability of wind transport and spread. In continental regions severe winters limit survival to those insects found in areas of bark at the base of trees and below the snowline. Infestations are localized, and tree mortality and gouting are negligible. Overwintering mortality in the upper crown is substantial enough to allow only minimal dispersion (McManus, 1976).

Climate and Distribution

The climate of forested areas can play a role in limiting the distribution of a species. The concurrent effects of physical factors such as temperature, humidity, wind, and light on survival, activity, rates of physiological processes, reproduction, and behavior restrict an insect species to those areas with a specific optimal range of physical factors. In addition to differences among species, there are differences in the optimal range of temperature, moisture, light, and so on among the developmental stages of an insect's life cycle. Thus, the distribution of immature forms may not always coincide with that of adults.

Distribution limitations may occur not only during the active growing season but also during overwintering periods when insects are exposed to the extremes of physical factors. The northern edge of the European pine shoot moth's habitat is confined at approximately the $-15°F$ ($-26°C$) isotherm where total annual snowfall is less than 80 in. Based on records of annual snowfall and winter temperatures, predictions indicate that continuous infestations of *R. buoliana* should not occur in the western pine zone. Areas least likely to suffer damage include interior British Columbia, Alberta, Montana, Idaho, parts of Washington and Oregon, Wyoming, Nebraska, and the Dakotas, Conversely, southern Oregon and northern

California (north of 40°N latitude) represent areas of extreme susceptibility (Daterman and Carolin, 1973).

It is not surprising that climate determines the distribution of a species. Even in the microcosm of an individual microhabitat, species distribution is affected by differences in factors such as temperature and humidity. For example, although the temperature of a felled tree is affected by various factors (see Tables 14.14, 14.15, and 14.16), the distribution of insects throughout the log still reflects minute spatial variation in factors such as temperature and humidity (Table 14.17 and Fig. 14.22) (Graham, 1924, 1925). Even small temperature differences in a log can have a signifi-

Table 14.14.   Noonday summer subcortical temperatures in white pine logs[a,b]

	Temperature (°C)		
Degree of Shade	Maximum	Minimum	Average
0	65.0	17.5	45.6
1/3	54.0	15.5	36.5
1/2	48.0	14.0	34.4
3/4	39.0	14.0	27.7
Complete	28.0	11.0	18.4

[a] From Graham (1924).
[b] Logs lying in a north and south direction, with north end slightly elevated. These figures are based upon data collected over four seasons.

Table 14.15.   Influence of bark character upon subcortical noonday temperature[a]

	Bark		Temperature (°C)		
Species	Character	Thickness (mm)	Maximum	Minimum	Average
White pine	Rough	13	47.0	18.0	35.1
White pine	Rough	12	48.0	17.0	36.4
White pine	Smooth	5	53.0	17.0	40.6
Norway pine	Scaly	13	45.0	18.0	32.7
Norway pine	Scaly	5	45.0	17.0	32.8
Norway pine	Scaly	1.5	46.0	17.0	33.8

[a] From Graham (1924).

Table 14.6.   Variations in noonday subcortical temperatures between logs of different species[a,b]

Species	Bark		Temperature (°C)		
	Character	Thickness (mm)	Maximum	Minimum	Average
White pine	Smooth	5	55.0	17.0	40.0
Jack pine	Scaly	4	51.0	17.0	38.0
Spruce	Scaly	2	49.0	17.0	36.0
Norway pine	Scaly	5	45.0	17.0	32.0

[a] From Graham (1924).
[b] Logs lying in a north and south direction with north end slightly elevated. These figures are based upon data collected over four seasons.

cant impact on resident insect species, depending on their spatial distribution (Table 14.18). When *Monochamus scutellatus* larvae develop in an area exposed to the sun, they complete their development in 1 yr. However, in heavily shaded areas complete development may require 3 yr (Graham, 1925).

Indirect Effects of Climate

It is obvious that most major climate-related changes in host plants can be important to the dendrophagous species associated with the affected host tree. Beyond the obvious there are subtle effects of weather that act through influences on host trees. Other sections of this book have already discussed the importance of host vigor and growth rate, resin pressure, and rate of recovery from injury on host selection and colonization by forest insects. Weather can affect host plants and natural enemies, which may be more important than direct physical effects (Martinat, 1987). For example, weather can cause host-plant stress that can affect herbivores through changes in plant anatomy, color and spectral qualities, plant temperature, nutritional content and allelochemicals (Mattson and Haack, 1987). In general, the mechanism by which environmental factors influence insect population dynamics is poorly understood.

Phenology and Physical Factors

Physical factors of the environment influence the phenology of insect and host development. The effect of temperature on insect development has been discussed previously (Temperature and Insect Development, this chapter). Temperature and other environmental factors can also influence the phenology of the host (Beckwith and Kemp, 1984; Cameron *et al.*, 1968;

Table 14.17. Density (no./ft^2) distribution of subcortical insects under various conditions.[a]

Host Species	Side of Log	Degree of Shade					Side of Log	Degree of Shade				
	Log	0	1/3	1/2	3/4	Full	Log	0	1/3	1/2	3/4	Full
						Ips pini						
White pine	Top	1.1	40.2	77.1	131.4	42.1	West	67.4	56.1	98.8	91.8	50.0
Jack pine	Top	0.4	9.2	33.5	64.4	21.6	West	21.7	36.1	43.7	33.6	23.5
Norway pine	Top	73.4	69.4	164.2	125.9	41.5	West	86.5	60.0	99.0	84.4	37.5
White pine	East	69.9	74.9	90.5	64.0	39.3	Bottom	23.3	21.2	23.1	23.2	35.5
Jack pine	East	39.7	48.0	63.1	33.0	27.0	Bottom	17.4	19.3	12.5	14.7	19.6
Norway pine	East	80.6	59.3	95.1	81.5	42.1	Bottom	40.0	16.5	31.4	35.8	44.1
						Monochamus scutellatus						
White pine	Top	0.5	2.0	5.2	2.2		West	6.7	7.0	10.5	6.0	
Jack pine	Top	1.6	0	1.3	0.6		West	2.6	4.3	1.3	2.0	
Norway pine	Top	1.0	1.0	0.6	1.0		West	4.0	2.0	3.0	3.0	
White pine	East	5.2	4.5	11.2	4.5		Bottom	5.5	4.2	11.6	2.2	
Jack pine	East	1.3	4.0	1.3	0.3		Bottom	3.0	2.3	2.3	1.0	
Norway pine	East	1.3	1.6	4.3	1.3		Bottom	4.6	3.3	1.3	2.0	
						Chrysobothris dentipes						
White pine	Top	19.7	17.2	2.5	0	0	West	2.7	0.2	0	0	0
Jack pine	Top	17.0	5.0	0.6	0	0	West	0	0	0	0	0
Norway pine	Top	12.0	1.0	0.3	0	0	West	0	0	0	0	0
White pine	East	4.0	2.5	0	0	0	Bottom	0	0	0	0	0
Jack pine	East	3.0	0.6	0	0	0	Bottom	0	0	0	0	0
Norway pine	East	0.3	0	0	0	0	Bottom	0	0	0	0	0

[a] From Graham (1924).

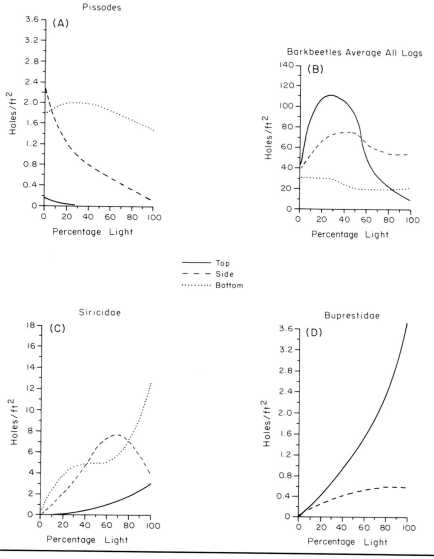

Figure 14.22.   Effect of insolation on colonization of logs, where —— denotes top of log, --- denotes side of log, and ... denotes bottom of log. (a) Distribution of *Pissodes* sp. based on exit holes per square foot; (b) distribution of bark beetles (average for all logs) based on exit holes per square foot; and (c,d) distribution of Siricidae and Buprestidae, respectively, based on exit holes per square foot. Note that minimum and maximum percent light represent full shade and full sun, respectively. (From Graham, 1925.)

Table 14.18.   High temperatures fatal to coleopterous species attacking logs[a]

Species	Stage	Minimum Death Temp. (°C)	Average Death Temp. (°C)	Maximum Death Temp. (°C)
Monochamus scutellatus	Adult	43	47	50
	Larvae	45	48	50
Ips pini	Adult	44	48	50
	Larvae	44	46	49
Chrysobothris dentipes	Adult	45	50	52
	Larvae	—	—	—

[a] From Graham (1924).

Wickman, 1976a, 1977). The synchrony between the host and the insect is especially important for the western spruce budworm (Thomson *et al.*, 1984). In the southwestern United States, Engelmann spruce is less damaged by the western spruce budworm primarily because budbreak and shoot elongation are phenologically later than in other budworm hosts. It has been suggested that delaying the host phenological development may be an effective control of spruce budworm (Eidt and Little, 1966, 1968).

The seasonality of insect activities is reflected in a large number of aspects of their life history. During the growing season, physical factors play an important role in determining the number of generations a species can complete in a year. Although the heat required (degree-days) for growth from egg to adult (one generation) is accumulated over the entire growing season in the northerly limits of a species' distribution, the heat summation period along the southern limits may allow two, three, or more generations per year. During a nongrowing season, characterized by extremes of temperature, humidity, or both, dormancy enables insects to survive. This period may be spent in the soil, under the bark, in a bud, or in a variety of sites. Physical factors also play a key role in the onset of dormancy duration and termination.

Diapause is a state of developmental or reproductive dormancy controlled by neuroendocrine centers in the brain. Once initiated, it persists even when conditions are favorable until certain physiological changes have occurred (so-called diapause development). The diapause state confers tolerance to harsh overwintering conditions and usually occurs in a specific developmental stage for a given species (i.e., larva or nymph, pupa, or adult). Diapause also allows synchronization of development, which in many species enhances the probability of successful mating.

Diapause is usually induced by temperature, food quality, and perhaps most often by photoperiod. The action of these factors is not mutually exclusive. Photoperiodic induction of diapause is often a temperature-sensitive process. In univoltine (single generation) species, diapause is obligatory; essentially all individuals go into diapause. Some degree of facultative diapause occurs in multigeneration species. Diapause may be induced primarily in the individuals of the last generation of each growing season, or the proportion of diapausing individuals may increase in each succeeding generation.

Species often are categorized as long-day or short-day insects. Long-day species tend to occur in temperate areas; for these insects, exposure to long days (long photoperiods) can terminate and prevent diapause. Conversely, short-day species occur at southern latitudes; and when exposed to short days (i.e., short photoperiods), they terminate diapause or maintain growth. The two prime factors in the termination of diapause are temperature and photoperiod (Bradshaw, 1974). These act by regulating the rate of diapause development (Tauber and Tauber, 1976).

Photoperiod also regulates the timing of many other activities of forest insects. Many behaviors exhibit a periodicity or rhythm that is triggered or terminated by specific photoperiods. For example, many species such as the spruce budworm exhibit flight and oviposition periodicity. Females show a very pronounced flight peak shortly before dark. Their activity, however, is complicated by mating, oviposition, and food quality (Sanders and Lucuik, 1975) (Figs. 14.23, 14.24, and 14.25).

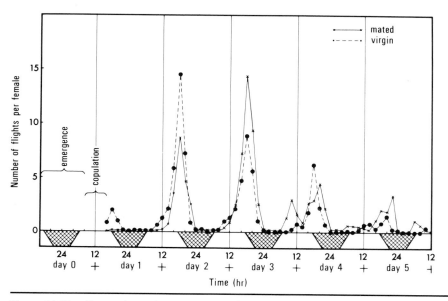

Figure 14.23. Comparison of mean activity of mated (——) and virgin (---) female *Choristoneura fumiferana* under artificial illumination. Shaded areas denote scotophase. (From Sanders and Lucuik, 1975.)

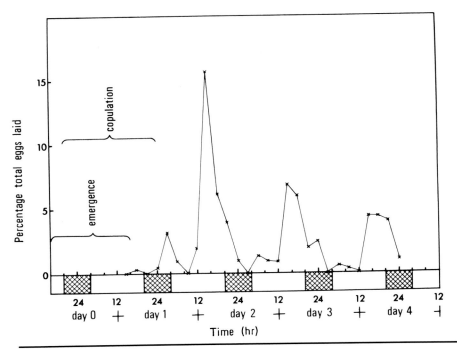

Figure 14.24. Pattern of oviposition of mated female *Choristoneura fumiferana* during the first five days of life. Shaded areas denote scotophase. (From Sanders and Lucuik, 1975.)

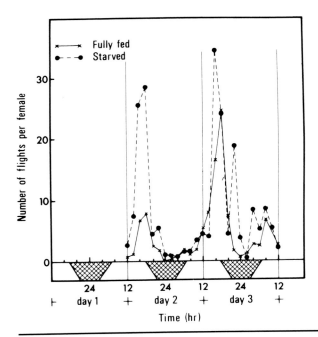

Figure 14.25. Comparison of mean activity between fully fed (——) and partially starved (---) mated female *Choristoneura fumiferana*. (From Sanders and Lucuik, 1975.)

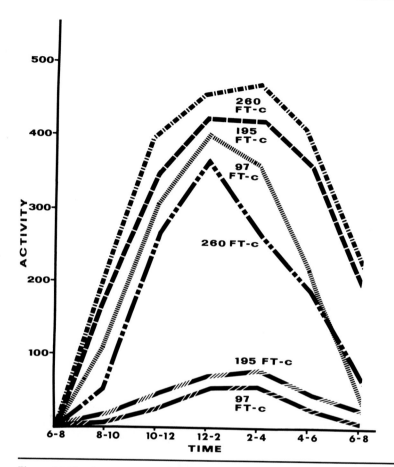

Figure 14.26. Locomotory and flight activity of adult *Brachymeria intermedia* exposed to 97 fc, 195 fc, and 260 fc of light. The top three curves represent locomotory activity; the bottom three represent flight activity. Temperature was held constant. (From Barbosa and Frongillo, 1977.)

Other species also exhibit periodicities in their feeding, general activity, and mating.

In addition to the duration of the light period, light intensity can be an important factor for many species (Fig. 14.26). Flight behavior of the scolytids *Gnathotrichus retusus* and *G. sulcatus* depends on the interaction of temperature and light intensity (Fig. 14.27). Above a temperature threshold of 58 to 60°F (14 to 16°C) a light intensity of 2000 footcandles (fc) retards flight activity. When light intensity is greater than 2000 fc, flight is terminated. An intensity of 1–200 fc appears to be optimal. In dense coniferous stands of the Pacific Northwest or on overcast days, flight continues throughout the day. But in areas with greater solar exposure,

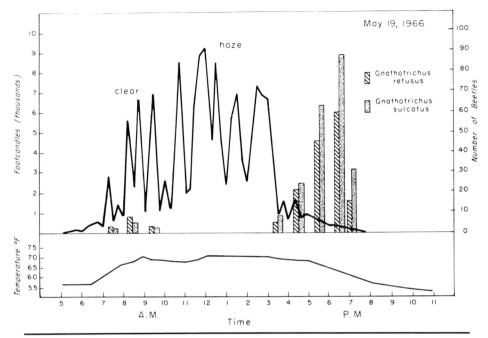

Figure 14.27.   Flight of *Gnathotrichus retusus* (◨) and *G. sulcatus* (▦) in relation to light intensity, temperature, and time of day on a hazy day. (From Rudinsky and Schneider, 1969.)

flight begins in the morning when temperatures reach 58°F (14°C) and ceases when intensity surpasses 2000 fc (Rudinsky and Schneider, 1969).

Chapter Summary

The importance of physical factors to the abundance, population dynamics, and overall survival of forest and shade tree insects has long been recognized. Physical factors such as temperature, relative humidity, light, and wind operate on different scales or on micro-, meso-, and macroenvironmental levels. These levels have distinct effects but also interact. Temperature is a major physical factor that varies seasonally, diurnally, within the canopy, and by microsite. Weather station temperatures are of little relevance outside of the context of individual forest microsites. Humidity also varies substantially in a forest ecosystem and over time. Many behavioral modifications exist in forest insects that allow them to adapt to varying humidity levels. The ecological niches forest insects occupy are often a function of their ability to adapt to variations in wind, temperature, and humidity, all of which affect forest and shade tree insects in complex and interesting ways.

CHAPTER 15

Population Growth and Outbreaks

Introduction

A forest insect population is defined as a group of organisms of the same species that occupies a given space at a given time. We may be interested in the population on oak trees, in a particular state park or a state or geographical region. Many studies involve the evaluation of localized populations, that is, the smallest group of interbreeding individuals. These sub-populations are often called demes. The spatial and temporal boundaries of a population are often arbitrary. The most important reason for considering the concept of a population is that populations manifest characteristics that are more than the sum of individual responses to the environment. Measureable traits that are associated with populations but not with individuals include density, natality (birth rate), mortality (death rate), immigration, emigration, age distribution, and dispersion. Other traits are correlates of those expressed in individuals; that is, they are similar but take on a different form or significance in populations. These include biotic potential, growth form, and genetic characteristics.

Population Density

Density is generally measured or expressed as the number of individuals per unit area or volume. The ability to measure density and the accuracy of the measurement are often functions of species-specific characteristics involving behavioral, physiological, and ecological optima, as well as the technique or procedure used for measurements. Estimates of density can be absolute or relative. Absolute density measurements are counts of the total number of a species per unit area. Not only are absolute measurements extremely difficult to obtain; but since the count is a statistic for time t (one point in time), and population numbers often change rapidly, the effort associated with such measurements may not meet desired objectives. Relative density measurements provide comparative measures of abundance by measuring numbers per unit of area or volume that can actually be colonized (habitat space). Thus, one can count the actual number of elm leaf beetles (*Pyrrhalta luteola*) in a given area, or one can develop an estimate of the average number of beetles per tree in area A versus an average per tree in area B. Given that other factors are equal, this index of relative abundance can serve as an indicator of the severity of infestation.

Dispersion and Dispersal of Populations

Individuals of a population distribute themselves in their habitat; this internal distribution generally forms three basic patterns: random, uniform, or clumped. Forest and shade tree insects rarely exhibit a random or scattered distribution. The specific pattern for any given species reflects the distribution of the factors it requires to survive, develop, and reproduce. Solitary general predators such as spiders may exhibit random dispersion if they occur in a forest floor with a relatively uniform covering of leaf litter. When individuals occur in situations in which some environmental interaction (e.g., intraspecific competition) promotes even spacing, they are uniformly dispersed. If the interactions among host trees are such that individual susceptible trees are relatively evenly spaced, their insect associates may show a similar uniform dispersion.

Perhaps the most common dispersion pattern for forest insects is the clumped dispersion. This occurs when aggregates of individuals are irregularly or nonrandomly distributed. The occurrence and intensity of this dispersion pattern is a function of the species, time of year, age of the individuals, sex, and so on. The clumping pattern may be a response to (a) the habitat mosaic created by structural diversity of the tree, (b) microclimatic differences in the habitat, (c) clumped distribution of host trees, (d) differential predation by natural enemies, or (e) a response to interspecific or intraspecific chemical cues (e.g., clumping as a result of aggregation pheromones produced among bark beetles).

The patterns just discussed are maintained by movements of individuals on a limited scale within the confines of a deme (or local population). Widespread movements that take individuals into and out of a population area are generally referred to as population dispersal, which takes the form of emigration (one-way outward movement), immigration (one-way inward movement), and migration (periodic departure and return of individuals). This is important because, along with natality (birth rate) and mortality (death rate), these are components of population growth. In addition, these are the primary means by which new, uncolonized areas are populated.

Local areas invariably become unfavorable to forest insects over time. Dispersal insures that some individuals of the population reach new areas before the collapse of a local population. Dispersal also insures gene flow among separate populations of a species. Although some dispersal may occur in all species, the amount of dispersal is a function of its vagility (i.e., the inherent capacity or power for movement of individuals of a population). The propagule or stage that is capable of dispersing varies. In most species the adult is the dispersal stage; however, in many Lepidoptera and scales, the first instar (so-called crawlers in scales) is the dispersal stage. The wind is often of primary importance in the dispersal of forest and shade tree insects. The pattern of establishment of dispersants often mirrors wind patterns and may be the areas where the greatest insect damage occurs.

Life Tables and Population Dynamics

One of the most useful ways of understanding the population dynamics (factors affecting numerical change) of a dendrophagous species is to construct a life table. By providing reproductive and survival parameters, a life table enables a scientist to calculate the growth performance of a population. Actually, the term *life table* is a misnomer. It is really more of a death table because it records mortality factors and age distribution at mortality. A single life table indicates stages at which high mortality occurs, whereas a series of life tables, each considering specific population levels (Table 15.1) and environmental conditions, can provide a substantial understanding of population dynamics over time.

The reliability of life tables is directly correlated to the reliability of sampling techniques. In general, two basic types of life tables are used in the evaluation of mortality in dendrophagous species. Time-specific life tables evaluate the fate of an imaginary cohort based on determinations of the age structure at a given time of a sample population with substantial overlapping of generations. The time-specific life table is used only if the population has a stable age distribution. It has limited usefulness in studies pertaining to the influence of environmental factors on numerical trends.

For most insects the age-specific or cohort life table is more appropriate because numerical levels can vary greatly between generations (Table 15.2). Such tables are based on the fate of a real cohort of individuals for which deaths are recorded in each of several age classes. The actual population in different age classes is listed and the action of various mortality factors recorded. Thus, life tables can be developed for each of several generations or for each of various life history conditions. Alternatively, data in age-specific life tables can be corrected so as to start with a uniform size cohort, such as 1000 (Southwood, 1966).

To understand the influence of a given environmental factor, a series of age-specific life tables is required for a number of generations. Summaries of average effects can be made by the construction of survivorship curves or by key-factor analysis. Key-factor analyses are particularly useful

Table 15.1. Life table for 1973–1974 generation of Douglas-fir tussock moth in moderate defoliation class plots (average of six plots)[a]

Age Interval (x)	Number Alive at Beginning of x (lx)	Factor Responsible for dx (dxF)	Number Dying during x (dx)	dx As a % of lx (100 qx)
Instar I (N₁) (1973)	313.5	Virus disease	21.9	7.0
		Predation, dispersion	65.4	20.8
			87.3	27.8
Instars I, II, III	226.2	Virus disease	16.1	7.1
		Hymenopterous parasites	0.9	0.4
		Dispersion, predation	185.6	82.1
			202.6	89.6
Instars III, IV	23.6	Virus disease,	2.1	8.9
		Hymenopterous parasites	0.2	0.8
		Predation, starvation	14.5	61.4
			16.8	71.1
Instars IV, V	6.8	Virus disease	1.5	22.1
		Hymenopterous parasites	0.6	8.8
		Other	1.0	14.7
			3.1	45.6

362

Stage		Mortality factor		
Instars V, VI	3.7	Virus disease	0.7	18.9
		Hymenopterous parasites	0.5	13.5
			1.2	32.4
Pupae	2.5	Virus disease	0.3	12.0
		Hymenopterous parasites	0.4	16.0
		Dipterous parasites	0.7	28.0
		Unknown parasites	0.2	8.0
		Other	0.4	16.0
			2.0	80.0
Moths	0.5	Sex ratio (38% female)	0.12	24.0
Females $\times$ 2	0.38	Reduction in fecundity	0.10	26.2
Normal females $\times$ 2	0.28	End of longevity	0.28	100.0
Eggs	20.9	Hymenopterous parasites	2.9	13.9
		Infertility	11.8	56.6
		Overwintering stress and predation	4.0	19.1
			18.7	89.6
Expected instar I	2.2			
Actual instar I (N_2) (1974)	1.4			
Generation Totals $I = N_2/N_1 = 0.004$			332.2	99.34

[a] From Mason (1976).

Table 15.2. Life table for the 1968–1969 generation of the forest tent caterpillar in northern Minnesota[a]

Age Interval(x)	Number Alive per Tree (lx)	Factor Responsible for dx (dxF)	Number Dying during x (dx)	dx as % of x (100 qx)	Survival Rate within x (Sx)
		1968 Generation			
		Absolute population measurement			
Egg	3105	Pharate larval[b] Parasitism Infertility	1204 230 83 1517	38.8 7.4 2.4 48.9	0.51
Larval, 10 days	1588	Weather, starvation, predation, and/or disease	458	28.8	0.71
Larval, 20 days	1130	Weather, starvation, predation, and/or disease	422	37.3	0.63
Larval, 30 days	708	Weather, starvation, predation, and/or disease	365	51.6	0.48
		Relative population measurement			
Larval, late 4th stage	343	*Rogas* Tachinids (Diptera) Nematodes	7 5 1 13	2.2 1.6 0.2 4.0	0.96
Larval, late 5th stage	330	Tachinids (Diptera) Hymenoptera Nematodes	31 1 1 33	9.2 0.2 0.1 9.5	0.90
Pupal	297	Parasitism Starvation[c], predation, diseases, and/or migration out of study area	219 60	73.7	0.26
Adults	18[d]				
Generation			3087	99.3	0.01

1969 Generation

	Absolute population measurement				
Egg	1770	Pharate larval[b]	44	2.5	
		Parasitism	170	9.6	
		Infertility	27	1.5	
			241	13.6	0.86
Larval, 10 days	1529	Weather, starvation, predation, and/or disease	601	39.3	0.61
Larval, 20 days	928	Weather, starvation, predation and/or disease	515	55.4	0.45
Larval, 30 days	413	Weather, starvation, predation, and/or disease	185	44.8	0.55
		Relative population measurement			
Larval, late 4th stage	228	*Rogas*	3	1.4	
		Tachinids (Diptera)	1	0.6	
		Nematodes	0	0.0	
			4	2.0	0.98
Larval, late 5th stage	224	Tachinids (Diptera)	4	2.9	
		Hymenoptera	1	0.1	
		Nematodes	1	0.1	
			6	3.1	0.97
Pupal	218	Parasitism	110	50.4	
		Starvation[c], predation, diseases, and/or migration out of study area	70		0.50
Adults	38[d]				
Generation			1732	97.8	0.02

[a] From Witter *et al.* (1972).
[b] Refers to mortality of fully developed first-stage larvae within the egg that did not hatch.
[c] Includes mortality other than parasitism that also occurred during the late larval, pupal, and adult stages.
[d] Count inferred by number of egg masses ×2.

because they can provide a method of pinpointing the key factor(s) determining numerical changes and the existence of density-dependent relationships. For example, key values can be calculated for each age class using data from an age-specific life table. The total generation mortality K is the sum of successive k-values. Curves are plotted representing the k-values over a number of years (or generations) for each age class as well as the K values for all years or age classes studied. The identification of the key factor is achieved when the curve for a particular age class is significantly correlated to the curve for K (Fig. 15.1). Mortality in a particular age class is generally due to a dominant factors(s). This is identified as the key factors(s). Key factors may have useful predictive values. However, these factors may provide a cause-and-effect relationship for only one set of circumstances since their effect may vary from time to time or place to place.

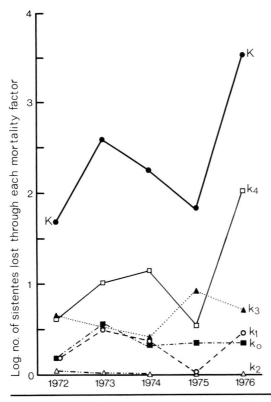

Figure 15.1. Comparison of total annual mortality K with individual mortality levels k for *Adelges cooleyi* (Cooley's spruce gall aphid) sistentes on needles of Douglas-fir. (From Parry, 1978.)

Population Fluctuations

Regular patterns of abundance are commonly observed in response to the annual or seasonal variations of limiting physical factors, primarily temperature and rainfall. Other environmentally modified factors such as flowering or seed set and host quality are similarly affected by physical factors.

In temperate forests, insect populations often respond to seasonal temperature changes. Leaf flush in the spring and leaf drop in the fall delineate the active season for both tree host and dendrophagous insect species. Mortality and various forms of dormancy keynote the inactive, nongrowing period of depressed population density. In tropical forests, rainfall is commonly the key factor governing seasonal changes in population abundance. Leaf production and leaf drop in tropical deciduous forests are closely correlated to wet and dry seasons.

In all types of forests, periodic fluctuations of insects occur that are not related to any obvious seasonal fluctuations. These long-term oscillations sometimes appear unusually regular. The periodicity of oscillations of many defoliators ranges between 4 and 10 yr. For some dendrophagous species these cycles may reflect cyclic changes in host species, such as flowering, seed reduction, or changes associated with aging. Perhaps the most regular and fascinating population cycle is that of the 17-yr cicada. The 17-yr interval between adult populations is so regular that exact predictions of the emergence of specific broods can be made. Adults emerge synchronously in early summer, live 4–6 weeks, and deposit eggs in host twigs.

The larch budmoth, (*Zeiraphera griseana*) is a univoltine defoliator of alpine larch (*Larix lyallii*) forests of Europe. Extensive defoliations by this insect in the forests of the European alps have been recorded at intervals of 6 to 8 yr. Favorable weather conditions and a constant food source combine to allow yearly population increases of the budmoth, which are ultimately halted by exhaustion of the food base and subsequent intraspecific competition during the larval feeding period. Ultimately the size of the population is reduced. The interval of oscillation is determined by the rate of increase of the insect population and the carrying capacity of the host plant. In the case of the budmoth, outbreak levels are reached within four generations, and the regression or collapse phase takes 2–4 yr (Baltensweiler, 1964).

Various theories have suggested factors ranging from climate to sun spots as the causes for the periodicity of long-term cycles of abundance. Although correlations can be made between cycles of abundance and a variety of factors, these do not necessarily represent a cause-and-effect relationship.

Population Growth Forms

Density measurements of a population represent the numerical level at a given time. Numerical levels of populations are not static but dynamic. There are incremental increases due to birth and immigration as well as losses from death and emigration. The population growth rate is essentially the change in the number of individuals ΔN divided by the time elapsed Δt;

$$\frac{\Delta N}{\Delta t}$$

The mathematical description that fits the other side of this equation has been the source of considerable research and controversy for many years. The simplest description of population growth is the geometric growth curve, also referred to as a *J*-shaped curve or an exponential growth curve. This growth form suggests that rate of change in number per unit time $\Delta N/\Delta t$ is equal to the innate capacity for increase r times the population size:

$$\frac{\Delta N}{\Delta t} = r_m N$$

where N is the population size, t the time, and r_m the innate capacity for increase at given environmental conditions. The unrestricted growth described by these equations is unlikely to continue because real populations are checked by environmental restraints such as predators, pathogens, parasites, weather, and competition. These limitations tend to dampen growth to a level that can be sustained by available resources. This maximum population is referred to as the carrying capacity K. The equation can be rewritten to describe the S-shaped (or logistic) curve:

$$\frac{\Delta N}{\Delta t} = rN \left(\frac{K - N}{K} \right)$$

where $(K - N)/K$ is a factor that reduces growth rate as population density increases and that can be considered the nonutilized opportunity for population growth.

This S-shaped or logistic curve is limited in its ability to reflect the growth of natural populations because the following assumptions must be made: (a) At all times the same proportion of individuals is reproducing (stable age distribution), (b) the carrying capacity is constant, (c) the environmental resistance provided by competition increases linearly with density, and (d) reproduction is constant and is not affected by environmental factors. For the most part, this growth form holds for species that have a relatively simple life history and for those under constant laboratory conditions in which food is maintained and the life cycle is short. However, natural populations exhibit a specific time interval in which population levels increase or decrease due to environmental pressures. Populations also exhibit interrupted activity during favorable periods as well as delayed responses to environmental changes. All of these factors can alter the growth form. Thus, although the growth form is a useful tool, it has limitations as a predictive tool (Odum, 1971; Price, 1975).

Time-lag models of population growth have incorporated a reaction time-lag that occurs between a change in the environment (favorable or unfavorable) and the associated change in population growth rate. This is described as

$$\frac{\Delta N}{\Delta t} = r_m N \frac{K - N_{(t-r)}}{K}$$

where r is the reaction time-lag. The major change introduced by the time-lag growth form is that the stable K is replaced by converging oscillation toward a stable asymptote, a stable oscillation around an equilibrium level, or a smooth climb toward an equilibrium level (Krebs, 1972).

These models provide growth forms that predict singular outcomes for a set of initial conditions (deterministic models). Stochastic models, on the other hand, add the element of biological variation by considering probability. Thus, instead of a single prediction, one has several predictions of potential population growth, which may range above or below the level predicted by the deterministic models. In general, three basic trends in the growth of most natural populations must be accounted for by models (Krebs, 1972). Some populations tend to remain at or near constant equilibrium levels and are kept there by stabilizing environmental influences. Other populations give no appearance of stability; nonstabilizing influences induce changes in abundance that appear to have no relation to an equilibrium level. Still other populations exhibit regular oscillations. One would expect that models that account for these situations and their variations would be extremely difficult to develop. Ultimately, the form of the population model is strongly influenced by its intended purpose(s). Perhaps the greatest utility is achieved, not necessarily by comprehensive or holistic models, but by models designed to meet the objectives of a limited scope.

Theories of Population Regulation

The question "What prevents unlimited growth?" has been the impetus of innumerable studies and debates on the causes of fluctuations in abundance. The controversies generated from the issue of natural regulation of populations is almost as old as the discipline of ecology.

Biotic School

Proponents of this viewpoint suggest that populations are maintained at or near an equilibrium level because of the activity of so-called facultative agents. These agents are responsible for mortality, which increases as insect density increases and decreases as pest density decreases (density dependence). Such agents include parasites, pathogens, predators, and lack of food. The effects of other controlling factors are independent of the abundance or scarcity of the insect. Among these so-called catastrophic agents are storms and various other weather extremes. Early proponents of this viewpoint were L. O. Howard and W. F. Fiske whose research was very important in the first biological control program against the gypsy moth.

The theoretical components of the biotic school were further developed and modified by the theories of the balance of nature as proposed by A. J. Nicholson. He suggested that there is a state of balance or dynamic equilibrium between population size and environmental conditions. This balance is maintained by the mutual influences of populations and environment. Nicholson concluded that competition (for food, space, etc.) is the only factor that could control populations. Concepts of density-independent and density-dependent factors summarize the tenets of the biotic school. These are, in effect, that a balanced state is maintained in nature by the effects of density-dependent factors (Smith, 1935).

Other researchers continue to emphasize the synergistic interaction of density-dependent and density-independent factors (Milne, 1957, 1962). They suggest that only perfectly density-dependent factors could control numerical increase indefinitely. Most factors (competing species, parasites, predators, pathogens, etc.) are imperfectly density dependent. For example, some parasites are effective at low host density but not at high host density. However, when combined with density-independent factors, these so-called imperfectly density-dependent factors should control population increases.

Climatic School

The central theme of the proponents of the climatic school is that insect population size or density is controlled by the effects of weather (Uvarov, 1931; Bodenheimer, 1928). The basic hypothesis suggests that extremes of temperature and humidity are responsible for changes in growth, fecundity and, ultimately, survival. The proponents of this theory emphasize the instability of natural populations and correlations between specific weather patterns and population fluctuations or outbreaks.

Wellington's theory of climatic release, though not proposing a mechanism of population regulation, presents a recent variation on the early climatic school theme. The climatic release concept emphasizes the conditions that enhance the establishment and development of outbreaks and attempts to explain why epidemic populations occur in a particular place and at a particular time (Wellington, 1952, 1954b; Wellington *et al.*, 1950). The concept of climatic release states that outbreaks of a species occur when a specific set of environmental conditions or a sequential pattern of climatic events occur. An evaluation of the theory in relation to outbreaks of the spruce budworm and the forest tent caterpillar in the boreal forest region indicates that epidemic populations of these two species do not occur simultaneously because of the different environmental requirements of their larvae. Dry and clear weather precedes spruce budworm outbreaks, whereas warm, humid, cloudy weather precedes tent caterpillar outbreaks (Wellington *et al.*, 1950; Wellington, 1950). Similarly, analysis of available historical data demonstrates that specific climatic patterns have preceded outbreaks of other species such as the South African looper (*Neocleora herbuloti*), the Columbian timber beetle (*Corthylus columbianus*), the pine looper caterpillar (*Bupalis piniarius*), the phantom hemlock looper (*Nepytia phanatasmaria*), the brown locust (*Locustana pardalina*), the migratory locust (*Locusta migratoria*), the desert locust (*Schistocerca gregaria*), and the southern pine beetle (White, 1974, 1976; King, 1972; McManus and Giese, 1968; Wellington *et al.*, 1950). Most analyses undertaken to demonstrate climatic release are correlations and do not experimentally demonstrate cause and effect. Indeed, whether climatic release is a valid concept at all has recently come under scrutiny and has not been adequately supported (Martinat, 1987).

Comprehensive School

Proponents of this hypothesis conclude that the factors affecting a population comprise a complex that cannot be split, or at least does not act as distinct, separate units (Thompson, 1929). That is, no specific factor limits a species, but outbreaks occur when the overall environment is optimal for a given species. Typically, evaluations of data on the fluctuations of insects, such as lepidopteran forest species that feed on conifers, indicate that fluctuations arise from the combined effects of many factors (Schwerdtfeger, 1941).

A variation on this theme focuses on the factors influencing an individual organism's capacity to survive and multiply. It rejects previous attempts to categorize the environment into biotic and abiotic or into density-dependent and density-independent factors. Instead, the concept considers four components of the environment: (a) other animals and pathogens, (b) weather, (c) habitat, and (d) food (Andrewartha and Birch, 1954). In most situations one or two factors (e.g., weather or other animals) would be expected to be of primary importance. In actual fact, research led the leading proponents of these tenets to stress the influence of weather.

Perhaps this stems from the nature of the extreme habitats in which they worked.

Self-Regulation School

The theories in this school of thought deemphasize the role of extrinsic factors in the regulations of abundance and argue for the importance of intrinsic factors. In general, previously discussed theories ignored the role of intrinsic phenotypic or genotypic changes in populations. Although there are many variations on the basic theme of self-regulation, they all suggest that the individuals of an insect population may differ in their susceptibility to mortality factors. Variation among individuals may be environmentally controlled. Alternatively, individuals of a species may be genotypically different throughout their distribution.

Self-regulation may occur through some social-feedback mechanism that acts as a buffer to stop population growth (Wynne-Edwards, 1962). These mechanisms might include territoriality or hierarchical organization of a group. This type of self-regulation would most likely be effective in the core areas of a species range and in species that are food limited.

Other researchers propose a general adaptive syndrome that limits abundance through the pathological stress condition of individuals. This conditions results in reduced fertility and survival even in the presence of excess requisites (Christian, 1950, 1959). Similar hypotheses propose that under appropriate conditions unrestricted numerical increase is prevented by a deterioration in the quality of a population (Chitty, 1960). Mortality factors such as weather also tend to be more severe as quality is reduced.

A related hypothesis is based on feedback mechanisms and does not require any degree of social organization (Pimentel, 1961). It suggests that changes in abundance are the result of alterations in the genetic characteristics of individuals in a population. This genetic-feedback mechanism results in a series of coevolutionary adaptations between a defoliator, for example, and its host. Perhaps this theory of self-regulation is in a special category, since it deals less with the "routine" regulation of populations and more with the acyclic fluctuations of introduced species or species that have gained some evolutionary (or adaptive) edge on their hosts. Thus, the hypothesis is very similar to the earlier fundamental concept that proposed self-regulation as a consequence of selection in favor of certain genotypes (Chitty, 1957, 1960, 1971).

Finally, behavioral and physiological polymorphism for dispersal activity, as well as other characteristics, may be the factors that influence the population dynamics of a number of species (*Malacosoma pluviale*, *Hyphantria cunea*, *Lymantria dispar*, etc.) (Wellington, 1957, 1960; Barbosa and Capinera, 1978; Morris, 1971a). In populations of these and other species, individuals differ in their ability to survive and reproduce under similar environmental conditions. This variability allows for survival in environments that are variable in time and space.

Outbreaks

It is ironic that the insect species that occur in the highest numbers and attract the most interest and/or research represent probably less than 1% of all forest insects. Because we have inadequate procedures for the study of sparse (endemic) populations, we often study the causes of outbreaks by looking at the after-the-fact epidemic phase. Nevertheless, in addition to understanding the factors responsible for the release of endemic populations, it is also necessary to understand the factors that induce the collapse of peak populations.

Although outbreaks are varied and difficult to categorize, two types of outbreaks are important among forest and shade tree insects, that is, sporadic and periodic outbreaks (Graham, 1939). Sporadic outbreaks occur in species that remain at relatively low numerical levels for many years and erupt to epidemic levels as a result of some temporary change in the environment. Events that drastically weaken or in some way deletriously affect the condition of trees or the accumulation of breeding material as a result of procedures such as thinning can trigger this type of outbreak. Windthrown trees, ornamental trees whose roots are injured by compaction, and trees injured as a result of construction activities may all be the source of attraction for secondary insects, such as bark beetles. Swaine jack pine sawfly outbreaks often appear as foci in the large distribution of jack pine in Canada. Occasionally, however, these small isolated outbreaks occur following salvage cuts of jack pine (McLeod, 1970).

Other examples of sporadic outbreaks cannot be correlated in any obvious way with a temporary change but are equally aperiodic. In the first 61 years of the 20th century, 18 major infestations of the elm spanworm have been recorded in the United States, from Texas to Maryland. Repeated spanworm outbreaks in the same locality have been relatively uncommon or widely separated by periods of extreme scarcity (Fedde, 1964).

Periodic outbreaks are perhaps the most dramatic of epidemic populations and among the most important. Often occurring at intervals of several decades, outbreaks of species such as the spruce budworm (Elliot, 1960; Blais, 1961), larch sawfly, European spruce sawfly, hemlock looper, and mountain pine beetle are characterized by severe defoliation and/or tree mortality over widespread areas. Most periodic outbreaks exhibit these same characteristics as well as apparently sudden, large concentrations of individuals that disappear with comparable suddenness. This latter characteristic is highly variable from species to species and is often based more on the visibility of the signs of damage (e.g., defoliation and dieback) than on actual number of insects. Periodic outbreaks may also occur at protracted and sometimes relatively regular intervals over extensive forest areas where single tree species predominate.

Another method of categorizing outbreaks is based on continuity of an area. The first category is the *spreading-out* type, in which an infestation occurs first in one place, then gradually spreads over the course of several years, and ultimately collapses after reaching a peak. In the second or *scattered* type of outbreak, infestations occur concurrently in various places scattered over a vast area (Watt, 1968). Further details of this and other schemes for categorizing outbreaks are discussed by Berryman (1987).

The epidemiology of insect outbreaks remains an enigma to forest entomologists. Despite the publication of well over 500 reports and scientific paper on the spruce budworm in the last 27 yr, the issue of whether outbreaks develop from discrete centers or whether populations peak simultaneously over large areas remains debatable. However, studies have provided insight into the outbreak phenomenon in other species. For example, evaluation of the 3-yr outbreak (which occur in 7–10 cycles) of the Douglas-fir tussock moth suggests that 3-yr outbreaks develop from a resident population for at least 2 yr (Mason, 1974).

Typically, outbreaks have three identifiable phases: a buildup phase, an outbreak phase, and a declining or collapse phase. The duration of each phase is sometimes consistent from outbreak to outbreak but is often affected by key environmental factors. In some species such as the Douglas fir tussock moth, the outbreak cycle is a short 3 yr from release to collapse (Mason, 1974). Distinguishing among the phases may be important because the factors regulating insect numbers during each phase may differ. A hypothesis on the regulation of the phasmid *Didymuria violescens* in the forests of Australia suggests that, at low density, population increase is limited primarily by egg parasites and/or birds and that outbreak release is initiated by very cool summer weather. At high density, population increase is ultimately halted by intraspecific competition for food (Readshaw, 1965).

Finally, recent analyses of forest insect populations have prompted a new hypothesis that proposes that insect populations are regulated by processes that produce a numerical bimodality. Forest insects may exist at two major, relatively stable levels (outbreak and innocuous equilibria). These two modes may be linked by two transient phases (release and decline). This type of discontinuous stability concept views the onset of outbreaks as changes in the population level from an endemic stable density, across the unstable (transient phase) equilibrium, to the epidemic stable density. Conversely, outbreak collapses are viewed as movement from the high-density stable equilibrium, across the unstable decline phase, to the lower stable equilibrium (Campbell and Sloan, 1978; McNamee, 1977).

Forest management practices may interfere with normal host plant—insect cycles. Successful spray operations may shorten the intervals between outbreaks of insects such as the spruce budworm, which thrives in mature balsam stands (Blais, 1974). The reasoning behind the theory suggests that in the absence of insecticidal treatment, a considerable amount of the mature stands of balsam fir would die. For example, the use of insecticide spray applications to save an estimated 13 million cords (47 million m^3) of balsam fir in Quebec, Canada, prevented the completion of the normal sequence of events, that is the selective destruction of overmature trees due to spruce budworm defoliation. Had this occurred, it would have taken many years to grow another generation of overmature trees (Blais, 1974). Regardless of the effect of spraying, proponents of this idea point out that no treatment would mean the loss of a considerable amount of an important economic resource.

Food and Outbreaks

The role of food in the survival, development, and fecundity of forest insects is discussed elsewhere. However, the general patterns of variation in nutrients and secondary substances are hypothesized to parallel patterns of abundance of some dendrophagous species. The occurrence of improved development and vigor of insects, in addition to the increased nutritional quality of their food, has led to possible correlations between outbreak development and major shifts in food quality. For example, the mountain pine beetle infests a greater proportion of trees in each year of an epidemic infestation. Adult emergence is closely correlated to tree diameter and bark thickness, which in turn is correlated to phloem thickness. The relationship between gallery starts, beetle survival, and phloem thickness serves as the basis for a theory that suggests that epidemics of the species are strongly dependent on the occurrence of large trees which have thick phloem. As these trees are killed, larval survival and percent emergence from remaining trees are reduced; and the population returns to endemic levels (Cole *et al.*, 1976).

These and other data provide evidence supporting theories of nutritional involvement in insect population regulation. However, these data have not always demonstrated causal relationships. One of the most important reasons for the lack of clear-cut cause-and-effect relationships may be the close association between changes in food quality and changes in other environmental factors.

There is often a hierarchy among the acceptable hosts of polyphagous species. Similarly, the suitability of these hosts varies such that the fecundity of adults reared on one host species differs from that of adults reared on other hosts. Thus, host species composition in a given habitat may have a major impact on numerical increases in insect populations. Not only may the preference for different hosts vary; but also the preference for certain types of foliage may not be constant among instars of an insect species. Oak is favored by all instars of gypsy moth; maple is not favored by any instar but will be fed on in the absence of more favored food (Barbosa *et al.*, 1979). White pine is refused by early instars but is palatable to later instars. Thus, in stands that contain both preferred and non-preferred hosts and that are subjected to heavy defoliation, the non-preferred hosts are sometimes defoliated by older larvae after the earlier instars have completely defoliated the preferred host species. Results of experiments evaluating the role of host plant composition in survival and numerical increase indicate that concentrations of preferred species can account for some differences in the degree of defoliation from stand to stand.

Chapter Summary

Numerical change in the total population of forest insects is referred to as population dynamics. Population dynamics is basically the study of the interaction between natality, mortality, immigration, and emigration that leads to changes in insect abundance. Population density is generally expressed as the number of individuals per unit area. Forest insect populations tend to be clumped in their distribution due to the mosaic nature of forests. Life tables (stage-specific mortality records) are the most common and useful sources of information on the population dynamics of forest insects. Dramatic increases in population (outbreaks) occur sporadically or periodically, depending on insect species. Why populations change (in some cases, 1-million-fold) is still an important unanswered question. But there are several schools of thought or hypotheses that may explain why populations fluctuate.

PART IV

Control of Forest and Shade Tree Insects

CHAPTER 16

Forest and Shade Tree Insect Pest Management

Introduction

Pest management is part of a crop production (or resource maintenance) system that aims, through the use of suppressive, preventive, and regulatory tactics, to reduce or maintain pest populations at tolerable economic levels. This comprehensive approach strives to use techniques and methods that are based on biologically and ecologically sound data, to exploit existing natural controls, and to organize informational inputs, predictive models, and decision-making outputs using systems science. Most pest management systems include several critical components including (a) pest population dynamics, (b) stand dynamics, (c) economic ramifications of injury, (d) ongoing monitoring of the abundance of natural controls and the abundance of pests in relation to both economic thresholds and economic injury levels (see Stark and Waters, 1985), and (e) evaluation of treatment options.

Integrated pest management (IPM) is a philosophical way of approaching pest problems that has its roots in systems thinking and concern for the impact of environmental manipulations. It attempts to bring together in a holistic way all the available information and experience about forest ecosystems to solve pest problems in an environmentally compatible way. Important aspects of IPM discussed in this chapter include the concept, its structure, and its implementation.

The Concept of Integrated Pest Management

The historical development of the concept of integrated pest management as applied to forestry has been discussed by Coulson (1981) and will not be reviewed here. The concepts of forest pest management are best discussed by Waters (1974), Waters and Stark (1980), and Stark and Waters

(1985). The concept of pest management as developed by these authors is based on two fundamental relationships: Forest insects are part of the forest ecosystem, and forest insects can have a negative impact on forest productivity and consequently can interfere with management objectives.

Because forest insects are a part of the total ecosystem and influence management objectives, the forest manipulations directed at pest regulation become an integral part of the forest management process. Waters (1974) defines IPM as follows:

> Integrated pest management is the maintenance of destructive agents, including insects, at tolerable levels by the planned use of a variety of preventative, suppressive or regulatory tactics that are ecologically and economically efficient and socially acceptable. It is implicit that the actions taken are fully integrated into the total resource management process—in both planning and operation.

Waters' definition includes two important aspects of integration: the integration of all available strategies to manage pest populations and the integration of these strategies into the overall forest management process. These two aspects of integration are worthy of additional discussion.

Integration of Pest Management Strategies

The blossoming of forest insect pest management can be characterized by the recognition that the complexities of the forest and shade tree ecosystems and the multiple objectives of the forest manager require varied and integrated procedures. The procedures involved in selecting a control strategy have many interrelated components (Table 16.1). The holistic approach called pest management includes a firm basis in the ecology and biology of interacting insects and trees, an understanding of economic theory and management by objectives, the appropriate use of a multitude of suppression tactics, and the use of the predictive and information-handling capacities of the computer. Ecological studies provide an evaluation of (a) the potential of any given species for numerical increase, (b) the key triggers or suppressors of population fluctuations, (c) the relative consistency of population trends, (d) the impact of insect populations on their hosts, and (e) the interdependency of the various dendrophagous species on a given host. Having determined the components of the forest insect community (the system) and how they interact to impede or perpetuate the community, particularly the target species, one can develop a model. The goal of the model is to reflect the real ecosystem and thus may require added research and subsequent changes in the model. Once a model is developed, it can then be used to predict the feasibility or outcome of the application of specific pest management tactics. The model is tested when strategies are applied in the field. Deficiencies in theory become evident, and additional investigations become necessary to deal with miscalculated interactions or to evaluate

Table 16.1. Factors contributing to efficacy of controls used against forest and shade tree insects[a]

I. Artificial Factors (factors controlled by humans)

Economic Considerations
Value of resource
Impact of specific damage on economic value of product
Cost of control
Cost of alternatives
Benefits from control

Socio-Environmental Considerations
Impact of control
Impact of alternative control projects
Impact of no control

Application Considerations
Coverage (ground or air):
Type of aircraft
Nozzles (number, type, position)
Airspeed
Application height
Droplet size
Spray pattern
Canopy penetration
Swath width
Weather, e.g., wind
Uniformity in distribution of deposit
Gallonage
Formulation:
Sprayability, uniformity
Diluent
Compatibility of additives
Spreaders (aids in foliar coverage)
Stickers (aids in adherence to foliage)
Anti-evaporants
Feeding inhibitors
Feeding stimulants
UV protectants
Weather
Dosage
Timing
Number of treatments
Interval

Objectives of Control Programs
Foliage protection:
Preserve aesthetic values
Prevent tree mortality
Prevent tree decline and growth loss
Prevent tree deformation
Prevent need for tree to refoliate
Alleviate entomophobia
Forest insect population reduction:
Remove nuisance
Remove environmental contaminants
Prevent recurrence of damaging populations
Population eradication
Pest management

Table 16.1. (*continued*)

Methods of Evaluation
    Standard data:
        Pre- and postapplication egg mass or larval counts, frass production
        Larval population reduction
        Degree of defoliation
    Control plots:
        Untreated
        Insecticide standard
    Monitoring of coverage, penetration, viability
    Ground observations during defoliation period

Strain Selection of Chemical Purity

II. Biotic Factors

Susceptibility to Pathogen Infection or Chemical
Laboratory bioassays
    Age or stage of development
    Dosage
    Pathology
    Feeding behavior
    Sublethal dosage recovery
    Fecundity
    Rate of development

Population Considerations
    Distribution over host trees
    Synchronization of life stages (e.g., hatching)
    Synchronization with tree host
    Size, density
    Quality, previous stresses
    Behavior

Compatibility with Natural Control Factors or Chemicals
    Parasites, predators, disease, insecticides

Host Tree Consideration
    Forest composition
        Canopy structure
        Percentage of susceptible hosts
        Antimicrobiosis-bacteriocidal foliage components

Residential Trees; Isolation, Value

III. Climatic and Physiographic Factors

Effects on Forest Insect Populations
Effects on Host Trees
Effects on Insecticide
Effects on Spray Operations

[a] Modified from Harper (1974).

unexpectedly significant causal agents. The cycling process of refinement and readjustment thus continues.

Integration of Pest Management into Forest Management

The terminology used to describe the manipulation of pest populations in forestry has had a major influence on foresters' understanding of pest management principles. Hawley and Stickel (1956) published a textbook entitled *Forest Protection*; and many forestry schools in the United States still teach courses in forest protection. The term *protection* probably has its historical roots in the early concerns for tremendous loss of forests due to fire. For many years the U.S. Forest Service had the policy of protecting all forest lands from fire. The use of this term led many foresters to believe that fire and insect outbreaks are unnatural events. Today's foresters, particularly in western and southeastern United States, have modified their view of fire, recognizing it as an important part of some forest ecosystems. Fire, prescribed or natural, can be an important management tool. Pest outbreaks are also a very natural part of some forest ecosystems, though many foresters still believe the solution to insect problems is to call a forest entomologist and ask that the problem be *solved*. Catastrophic insect outbreaks may play an important role in maintaining some forest communities (Peterman, 1978). The replacement of the old terminology with *pest management* has inclined foresters to accept the fact that insects and their management are an integral part of forest management.

This view, however, often makes it difficult to establish decision-making thresholds crucial to pest management. Control decisions in IPM may be based, in part, on knowledge of economic thresholds and economic injury levels. *Economic threshold* is defined as the density at which control measures should be applied to prevent an increasing pest population from reaching the *economic injury level*. The latter can be defined as the lowest pest population density that will cause economic injury. For forest ecosystems, the consideration of these concepts in IPM decisions is difficult for a variety of reasons, including the long maturation period of the crop, the economic restraints of forest production, and the simultaneous multiple objectives of forestry. Thus, the development of economic thresholds and economic injury levels is rare, except in special settings.

Pest management must be integrated into forest management because essentially all of the available methods for insect population manipulation require activities that have an impact on other forest uses. The Multiple-Use Sustained-Yield Act of 1964 requires that all public forest land be managed for multiple uses. Included in these uses are timber, wildlife, water, recreation, and range. Manipulation of forest insects will affect some or all of these multiple uses. Decisions about pest management must be made in the context of overall forest management objectives. Indeed, the blend of forest management uses that are proposed for a given area has a major influence on which pest management strategies can be applied.

Forest planning constraints also require that pest management be integrated into forest management; and pest management activities must be consistent with time, personnel, and monetary resources. Many pest control options are limited by these constraints. Most pest management approaches require manipulation of the standing forest crop, which may have a major impact on long-term planning of timber harvests. Major pest outbreaks may require timber salvage operations, which result in an oversupply of timber in the short term and an undersupply at some future date.

The Structure of Pest Management Systems

A basic component to IPM is its holistic or systems approach. Because forest insects are a part of the forest ecosystem, manipulations directed at pests will influence a multitude of other components in the system. The complexity of these interactions make them difficult to conceptualize. An important tool for enhancing our ability to understand complex interactions is the mathematical model. Models enable us to incorporate numerous well-understood interactions into a complex mathematical equation that integrates the various interactions. Models in no way increase our knowledge of systems, nor are they the only component to IPM; they merely enable us to understand the relationships among the components. Important components of a pest management model include (1) an insect population dynamics model, (2) a stand dynamics model, (3) a pest impact model, (4) a treatment strategy model, and (5) a management decision model.

Population dynamics models have been developed for many major forest pests, including the Douglas-fir tussock moth (Brookes *et al.*, 1978), the southern pine beetle (Hines *et al.*, 1980; Feldman *et al.*, 1981), and the western spruce budworm (Sheehan *et al.*, 1986). Population dynamics are discussed in some detail in Chapter 15. It is an area for which large amounts of data exist. The population dynamics models have been very helpful in understanding the relative contributions of various aspects of the overall population changes of forest insects. Population dynamics models of forest defoliators often include estimates of defoliation that can be used in stand dynamics models. The western spruce budworm model developed by Sheehan *et al.* (1986) consists of two components: BWMOD (Fig. 16.1) and BWFLY. BWMOD simulates the complete budworm life cycle and predicts defoliation. BWFLY predicts adult dispersal and oviposition and therefore the beginning population of the following year. Many years of research from various scientists are incorporated into the BWMOD and BWFLY simulation models.

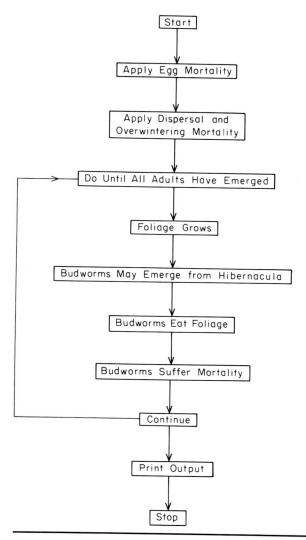

Figure 16.1.   Flowchart describing the sequence of events for the western spruce budworm population model (BWMOD). (From Sheehan *et al.*, 1986.)

Stand dynamics models are usually linked to population dynamics. This linkage involves a two-way interaction in which the insect pest influences the stand conditions and stand condition influences the population dynamics. The Prognosis Model (Stage, 1973) is probably the best known stand dynamics model. Prognosis predicts the future growth of forest stands under unmanaged conditions and for a variety of management regimes. Prognosis can receive data directly from the two budworm models (Fig. 16.2). This allows for the comparison of forest growth under a

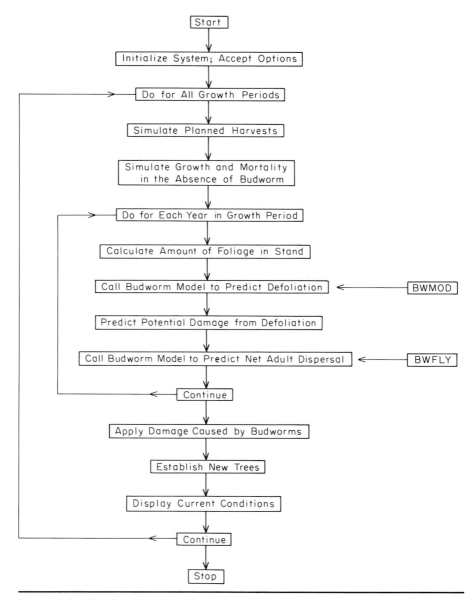

Figure 16.2. Simplified flowchart illustrating the simulation of a single stand using the Prognosis-Budworm model. (Adapted from Sheehan *et al.*, 1986.)

variety of budworm population scenarios. Stand Prognosis is also capable of receiving input from the mountain pine model and the Douglas-fir tussock moth model (Crookston and Stark, 1985). The linkage of the stand dynamics model and the population models allows prediction of pest impact.

Treatment models are relatively informal models that compare the effects of various treatments on stand dynamics and population dynamics models. These models usually involve a series of user-specified possible alternatives. Each alternative is independently evaluated. The effect of each alternative and the derived resource outputs are considered in the choice of a preferred alternative. The details of the control alternatives available are the subject of several sections of this text. Historically, considerable attention has been given to an evaluation of the various control tactics available to the forest resources manager.

The final structural component is the management decision model. Decision models are normally designed to compare a series of predetermined options. In some ways they are similar to the tactics model except that the alternatives are often optimized. Management decision models rely heavily on economic assumptions and benefit/cost analyses. Perhaps the best-known decision model is FORPLAN (Forest Planning Model) (Johnson *et al.*, 1980). FORPLAN incorporates information provided by stand dynamics models such as Prognosis and attempts to optimize the allocation of resources within specified economic constraints. The output from FORPLAN usually indicates the number of areas allocated to any specific alternative.

Implementing Integrated Pest Management

The concept and structure of integrated pest management may lead one to believe its implementation is a difficult task. The accomplishment of IPM does indeed require considerable skill in a variety of areas. To assist forest managers in implementing IPM, additional models have been developed. These so-called decision support systems have been developed for several important forest pests. The southern pine beetle decision support system (Fig. 16.3) outlines the basic steps required of a forest resource manager in making a pest management decision. The model also indicates personnel inputs and inputs from the various other structural components of the IPM system.

The use of integrated pest management principles in forestry is in the early stages of development. IPM procedures have been developed for only a very few of the important forest pests. Many of the models developed to date will require substantial revision and testing in the future. The links between different components of the IPM system will also require considerable modification and improvement. However, it is clear that pest management systems including mathematical models have been

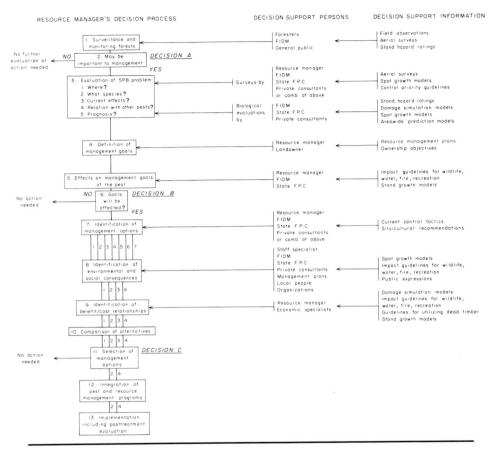

Figure 16.3.   Decision process and decision support system for southern pine beetle management. (From Coster, 1980.)

a tremendous tool for land managers and will remain an important component of decision making for years into the future.

Chapter Summary

Integrated pest management (IPM) is a philosophical approach to managing damaging forest insects within the context of understanding their role in forest ecosystems. Successful IPM depends on integrating the available technologies of pest suppression into the forest management decision process. Components of IPM include population-dynamics, stand-dynamics, and pest-impact models. The components of IPM are integrated into forest management through the use of decision support systems that have been developed for several major forest and shade tree pests.

CHAPTER 17

Survey, Detection, and Evaluation of Insect Pests

Introduction

The success of any strategy for the management of pest insects is based on the ability to assess their presence, location, density, and impact. The detection, surveillance, and evaluation of pest populations usually involve some sort of assessment of the insect, but also may involve evaluations of tree damage. This assessment can include an evaluation of the presence, type, location, abundance, and severity of damage to the host tree. Similarly, surveys may attempt to predict damage by assessments of host tree susceptibility. The latter type of survey is important because the greater the management inputs in a forest, the greater is the need for accurate information on pests. That is, as the intensity and costs of resource management increase, there is an increasing need for the accurate prediction of the development and course of infestations. This is particularly important in high-value stands in recreational areas, near population centers, and in plantations and nurseries.

A multitude of sampling methods are useful for conducting surveys and assessing pest activity. Often, available methods must be modified to reflect the peculiarities of the target insect's life history or the objective of the assessment. Therefore, sampling methods for individual species will not be discussed in detail. However, the numerous factors that must be considered in sampling forest and shade tree insects will be examined.

The design of a sampling program involves a series of decisions about how and when to sample. For example, among the more important factors in a sampling program is timing of the sampling in relation to the events in a species' life cycle. Other key data include the location of each stage and its distribution and accessibility. Here lies the ultimate incongruity of sampling; the more thorough the knowledge of the life history, the more efficient and accurate the sampling procedure must be. Yet, proper sampling procedures usually are prerequisites for the efficient investigation of an insect's life history.

Sampling

At its simplest, sampling involves collecting and recording data on a portion of a population that is representative of the entire population. Very little about sampling, beyond this brief description, is as simple. The diversity in application of the principles of sampling often parallels the diversity in life history and habitats of forest insects. If we add the multitude of objectives for the use of sampling, the complexity of the issue can become intimidating. The size of the host tree and the frequently vast contiguous area occupied by trees present obvious problems in forest insect sampling. Conversely, the relative permanence of tree ecosystems, compared with an agroecosystem, provides an excellent opportunity to develop sophisticated sampling methods or procedures.

The complexity of sampling methods is related, in part, to the objectives of the sampling. Population sampling can be extensive (e.g., for the prediction of damage) or intensive (e.g., to evaluate population dynamics within a small area). Extensive surveys employ rapid, simple methods to determine the presence or absence of a species, to assess relative population levels, or to map the extent of an outbreak. Intensive surveys generally provide data on interactions such as the effect of one or more factors on population dynamics.

Sampling of forest and shade tree insects is often difficult because their distribution may reflect the distribution of a key resource or may result from the response to a specific set of stimuli. In the case of the European spruce sawfly, the ovipositing female responds to needle size of its host foliage. Eggs are concentrated on 1-yr-old needles at the periphery of the upper crown of Sitka spruce (Billany *et al.*, 1978). Oviposition preferences have also been described for the spruce budworm (upper crown branch tips) (Morris, 1955; Wilson, 1959; in Billany *et al.*, 1978), the fall webworm (branch tips), *Neodiprion swainei* (near the tops of trees) (Lyons, 1964; in Kaya and Anderson, 1973), the Douglas fir tussock moth (bottom half of the crown) (Luck and Dahlsten, 1967; Mason, 1970; Kaya and Anderson, 1973), and the elm spanworm (primarily on the bole, 15–40 ft above the ground) (Kaya and Anderson, 1973).

In summary, sampling may be a simple tabulation of biological events; this includes detection of pest insects and their damage, species composition, dispersion, and parasitism. Sampling may also provide post-injury assessments of insect activity and may be used to assess specific and dynamic attributes of populations.

Distribution of Insects

The interrelationship between sampling and distribution is most clearly illustrated in the relationship between sample unit size (area or volume) and the distribution of the insect. Stated in its simplest form, often the larger the sample unit size, the more likely it is that the distribution of the insect will appear random. Conversely, the smaller the unit size, the more likely it is that a clumped distribution will be observed (Southwood, 1966, 1978).

The selection of a sampling unit is critical to the successful assessment of a population. The unit must be characterized by the following traits: (a) the sampling unit must be such that all units have a comparable chance of selection; (b) the unit must be stable, that is, unaffected by changes in the host tree or in insect activity; (c) the proportion of the insect population using the unit must remain constant; (d) the unit must lend itself to conversion to unit areas; and (e) the sample unit must be relatively small and easily delineated in the field to enhance the feasibility of sampling. Finally, although sampling precision is a desirable goal, the best approach also incorporates cost-benefit analyses, which strive for maximum precision of results per unit of cost (Morris, 1955).

Mathematical distributions of insect populations are mathematical models of the biological patterns of insect occurrence. Insects can be randomly, uniformly, or contagiously distributed. These various patterns of distribution result in binomial, negative binomial, logarithmic, Poisson, and normal distributions (to mention a few). The mathematical nature of these distributions is described in detail in Southwood (1978). Population distributions for many forest insects are of the negative binomial type or are contagious (i.e., clumped or aggregated). These distributions can arise from the nonrandom distribution of essential resources. Clumped distributions may be explained by the clustering of eggs when they are laid or by the tendency of larvae to feed gregariously. In these life histories the presence of one insect increases the probability of finding another (Morris, 1955). Apparent distributions of certain species actually may be artifacts of the sampling program. For example, in very sparse populations, the probability of individuals occurring in a sampling unit might be so low that the populations may appear to be randomly distributed. However, the distribution of populations of several forest insects do actually tend toward the Poisson or random distribution (Waters, 1959; in Southwood, 1966).

In general, variation in the distribution and abundance of forest insects between trees is greater than the variability within a tree, that is, between areas of any given tree. This generalization is true for the spruce budworm, large aspen tortrix (*Choristoneura conflictana*), lodgepole needle miner, larch sawfly, winter moth, European spruce sawfly, and others (Morris, 1955; Morris and Reeks, 1954).

Detection Surveys and Sequential Sampling

Insect surveys are the first line of defense against tree pest species. Although sometimes superficial, surveys nevertheless provide current information on (a) the status of known pests, (b) the occurrence of an unusual species, (c) where potential problems may arise, (d) the extent and severity of outbreaks, and (e) the effectiveness of applied control measures (Waters, 1955). A great many issues must be considered in designing an appropriate survey, which should include questions such as, What degree of precision is required?, How many samples should be taken?, and What are the cost variables (e.g., the person-hours per plot)? The answers to these questions depend to some degree on the objectives of the sampling.

A variety of approaches deal with key sampling questions and allow the development of effective surveys. One of the most successful and widely used approaches is sequential sampling, in which the total number of samples taken is variable. A sequential plan usually allows one to stop sampling as soon as adequate data have been gathered and often provides substantial savings of time, labor, and cost. Thus, this type of sampling is frequently used in forest insect surveys.

Substantial preliminary data must be generated before applying a sequential sampling plan. The type of distribution of the target insect must be established. The distribution is, in part, described by a statistic that is a function of the relationship between the mean and the variance among samples. Secondly, sufficient data must be collected and analyzed to set meaningful class limits or degrees of infestations. Generally, these are densities that are allowable or that are correlated with extensive damage (Waters, 1955) (Table 17.1).

Two parallel decision lines are used. These are calculated using the equation for a straight line with slope and intercept values generated from the general formula for a binomial distribution (or other appropriate distribution). The lines are plotted, and the resultant graph represents the sequential plan for decision making. As shown in Figure 17.1 for tussock moths, if the cumulative number of infested sample units drops below the lower line, the population is classified as low level. Conversely, if the cumulative number of infested sample units is above the upper line, the population is considered suboutbreak. Values falling between the lines represent intermediate levels of infestation and require that sampling of trees continue until a decision line is crossed. If the lines are not crossed after extensive sampling (an arbitrary limit), the population may be simply classified as intermediate. Finally, for the sake of convenience, a tally sheet can be prepared for use in the field (Fig. 17.2). Sequential sampling systems have been developed for various stages of many forest and shade tree insects including the larch sawfly (Ives and Prentice, 1958; Ives, 1954), spruce budworm (Cole, 1960), winter moth (Reeks, 1956), forest tent caterpillar (Shepard and Brown, 1971), lodgepole needle miner (Stark, 1952),

Table 17.1.   Sequential table for sampling spruce budworm larval populations (fourth through sixth instars) on 15-in. balsam fir twigs (90% confidence level)[a,b]

Number of Twigs Examined (1)	Cumulative Number of Budworms			
	Light (2)	Medium (3)	Medium (4)	Heavy (5)
1	—	—	—	29
2	3	—	—	39
3	8	—	—	50
4	12	22	23	60
5	16	26	33	71
6	20	31	44	81
7	25	35	54	92
8	29	39	65	102
9	33	43	75	112
10	38	48	85	123
11	42	52	96	133
12	46	56	106	144
13	50	60	117	154
14	55	65	127	164
15	59	69	138	175
16	63	73	148	185
17	67	77	158	196
18	72	82	169	206
19	76	86	179	217
20	80	90	190	227
21	84	94	200	237
22	89	99	210	248
23	93	103	221	258
24	97	107	231	269
25	101	111	242	279

[a] From Waters (1955).
[b] This table is used as follows: Sampling is continued only as long as the cumulative total count of budworms is between the listed values for light and medium (columns 2 and 3) or between those of medium and heavy (columns 4 and 5) or where a dash is shown. For example, if counts of two, three, five, four, and one budworms on successive twigs occurred, sampling would cease after the last (the fifth) twig because on that twig the count (15) fell below the limit value for light, and the sample would be so labeled. Similarly, to distinguish between a medium and heavy population (columns 4 and 5), sampling is continued until the cumulative count of budworms for any given number of twigs is equal to or less than the count listed medium (column 4) or equal to or higher than that listed under heavy (column 5). A medium population is determined when the cumulative budworm count for any number of twigs (that is, four or more) is equal to or higher than the count listed for medium (column 3) and equal to or lower than that listed for medium (column 4). As indicated, it is possible to identify a light population after examining just two twigs, a medium population after four twigs, whereas a count of 29 or more budworms on the first twig would immediately specify a heavy population.

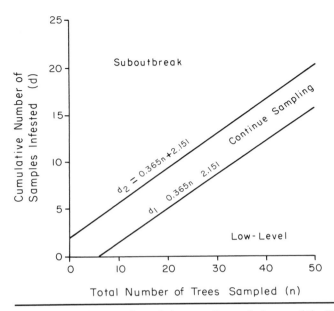

Figure 17.1. Sequential graph for sampling early larvae of the Douglas-fir tussock moth (*Orgyia pseudotsugata*) in the lower tree crown. (From Mason, 1978.)

Plots	Cumulative Number of Trees																				Class
	1	2	3	4	5	6	7	8	9	10	11	12	13	14	15	16	17	18	19	20	
	Low / Sub			−/4	−/4	0/5	0/5	0/6	1/6	1/6	1/7	2/7	2/7	2/8	3/8	3/8	4/9	4/9	4/10	5/10	
A	0	0	0	0	0	0															L
B	0	0	0	1	1	1	2	2	3	3	3	3	3	3	3						L
C	1	2	3	4																	S
D	0	1	1	2	3	3	3	4	4	5	6	6	7								S
E	0	0	1	1	1	1	2	2	3	3	3	3	4	4	5	5	5	5	5	6	I

Figure 17.2. Field tally sheet with examples for sequential sampling of early larvae of the Douglas-fir tussock moth (*Orgyia pseudotsugata*). Note that class limits for decision making are found across the second row. The top left corner provides the upper limit of the low-level population classification, and the lower right corner provides the lower limit of the sub-outbreak classification. The cumulative totals in sample units are recorded across the sheet. When the cumulative total is equal to or less than the top value (second row), sampling is terminated, and the plot classified as low-level (i.e., L). When the cumulative total is equal to or greater than the bottom value, the plot is classified as sub-outbreak (i.e., S). If a decision is not achieved after an arbitrary number of trees (e.g., 20), sampling is terminated, and the plot classified as intermediate (i.e., I). (From Mason, 1978.)

Swaine's jack pine sawfly (Tostowaryk and McLeod, 1972), saddled prominent (Grimble and Kasile, 1974), Dougls-fir tussock moth (Mason, 1969), pine leaf chermid (Dimond, 1974), and red pine sawfly (Cannola *et al.*, 1959).

Direct Methods of Assessment

The detection and survey of forest and shade tree insects, as well as sampling for intensive studies of tree insect populations, can be achieved using various techniques. Methods of making relative assessments of population density include direct counts (e.g., hand-picking, visual counts, trapping, and larval drop catches) and indirect methods (e.g., frass drop catches or tent counts) (Fig. 17.3). Although it is difficult to provide a comprehensive overview of all available methods, some common procedures and techniques are outlined below.

Estimates of Resting Stages: Egg and Pupal Counts

Although there are obviously a multitude of differences in the distribution, location, and accessibility of pupae and insect eggs, they have several features in common that are important in sampling. The other life stages of most tree insects are highly mobile over time and space and in moderate or light infestations may be relatively inaccessible. Attempts to sample the mobile stages of many insect species often cause disturbances that influence the validity of the assessment. For these and other reasons, the so-called resting stages (pupae or eggs) are sampled instead of other life cycle stages. Because these stages often overwinter, sampling these stages allows sufficient time to carefully evaluate existing conditions and design and implement a management strategy before injurious larval stage becomes active (De Gryse, 1934).

The choice of a sampling method or procedure for eggs or pupae, as in the sampling of other stages, is influenced by a multitude of factors. The methods of assessing insect density based on egg counts, which are discussed here, are merely illustrative and not necessarily optimal for all circumstances.

Often, eggs (or ovipositions) are sampled because of biological peculiarities of the target insect. The larch sawfly, an important pest of larch, lays its eggs in the shoots of current-year growth. The ability to count egg-scars simplifies the task of making population estimates. The direct estimation of eggs, in this and other situations, often is unfeasible; so estimation must be indirect. Because oviposition by the larch sawfly causes shoots to curl, counts of curled tips provide a useful population index of sufficient accuracy for survey purposes (Ives, 1955b). The eggs of the western hemlock looper (*Lambdina fiscellaria lugubrosa*) are deposited on all parts of a tree. However, in the years before heavy defoliation most eggs are located on the bole in preferred oviposition sites (i.e., bark crevices,

Figure 17.3.   Assessment of the activity of defoliators using frass collectors. (Courtesy USDA Forest Service.)

moss, and lichen). Winter estimates of the number of eggs per half square foot of bark at the mid-crown can be a useful predictive index of defoliation the following season (Thomson, 1985).

In the life cycle of the southern pine beetle and other similar species, the overlapping periods of attack, oviposition, and hatching make it difficult to pinpoint a time period when a direct count of total egg production can be made. In addition, counting eggs is difficult and time-consuming. Alternatively, egg densities of *D. frontalis* can be estimated from measurements of completed egg galleries (Foltz *et al.*, 1976).

Although egg or pupae sampling has a number of advantages, it is subject to certain problems or sources of error. Overestimations as well as underestimations can result from the inclusion of old egg masses or pupae retained from the previous season or from the loss of egg masses or pupae between the time they are produced and the sampling date due to predation or abiotic factors. The inaccessibility of these stages may also cause a low count in samples and thus, result in underestimations. Often, however, the texture and condition of egg masses or pupae or the changes in color due to exposure enable observers to distinguish each season's egg masses or pupae. Thus, although some characteristics associated with these life cycle stages make them appropriate to sample, other characteristics reduce sampling accuracy.

Larval and Adult Counts

The mobility of larval stages increases the difficulty of the accurate assessment of relative density. However, sections of the host tree can often be selected as sampling units for larval sampling because the requisites provided by the host tree may tend to diminish activity and/or restrict larvae to certain areas of the tree (Morris and Reeks, 1954). These requisites may include nutrition, shelter, or appropriate microenvironment. Thus, the area to be sampled may be substantially limited and relatively easy to sample if the insect is a specialist or monophagous, particularly if the host tree is not abundant.

Units commonly used in sampling forest insects include whole trees, twigs of specific lengths, buds, leaf clusters, single leaves, and unit areas of bark. The type of unit and the approach chosen are dependent on the habitat, sampling objectives, species being sampled, and availability of resources (economic and personnel). The tendency of certain species to respond to specific stimuli by becoming less active may make the choice of larval sampling appropriate. For example, the use of burlap bands (Fig. 17.4) to assess larval numbers is essentially the use of a surrogate shelter normally provided by the host tree or resting sites in the litter. Larvae or adults may be counted by direct visual observations or by observations after mechanical displacement of insects off their host tree (e.g., by shaking or beating tree or branches).

As previously mentioned, the choice of a method or approach is affected by a multitude of biotic and abiotic factors. The behavior of larvae may determine the method of sampling. For example, red pine scale (*Matsucoccus resinosae*) immatures (crawlers) are dispersed by wind and/or drop to the ground. The relative population density of this immature stage obviously is not achieved by direct counts, but may be better assessed by

Figure 17.4.   The use of bands for larval sampling.

the use of plastic Petri dishes that are coated with a diluted layer of sticky resin and placed near ground level (Grimble and Palm, 1976).

Seasonal changes may also determine both the location of the target insect stage and the type of sampling unit required. Approximately 94% of birch casebearer larvae (*Coleophora fuscedinella*) abandon the foliage in the fall and attach their cases to bark surfaces at the crotch of twigs and the base of buds. The within-crown distribution of overwintering birch casebearers also changes. Approximately 50% of the insects concentrate in the outer 50 cm of the middle half of the tree crown (Raske and Bryant, 1976). In the latter part of August, first instar spruce budworms spin hibernacula, molt, and then overwinter as second instars. The following spring, second instars emerge and begin to feed. Assessment of over-wintering populations can be made by evaluating larval hibernaculae. Losses due to wind-borne dispersal occur among both first and second instars. The relative effects of fall and spring dispersal on a population can be determined by assessments of first and second instar populations over time (Miller, 1958).

Occasionally, overwintering adults and larvae are disassociated from the host tree. Thus, by necessity, the appropriate sampling unit may be a unit area of forest floor. Most often this approach is used for pupal stages. For example, 1-ft^2 samples of forest floor litter and soil to a depth of 2–3 in. into the mineral soil have been used for the assessment of pupal popula-tions of the saddled prominent (Grimble and Newell, 1972b). Similar methods can be used for sampling adults. The risk of damage due to the striped ambrosia beetle can be determined by assessing the size of over-wintering populations. This is achieved by taking 10 pints of litter for each of 10 sampling sites at approximately 200-ft intervals in areas that are most suitable for beetle hibernation (Chapman, 1974) (Fig. 17.5).

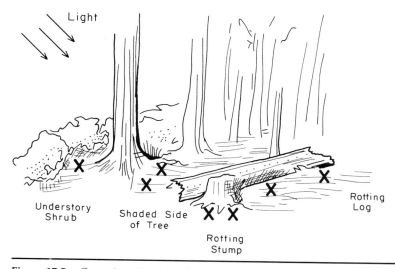

Light

Understory
Shrub

Shaded Side
of Tree

Rotting
Stump

Rotting
Log

Figure 17.5. Overwintering sites of ambrosia beetles. Litter is collected to sample the population at understory shrub, shaded side of tree, rotting stump, and rotting log. (Redrawn from Chapman, 1974.)

Traps

Although a great deal of research is currently being directed at developing traps [see Ruesink and Kogan (1975) and Southwood (1966) for detailed discussion of traps] that can be used for the control of forest insects, to date their most efficient use has been in the detection and monitoring of pest populations (Fig. 17.6). The Douglas-fir tussock moth is a defoliator in western North America that has reached outbreak levels approximately every 7–10 yr since 1900. The tussock moth is probably always present in certain forests, but like other forest insects it usually occurs in low numbers. As with other economically or socioeconomically important species, the inability to detect low level populations before they become epidemic has been a long-standing problem. An early-warning system would be the most efficient way of dealing with outbreaks of this and other species (Fig. 17.7)

For many tree insect species, progress is being made toward the development of such a trapping system that uses insects' natural sex attractants or synthetic versions of these compounds. In their simplest forms these pheromone traps could tell us whether the insect is present or absent (Fig. 17.8). If procedural methods are consistent, trap catches can also act as indicators of relative population trends. Ultimately, it may be possible to routinely determine population densities based on trap catches of individuals at low population levels. The use of pheromone-baited traps to indicate the presence or absence of insects at low densities, as well as for

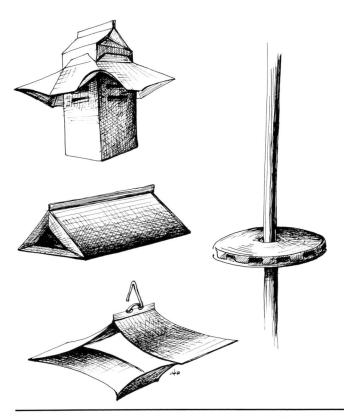

Figure 17.6.   Selected examples of traps used for the detection and
monitoring of forest and shade tree insects.

relative population trend assessments, has been demonstrated for the
spruce budworm, Douglas-fir tussock moth, gypsy moth, *Rhyacionia* tip
moths, pine sawflies, bark beetles, and others. Where feasible, virgin
individuals may be incorporated in monitoring traps (Fig. 17.8). In many
cases commercial synthetic pheromones such as sex attractants are available for trapping programs. These compounds and other similar attractants
are used for species such as the smaller European elm bark beetle, gypsy
moth, eastern spruce budworm, southern pine beetle, two-lined ambrosia
beetle, five-spined ips, oak leaf rollers (*Archips semiferanus*), European
pine shoot moth, and carpenter worms (Tagestad, 1975; McLean and
Borden, 1975).

A great deal of research is needed to increase the effectiveness of
traps. Most obviously, detailed data are required on the orientation behavior of the target insect to traps, the biochemical and visual components
of orientation, and the influence of environmental factors. Information on
trap design (size, shape, and color), placement, and density is also critical

Figure 17.7. Trap showing pheromone bait, suspended on pine (arrow), and captured European pine shoot moth (*Rhyacionia buoliana*) males. (From Daterman, 1974.)

Figure 17.8.   Sex-lure trap baited with one virgin southwestern pine tip moth (*Rhyacionia neomexicana*) female (inside cage) and the attracted males, which became stuck on board. (From Jennings, 1975.)

Table 17.2. Effect of trap color on response of male European pine shoot moth to traps uniformly baited with synthetic pheromone[a]

Item	Trap Color		
	White	Fluorescent Yellow	Dark Green
Mean no. males per trap[b,c]	39.13 (±9.2)	3.25 (±0.8)	17.38 (±3.7)

[a] From Daterman (1974).
[b] Each treatment replicated eight times.
[c] Standard error of mean in parenthesis.

(Table 17.2). For monitoring programs such as those designed for low-density populations certain obvious, simple but basic trap requirements must be met, including (a) the ability to capture the target insect at low densities, (b) the capacity to trap insects without having the trap become saturated (i.e., so fully covered that new arrivals escape), (c) the capability for trapping consistency, (d) the ability to exclude extraneous debris, (e) the ability to exclude living material, including nontarget organisms or predators (such as birds, which might remove trap catches), (f) the ability to withstand prevailing weather conditions, (g) the ease of handling, and (h) reasonable cost (Sanders, 1978). A wide variety of monitoring traps have been designed to meet the requirements necessary for efficient trappings (Fig. 17.6).

In addition to the use of traps in low-density populations to detect incipient outbreaks, traps may be used in quarantined areas to detect invasions by pest species. Traps may also be used to monitor specific life history events such as seasonal fluctuations, emergence, flight or dispersal periods, and oviposition periods.

Radiography

Although the use of radiography (X-rays) for detection and sampling requires specialized equipment, it does have unique features that make it particularly effective in sampling small and cryptic forest pests. Radiography has been used to study wood borers such as *Oberea schaumii* (poplar branch borer) and *Saperda inornata*, the white pine weevil, the western pine beetle, the southern pine beetle, and various seed-infesting insects (Amman and Rasmussen, 1974; Zerillo, 1975).

Radiographic procedures have been used for counting and quantifying development and mortality of various parasitoids, including *Blepharipa scutellata*, *Parasetigena silvestris*, and *Compsilura concinnata*, in their gypsy moth hosts (Fig. 17.9). Similarly, *Neodiprion sertifer* cocoons and gypsy moth pupae have been classified as healthy, parasitized, dead, or diseased by using radiographic techniques (Odell *et al.*, 1974; Ticehurst, 1976).

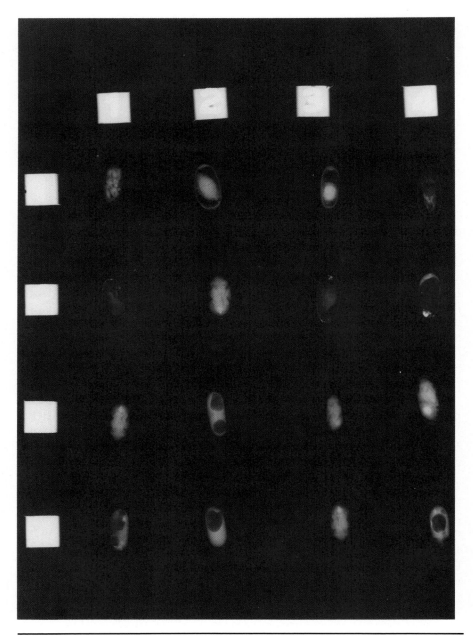

Figure 17.9.   Radiograph of *Blepharipa pratensis* puparia. *B. pratensis* is a tachinid parasitoid of the gypsy moth. Visible in the radiograph are hyperparasitoids, polyembryony, dried *B. pratensis* prepupae, and fully formed pupae. (From Odell *et al.*, 1974.)

Insecticidal Techniques

A method that has been applied to sampling is the use of insecticides to kill the insects in a tree, after which one counts them. This convenient method of whole tree sampling has been used to assess a variety of species including spring cankerworm larvae (*Paleacrita vernata*) in midwestern shelter belts and insects associated with pine plantations (Stein and Doran, 1975) (Fig. 17.10). The insecticidal sampling technique has been used to evaluate species composition and density, annual and seasonal fluctuations of a variety of species, and the succession of arthropod fauna in, for example, the crown strata of red pine stands (Martin, 1966).

Often materials such as pyrethrum are used because they provide a rapid knockdown that allows for minimal collection time, low mammalian toxicity, no phytotoxic effects, and no long-lived residues. Whole-tree sampling with insecticides can also be used to measure the efficiency of other sampling techniques. This sampling method is often selected because it is a nondestructive method that avoids any aesthetically unfavorable changes in shade trees, which often results when portions of a tree are removed during sampling (Stein and Doran, 1975). A similar procedure involves the use of systemic insecticides. The problems often associated with the use of systemics, including phytotoxicity and effective transport, must be considered before they are used (Satchell and Mountford, 1962).

Indirect Methods of Assessment

Insect Excrement

In an effort to devise practical methods of securing reliable quantitative data on insect populations, entomologists have occasionally turned to indirect assessments of the presence of insects and their distribution and abundance. The measurement of frass that drops from trees during feeding by larvae can be used to determine population trends once the relationship between frass quantity and number of larvae is established.

This method has been used to assess the effects of insecticidal control attempts (Green and DeFreitas, 1955; Doane and Schaefer, 1971) (Table 17.3), as well as in intensive studies of the influence of weather, foliage type, and so forth (Table 17.4) on factors such as feeding and development. Trends of daily frass-drop over a season provide a relative index of the time and extent of larval mortality but not the actual percent mortality. For those species that are univoltine and have a relatively short emergence period, an estimate of mortality may be generated when frass-drop is converted to larval numbers. If the relationship between frass-drop and foliage consumption is known, then the measurement of frass-drop can be used to determine foliage loss. This might be particularly useful at low to moderate levels of infestation when visual estimates are difficult.

Figure 17.10.   Mistblower application of a pyrethrin spray to sample Lepidoptera larvae. (From Stein and Doran, 1975; U.S. Forest Service photo.)

Table 17.3. Relative amounts of gypsy moth frass falling into drop nets in treated and untreated woodland[a]

Plot No.	Insecticide	Mean Weight in Grams of Frass per Net[b]	
		June 9	June 16
3	Gardona	0.04	0.26
	Control	9.19	67.20
2	Sevin	0.31	1.12
	Control	4.00	25.33
4	Sevin	0.00	0.03
	Control	3.90	21.50
1	Dylox	0.37	2.18
	Control	7.75	49.00
6	Dylox	0.06	0.51
	Control	6.25	53.90

[a] From Doane and Schaefer (1971).
[b] The June 9 and June 16 collections represent 1-day and 7-day accumulations, respectively.

Table 17.4. Comparison of the hourly frass-drop from colonies of 20 larvae of *Neodiprion lecontei* feeding upon either current- or previous-year foliage of red pine at two different temperatures[a]

	Instar IV (32.5°C)		Instar V (25.5°C)	
	New Foliage	Old Foliage	New Foliage	Old Foliage
n[b]	12	12	12	12
$\bar{x}$	122.5	91.6	88.3	75.7
$s\bar{x}$	1.47	1.04	1.25	1.39
t		11.58		6.75
p		<0.001		<0.001

[a] From Green and DeFreitas (1955).
[b] n = number of hours of collection, $\bar{x}$ = mean hourly frass-drop (frass pellet drop), $s\bar{x}$ = standard error of the mean, t is the value of t-statistics, and p is the level of statistical significance.

There are a number of advantages to this technique. Because larvae are not disturbed, the same trees can be sampled repeatedly. This approach can reduce the normal intertree variation that is commonly associated with other methods (Morris, 1949).

Several caveats must be considered. There may be, for example, an inherent difference in frass yield by larvae on different host species or different age foliage, which must be determined before the method is used. Water-soluble constituents of frass may be lost. As much as 25% of the weight of the frass of some species may be lost on exposure to rainfall. Finally, pellets may lodge on foliage and introduce another source of error (Morris, 1949). One of the problems with measuring frass-drop is that often a number of defoliator species are actively feeding. However, the frass of several defoliators and wood borers of forest trees in the United States and Canada can be easily differentiated (Hodson and Brooks, 1956; Solomon, 1977) (Fig. 17.11).

Shelters

Forest and shade tree insects often form shelters that are easily visible and can be readily surveyed. Although insects may not use these shelters throughout the season, empty shelters may withstand ambient conditions and can be used as a relative index of population trends. Casebearers, webformers, and tentmakers can all be surveyed by counting their shelters. For example, extensive roadside surveys of fall webworm tents can provide reliable estimates of seasonal population trends (Morris and Bennett, 1967).

Assessment of the Degree of Damage

There are occasions when it is essential to determine the degree of damage inflicted by dendrophagous species. Forest and shade tree entomologists are concerned with whether pest populations are (a) increasing or decreasing (based on the extent of damage), (b) spreading over wider areas, (c) selectively injuring only particular host species, or (d) injuring only selected host tissues. In other situations the prime concern is the degree of damage and/or the specific effects on tree growth or stand growth and composition.

Perhaps the most common indirect measure of a tree insect population is the extent (or percentage) of defoliation. It is a useful tool in surveys and extensive studies. The obvious advantages of low cost, high speed, and overall convenience are countered by equally obvious disadvantages of observer subjectivity in the estimates and difficulty in making valid comparisons between observers, areas, seasons, and so forth. In addition, certain trees, such as most coniferous species, may suffer considerable defoliation before the damage becomes noticeable (Morris, 1949).

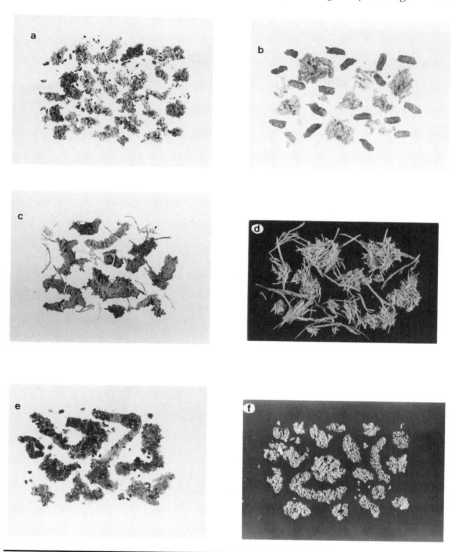

Figure 17.11.   Frass characteristics for identifying insect borers: (a) oak clearwing borer (*Paranthrene palmii*), (b) carpenter worm (*Prionoxystus robiniae*), (c) hickory borer (*Goes pulcher*), (d) poplar borer (*Saperda calcarata*), (e) peach tree borer (*Synanthedon exitiosa*), and (f) cottonwood clearwing borer (*Paranthrene dollii*). (From Solomon, 1977.)

Selective feeding by some species on old or young needles may account for some of the difficulty in accurate assessments of defoliation. Nevertheless, the advantages that are associated with this technique are such that percent defoliation estimates are widely used (Fig. 17.12). In addition to defoliation estimates, other forms of damage in the crown can

Current Budworm Defoliation (twigs)

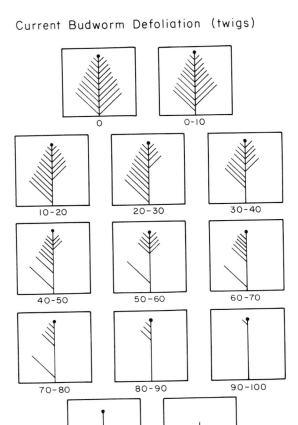

Figure 17.12. Schematic representation of spruce budworm defoliation of balsam fir current growth (twigs), expressed in percent. (From Ashley and Stark, 1976.)

provide a relative index of the degree of infestation. These may include the degree of dieback (branch mortality), top kill (mortality of upper crown branches), or flagging.

Changes in the crown and trunk of trees that can be visually tabulated are often used to assess the degree of infestation by bark beetles. Although the very early stages of infestation often are not easily observable, as the infestation develops the observed symptoms can be used to fairly accurately determine the stages of development of the infestations. Symptoms are manifested first in the crown, then in the bole, and finally in the inner bark and sapwood. When green trees are attacked by the mountain pine beetle in midsummer, the first signs of injury occur when foliage color changes in the spring of the following year. Environmental conditions may also accelerate host tree changes. An early attack followed by a hot, dry summer may result in foliage fading in the fall. Foliage usually changes from green to yellow, to yellow-brown, and finally to red-brown. At this point, needles drop off the tree, although red-brown needles may be retained for as long as 3 yr. For mountain pine beetle-infested trees, the yellow-green or yellow-brown phase may occur in June and July of the year after attack. Foliage color changes are convenient symptoms because they can be viewed from the ground or from planes. However, because trees killed by other agents may show similar color changes, ground checks are essential. Although the exact timing of color changes may vary among different bark beetle species, the overall sequence of events is what is important in many population surveys (Safranyik *et al.*, 1974) (Figs. 17.13 and 17.14).

Inspection of tree boles in suspected infestation areas can provide valuable information on the extent of beetle attack and the intensity of infestations on individual trees. Tree inspection should occur just after the beetle emergence and attack periods (e.g., from late August to September for the mountain pine beetle). The first signs of attack, boring dust or pitch tubes extruding from entrance holes, can be useful indices. Boring dust is often distinctive in color (e.g., red-brown) and similar in texture to fine sawdust. This dust is extruded in the early phase of gallery construction but can easily be lost through the action of wind and rain. In some cases pitch tubes are formed (Safranyik *et al.*, 1974).

The susceptibility or vulnerability of trees or stands to attack or damage by dendrophagous species is often evaluated using risk-rating or hazard-rating systems. Risk-rating systems predict the probability of attack or outbreak (susceptibility) and are most frequently used when considering immediate action. Hazard-rating systems predict the probability of damage or impact (vulnerability). Hazard rating is most frequently used when considering preventative and cultural treatmens over a longer time scale than risk-rating. Hazard- and risk-rating schemes are used for general surveys of forest stands as well as base for the implementation of preventative silvicultural tactics such as thinning. These schemes are arbitrary; but when applied to meet specific objectives, they can be very useful. The criteria used in risk-rating systems usually reflect tree susceptibility or vigor and include many of the symptoms described above (i.e., bark characteristics, crown length, width and density, crown conformation, age–diameter ratios, needle length and color, and tree dominance) (Keen and Salman, 1942). Hazard-rating systems normally involve stand conditions, stand location, and species composition. Risk-rating and hazard-rating systems have been developed for ponderosa pine stands attacked by western pine beetle (Smith *et al.*, 1981; Salman and Bongberg, 1942; Keen, 1936, 1943; Miller and Keen, 1960) and spruce–fir stands attacked by

Figure 17.13.   Crown symptoms of lodgepole pine attacked by the mountain pine beetle (*Dendroctonus ponderosae*). (a–c) Foliage color changes from green in May and June of the year of attack to a red-brown color, usually in July and August the year after attack. (From Safranyik *et al.*, 1974; Canadian Forestry Service.)

Figure 17.13. (*continued*)

Figure 17.13. (*continued*)

Figure 17.14. Crown characteristics of ponderosa pine trees symbolizing each of the four ratings of the ponderosa pine risk-rating system and the relative risk, or susceptibility, of trees in each rating to attacks by the western pine beetle (*Dendroctonus brevicomis*): (a) Risk 1, low risk; (b) risk 2, moderate risk; (c) risk 3, high risk; and (d) risk 4, very high risk. (From Johnson, 1972.)

Figure 17.14. (*continued*)

Figure 17.14. (*continued*)

Figure 17.14. (*continued*)

Table 17.5. Risk categories for potential spruce beetle outbreaks for
each stand characteristic[a]

Risk Category[b]	Physiographic Location	Average Diameter of Live Spruce above 10 in. dbh	Basal Area (ft²)	Proportion of Spruce in Canopy (%)
High (3)	Spruce on well-drained sites in creek bottoms	>16	>150	>65
Medium (2)	Spruce on sites with site index of 80 to 120	12–16	100–150	50–65
Low (1)	Spruce on sites with site index of 40 to 80	<12	<100	<50

[a] From Schmid and Frye (1976).
[b] Number in parentheses indicates arbitrary value to be used in calculating stand priority, and is used only for convenience.

spruce budworm (McCarthy and Olson, 1983; Westveld, 1945, 1954; Balch, 1946; McLintock, 1949; Witter and Lynch, 1985). Systems have also been developed for the southern pine beetle (Lorio, 1978; Hicks *et al.*, 1980), spruce beetle (Schmid and Frye, 1976) (Table 17.5), fir-engraver (Schenk *et al.*, 1977), gypsy moth (Valentine and Houston, 1981), and Saratoga spittlebug (Heyd and Wilson, 1981), to mention a few.

Assessments of injury to large forest stands can often be accomplished via airplane or satellite evaluations. A more sophisticated version of visual assessment of color changes in the foliage involves the use of infrared photography as a survey tool. Remote-sensing techniques have been used to evaluate current feeding, tree and forest condition as affected by previous feeding, forest recovery after spraying, and location of specific tree mortality. From the broadest perspective remote-sensing techniques can be used in the ultraviolet, visible, infrared, and microwave regions of the electromagnetic spectrum, although the form most commonly used in survey and detection is infrared and regular color photography.

Often the objective of a survey is not to determine the presence, abundance, and distribution of pest insects but to assess the potential quality of lumber, the extent of aesthetic damage, or the extent of defects in the tree (or wood). For example, the Nantucket pine tip moth inhabits conelets, buds, and new shoots of various pine species. Tip moth damage is manifested by browning, curling, and dieback of shoots. Surveys are most often made by counting the number of damaged tips in the top whorl. This approach provides a relative index since not all apparently damaged tips are infested by tip moths (Stephen and Wallis, 1978). Depending on the species of insect, length of infestation, susceptibility of the stand, and site conditions, such ground surveys can characterize the nature and extent of damage while uncovering bole distortions (e.g., forks and crooks), flagging, dieback, and top-kill (Collis and Van Sickle, 1978).

Surveys or assessments of damage (or defects) are not only undertaken in forest stands of living trees or among shade trees. Defects in wood can also be caused by insects active in living trees, freshly felled trees, green saw logs, unseasoned lumber, and seasoned rough or finished wood products. Injury may vary from slight defects to extensive damage due to insects or secondary invaders such as fungi. Use of timber for purposes that provide the greatest financial return depends on accurate grading or determination of the extent of pest-induced damage.

Chapter Summary

Many methods are used to survey and detect forest and shade tree insects, depending on the insects' life history and the objectives of the survey. All survey methods involve sampling, or estimating the population based on a representative sub-unit. In sequential sampling, one of the most useful sampling techniques, a variable number of samples is taken, depending on the level of the population. Methods of sampling include direct approaches (e.g., visual counts, traps, and hand picking) and indirect approaches (e.g., frass drop catches and tent counts), which are often developed for specific insect species. Predicting and assessing damage to trees by dendrophagous insects are important aspects of forest pest management. Assessment is often done by determining tree mortality or level of defoliation. Hazard and risk rating systems have been developed for many insect species. Hazard rating systems predict the probability of damage or impact and are usually based on stand level factors. Risk rating systems predict the probability of attack or outbreak and are often based on population level factors.

CHAPTER 18

Biological Control of Forest Insects

Introduction

Many biotic factors aid in the regulation of an insect population. Some of these are so-called natural enemies including pathogens, parasites, and predators. The type of insect control that uses and manipulates these natural enemies to reduce injurious pest populations and their damage is referred to as biological control.

The ecological diversity and temporal stability of the interactions between plants and insects in forests often afford enhanced opportunities for the successful application of biological control. The forest habitat also presents unique problems. In either case, biological control of forest pests exhibits characteristics that are distinct from those observed in agroecosystems. For example, dramatic manipulations that often affect the success of biological controls, such as soils preparation and harvesting, are much less prevalent in a forest ecosystem or even in a plantation, compared with a farm of comparable size. Although the lack of disturbances, the long periods between harvests, and the stability and diversity of the habitat may be conducive to the successful application of biological control, other factors may work against its success. For example, the rich indigenous complex of natural enemies that compete with introduced biological control agents may hinder their establishment and thus make biological control attempts difficult (Pschorn-Walcher, 1977).

In agroecosystems, variation in crop species, cultivar characteristics, and agricultural practices can lead to significant localized variation in natural enemy complexes. Thus, the importance of a pest species, either indigenous or introduced, may vary more widely from region to region than in comparable forested areas. The differences between forests and agroecosystems help determine the feasibility of certain approaches to biological control. For example, although inundative releases (release of large numbers of a species designed to achieve prompt mortality of a host) may be feasible for many agricultural situations, such releases in the frequently vast stretches of forests are often unfeasible.

Biological Control Strategies

The classical approach is perhaps one of the most widely used strategies in the biological control of undesirable forest insects. Many pest species are exotic; that is, they originated in another country. The classical approach is geared primarily toward the control of these exotic invaders. After the identification of the country of origin, an appropriate program is initiated involving foreign exploration for natural enemies, introduction of potential candidate species, quarantine, mass rearing (Fig. 18.1), and release of individuals.

Other strategies include the conservation, augmentation, and/or redistribution of established natural enemies. Conservation essentially involves the manipulation of the environment to favor existing natural enemies. Augmentation strategies seek to increase the effectiveness of natural enemies through several methods, which may include mass production, periodic colonization, and genetic improvement. This approach is often particularly useful when a natural enemy has been proven to be relatively effective; however, some component of its biology or life history, such as emergence that is asynchronous with that of its host, inadequate numerical increase, or inability to overwinter, may prevent the biological control agent from meeting its full potential. Redistribution involves purposeful movement of natural enemies from one area to another area in which they do not occur. This approach is useful, e.g., for natural enemies that are slow moving.

A fourth strategy suggests that undesirable species can be regulated by using natural enemies of closely related species or genera (Pimentel, 1963). Recently, the first example of a successful application of this approach against a forest pest was demonstrated. A hymenopteran parasitoid (*Telenomus alsophilae*) of the fall cankerworm (*Alsophila pometaria*) was used in a successful suppression program against a South American forest pest in the same family but of another genus (*Oxydia trychiata*) (Drooz *et al.*, 1977)

Regardless of the strategy being implemented, natural enemy populations may be established by inoculative or inundative releases. Inoculative releases are repeated colonizations of relatively small groups of natural enemies designed to build up a self-perpetuating population that will continue to control pest populations over time. Inundative releases are the introduction of large numbers of a species to achieve prompt mortality of the host without any expectation of long-term persistence or control. Biological control of forest insects has been typified by inoculative releases of introduced natural enemies. The release of parasites and predators against the gypsy moth is a good example.

Figure 18.1.   Mass-rearing of *Brachymeria intermedia,* a pupal parasitoid of the gypsy moth. (Courtesy of USDA; photograph by Murray Lemmon.)

There remain many controversial issues associated with the relative effectiveness and requirements of the various strategies. It is still unclear, for example, how much, if any, preliminary investigation is needed in the country of origin or in the area of introduction before a decision is made on which species should be introduced or released. Is it best to introduce only one parasitoid species at a time? Should several be introduced at once? What characteristics describe a good biological control agent? Are polyphagous parasites better than monophagous species? These are but a few of the questions that need to be considered in the application of any biological control strategy; yet they are questions for which we do not have consistent, widely accepted answers. Clearly biological control has been used successfully in the past (Table 18.1), but more research is needed.

Predation and Natural Regulation

Numerical and Functional Responses

Regulation of forest insect populations occurs, in part, as a result of the actions of vertebrate and invertebrate predators. The effectiveness of vertebrate or invertebrate predators (or parasitoids) may be partially determined by the change in their response to changes in prey density. An increase in the number of predators (or parasitoids) over time, as the prey population increases, is referred to as a *numerical response*. An increase in the number of attacks on hosts or prey by predators as the host or prey numbers increase is referred to as a *functional response* (Figs. 18.2 and 18.3).

Among birds in general, both active response and lack of response have been demonstrated. In outbreaks of spruce budworms and the larch sawfly, numerical and functional responses have been demonstrated in avian predators (Morris *et al.*, 1958; Buckner and Turnock, 1965). Increased clutch size occurs in certain wood warblers in response to increases in spruce budworm populations. The numerical response that has been observed in the bay-breasted and Cape May warblers is partly due to larger clutch size in abundant spruce budworm years. The response of the ovenbird to a spruce budworm outbreak can be a functional response that includes atypical arboreal foraging by this ground nester. The numerical response is caused by smaller, more tightly packed territories and greater frequency of second and third nests (Zach and Falls, 1975).

Numerical responses in mammals appear to be relatively rare. The length of the life cycle of most mammals is such that the temporary event represented by an insect outbreak is of too short a duration to have an impact on the abundance of subsequent generations of predators.

Often various habitat traits and behavioral characteristics of birds and other predators can have a significant influence on their numerical levels. Variable habitats may be more likely to produce variations in bird populations than stable ones. The territorial habits of certain species of birds and

Table 18.1. Selected examples of biological control programs initiated against tree pests

Region	Time Period	Pest Species	Major Biological Control Agents Released	Result[a]
Canada	1959–1968	*Adelges piceae* (balsam woolly aphid)	*Aphidecta obliterata* (Col: Coccinellidae)	–
			Aphidoletes thompsoni (Dipt: Cecidomyiidae)	
			Cremifania nigrocellulata (Dipt: Chamaemyiidae)	
			Leucopis (Neoleucopis) sp. (Dipt: Chamaemyiidae)	
			Laricobius erichsonii (Col: Derodontidae)	
			Pellus impexus (Col: Coccinellidae)	
Canada	Mid-1930 to mid-1940 & early 1950, and 1969–1980	*Coleophora laricella* (larch case bearer)	*Agathis pumila* (Hym: Braconidae)	++++
			Chrysocharis laricinellae (Hym: Eulophidae)	
			Diadegma laricinellum (Hym: Ichneumonidae)	
			Dicladocerus japonicus (Hym: Eulophidae)	
			Drino bohemica (Dipt: Tachinidae)	
Canada	Mid-1940s to mid-1950s	*Diprion hercyniae* (European spruce sawfly)	*Exenterus vellicatus* (Hym: Ichneumonidae)	++++
			Exenterus spp. (>5 species)	
			Borrelinavirus hercyniae (NPV)	
Canada	1959–1968	*Malacosoma disstria* (forest tent caterpillar)	Nuclear polyhedrosis virus	+
			Cytoplasmic polyhedrosis virus	
Canada	1959–1968	*Neodiprion lecontei* (redheaded pine sawfly)	Nuclear polyhedrosis virus	+++
Canada	1959–1968	*Neodiprion sertifer* (European pine sawfly)	*Monodontomerus dentipes* (Hym: Torymidae)	++
			Dahlbominus fuscipennis (Hym: Eulophidae)	
			Exenterus amictorius (Hym: Ichneumonidae)	
			Pleolophus basizonus (Hym: Ichneumonidae)	
Canada	1959–1968	*Neodiprion swainei* (Swaine jack pine sawfly)	*Lophyroplectrus luteator* (Hym: Ichneumonidae)	++
			Exenterus amictorius (Hym: Ichneumonidae)	
			Pleolophus basizonus (Hym: Ichneumonidae)	
			Dahlbominus fuscipennis (Hym: Eulophidae)	
			Drino bohemica (Dipt: Tachinidae)	
			Borrelinavirus (NPV)	

(continued)

Table 18.1. (*continued*)

Region	Time Period	Pest Species	Major Biological Control Agents Released	Result[a]
Canada	1959–1968	*Operophtera brumata* (winter moth)	*Agrypoon flaveolatum* (Hym: Ichneumonidae) *Cyzenis albicans* (Dipt: Tachinidae)	+ + + +
Canada	1959–1968	*Pristiphora erichsonii* (Larch sawfly)	*Mesoleius tenthredinis* (Hym: Ichneumonidae) (Bavarian strain) *Olesicampe benefactor* (Hym: Ichneumonidae) *Myxexoristops stolida* (Dipt: Tachinidae) *Hypamblys albopictus* (Hym: Ichneumonidae)	+ +
Canada	1959–68, 1969–80 1959–68, 1969–80 1959–1968 1969–1980 1969–1980	*Rhyacionia buoliana* (European pine shoot moth)	*Lypha dubia* (Dipt: Tachinidae) *Orgilus obscurator* (Hym: Braconidae) *Temelucha interruptor* (Hym: Ichneumonidae) *Parasierola nigrifemus* (Hym: Bethylidae) *Agathis binominata* (Hym: Braconidae)	+ (1959–68) + + + (1969–80)
Canada	1959–1968	*Stilpnotia salicis* (satin moth)	*Apanteles solitarius* (Hym: Braconidae) *Compsilura concinnata* (Dipt: Tachinidae) *Meteorus versicolor* (Hym: Braconidae)	+ + +
Caribbean and Bermuda	Up to 1982	*Oligonychus milleri* (pine mite)	*Stelhorus salutaris* (Col: Coccinellidae) *Amblyseius californicus* (Phytosiid mite) *Typhlodromus occidentalis* (Phytosiid mite) *T. citri*	–
		Hypsiphyla grandella (mahogany shoot borer)	*Apanteles* sp. *vitripennis* (Hym: Braconidae) *Antrocephalus vitripennis* (Hym: Braconidae) *Phanerotoma* sp. (Hym: Braconidae) *Tetrastichus spirabilis* (Hym: Eulophidae) *Trichogrammatoidea robusta* (Hym: Trichogrammatidae)	–
Australia	1902	*Eriococcus araucariae* (Araucaria mealy bug)	*Cryptolaemus montrouzieri* (Col: Coccinellidae)	+ + + +
Australia	1932–1938	*Chermes boerneri* (pine aphid)	*Leucopis atrifacies* (Dipt: Ochlhiphilidae) *Lipoleucopis praecox* (Dipt: Ochliphilidae) *Wesmaelius concinnus* (Neupt: Hemerobiidae) *Exochomus quadripustulatus* (Col: Coccinellidae)	+ + +

Region	Years	Pest	Natural enemy	Result
Australia	1931–1932	*Asterolecanium variolosum* (golden oak scale)	*Habrolepis dalmani* (Hym: Encyrtidae)	+++
Eastern United States	1932–1936	*Coleophora laricella* (larch case bearer)	*Agathus pumila* (Hym: Braconidae); *Chrysocharis laricinellae* (Hym: Eulophidae)	S
Eastern United States	1933–1960	*Adelges piceae* (balsam woolly aphid)	*Leucopis obscura* (Dipt: Chamaemyiidae); *Laricobis erichsonii* (Col: Derodontidae); *Scymnus impexus* (Col: Coccinellidae)	S
Eastern United States	1932–1940	*Diprion hercyniae* (European spruce sawfly)	*Dahlbominus fuscipennis* (Hym: Eulophidae); NPV	S
Eastern United States	1929–1930	*Cnidocampa flavescens* (oriental moth)	*Chaetexorista javana* (Dipt: Tachinidae)	+++
Western United States	1934–1939	*Pyrrhalta luteola* (elm leaf beetle)	*Erynniopsis rondanii* (Dipt: Tachinidae); *Tetrastichus brevistigma* (Hym: Eulophidae)	+++
Western United States	1957–1968	*Adelges piceae* (balsam woolly aphid)	*Leucopis obscura* (Dipt: Chamaemyiidae); *Cremifania nigrocellulata* (Dipt: Chamaemyiidae); *Laricobius erichsonii* (Col: Derodontidae); *Aphidecta obliterata* (Col: Coccinellidae); *Aphidoletes thompsoni* (Dipt: Cecidomyiidae)	F

[a] −, no control; +, slight pest reduction or too early for evaluation of control; ++, local control (distribution restricted or not fully investigated); +++, control widespread but local damage occurs; ++++, control complete; F, failed; S, successful control.

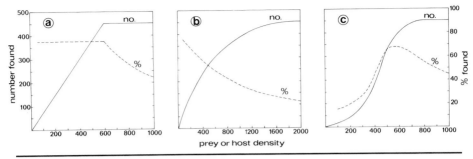

Figure 18.2.   Three types of functional responses: (a) The number of prey (hosts) found or attacked per predator (parasitoid) is directly proportional to prey (host) density, but at a certain prey (host) density the predator is satiated (or the parasitoid has depleted her eggs); (b) the number of prey (hosts) found also increases with prey (host) density but at a continually diminishing rate (due to increasing handling time); and (c) the increase in prey (host) found or attacked at low density is larger than the increase in prey (host) density but falls off after reaching a maximum (due to increased handling time). (From van Lenteren and Bakker, 1978.)

small mammals may strongly influence any potential numerical response. Large population movements of predators in response to localized changes in prey population have been referred to as a behavioral numerical response. For example, small mammalian cocoon predators and birds that normally do not inhabit bogs sometimes are found in this habitat during times of larch sawfly abundance (Buckner and Turnock, 1965). Woodpeckers often move into and remain in areas with an abundance of bark beetles, wood borers, or carpenter ants. Thus, real or apparent numerical responses may result from changes in reproductive success and/or behavior (Fig. 18.4). Regardless of the type of numerical response, the result is the same: increased predatory activity. Those concerned with tree insect pest management can capitalize on the existence of both types of numerical response. For example, when provided with nest boxes, broods of the mountain chickadee in areas infested with lodgepole pine needle miners outnumbered those in noninfested areas by 13:1 (Coppel and Sloan, 1971).

The capacity for learning greatly affects the functional response of vertebrate predators (Fig. 18.5). Many aspects of predator behavior determine the nature of the functional response over time. These include the importance of searching images in feeding behavior, feeding patterns and handling time of prey by the predator, and the tendency to vary diet (Buckner, 1967). The complete response of a predator to changing prey density is the summation of numerical and functional responses.

The Type 3 curve (Fig. 18.2c), which often is presumed to describe the response of vertebrates, can also represent the response of invertebrates (such as species in the Hymenoptera) (Holling, 1965). This speculation has

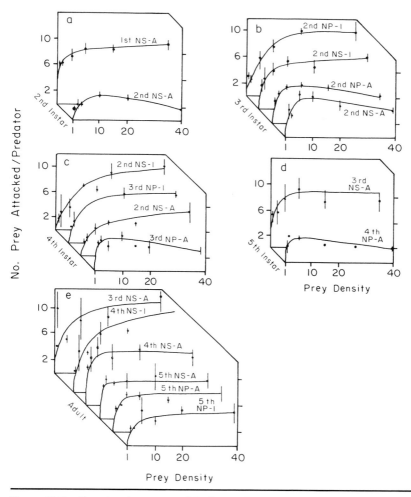

Figure 18.3. Functional response of *Podisus modestus* (nymphal and adult stages) to the density of prey, *Neodiprion swainei* (NS) and *N. pratti banksianae* (NP). Instars one through five of the prey were provided in two forms: A, active (i.e., free to defend themselves); I, inactive (i.e., unable to defend themselves). Vertical lines represent ±1 SE of the mean. (From Tostowaryk, 1972.)

been confirmed by various researchers who have suggested that experimental laboratory artifacts and/or lack of detailed behavioral observations have obscured the fact that many invertebrates can exhibit a Type 3 response (van Lenteren and Bakker, 1976, 1978). This possibility is important because a predator (or parasitoid) exhibiting a Type 3 response may have a regulating influence on its prey/host over the full range of the host population's growth.

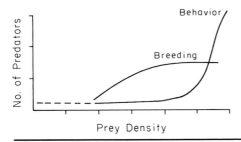

Figure 18.4. Breeding and behavioral numerical responses. (From Buckner, 1967.)

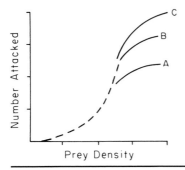

Figure 18.5. Functional response curve of the masked shrew attacking larch sawfly cocoons: Curve A, feeding response; curve B, feeding plus indiscriminate attack response; and curve C, feeding plus indiscriminate attack plus hoarding response. (From Buckner, 1967.)

In general, a large number of variables affect the outcome of all predator–prey interactions, including (a) the density of prey and/or predator populations, (b) behavioral and physiological characteristics of prey and/ or predator, and (c) density and quality of alternate food available to the predator. Although both vertebrate and invertebrate predators exhibit characteristics that define them as predators, there are advantages in considering them separately. Each of them acts in a fashion that reflects their size, life history, and physiological differences.

Vertebrate Predators: Birds and Small Animals

Although references to vertebrate predators of forest insects are relatively rare, reports have been published of predation even by amphibians (e.g., frogs) and small forest reptiles. However, the energy demands of their high metabolic rates, as well as other traits, make bird and mammal predators more important than frogs or snakes (Buckner, 1966). Birds have been recorded as predators of all stages of insects for many decades. Forbush and Fernald (1896) in their comprehensive treatise on the gypsy moth listed 38 species of birds known to feed on the gypsy moth. Flocking birds were observed to exert considerable pressure on gypsy moth populations. Avian predators, including nuthatches (*Sitta carolinensis carolinensis*), downy woodpeckers (*Dryobates pubescens medianus*), cuckoos (*Coccyzus americanus americanus* and *C. erythrophthalmus*), scarlet tanagers (*Piranga erythromelas*), black-capped chickadees (*Parus atricapillus*), and kingbirds (*Tirannus tirannus*), were observed feeding on gypsy moths. However, these predators apparently did not impose a significant degree of regulation on the pest population because they were only occasional visitors and were concentrated in heavily infested sites, thus reducing their impact. Repeated visits by flocks of birds, such as the grackle (*Quiscalus* sp.), red-winged blackbird (*Agelains phoeniceus phoeniceus*), and starling (*Sturnus vulgaris vulgaris*), appeared to have a greater effect. Similarly, over 45 species of birds may feed on the eggs, larvae, pupae, and adults of spruce budworm populations. Approximately 84–89% of the stomach contents of just one of those species, the cedar waxwing, consisted of spruce budworm larvae and pupae (Mitchell, 1952).

Repeated reference has been made in the scientific and lay literature to the beneficial role of birds in dampening fluctuations of forest insect numbers. Experiments in which birds have been excluded from an area have demonstrated their high potential for destroying forest insects. For example, Takekawa and Garton (1984) found that when birds were excluded from host trees, western spruce budworm populations increased by as much as 72%.

Evaluations of the effectiveness of predaceous birds have ranged from a low of 3.5–7% (Kendeigh, 1947; George and Mitchell, 1948) to a high of 48–71% (Dowden *et al.*, 1953) in spruce budworm outbreaks. Similarly, others (Sloan and Coppel, 1968; Coppel and Slogan, 1970, 1971) have reported avian predation on all but 5% of overwintering pine sawflies above the snowline and all but 23.5% of larch casebearers (*Coleophora laricella*) (McFarlane, 1976). Tits may consume 50% of *Ernamonia conicolani* attacking pine woods (Tinbergen, 1960), and 0.5–2.6% of larvae and 10% of adults of the winter moth attacking oak trees (Betts, 1955).

In northern and eastern Europe, particularly in Russia and Germany, birds are often viewed as effective predators and are artifically colonized in some forests. Five- to ten-fold increases in the populations of insectivorous birds have been accomplished by the use of nest boxes and similar techniques. Although studies to compare their effectiveness relative to that of other agents are rare, there is some evidence that in plots with augmented bird populations outbreaks do not occur or are of moderate intensity (Franz, 1961).

Woodpeckers have been credited as being efficient predators of wood borers such as cerambycids, buprestids, wood wasps, and weevils (Marshall, 1967) as well as being major influences in the reduction of important bark beetle species such as the Engelmann spruce beetle and the western pine beetle (Shook and Baldwin, 1970). Woodpeckers have been responsible for 45–98% reductions in Engelmann spruce beetle populations (Knight, 1958). The downy, hairy, and white-headed woodpeckers (*Dendrocopus pubescens*, *D. villosus*, and *D. albolarvatus*, respectively) are the most important avian predators of *Dendroctonus brevicomis*. There can be a 50% or more reduction in *D. brevicomis* brood in woodpeckered areas. After woodpecker predation and consumption of approximately 32% of two host generations, there can be an increase in the number of parasitoids per square foot and a three- to ten-fold increase in percent parasitism. This may be caused by the greater access of hosts to parasitoids with short ovipositors, resulting from bark flake removal by woodpeckers and the occurrence of near uniform bark thickness. Conversely, predator abundance may be reduced due to woodpecker activity and/or actual consumption of predators by the woodpeckers (Otvos, 1965). Finally, in hardwood forests woodpeckers may remove as much as 39% of the white oak borers (*Goes tigrinus*), 39% of the living beech borer (*G. pulverulentus*), and 65% of the poplar borer (*Saperda calcarata*) (Solomon, 1969).

Although predators can strongly influence prey numbers, it is still a matter of controversy whether their activities can be considered regulatory (Buckner, 1966). Since most birds do not appear to show significant numerical and/or functional responses, their impact as predators is likely to be most important only at low prey densities. The impact of predation by small mammals on insects such as the European pine sawfly, larch sawfly, European spruce sawfly, and others have been assessed in various studies (Tables 18.2 and 18.3). Many other studies have not been able to demonstrate an effective regulatory role of avian or mammalian predators. Therefore, their impact as predators remains controversial (Buckner, 1966).

The relative effectiveness of birds and small mammals is often a function of variables such as prey behavior or habitat characteristics. The proportion of a bird's diet that consists of a given insect may be dependent on prey density, size, conspicuousness, and palatability (Tinbergen, 1960). In addition, the acceptance of prey into a bird's diet may be delayed until a specific search image of the prey is established, based on the frequency of encounters. In areas where insects such as the spruce budworm reach epidemic levels, not only do insectivorous birds freely use this abundant food source, but birds that normally feed on other material will often feed chiefly on budworms (George and Mitchell, 1948).

Predation on the gypsy moth is a good illustration of the role various types of vertebrate predators can play at various densities and the influence of habitat characteristics and prey behavior. Small mammals such as deer mice (*Peromyscus leucopus*), short-tailed shrews (*Blarina brevicauda*),

Table 18.2 Weight, food, and activity of certain small mammals in northeastern United States[a]

Species	Avg. Weight (g)	Daily Ration (g)	Insect Food Consumed (%)	Time of Activity	Season of Activity
Deer mouse	20	6	20+	Nocturnal	Throughout year
Jumping mouse	22	6	20+	Chiefly nocturnal	Mid-Apr–Oct
Red-back mouse	21	6	10+	Diurnal-nocturnal	Throughout year
Lemming mouse	25	12	5	Diurnal-nocturnal	Throughout year
Chipmunk	87	15	10	Diurnal	Apr–Nov
Flying squirrel	60	12	20	Nocturnal	Throughout year
Short-tailed shrew	18	9	50	Diurnal-nocturnal	Throughout year
Hairy-tail mole	54	36	75	Diurnal-nocturnal	Throughout year

[a] From Hamilton and Cook (1940).

Table 18.3. Number of gypsy moth egg masses in untreated sites and in sites affected by selective manipulation of predators[a]

| Forests | Egg Masses Found for Various Treatments[b] | | | |
	Mammals Removed	Birds Excluded	Mammals Removed and Birds Excluded	Control
Site A	124	7	357	14
Site B	215	10	186	35
Site C	44	0	129	21
Site D	8	0	23	3
Site E	1	4	7	0
Mean[c]	96.5 ($\pm$37.7)	8.2 ($\pm$4.3)	128.8 ($\pm$53.1)	14.6 ($\pm$6.3)

[a] Modified from Campbell and Sloan (1977a).
[b] Number of egg masses found Fall 1971 in treated one-acre plots where 100 egg masses had been placed in Spring 1971.
[c] Standard error in parenthesis.

and gray squirrels (*Sciurus carolinensis*) appear to be important predators in sparse larval populations, whereas birds are important predators on larger larvae in intermediate and dense populations. These differences may be due to larval behavior at epidemic levels. At these levels, larvae are found more often on trees than in the litter (Campbell, 1961).

Similarly, because soil serves as a hibernaculum for European spruce sawfly cocoons, small mammal predation is a key factor in the population dynamics of this insect (Buckner, 1966). On the other hand, in jack pine forests of Michigan, bird predation on the jack pine budworm (*Choristoneura pinus*) accounts for 40–65% of the mortality of late instars and pupae. In spite of the fact that there are more small mammals per unit area of forest than there are birds, the importance of these mammals is probably minimal because few climb trees to find food (Mattson *et al.*, 1968).

One of the reasons that predators are not effective regulators of dendrophagous insects may be their relative lack of host specificity. The abundance of alternate prey can cause a reduction in predation on the pest insect. On the other hand, the availability of alternate hosts may allow the predators to survive when the pest population is low.

The cumulative effect of vertebrate predators and the sequence of predation may be ecologically meaningful even though the activities of a single predator species are relatively trivial (Table 18.3). In general, mammals prey primarily on ground-inhabiting stages of forest insects, that is, the pupal and cocoon stages. Birds prey chiefly on late larval stages and adults. Birds and mammals do not usually prey on eggs or early larval stages. The sequence of predation has important implications to the population dynamics of insects because the elimination of later stages would likely have a more pronounced effect on subsequent population density than the mortality of earlier stages. Thus, one would have bird predation on late instars, followed by mammalian predation on pupae or cocoons, and by bird and occasionally mammalian predation of adult insects. In any given forest habitat, 50 avian and 12 small mammalian species may be acting as predators of one insect species. The resultant pressure on the population may be substantial. However, in addition to habitat complexity, the number of predator species, their complex behavior, and the potential independence of their response to prey density make it difficult to assess the nature or extent of insect regulation by predator complexes.

Use of Vertebrate Predators as Biological Control Agents

Two basic approaches to biological control by vertebrates predators have been applied: Exotic species have been imported and used, and native species have been augmented to increase their potential as biological control agents. Encouragement of native species is often the most practical approach, particularly in habitats that can be manipulated, such as plantations or nurseries. Greater success is probable with introduced species in large areas or in habitats where manipulation is less feasible.

The provision of nest boxes for birds or small mammals, such as deer mice, may increase populations of these insectivores (Buckner, 1966). In certain experiments, bird populations have been increased by using nest boxes, thereby resulting in a 100-fold reduction in the number of pine looper larvae per tree (Bruns, 1960). However, the benefits accrued from the activities of some vertebrates may be countered by other behaviors. For example, although woodpeckers can be voracious feeders on dendrophagous insects, they may also produce holes which reduce the value of the tree. Similarly, although small mammals may consume a variety of insects, they may also feed on seedlings, saplings, and seeds.

Invertebrate Predators

Insect Predators

In many predator species, both adult and immature stages are predaceous and must actively search out their prey. In other species, the larval stage is predaceous, but the adult needs only a high-protein food source such as pollen. Commonly, predators and their hosts show little close physiological and bionomic synchronization. Nevertheless, some of the more spectacular successes of biological control have involved the use of invertebrate predators (e.g., coccinellid beetles, syrphid flies, and mites) against insect pests and their relatives with piercing-sucking mouthparts, such as aphids, scales and mites.

Among the many important dendrophagous groups, aphids, scales, and mites are often considered to be of secondary importance in rural forests. However, with only a few exceptions, their impact can be substantial in urban forests and on individual shade trees. In tree plantations, seed orchards, christmas tree plantations, and urban settings the effect of these insects can be pronounced on growth and the aesthetic value of trees. Presently, little biological control is being applied in these situations, although there is great potential for this approach.

The balsam woolly aphid is not only a homopteran of great importance in forests, but the attempts at its biological control provide an interesting illustration of the use of invertebrate predators. The introduction of predators of the balsam woolly aphid from Europe into eastern Canada began in 1933. Liberations of predators were made from 1933 to 1937, in 1941, and from 1951 to 1955. Of the species released between 1933 and 1941, one dipteran predator species (*Leucopis obscura*) became established (although recent unpublished information indicates that initial releases were made of several *Leucopis* species). Of all liberated species approximately 12 have become established; they include neuropteran, coleopteran, and dipteran predators. In addition, a number of native predators were recorded feeding on the balsam woolly aphid.

The introduction of *L. obscura* into Canada was somewhat successful. It was infrequently attacked by any other species (e.g., hyperparasites) and became numerous at high host density. This dipteran also dispersed over the range of its host. The density of the predator increased at a slower rate than that of the host. It appeared that *L. obscura* could play a significant role in limiting host populations if the predator complex included a species that fed on host eggs or on early nymphal stages (whereas *L. obscura* fed primarily on the intermediate stages and adults).

A small beetle, *Laricobius erichsonii*, which is a common predator in Europe on *A. piceae*, is one of several species introduced into Canada between 1951 and 1955. Larvae feed on the developing stages and on adults of the early-season prey generation, but prefer the eggs of the late-season generation. A comparison of prey population trends in the absence and presence of *L. erichsonii* (including the activities of *L. obscura*) indicates that the added pressure of *L. erichsonii* predation is sufficient to alter the normal upward trend between generations (Fig. 18.6). *L. erichsonii* increases the overall effectiveness of predation, because its peak larval feeding occurs earlier than that of *L. obscura*, and it feeds on host eggs (Brown and Clark, 1956a,b, 1957; Clark and Brown, 1958). In summary, the balsam woolly aphid biocontrol program illustrates not only the success that can be achieved but also the subtleties that must be considered to insure success with a well-executed program.

Among the most numerous, widely distributed, and efficient predators in the forest are the ants (Table 18.4). Their social or colonial life-style has played an important part in their ability to invade and exploit a large number of habitats and to reach relatively high numbers. The study and use of ants as biological control agents of forest insects has been

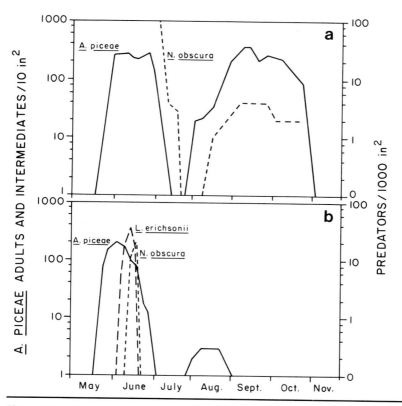

Figure 18.6. Seasonal development of *Adelges piceae* on trees with and without the predator *Laricobius erichsonii*: (a) Tree 12–7, without *L. erichsonii*; (b) tree 8–24, with *L. erichsonii*. (From Clark and Brown, 1958.)

concentrated in Europe, particularly in Germany. The most frequently used and most efficient species are those in the *Formica rufa* group, commonly referred to as the red wood ants. In northern European countries (e.g., Russia) the density of species such as *Formica rufa* can reach 180 colonies/ha, or approximately 4237 ants/m². Estimates of the effectiveness of ant predators suggest that ants from the one million ant nests of the *Formica rufa* group in the Italian Alps consume approximately 14,000 tons of live insects in a 200-day growing season (Fig. 18.7). Three to four of the eight species in the *F. rufa* group in Europe are considered beneficial. *Formica polyctena* is generally recognized as having great potential as a biological control agent of forest insects. Although there is a comparable number of ant species in North America, little is known of their potential as predators or as biological control agents.

A number of bionomic and behavioral characteristics of ants suggests greater efficiency compared with other predators, both vertebrate and

Table 18.4.   Species of forest pests against which ants have been reported to be more or less useful[a]

Host Plant	Herbivore
Spruce	*Cephalcia abietis*
	Pristiphora abietina
	Epinotia tedella
	Lymantria monacha
Pine	*Diprion pini*
	Neodiprion sertifer
	Dendrolimus pini
	Thaumetopoea
	pityocampa
	Bupalus piniarius
	Palolis flammea
	Lymantria monacha
Larch	*Pachynematus scutellatus*
	Coleophora laricella
	Eucosma griseana
Broadleaf trees	*Tortrix viridana*
	Operophtera brumata
	Dasychira pudibunda

[a] Adapted from Adlung (1966).

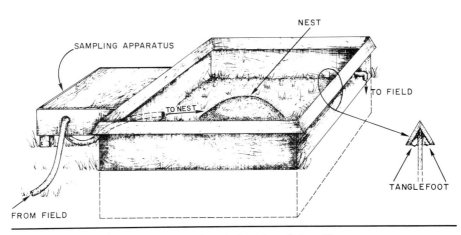

Figure 18.7.   Sampling apparatus used to study ant predation. (From Finnegan, 1969.)

invertebrate. They can attain a very high population density, and even during periods of live-prey scarcity, density can be maintained by the use of alternate food such as aphid honeydew. In addition, there is very little numerical lag behind an increasing dendrophagous prey population. The predaceous activities of ants occur in all forest strata from the floor to the tree crown, unlike other predaceous animals. Activity periods of ants are also quite extensive. Although nocturnal activity is slightly reduced, ants are active for the entire day (Fig. 18.8). The seasonal period of ant activity is approximately 200 days/yr in central Europe and only slightly less in North America. The area protected by an average nest varies, but it is estimated that the beneficial value of ants extends 25–35 m from the nest in all directions (Adlung, 1966).

In many of the more important species the presence of multiple queens, as many as 2000 in a colony with 300,000 workers, assures an extended nest life. Other useful behavioral traits include their lack of prey-stage specificity. Ants attack eggs, larvae, pupae, or adults. Their colonial existence and the need for individuals to forage for the queens, brood, and other workers causes ants to maintain a high level of predaceous activity. *Formica lugubris*, *F. polyctena*, and *F. aquilonia*, which occur widely across across Italy and other areas in Europe, typically display the above characteristics and are considered outstanding predators. *F. lugubris*,

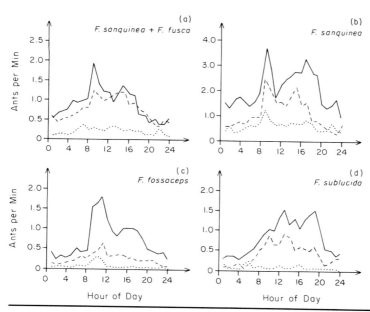

Figure 18.8.   Average daily foraging activity (May to October for 3 yr) of four ant nests kept in the laboratory: (a) *Formica sanguinea* and *F. fusca*, (b) *F. sanguinea*, (c) *F. fossaceps*, and (d) *F. sublucida*. Curves are for 70°F (——), 60°F (---), and 50°F (. . .). (From Finnegan, 1973.)

a species that occurs in areas of Europe where climatic conditions are similar to that of the boreal forest region of eastern Canada, was selected for importation and release in a 35-yr-old predominantly mixed conifer plantation of Canada (Finnegan, 1975). Establishment and acclimatization of this species in Quebec has been successful (Finnegan, 1971) (Fig. 18.9).

Like most tools used to control insects affecting the resources we use, not all aspects of the use of ants are beneficial. Their prey may include not only pests but also other species whose activities are not construed as harmful. Some ants actually protect Homoptera species that provide them with food in the form of honeydew. These protected homopterans may be directly responsible for damage to trees or may promote invasion by tree pathogens. Trophobiotic relations with various *Cinara* aphids and other aphids, such as *Pterochlorus exsiccator* on beech and woolly aphids on pines and spruce, may result in heavy infestations (Adlung, 1966).

Figure 18.9.   Importation and use of *Formica lugubris* for the biological control of the larch sawfly (*Pristiphora erichsonii*): (a) a 2-yr-old nest (measuring 120 cm) introduced from northern Italy to Quebec, Canada; (b) arrival of a colony in Quebec from northern Italy; and (c) adult *F. lugubris* attacking a larch sawfly larva. (From Finnegan, 1975.)

One might also expect difficulties with the use of ants in recreational or other areas of public use. The presence of predaceous ants would have a negative impact on public recreation; this would be an important consideration in their use as biological control agents (Finnegan, 1974). The qualities deemed desirable for an effective predator (i.e., aggressiveness, abundance, and nest size) are just the traits that are likely to escalate negative interactions in high-use recreational areas.

The unresolved problems of the taxonomy of many ant genera add to the problems that must be solved before an appropriate evaluation of ants as biological control agents can be made. Part of the problem is the confusion over proper identification of species. Although recent work has clarified the situation for European species, much work remains to be done in North America (Finnegan, 1971).

Some predaceous ants may be considered direct pests of trees as a result of their normal behavior. In the eastern and northeastern United States, the Allegheny mound ant (*Formica exsectoides*) feeds on living and dead insects as well as honeydew. However, to prevent shading of the nest the ant kills surrounding conifers by injecting formic acid in their cortical tissues. Young white pine, red pine, Scots pine, and spruce are most susceptible. Damage in plantations may be severe (Peirson, 1922; Baker, 1972). Other ant species are involved in similar disruptive activities (Peirson, 1922).

Coleoptera

Predaceous species in the order Coleoptera make up the greatest proportion of predators in forests. Many of these are predators of bark beetles and thus take on great significance as regulatory agents. Species of the Cleridae (checkered beetles) and the Nitidulidae are important bark beetle predators in Europe and North America. *Thanasimus* and *Enoclerus* are important North American genera in the family Cleridae (Berryman, 1967). Other families, such as the Staphylinidae (rove beetles), Cucujidae (flat bark beetles), and Elateridae (click beetles), prey on various life stages of bark beetles.

The predatory activity of clerids represents an important natural control of many species of bark beetles. Species in the major genus *Enoclerus* attack prey in both deciduous and coniferous forests. The black-bellied clerid (*Enoclerus lecontei*), for example, is an important predator of the western pine beetle and is also associated with several other *Dendroctonus* species including *D. frontalis* and *D. ponderosae* (Berryman, 1966). Clerid adults are aggressive and voracious feeders, averaging approximately 0.68–1.31 bark beetles/day, depending on whether they are reared individually or in pairs. An adult clerid lives from 36 to 114 days and may consume 44 to 158 bark beetle adults. Adult *Enoclerus sphegeus*, a predator of *D. ponderosae*, also averages approximately one prey per day. Each larva,

Table 18.5 Maximum number of larvae of the clerid predator *Enoclerus sphegeus* per ft of bark for four sampling heights[a]

Height of Sample[b] (ft)	Number of Larvae/ft² of Bark		
	1965	1966	1967
1.5	32	34	22
5.0	24	16	8
10.0	12	8	10
15.0	12	6	12

[a] From Schmid (1970).
[b] Samples not necessarily from the same tree.

however, kills approximately 25 larvae or pupae of *D. ponderosae*. The impact of adults and larvae of this clerid species on *D. ponderosae* populations has been estimated at approximately 1% and 5–11% of the adult and larval prey population, respectively (Schmid, 1970). The activity of these predators is sometimes concentrated at certain heights on the tree bole (Table 18.5).

The predaceous ground beetles (Carabidae), in both larval and adult stages, serve as effective consumers of all types of animal life on the forest floor, including soil insects, caterpillars, pupae, and phytophagous Coleoptera. A few species are arboreal. Species in the genus *Calosoma* are major predators of lepidopterous larvae. Another group whose species are widely distributed throughout the forest strata is the Coccinellidae. Both larvae and adults are vigorous predators of aphid and scale insects (Doane *et al.*, 1936).

Miscellaneous Insect Predators

Other groups of predaceous invertebrates that merit mention include species in the orders Diptera, Hemiptera, and Neuroptera. Among the dipterous predators, for example, are three families that include commonly encountered species. The families Xylophagidae, Lonchaeidae, and Dolichopodidae (long-legged flies) contain important predators. *Medetera aldritchi* is an important predator of scolytids, such as *Dendroctonus ponderosae*. Others, such as the syrphids, are important aphid predators. Only the larvae of most Diptera are predaceous, although there are a few exceptions, such as the robber flies (Asilidae). Among the most important hemipterous predators are the Pentatomidae (stink bugs), Reduviidae (assassin bugs), and Anthocoridae (minute pirate bugs), which feed on bark beetles. The Raphidiidae (snake flies) and lacewings represent the most common neuropteran predators (Thatcher, 1961; Doane *et al.*, 1936).

Other Predaceous Arthropods

Spiders and mites are the most important noninsect predaceous arthropods (Figs. 18.10 and 18.11). In spite of their importance, very little is known of their ecological significance and potential as effective biological control agents for forest insects. A study of the impact of spiders in a mesic *Liriodendron tulipifera* forest litter habitat indicates that spiders annually

Figure 18.10.   Representative of a major group of invertebrate predators, the spiders.

Figure 18.11.   The spider *Oxyopes scolaris* guarding egg sac and feeding on *Ips pini*. (From Jennings and Pase, 1975.)

consume 43.8% of the mean annual standing crop of all cryptozoans on the forest floor. The impact of large spiders (>10 mg) is less than that of small spiders (<1 mg) (Moulder and Reichle, 1972).

Mites

Mites have been known to be associated with bark beetles for some time. Their size, cryptic behavior, and reluctance to feed when exposed to light has made it difficult to characterize their role in natural control. Therefore, their potential as biological control agents is also uncertain. Surveys of mites associated with the galleries of the southern pine beetle have demonstrated that 32 of the 51 mite species collected will prey on one or more brood stages. Based on feeding preferences and relative frequency, at least eight species seem to be prime candidates for biological control (Moser, 1975). With greater efforts to develop the required data base, mites should provide effective biological control in urban and rural forest ecosystems.

Insect Parasitoids

All forest insects are hosts to a variety of parasitic insect species, often referred to as parasitoids. This term reflects a behavior that is intermediate between that of a predator and a classical parasite. Whereas an adult parasitoid is free-living, the larvae are parasitic on a host, which generally dies after the parasite has completed feeding (Fig. 18.12). Most hosts are other invertebrates and are generally the eggs, larvae, or pupae of other insects (Askew, 1971).

Parasitoids can be characterized by the ways in which they differ from predators. For example, although many predators search for prey both as immatures and adults, most parasitoids search for hosts only as adults. Parasitoid eggs or larvae may be placed on leaves to be consumed by the host, or placed on or in the appropriate host stage. A single host is

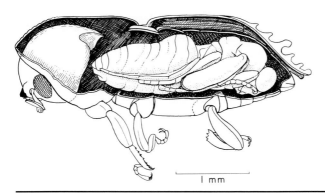

Figure 18.12.   Pupa of the parasitoid *Tomicobia tibialis* in its normal position within cutaway host, *Ips paraconfusus*. (From Bedard, 1965.)

sufficient to complete the development of an immature parasitoid. Parasitoids are frequently host specific, relative to predators. Direct evidence of behavioral life-history characteristics, as well as theory, suggests that parasitoids of forest insects are likely to be effective regulators of forest pests. However, historical data and the present lack of conclusive evidence indicate that the ultimate success of a natural enemy or a given species cannot always be predicted (Askew, 1971).

Host defenses against parasitoids vary from passive to active, as well as in their effectiveness. Pupae of many forest species have hard, smooth cuticles, which make grasping and ovipositor penetration difficult. Many aphid and scale species have a woolly covering of wax that may make the exact location of the insect difficult to determine. Some protection may be gained by habitation in protected places, such as galls, mines, wood, and soil. Violent movements, wriggling, or defensive postures are often observed in larvae or, more commonly, in pupae.

Physiological defense reactions are important protective mechanisms after oviposition has occurred. The most common reaction is encapsulation (aggregation of host blood cells around the parasitoid). An envelope of blood cells is formed that kills the parasitoid. This type of physiological reaction has been observed in 14 orders of insects and in all stages except the egg. In general, hymenopterous parasites are not encapsulated in their normal hosts, although they may be in unusual hosts (Askew, 1971).

Although parasitoid species occur in a number of insect orders, by far the most important parasitoids of forest and shade tree insects occur in the orders Diptera and Hymenoptera. Several families in these orders, such as the Tachinidae, Acroceridae, Pipunculidae, Chalcidae, Braconidae, and Ichneumonidae, include important parasitic species (Askew, 1971).

Diptera

Dipteran parasitoids lack the strong, protected ovipositor of the Hymenoptera. Therefore, they generally do not attack hosts in wood galleries, leaf mines, galls, or other protected habitats. The absence in adults of chewing mouthparts for emergence from those host habitats adds to the difficulty. However, in some dipterous families, accessory structures have evolved for piercing tissues and allowing larviposition or oviposition inside of hosts. Although life histories of dipterous parasitoids are less diverse than those of hymenopterous parasitoids, their host range is much more diverse and includes slugs, snails, spiders, earthworms, and centipedes, as well as insects.

The major dipteran family of parasitoids is the Tachinidae. Larvae of Lepidoptera and Coleoptera comprise the majority of hosts. Tachinids tend to be more polyphagous than parasitic Hymenoptera. One species, *Compsilura concinnata*, a parasitoid of the gypsy moth and other forest insect species, parasitizes well over 200 host species (Clausen, 1956).

The reproductive strategies represented in the tachinidae are quite diverse. Species are capable of larviposition (deposition of larvae), oviposition (deposition of eggs), and ovoviviparous behavior (deposition of fully incubated eggs in which larvae are at the point of hatching). Eggs or larvae are laid on or in the host, as well as on vegetation or the soil surface. The reproductive potential of tachinids generally varies inversely with the likelihood of a larva gaining access to a host (Askew, 1971). The tachinid parasitoid guild of larval gypsy moths is a good illustration. *Blepharipa pratensis* oviposits on leaves and has a reproductive capacity as high 5000 eggs per female. *Parasetigena silvestris* females lay approximately 200 eggs on the surface of larval gypsy moths, and *Compsilura concinnata* larviposits approximately 90 to 110 progeny within the host per female.

Hymenoptera

The majority of insects in this order are parasitoids of other insects. Thus, the Hymenoptera can be considered among the most important orders that include agents of natural and biological control. The ovipositor of parasitic Hymenoptera has evolved as a key structure, aiding in their success. Mechano- and chemoreceptors occur on the ovipositor sheath and terebra (the structure through which the eggs pass) (Fig. 18.13). Cutting ridges on the terebra pierce the host cuticle. The ovipositor may be used to inject a toxin and paralyze the host, as well as to lay eggs.

Unlike dipterans, many hymenoterous species are hyperparasitic; that is, they are parasitoids of parasitoids. Hyperparasitic behavior is by no means obligatory, and some species are both primary parasitoids of a phytophagous host and secondary parasitoids (hyperparasitoids). Very often two species lay their eggs on or in the same host individual, with only one species generally being successful. This type of parasitism is called multiple parasitism. Many eggs or larvae may be placed on or in the same host individual by one female or by several females of the same species. This phenomenon is termed superparasitism. In solitary species only one progeny survives. In gregarious species some or all progeny survive (Askew, 1971).

Chalcidoidea

This group of Hymenoptera includes a large number of species, which average 2–3 mm in length. Most of these parasitoids are robust insects with a well-developed flight capacity. Although their armored ovipositors are capable of penetrating hard vegetative or host tissues, the depth of penetration is limited by the overall size of their ovipositors. Hosts of chalcids occur in several orders including Hymenoptera, Coleoptera, Lepidoptera, Diptera, and Hemiptera.

Feeding requirements of these adult parasitoids vary. Many females, after stinging the host, feed on fluids that exude from the wound. Adults may also feed on nectar or honeydew. This latter type of feeding behavior is found in many different hymenopterans. Flowers of Umbelliferae are particularly attractive to many parasitic species (Askew, 1971).

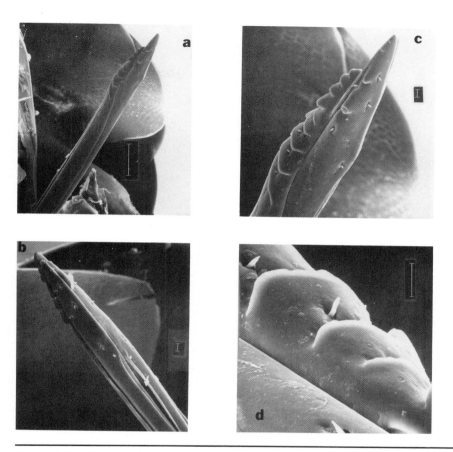

Figure 18.13.   Views of the ovipositor of *Brachymeria intermedia*, a pupal parasitoid of the gypsy moth (*Lymantria dispar*); scales represent (a) 100 μm and (b–d) 10 μm.

Ichneumonoidea

This group includes species that are larger than the chalcids. They are relatively slender and long-legged and have a narrow, stalklike connector between the thorax and abdomen. Many of the species in this group are found searching for hosts along woodland edges or in the woodland canopy. The two major families in this group, the Ichneumonidae (Fig. 18.14) and the Braconidae, attack insect species in various orders, as well as spiders and pseudoscorpions.

Pathogens

Insect pathogens play a central role in the regulation of forest insect populations and the collapse of outbreaks. Their compatibility with other chemical control agents, biotic control agents, and silvicultural practices

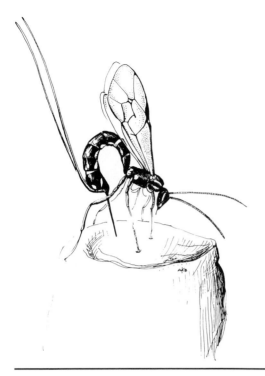

Figure 18.14.   Oviposition stance of an *ichneumonid* parasitoid. (Redrawn from Arthur, 1963.)

make them potentially valuable tools in forest insect pest management. Pathogens commonly occur in field populations of forest insects. Studies of the spruce budworm, the European spruce sawfly, the winter moth, and the gypsy moth have shown that pathogens such as viruses, microsporidia, and bacteria are continually present in populations at a low level (enzootic). Outbreaks (epizootics) of disease occur in a large number of insect species and are usually associated with high host densities. Epizootics of nuclear polyhedrosis viruses (NPV) have led to the collapse, in the second or subsequent generation, of major defoliators such as the gypsy moth, the tent caterpillar complex (*Malacosoma* spp.), the Douglas-fir tussock moth, the pandora moth, and the red-belly tussock moth (*Lymantria fumida*). Although bacterial epizootics are uncommon, epizootics of protozoan and fungal diseases are relatively frequent. *Nosema* spp. and *Entomophthora* spp. epizootics have been responsible for the collapse of several important forest defoliators.

Microbial control has exhibited the most potential against forest insect defoliators. Control with *Bacillus thuringiensis* (Bt) and NPVs have been particularly effective against various sawfly and lepidopteran species. Like parasitoids, pathogens can be used in two basic ways: as short-term biological insecticides or as long-term controls in which the pathogen is introduced in an inoculative release. A great deal of additional research on the technology of microbial formulations and application is required. Data on production, storage, application methodology, safety, persistence, and cost are needed for many potential microbial agents. In addition, the virulence of many of the known pathogen strains is inadequate and must be improved. We have been unsuccessful in mass cultivation of many microorganisms including nematodes, protozoans, and others. The economic feasibility of many microbial insecticides will be based, in part, on efficient mass production on artificial substrates (Maddox, 1975). In general, few investigations have been conducted on the use of fungi and protozoa as biological controls. Basic biological and ecological studies are required before preliminary field tests can begin.

The use of pathogens in forest insect pest management promises to be one of many tools available for the regulation of dendrophilous species. Like other tools, they are and will be very useful under certain circumstances, and ineffective under others. Microbial control has a great deal of potential because it is relatively host-specific, is compatible with other control procedures, and does not interact negatively with other organisms in the environment. The use of pathogens is particularly promising in forest habitats. The environmental stability of these organisms, which is enhanced by the long-term character of their habitat, and the relatively high economic threshold of forest and forest products enhance the possibility of successful introduction and colonization of pathogens. Conversely, a number of problems need to be solved before microbials can be efficiently used in forests. Most fundamentally, we lack a clear understanding of the factors that initiate epizootics of insect populations.

Generalizations are difficult to make because there are so many types of pathogens. Nevertheless, pathogens are considered to be most effective at high host densities, compared with parasites and predators, which tend to be most effective at low host densities (Tanada, 1967). Clearly, there are exceptions, since epizootics can occur at low host densities.

Use and Effectiveness of Pathogens

Viral diseases are common among forest and shade tree insects (Martignoni and Iwai, 1986) (Table 18.6). Polyhedral viruses as well as bacteria have been applied against several dendrophagous species [i.e., the great basin tent caterpillar (*Malacosoma californicum fragile*), the pine processionary caterpillar (*Thaumetopoea pityocampa*) (Tanada, 1959), *Neodiprion lecontei* (Kaupp and Cunningham, 1977), the gypsy moth, the Douglas-fir tussock moth (Stelzer *et al.*, 1975) (Fig. 18.15), and others] and have resulted in substantial larval mortality (e.g., 90, 95, or 100%) and excellent foliage protection (e.g., less than 25% defoliation).

Table 18.6.   Virus diseases of forest insects[a]

Insect Order, Family, Genus, and Species	Type of Disease[b]
HYMENOPTERA	
Diprionidae (conifer sawflies)	
Diprion hercyniae (Hartig), European spruce sawfly	N
Neodiprion abietis (Harris), balsam fir sawfly	N,P
Neodiprion excitans Rohwer, blackheaded pine sawfly	N
Neodiprion lecontei (Fitch), redheaded pine sawfly	N,P
Neodiprion pratti banksianae Rohwer, jack pine sawfly	N
Neodiprion pratti pratti, Virginia pine sawfly	P
Neodiprion sertifer (Geoff.), European pine sawfly	N,P,PV
Neodiprion swainei Middleton, Swaine jack pine sawfly	N
Neodiprion taedae linearis Ross, Arkansas pine sawfly	N
Neodiprion virginianus	N
Tenthredinidae (sawflies)	
Trichiocampus irregularis (Dyar)	N
Trichiocampus viminalis (Fallén)	N
LEPIDOPTERA	
Arctiidae (tiger moths)	
Hyphantria cunea (Drury), fall webworm	CP,G,N,P
Gelechiidae (gelechiid moths)	
Coleotechnites milleri (Busck), lodgepole needle miner	G
Geometridae (geometrid moths)	
Erranis tiliaria (Harris), linden looper	CP
Erranis vancouverensis Hulst	N
Lambdina fiscellaria (Guenée)	N,P
Lambdina fiscellaria lugubrosa (Hulst), western hemlock looper	N,P
Lambdina fiscellaria somniaria (Hulst), western oak looper	N,P
Lasiocampidae (tent caterpillar moths)	
Malacosoma americanum (F.), eastern tent caterpillar	CP,N,P
Malacosoma californicum (Packard), western tent caterpillar	N
Malacosoma californicum fragile (Stretch), great basin tent caterpillar	N
Malacosoma californicum pluviale (Dyar), western tent caterpillar (subspecies)	CP,N,P
Malacosoma constrictum (Henry Edwards), Pacific tent caterpillar	N
Malacosoma disstria Hübner, forest tent caterpillar	CP,N,P
Lymantriidae (tussock moths)	
Orgyia antiqua (I.), rusty tussock moth	CP,N,P
Orgyia leucostigma J. E. Smith, whitemarked tussock moth	CP,N,P
Orgyia pseudotsugata (McDunnough), Douglas-fir tussock moth	CP,N,P
Orgyia vetusta Boisduval, western tussock moth	N
Lymantria dispar (L.), gypsy moth	CP,D,IV,N,S

[a] From Maksymiuk (1979).
[b] CP, cytoplasmic polyhedrosis; D, densonucleosis; G, granulosis; IV, irridescent virosis; N, nucleopolyhedrosis; P, polyhedrosis; PV, presumed virosis; and S, spheroidosis.

(*continued*)

Table 18.6. (*continued*)

Insect Order, Family, Genus, and Species	Type of Disease[b]
Pyralidae (pyralid moths)	
Dioryctria abietivorella (Groté), fir cankerworms	G
Tortricidae (leafroller moths)	
Acleris gloverana (Walsingham), western blackheaded budworm	N
Acleris variana (Fernald), eastern blackheaded budworm	N
Archips argyrospilus (Walker), fruit tree leafroller	G
Archips cerasivoranus (Fitch), uglynest caterpillar	N
Choristoneura biennis Freeman, 2-year budworm	S
Choristoneura conflictana (Walker), large aspen tortrix	S
Choristoneura fumiferana (Clemens), spruce budworm	CP,G,N,S
Choristoneura occidentalis Freeman, western spruce budworm	N
Choristoneura pinus Freeman, jack pine budworm	N

Pathogens can effectively reduce high host densities by inducing significant mortality and may also have other qualitative effects on survivors, which are often physiologically weakened. Chronic Bt infections of the gypsy moth cause reduced feeding activities in larvae (Harper, 1974). Surviving larval spruce budworms affected by various pathogens also exhibit reduced feeding activity as well as reduced pupal size, survival, and fecundity in individuals that reach adulthood (Katagiri, 1969; Harper, 1974; Morris, 1976). Virus-infected fall cankerworm females fail to produce any egg masses (Neilson, 1965). Gypsy moth individuals infected with cytoplasmic polyhedrosis virus that emerge from apparently normal pupae show reduced pigmentation. Wing coloration of males is significantly lighter than that of disease-free individuals (Magnoler, 1974) (Fig. 18.16). In still other species, sublethal infections may cause the larvae to be sluggish and render them susceptible to attack by parasitoids and predators (Smirnoff, 1961).

The potential of microbial control was dramatically illustrated in Canada with the virus-induced collapse of the introduced European spruce sawfly. When the sawfly was first discovered, a large outbreak had already developed. However, the sawfly population declined rapidly over the next few years due to an accidentally introduced polyhedrosis virus (Balch and Bird, 1944; Tanada, 1959). Populations of the European pine sawfly have been controlled by a virus in several European countries (Bird and Whalen, 1953). In 1949, virus-killed larvae were sent from Sweden, and the pathogen was used very successfully in Canada and later in the United States (Bird, 1953).

Populations of the Swaine's jack pine sawfly, the Virginia pine sawfly (*Neodiprion pratti pratti*), and the redheaded pine sawfly have all been successfully controlled with viral suspensions in Canada and the United States. Virus-infected individuals found in the field were collected and the

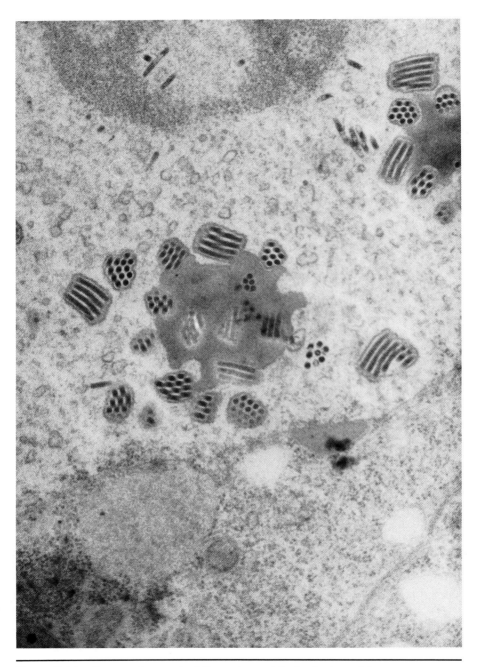

Figure 18.15.   Electronmicrograph of the nuclear polyhedrosis virus of the Douglas-fir tussock moth (*Orgyia pseudotsugata*). (Courtesy U.S. Forest Service; photograph by Kenneth M. Hughes.)

Figure 18.16. Healthy (upper) and diseased (lower) *Lymantria dispar* males showing normal and reduced pigmentation. (From Magnoler, 1974.)

pathogen was mass-produced for eventual use against the sawflies. Field populations of several species of Lepidoptera have succumbed to NPV infestation, including the gypsy moth relative *Lymantria fumida* (a pest of the Japanese fir), the Douglas fir tussock moth, the forest tent caterpillar, and the spruce budworm. Populations of other species, such as the pine processionary caterpillar of Europe and the spruce budworm, have been reduced with the use of cytoplasmic polyhedrosis virus (CPV) and granulosis virus (GV) (Stairs, 1971).

Epizootics are often associated with high host densities. Among the probable reasons for this association is the increased likelihood of disease transmission at high densities. However, in some situations viral diseases may be very effective at low density. These exceptions are sometimes related to host behavior and the spread of the disease within the population. For example, *Neodiprion sertifer* lays its eggs in a cluster, and the larvae feed gregariously. This results in a single focus of infection, to which the entire colony will likely succumb before any wandering behavior is triggered. On the other hand, *Diprion hercyniae* lays single eggs, and larvae feed singly. Thus, the progeny of infected adults establish many infection foci (Bird, 1955). In general, it should be noted that although pathogens can be very effective at regulating pest populations, not all virus diseases exhibit the dramatic effectiveness observed in many common forest insect–virus interactions. For example, many viruses such as the one affecting the jack pine sawfly are not highly pathogenic and cause only a low percentage of mortality (Bird, 1955).

Dispersal of Pathogens

The spread of infective agents occurs in various ways in nature. The prime methods are via the movement of the infected hosts, the phoretic transport of infected hosts, or contact with the fecal matter of other insects or animals that prey on diseased hosts. Physical factors such as wind and rain may also aid in the dispersion process. The infected insect can spread the pathogen in several ways. For example, females may deposit eggs contaminated with pathogens. Fecal deposits left behind by larvae are also infection foci. Ultimately, the dead infected host may burst, releasing infective stages of the pathogen.

Formulations of Pathogens

Currently, research is being directed at developing formulations that combine various types of pathogens or pathogens and synthetic insecticides. Chitinase, a hydrolyzing enzyme, can be combined with *Bacillus thuringiensis* to produce higher mortality than that produced by Bt alone. The chitinase affects the midgut peritrophic membrane and facilitates entry of ingested pathogens into the hemocoel. Substantial control of the spruce budworm and of *Lambdina fiscellaria* has been achieved with this formulation (Smirnoff *et al.*, 1973; Smirnoff, 1974a,b; Morris, 1976). Interestingly, when populations of spruce budworm are infested with a native microsporidian, the addition of chitinase to the Bt formulation is not required. The activity of the microsporidian breaks down the gut wall and thus facilitates penetration by Bt spores (Smirnoff, 1974a).

Viruses and other pathogens are compatible with most chemical insecticides, additives, fungicides, and chemical stressors. The selectivity of a pathogen may allow formulation with reduced amounts of chemical insecticides. Applications of *Bacillus thuringiensis*, entomopox virus, or nuclear polyhedrosis virus, combined with sublethal doses of the organophosphates fenitrothion or orthene, result in effective control of spruce budworm larvae. This approach may also be effective against other lepidopterous larvae. With advances in methodology this approach may provide a method of maintaining pest populations below economic injury levels while providing for minimum environmental disruption. An additional advantage is that protection by these Bt–insecticide combinations may carry over to succeeding generations (Morris *et al.*, 1974; Morris and Armstrong, 1975). *Bacillus thuringiensis* also has been combined with nucleopolyhedrosis virus and used against the Douglas fir tussock moth. The combinations vary, but the principle of potentiation is the same. This principle suggests that insects are more susceptible to disease when they are stressed, whether due to infection by pathogenic microorganisms or exposure to insecticides. More research and field testing must be conducted if we are to better understand the population dynamics of these pathogens and thereby increase their efficiency and achieve favorable cost–benefit ratios for their use.

Additives that increase the effectiveness of viruses include boric acid and silica powder. These additives have been used in formulations applied against species such as the gypsy moth and the pine caterpillar (*Dendrolimus spectabilis*). Inorganic compounds such as copper sulfide ($CuSO_4$), iron sulfide ($FeSO_4$), and zinc sulfide ($ZnSO_4$) reduce the time period needed for lethal infection of NPV in *Lymantria monacha*. Sodium silicate (Na_2SiO_3) and plant ash achieve the same result in NPV infections of the gypsy moth (Tanada, 1976). Other additives such as molasses increase feeding by the gypsy moth on virus-treated leaves. Finally, compounds called ultraviolet screens are incorporated to protect preparations against environmental degradation.

Persistence of Pathogens

Advantages of using microbial control include persistent effects, often beyond the year of application, and rapid spread of the disease. Infection of *D. hercyniae* on spruce trees sprayed with virus was found up to 2000 ft beyond the sprayed area, even though the area had previously been disease free. The spread of virus occurs most rapidly in heavily forested areas (Bird, 1955). One year after infestations of Swaine's jack pine sawfly were sprayed with a polyhedral virus, the disease had spread 2 miles from the sprayed areas.

Viruses persist in the field in several ways. The virus of the tent caterpillar (*Malacosoma californicum fragile*) persists through the winter on the food plant and infects the next generation (Tanada, 1973). Alternatively, after control of the Siberian silkworm (*Dendrolimus sibiricus*) with *B. thuringiensis* var. *dendrolimus* in forested areas of Russia, the bacteria persisted in the soil, in the remains of dead caterpillars, and in pupae (Falcon, 1971). Debris mats associated with dense gypsy moth populations consist of exuviae, disintegrated larval and pupal cadavers, and empty pupal cases. Emerging females often lay their egg masses near, under, or in such mats. The following spring, newly hatched larvae crawl over the debris, which can serve as a source of infection (Table 18.7). In addition, NPV persists in the field on the hair of the egg masses and on the eggs (Doane, 1975).

Table 18.7. Mean percent mortality of gypsy moth larvae hatching from disinfected eggs and emerging through layers of debris collected just prior to spring hatch in the field[a]

Collection Sites	No. Larvae Tested	Mortality (%)		
		NPV	Other Causes	Total
Site A				
Debris	107	90.6	9.4	100
Control, sterilized debris	114	4.4	0.9	4.7
Site B				
Debris	108	80.6	19.4	100
Control, sterilized debris	109	0.9	0	1.0
Site C				
Debris	415	89.3	10.7	100
Control, sterilized debris	160	0	1.2	1.2
Control				
Test larvae from disinfected eggs	100	2.0	1.0	3.0

[a] Modified from Doane (1975).

Pathogens and Environmental Factors

Physical factors such as rain, wind, temperature, and sunlight are important in determining the virulence, stability, persistence, and dispersal of pathogens, as well as the susceptibility of the host. In general, humidity appears to play a minor role in the persistence of most pathogens, except for fungi and certain nematodes. Humidity often affects the germination, survival, or persistence of fungal spores. The resistance and behavior of host insects may also be influenced by humidity and thus indirectly affect pathogen activity. Rain can increase the distribution of a pathogen because it may remove the pathogen from parts of a tree and redistribute or concentrate it. Temperature also is an important inactivator of viruses (Tanada, 1976) and fungi (Tables 18.8 and 18.9). Concurrent with weather factors are other physical and biotic factors (i.e., nutrition, parasites, crowding, and insecticides) that interact with both the pathogen and its host (Fig. 18.17 and Table 18.10).

In general, the physical environment plays a major role in the effectiveness of pathogens. The LT_{50} (lethal time for mortality of 50% of a population) of insects exposed to virus treatment in the field increases markedly with prolonged exposure of a virus to sunlight (Yendol and Hamlen, 1973). The ultraviolet component of sunlight can reduce the viability of some viruses in as little as a few hours to several days. Often,

Table 18.8. Some cardinal temperatures for various entomogenous fungi[a]

Species	Minimum (°C)	Optimum (°C)	Maximum (°C)
Entomophthora exitialis	6	24	30
E. coronata	6	27–33	33
E. virulenta	6	30	36
E. aphidis	5	21–24	28
Metarrhizium anisopliae	6	25–30	33
Beauveria bassiana	10	25–30	32

[a] From Yendol and Hamlen (1973). Reprinted by permission of the New York Academy of Science.

Table 18.9. Optimum temperatures for spore germination of some entomogenous fungi[a]

Fungal Species	Optimum Temperature for Germination (°C)
Beauveria bassiana	28
Entomophthora coronata	16–30
Metarrhizium anisopliae	20–30
Spicaria pracina	25–28
Spicaria rileyi	15–30

[a] From Yendol and Hamlen (1973). Reprinted by permission of the New York Academy of Science.

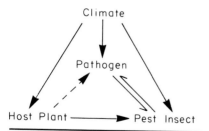

Figure 18.17. Interactions between the insect pathogen and the ecosystem. (From Tanada, 1973. Reprinted by permission of the New York Academy of Science.)

Table 18.10.   Environmental factors affecting short- and long-term microbial control [a]

Short Term	Long Term
Physical Factors	Physical Factors
Temperature	Soil temperatures
Sunlight	Sunlight
Rain	Rain
Humidity	Humidity
Application Methods	Biotic Factors
Additives	Predators
Coverage	Pathogens
Timing	Parasites
Pathogen virulence	Host plants
Formulation	Insect hosts
pH	

[a] Modified from Tanada (1973). Reprinted by permission of the New York Academy of Science.

attempts are made to add materials to virus preparations such as solar protectants, including activated carbon, molasses, and dried milk powder. The half-life of the gypsy moth NPV formulations can be extended from under 12 hr for purified virus to 36 hr when combined with egg albumin or peptonized milk (Yendol and Hamlen, 1973).

Biotic Factors and Pathogen Activity

Biotic factors, other than those just discussed, can be extremely important in pathogen activity. The structure and texture of the leaf may enhance viral persistence. When the NPV of the forest tent caterpillar is placed on the leaves of *Liquidamber styraciflua* or on a glass surface and is exposed to sunlight, it is inactivated at a faster rate on the glass than on the leaves. Midgut pH, which can be determined in part by the type of foliage consumed, can be a critical factor in the susceptibility of hosts. The activity of Bt toxic crystals is dependent upon the occurrence of an alkaline midgut in its lepidopterous hosts (Maddox, 1975). The leaf juices of conifers tend to inhibit Bt more than those of other plant foliage (Tanada, 1967). Sawflies, which have a low midgut pH, are susceptible to *Bacillus cereus*. Other insects that have a high pH, such as some Lepidoptera, are resistant to infection.

Predisposition to infectious diseases may be a function of nutrition. The quality and quantity of food often affects the incidence of disease. When the gypsy moth, the lackey moth, and the fall webworm are fed unfavorable food, the incidence of virus disease increases (Tanada, 1967). Similarly, forest tent caterpillars die most often from virus disease when they feed on maple foliage (Bird, 1955).

Thus, the relation between dosage, host survival, and pathogen dissemination can be influenced by the quality of host plants. The pest species may also differ in its susceptibility as it gets older. In the laboratory, first instar gypsy moths are approximately ten times more susceptible to viral mortality than second or third instars (Stairs, 1971).

Interactions of Biotic Agents and Pathogen Activity

Insects are often simultaneously infected with a number of pathogens. Multiple infection may enhance the probability of early death of the host. Interactions among types of pathogens, such as several viruses, in a host may lead to enhancement of the activity of the viral preparation (see Formulations of Pathogens). However, some pathogen mixtures may be antagonistic. The spruce budworm and the forest tent caterpillar may exist with mixed infections of NPV and CPV. When this occurs, the cytoplasmic polyhedrosis virus interferes with and retards development of the nuclear polyhedrosis virus. The pathogenicity of NPV is enhanced when it is separated from CPV. This suggests that the most effective use of these viruses could be achieved by sequential treatments, that is, the application of NPV followed by application of CPV (the most infectious virus affecting the budworm) after the NPV has become established (Bird, 1969). In another case, simultaneous infection by NPV and two species of microsporidia of the fall webworm results in interference with the NPV infection process by the *Nosema* species (Stairs, 1973; Smirnov, 1976).

A number of interactions take place between predators and parasitoids and their host's pathogens. There is no evidence to indicate that parasitoids or predators become infected by the pathogenic viruses or bacteria of their hosts; nevertheless, the development of a parasitoid in a diseased host may be jeopardized. Although parasitoids and predators may not be subject to infection by their host's pathogens, they may serve as vectors and transmit viruses mechanically, such as by means of ovipositor contamination.

A study of larval gypsy moths from six geographically distinct populations shows a positive correlation between the incidence of NPV and bacteria and the incidence of several parasitoids, including *Cotesia* (=*Apanteles*) *melanoscelus* and *Parasetigena silvestris* (with NPV) and *Compsilura concinnata* and *Blepharipa scutellata* (with bacteria) (Reardon and Podgwaite, 1976). Viral mortality is significantly greater among larvae exposed to contaminated *Cotesia melanoscelus* than among those exposed to uncontaminated parasitoids (Raimo *et al.*, 1977). Finally, release of *Cotesia melanoscelus* in areas treated with Bt results in additional foliage protection compared with either control agent alone. Larval populations are consistently reduced only where both the parasitoids and Bt are used (Wollam and Yendol, 1976).

The pathogen-laden feces of invertebrate predators can contaminate foliage and thus infect other hosts. The gypsy moth predator *Calosoma sycophanta* is capable of transmitting infective virus in its feces after consuming diseased prey (Capinera and Barbosa, 1975). Other predaceous insects also have the potential for inducing or spreading infection following unsuccessful attacks on prey. *Vespula rufa consobrina*, a predaceous wasp attacking *Neodiprion swainei*, feeds on virused larvae and may have viral polyhedra on its mouthparts (although not in the gut), which can be transmitted during unsuccessful attacks. A heteropteran predator, *Pilophorus uhleri*, also induces viral infections in healthy larvae (Smirnoff, 1959). Similarly, tree-to-tree dispersal of the nuclear polyhedrosis virus of the European pine sawfly occurs chiefly through the activities of parasitoids (Bird, 1961). Finally, adult *Sarcophaga aldrichi* feed on diseased larvae and can then contaminate foliage with its mouthparts and tarsi (Tanada, 1976).

Transfer of pathogens by vertebrate predators through the alimentary canal also occurs. Certain NPVs of sawflies remain infective after passage through the guts of birds (Bird, 1955). In fact, the stomach contents of catbirds and cedar waxwing that are fed virus-infected *Neodiprion sertifer* are highly infectious even after 8 months of storage (Bird, 1955). Feces of two gypsy moth predators, the white-footed mouse (*Peromyscus leucopus*) and the short-tailed shrew (*Blarina brevicauda*), contained active NPV. This suggests that small mammals also distribute infectious NPV in the wild (Lautenschlager and Podgwaite, 1977).

Viruses

In most of the known viruses of insects, virus particles are in protein crystals known as inclusion bodies. Nuclear polyhedrosis, granulosis, and cytoplasmic viruses are the three principal types of inclusion viruses. Most nuclear polyhedrosis viruses infect and attack the epidermis, trachea, fat body, and blood cells of insects such as lepidopterous larvae. Dead larvae infected with NPV often have a very characteristic drooping posture (Fig. 18.18). The cavader is essentially a sack of dark-colored fluid containing large numbers of polyhedra (protein crystals incorporating an envelope of one or several virus rods). Granulosis viruses primarily infect the fat body of lepidopterous larvae, although trachea and epidermis also serve as sites of infection. Cytoplasmic polyhedrosis viruses infect the cytoplasm of the cells of the midgut of larvae. The CPV are less pathogenic than NPVs and less host-specific than other viruses (Maddox, 1975).

Field tests have been conducted primarily with NPVs and to a lesser extent with GVs. The use of CPV and noninclusion viruses has been somewhat limited. Entry of viruses into insect hosts is generally either

Figure 18.18.   Typical *J*-shaped posture of dead, virused larva of the Douglas-fir tussock moth (*Orgyia pseudotsugata*). (Courtesy U.S. Forest Service; photograph by Paul J. Iwai.)

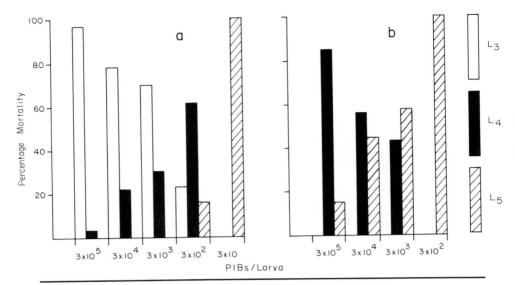

Figure 18.19.   Mortality of various *Malacosoma neustria* larvae after ingestion of different concentrations of nuclear polyhedra by (a) third and (b) fourth instars. (From Magnoler, 1975.)

through the mouth and midgut or transovarially (within the egg). In addition, access is possible through wounds made by predators and through ovipositor wounds made by parasitoids.

Innate resistance to virus infection has been shown in strains of various insect species. In some insects there is an age differential in susceptibility to infection; early instars are more susceptible than late instars (Fig. 18.19). Although the basis of resistance is yet to be established, it is probably caused by the inability of the virus to replicate and infect the cell.

Viruses hold great promise in microbial control of dendrophagous forest insects. Although the rate of research on various aspects of the biochemistry and physiology of viruses has increased in recent years, much more is needed. This need is particularly great for certain aspects such as virus ecology (Tanada, 1976).

Fungi

Hundreds of species in approximately 35 genera of fungi infect insects. Only a few have been studied extensively. Of these, very few are capable of regulating forest and shade tree insects because entomopathogenic fungi are not host-specific.

Transmission of fungi from host to host is accomplished during the spore phase. Entry is commonly through the cuticle, and eventually the insect body is filled with mycelia (threadlike vegetative filaments). Fungi, more than any other pathogen, require specific conditions for the development of epizootics. High humidity is perhaps the key environmental factor needed for development. One of the prime limitations to the use of fungi is our inability to mass-rear them outside of their hosts.

Although a great deal of research needs to be done before fungal agents can be effectively manipulated, the most promising species are in the genera *Beauveria*, *Entomophthora*, and *Empusa*, which attack many dendrophilous insects (Angus, 1962). Fungi such as *Entomophthora sphaerosperma* and *E. egressa* are sometimes found in late instar spruce budworms at a relatively low incidence. Occasionally, however, the incidence of disease can reach substantial levels (Vandenberg and Soper, 1975; Harvey and Burke, 1974). *Beauveria bassiana* and *Metarrhizium anisopliae*, the causal agents of white and green muscardine diseases of insects, have been reported in various forest insects such as *Dendroctonus frontalis* and *Ips typographus* (Moore, 1971). Yellow muscardine disease epizootics in the pine caterpillar, a major defoliator in Japan, are caused by the fungus *Isaria farinosa*. High-density pest populations collapse due to the action of this pathogen, but its pathogenicity to the economically important silkworm has limited its application (Katagiri, 1969). The effectiveness of entomopathogenic fungi against weevils in the genus *Hylobius* is limited by the influence of the antibiosis exerted by the microflora in the soil and leaf litter. However, experimentation indicates that with selected placement of appropriate doses and careful timing of application the use of this pathogen may be effective (Schabel, 1976) (Fig. 18.20).

Often, a complex of fungi interact to cause significant insect mortality; however, the key fungal pathogen may be difficult to identify. An example of an entomopathogenic fungal complex is that associated with bark beetle tunnels. Approximately one-third of the mortality of each year's brood of *Dendroctonus frontalis* may be attributable to its pathogenic complex (Moore, 1971).

Bacteria

Entomogenous bacteria are usually classified as obligate or facultative and as sporeformers or nonsporeformers. Sporeforming bacteria produce endospores, which persist in a dormant state outside the host. Upon ingestion by a host, these bacteria germinate in the gut. The crystalliferous sporeformers such as *B. thuringiensis* form both endospores and a proteinaceous crystal in the sporangium at the time of sporulation. The crystal contains an endotoxin that can paralyze the gut of most lepidopterous larvae. Other toxins may also be produced by crystalliferous bacteria (Falcon, 1971; Maddox, 1975). Although normally nonpathogenic, nonsporeforming bacteria can penetrate into the hemocoel and blood and

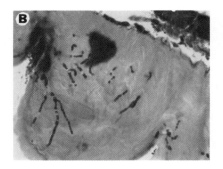

Figure 18.20.   (a) Scanning electron micrograph of various microflora on the pronotum of an adult *Hylobius pales*, including dormant conidium of *Metarrhizium anisopliae*; (b) *Metarrhizium anisopliae* hyphae penetrating through procuticular integument of larval *Hylobius pales* and melanotic reactions, 78 hr after inoculation; (c) appearance of adult *Hylobius pales* 2 weeks after death due to infection by *Metarrhizium anisopliae*. (From Schabel, 1976.)

become pathogenic when their host is stressed by temperature, poor food, parasites, or other pathogens. The greatest biological control potential and current use of these microorganisms (Table 18.11) lies with the sporeforming bacteria which are potent pathogens of many lepidopterous larvae and are perhaps the most important bacteria for insect regulation. Noncrystalliferous, sporeforming bacteria such as *Bacillus cereus* are widely distributed in the field and are often associated with Coleoptera, Lepidoptera, and Hymenoptera. However, their effectiveness is severely limited by the pH of the gut of most insects. With further research other bacteria may prove useful (St. Julian *et al.*, 1973).

Larvae of most lepidopterous species are the targets of bacterial pathogens, and when the pathogen is effective, infected individuals die as larvae. However, for some species, including the Siberian silkworm, the eye-spotted budmoth (*Spilnota ocellana*), the forest tent caterpillar, the mourning cloak butterfly, and the spruce budworm, mortality is delayed until after pupation or eclosion of adults (Falcon, 1971).

Table 18.11. Insects for which *B. thuringiensis* and nucleopolyhedrosis virus (NPV) are registered[a]

Insect	Insecticide Formulation	
	Bt[b]	Virus
Alsophila pometaria (Harr.), fall cankerworm	WP or aqueous conc.	
Choristoneura fumiferana (Clemens), spruce budworm	WP	
Heterocampa guttivitta (Walker), saddled prominent	WP	
Hyphantria cunea (Drury), fall webworm	WP or aqueous conc.	
Malacosoma americanum (F.), eastern tent caterpillar	WP or aqueous conc.	
Malacosoma californicum (Packard), western tent caterpillar	WP or aqueous conc.	
Malacosoma californicum fragile (Stretch), great basin tent caterpillar	WP or aqueous conc.	
Malacosoma disstria Hübner, forest tent caterpillar	WP or aqueous conc.	

(continued)

Table 18.11. (*continued*)

Insect	Insecticide Formulation	
	Bt[b]	Virus
Neophasia menapia (F. and F.), pine butterfly	WP	
Orgyia pseudotsugata (McDunnough), Douglas-fir tussock moth	WP	NPV[c]
Paleacrita vernata (Peck), spring cankerworm	WP or aqueous conc.	
Phryganidia californica Packard, California oakworm	WP or aqueous conc.	
Lymantria dispar (L.), gypsy moth	WP or aqueous conc.	NPV[d]
Schizura concinna (J. E. Smith), redhumped caterpillar	WP or aqueous conc.	
Thyridopteryx ephemeraeformis (Harworth), bagworm	WP or aqueous conc.	

[a] From Maksymiuk (1979).
[b] Follow information on label (dose, etc.; WP, wettable powder).
[c] TM Biocontrol-1 (trademark), polyhedral inclusion bodies of Douglas-fir tussock moth 70 million NPV units per gram; for population reduction aerially apply 1/2 oz. in 1–2 gallons spray/acre (14.175 g in 9.254–18.708 liters/hectare).
[d] Gypchek (trademark), polyhedral inclusion bodies of gypsy moth. For foliage protection make two applications 7–10 days apart at a rate of 25–125 million gypsy moth potency units per acre in sufficient water for thorough coverage.

Protozoa

In comparison with the rapid action of some bacteria and viruses, protozoan pathogens act slowly. Their effects are not rapid enough to prevent severe damage to trees, although a few are virulent and cause death a week or more after infection. Vital host functions are not often disrupted quickly; the result of infection is more often slow debilitation, reduced fecundity, reduced longevity, and lack of responsiveness. The primary route of infection is through ingestion. Transovarial transmission occasionally may be a method of dispersion (McLaughlin, 1973). The future of biological control with protozoa lies primarily in introductions that are designed to dampen impending or potential outbreaks.

The microsporidian *Perezia fumiferana*, a pathogen of the spruce budworm, does not generally induce a high level of mortality (Tanada, 1959). However, on occasion enzootic infections may vary from approximately a 5–20% infection rate and in some populations may increase to 81%. *Perezia* is transmitted to almost all progeny of diseased adult females. Spores can remain dormant in the egg or second instars (overwintering stage) for months. When larvae start to feed again in the spring, *Perezia* multiply and kill many second through fourth instars (Burges, 1973). Survivors have retarded larval and pupal development as well as reduced pupal weight, adult fecundity, and longevity. Thus, this pathogen is closely synchronized with its host's life cycle and can have a long-term impact.

Preliminary research has been conducted on a number of other protozoans. *Thelohania hyphantriae*, a microsporidian of the fall webworm, and protozoans of *Euproctis chrysorrhea* (the browntail moth) and the lackey moth show some promise. *Nosema bombycis* also appears highly infective against the fall webworm (Tanada, 1959). As with other microbials and insecticides, appropriate formulations incorporating protectants and enhancers are required in the use of protozoa. The key to the use of protozoa in an integrated pest control program may be its ability to debilitate individuals in a population and thus increase their susceptibility to other mortality agents.

Nematodes

Whether one considers nematodes to be pathogens, parasites, or pathogenic parasites, the central issue is that nematodes can cause mortality of their insect hosts in several ways. For example, phoretic bacteria may be introduced by the nematode when it enters the insect host. Ultimately, the bacteria in the dead insect may serve as food for the nematode. Nematodes sometimes kill their insect hosts when leaving the host to enter the soil for their transformation to the adult stage. A high rate of mortality in the beetle *Eutetrapha tridentata* results when the nematode

Diplogasteritus labiatus ruptures the host intestine to enter the hemocoel (Poinar, 1971). Nematodes infest the body cavities of many forest insects. Bark beetle species in several genera have been found to be infested with nematodes (Fig. 18.21).

Nematodes parasitic on bark beetles in the United States belong to two superfamilies, the Tylenchoidea and the Aphelenchoidea. Adult parasites range from 1/2 to 7 mm in size. As of 1963, 250 nematode parasites and associates (whose niches are unknown or are other than parasitic) of 60 species of bark beetles have been recorded worldwide (Nickle, 1963). Over 100 species and subspecies of nematodes have been associated with various bark beetle species in France and Germany.

Nematodes are well adapted to their parasitic existence. The life histories of nematodes and their hosts closely parallel each other. The effect of nematode activity is often mechanical rather than pathological, and infections are generally not lethal. Abdominal cavities of adult female hosts may be so filled with various nematode stages that normal egg production is greatly reduced. In Wisconsin, *Parasitaphelenchus oldhami* infestations of *Scolytus multistriatus* often average from 75 to 100% of the population. The number of nematodes per infested beetle ranges from 30 to 40, and observations of 100 to 200 per beetle are not uncommon in heavily infested populations (Saunders and Norris, 1961). In England, 40% of samples of *Scolytus scolytus* (*n* = 385) and *S. multistriatus* (*n* = 180) are sterilized by nematode infestation (Poinar, 1971). Castration by nematodes

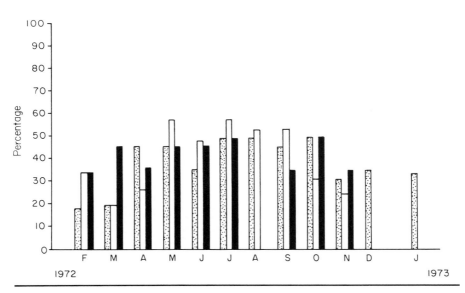

Figure 18.21. Percent of *Ips* beetles infected with nematodes, by months: *I. avalus* (stipple bars), *I. grandicollis* (white bars), and *I. calligraphus* (black bars). (From Hoffard and Coster, 1976.)

has also been reported in the wood wasp *Sirex noctillio*. The fact that nematode juveniles that have been removed from adult hosts can be cultured on the symbiotic fungus carried by the wood wasp enhances its potential as a biological control agent. However, the major drawback to the use of these nematodes is their wide host range, which includes hymenopterous parasitoids of wood wasps (Poinar, 1971). Still another impact of nematodes involves behavioral changes in their hosts due to infection. An example is the shorter-than-normal egg galleries produced by infected beetles (Massey, 1966) (Fig. 18.22). In summary, the outcome of specific host–parasite interactions varies, but the results of such encounters may include death, weakened hosts, reduction in egg-laying, sterilization, and behavioral changes.

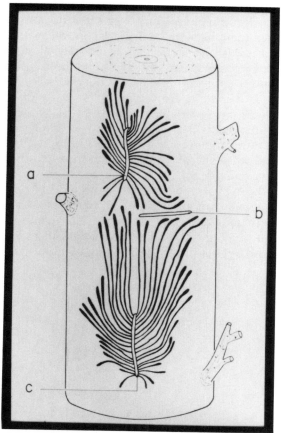

Figure 18.22. Illustration of altered gallery formation behavior of *Scolytus rugulosus* in peach trees caused by a nematode parasite: (a) and (c) show the normal female beetle gallery pattern; (b) female beetles parasitized by the nematode *Neoparasitylenchus rugulosi* form horizontal galleries and produce no eggs. (From Nickle, 1971.)

Intensive research on the ecology, behavior, systematics, and epidemiology of nematodes of dendrophilous insects is still scarce. In addition to the lack of a biological data base, technological problems must be solved, such as the development of artificial media for mass production (Massey, 1966). Until these data and techniques are available, evaluation of the effectiveness of nematodes for biological control is premature.

Chapter Summary

Biological control is the use of living organisms or their products for the suppression of pest insects. Classical biological control is an approach applied to exotic pests and involves introduction of control agents from the source of origin of the pest species. Biological control can also be accomplished by inundating pest populations with biological control agents. Vertebrate predators, including birds and small mammals, may play an important role in regulating pest populations but have not been used extensively in forest-wide control projects. Likewise, invertebrate predators, especially insects, ants, and spiders, are significant natural regulation factors. Insect parasitoids in the orders Hymenoptera and Diptera have been used in the classical biological control approach and to a lesser degree in inundative releases. Inundative release of parasitoids has limited use in North American forests because of cost but is used in forests of other parts of the world (e.g., China and Brazil). Pathogenic microorganisms including viruses, bacteria, and fungi have been extensively used in pest suppression projects. These agents are often applied with the same technology developed for insecticides and are sometimes referred to as biological insecticides. Numerous biological and physical factors affect the effectiveness of biological control agents in reducing populations of pestiferous insects.

CHAPTER 19

Insecticides and Chemical Control

Introduction

Although the numerical levels of dendrophagous insect pests are often impressive, the most extensive and intensive use of the 200 or more registered insecticides in the United States is against agricultural pests. Chemical control is also an important and useful tool for forest and shade tree pest management, though it is limited in comparison with agricultural uses. The approach of most forest managers is to incorporate practices that are in concert with fundamental ecological or silvicultural principles. This conservative focus is due to features unique to managed forests. Included among these is the economic reality that trees are a long-lived, relatively low-value crop. Therefore, costs of immediate controls must be measured against anticipated future yields. The use of chemical control can be impractical for several reasons, but most importantly because the costs of comprehensive chemical applications are high and the size of most forests is quite large.

The development of effective aerial methods of application (Fig. 19.1) and the development of the nonselective insecticide DDT made the use of insecticides economically practical and widespread after World War II. The extent of aerial application of insecticides ranged from a few hundred acres to millions of acres. However, an increasing trend toward the blanket coverage of vast areas heightened public concern over the environmental impact of insecticide applications (Morris, 1951b; Webb, 1959; Bevan, 1974).

A variety of chemicals are used for the control of forest and shade tree insects. The basis for classification of these insecticides is the nature of the chemical. Inorganic insecticides, which were the earliest used, include materials such as cryolite, elemental sulfur or phosphorus, and lead arsenate. They act primarily upon ingestion. Because they are toxic to

Figure 19.1. Helicopters capable of spraying 1000 acres/hr are used to treat large portions of infested areas. (From Mounts, 1976.)

mammals, persistent in the environment, able to induce insect resistance, and relatively ineffective compared with synthetic materials, inorganic insecticides have been replaced by organic insecticides. Organic insecticides encompass a range of materials including botanicals and their derivatives (e.g., nicotine, pyrethrum, and rotenone), oils, and synthetics. Synthetic insecticides are by far the most important and widely used chemicals in forest and shade tree pest management. They may be known by their common, chemical, or proprietary names (registered trademarks). For example, carbaryl is the common name of 1-napthyl N-methylcarbamate and is associated with the proprietary name Sevin.

Major Groups of Insecticides

Only the major groups of insecticides are discussed here. Many of the other groups are small or do not include materials commonly used in forest and shade tree insect control (Chater and Holmes, 1973).

Organophosphates

Organophosphates are insecticides that contain phosphorous; many are used systemically in plants. Some common representatives of the organophosphate group include diazinon, chlorpyrifos (Dursban), acephate (Orthene), dimethoate (Cygon), bidrin, ethion, and trichlorofon. This group also includes malathion, which is among the safest materials in terms of mammalian toxicity, and parathion, which is highly toxic (O'Brien, 1967; Pfadt, 1985).

Carbamates

The most common and effective compounds in this group are N-methyl aromatic carbamate esters. In general, carbamates degrade readily in vivo and in the environment and have low mammalian toxicity, but they are not broad-spectrum. Their selectivity is useful for incorporation into pest management schemes. The major exception to this selectivity is the compound Sevin, an insecticide commonly used to control various leaf-chewing forest insects (O'Brien 1967; Pfadt, 1985). Other carbamates such as bendiocarb are used on gypsy moth egg masses.

Organochlorines

This group includes perhaps the most famous and infamous of all insecticides, DDT. This and other compounds in the organochlorine group are composed of chlorine, hydrogen, carbon, and occasionally sulfur or oxygen. The relatively low toxicity of DDT, its low cost, and its broad spectrum of effectiveness made it appear to be a panacea. Its persistence in the environment, lipid-solubility, and apparent long-term hazards resulted in its elimination from forest pest control. Other insecticides in this group have also been widely used. These include methoxychlor, dienochlor (Pentoc) benzene hexachloride (BHC), and lindane (a purified form of BHC) (O'Brien, 1967; Pfadt, 1985).

Pyrethroids

The natural insecticide pyrethum has been used as a model to develop a variety of insecticidal compounds known as pyrethroids. These materials are stable in sunlight and effective against pests at low rates of application. Pyrethroids are important for the control of shade tree pests because they are highly toxic to insects and only slightly toxic to mammals.

Oils

These petroleum derivatives consist of a complex mixture of hydrocarbons. Oils are commonly sprayed on shade trees or plantation trees to control mites, scale insects, and mealybugs; they kill by suffocation. Highly refined oils, sometimes referred to as summer oils, are used on growing trees. Less highly refined oils (dormant oils) are frequently used before budbreak. Oils are usually applied as 1–3% emulsions. Often they are combined with an insecticide, such as ethion. The oil may enhance the action of the synthetic insecticide by increasing its ability to penetrate the target organisms. A major drawback of the use of oils is the potential for phytotoxicity (toxicity to plants). Low-viscosity oils are the safest to use on plants. New oil formulations that result from highly refined material allow application during both the dormant and the growing seasons and have greatly reduced problems of phytotoxicity. Certain characteristics of oils, such as their lack of mammalian toxicity, short residual activity, and apparent inability to induce resistance among insects, make oils important tools, particularly in urban forests (Chater and Holmes, 1973; Anonymous, 1973).

Miticides

Many important dendrophagous species are not insects but mites. The need to control these pests led to the development of chemicals (miticides) that are not effective against insects but are toxic to mites. These are generally relatives of DDT, organosulfur compounds, or nitrophenols. A miticide commonly used on shade trees is dicofol (kelthane).

Mode of Action of Insecticides

The majority of insecticides, including pyrethroids, organophosphates, organochlorines, carbamates, and some botanicals, exert their toxic effect on vertebrates and invertebrates by affecting the nervous system. Normally, a nerve impulse is conducted along an axon and ultimately reaches a synapse, which is essentially a space between two nerve cells (neurons) or between a nerve cell and an effector organ (i.e., muscle, gland, or sensory cell). The nerve impulse triggers the release of a chemical (transmitter) that diffuses across the synapse and triggers another nerve impulse or a response (if it is an effector organ). Common transmitter substances include acetylcholine and norepinephrine. To restore sensitivity and thus make the cell competent to transmit an impulse anew, the transmitter substance must be removed. This is accomplished enzymatically by acetylcholinesterase or monoamine oxidase (for norepinephrine) (O'Brien, 1967).

The insecticides just mentioned inhibit cholinesterase and thus disrupt nervous activity by causing, for example, the accumulation of acetylcholine at nerve endings. In vertebrates this results in a variety of symptoms including twitching, slowing of the heart, pupil constriction, urination, salivation, and paralysis. This type of toxicity is counteracted by either atropine or oximes (e.g., 2-PAM), depending on the specific symptoms. Insect poisoning is characterized by tremors, paralysis, and finally death.

With all the concern, interest, and intensive research that has been associated with organochlorines, particularly DDT, it is interesting that their exact mode of action is still unknown. The primary effect is believed to be on the nervous system. Symptoms such as tremors, hyperexcitability, convulsion (in mammals), and paralysis in poisoned organisms, as well as other experimental evidence, suggest that these materials act on the axon rather than the synaptic junctions (O'Brien, 1967).

Mode of Entry of Insecticides

Most insecticides must enter the body to cause mortality. Entry is generally achieved via the mouth, cuticle, or spiracles. Although an insecticide may have a dominant mode of entry, most current materials have the potential of penetrating insects in more than one fashion. The route by which a chemical enters an insect affects its toxicity in some cases. For example, the hierarchy of effectiveness from most to least toxic is DDT, lindane, and chlordane when they penetrate through the cuticle. When the insecticide is ingested, the toxicity hierarchy changes to lindane, DDT, and chlordane (Pfadt, 1985). This change can result from characteristic differences in each insecticide and/or their solvents, the feeding response of the insect to insecticide-coated food, or species-specific differences among insects.

Formulations of Insecticide

The form of an insecticide preparation can determine (1) its usefulness for the protection of particular crops, (2) the equipment with which it can be applied, (3) its effectiveness against the target organisms, (4) its effect on nontarget organisms, (5) its rate of degradation in the environment,(6) its ability to be deposited and to remain where it will be most effective, and (7) its mode of entry into the target insect. In general, a commercial insecticide consists of a toxicant (active ingredient) and one or more so-called inert ingredients. The functions of these inert ingredients may be quite varied and include dissolving or diluting the toxicant (so-called carriers or diluents), emulsifying two insoluble liquids with each other, or dispersing the insecticide (so-called wetting agents, spreader-stickers, or surfactants, which reduce the forces attracting like particles of the insecticide and allow better mixing of unlike particles). This last function also enables even coating of the plant surface. Finally, protective compounds such as so-called screens, which protect against degradation of the insecticide by ultraviolet radiation, are included in commercial insecticides.

The most commonly used formulations are granules, insecticide–fertilizer mixtures, wettable powders, emulsifiable concentrates, flowables, fumigants, and (most recently) insecticide–pathogen mixtures. The emulsifiable concentrate is the most important and widely used formulation. It is usually composed of a toxicant, a solvent, and an oil-soluble emulsifying agent. Water is added to form an emulsion, which is applied by spraying.

In certain situations in which relatively small trees are cultivated (e.g., nurseries, plantations, seed orchards, and shade trees), certain formulations are used routinely although they are rarely used in rural forests. Common among these are granule, wettable powder, and fumigant formulations. Granular insecticides (toxicants produced in granular form) are sometimes used on or in soil, in seedbed preparation, or in prophylactic procedures for young trees. Granular insecticides often are formulated (mixed) with fertilizers and thus control soil insects and provide additional nutrition to plants. Wettable powder formulations such as dusts are diluted and suspended in water and applied as sprays. This mixture includes a wetting agent and a spreader-sticker. Flowables are made from finely ground solid materials that are suspended in a liquid. In this form they can be mixed with water and applied. Among the advantages of flowables are that they do not clog nozzles and require only moderate agitation. Fumigants are insecticides that are formulated to act in gaseous form. These are used to disinfect insect-injured nursery stock or occasionally to control insects infesting seasoned wood (Pfadt, 1985).

Systemic Insecticides

Although a large number of dendrophagous species are effectively controlled by insecticidal sprays, control of cryptic species, such as leaf miners, rollers and tiers, bark beetles, wood borers, and several arthropods with piercing-sucking mouthparts, is less than adequate with conventional insecticides. However, these species are sometimes susceptible to systemic insecticides.

A systemic insecticide is absorbed by tree tissues and translocated via the vascular tissues (i.e., xylem and phloem) from the treated area to distant parts of the tree, including the stem, roots, and foliage. Systemics can be sprayed on foliage or trunks or broadcasted as granules on soil like other pesticides. In addition, they can be injected into tree trunks (into vascular tissues) or into soil in the root zone using, for example, bidrin or Meta-Systox R, respectively. Implants of systemic insecticides such as acecaps may be injected into the trunk, where they dissolve. Trunk treatment involves a variety of methodologies. Solutions may be introduced by gravity or pressure (at approximately 10–40 psi) into holes drilled in the trunk or major roots (Fig. 19.2). Insecticides and fertilizers can also be injected into trees via implants that dissolve and gradually release material into the xylem (Fig. 19.3). The most effective transport occurs when holes extend into the outer sapwood (Norris, 1967).

The use of systemics offers a number of advantages. Incorporation of insecticides within the tree frees the material from the degradation that generally results from exposure to weather. The environmental hazards that can result from drift and the spraying of nontarget organisms such as natural enemies can be avoided. Systemics are effective against a variety of chewing and sucking insects and mites. When well translocated, the toxicant continues to be effective even in the new growth.

As with any approach to control, there are also disadvantages to the use of systemics. Perhaps the major problem with systemics is the inability

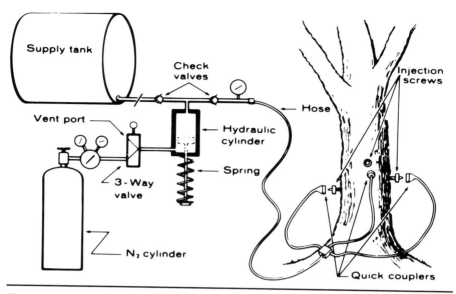

Figure 19.2. Essential parts for pressure-injection machine of the type used for the injection of systemic insecticides. (From Reil and Beutel, 1976.)

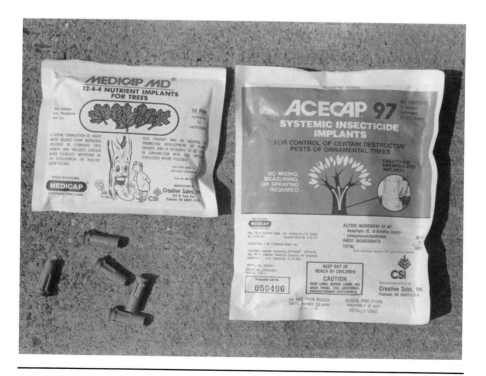

Figure 19.3. Plastic implants of fertilizer and insecticide that are placed in predrilled holes at the bases of trees.

to achieve uniform distribution of the toxicant throughout the physiologically functional tissues of the tree. Phytotoxicity is another potential hazard that can limit the effectiveness of systemics and requires a very accurate regulation of dosage. Phytotoxicity often results in yellowing or browning of needles, needle drop, marginal browning (burning) and stunting of leaves, lesions on the outer bark, cortical necrosis, and foliage dysfunctions. As with conventional insecticides, phytotoxicity caused by systemics is related to tree vigor, tree species, tree age, site conditions, and seasonal conditions (Norris, 1967).

The cost of systemic use can be high compared with the use of other insecticides. Systemic treatments of trees at present are practical only for high-value trees. Thus, their use is probably most appropriate in urban forests and for shade and ornamental trees. Although more research is needed into various aspects of their use, the future for systemics holds great promise.

Equipment for Chemical Applications

The equipment used for chemical control of dendrophagous species varies according to the target pest and the nature of the area to be treated. Accessibility and size of trees, woodlots, or forests can determine the type of equipment to use. Aerial sprays are commonly used to control insects in the forest, whereas pests in forest nurseries, seed orchards, and urban forests are more frequently controlled with mist blowers, high- and low-pressure sprayers, soil injectors, granular formulations, and soil fumigation (Fronk, 1971) (Table 19.1 and Fig. 19.4).

Plantation or urban trees can be sprayed with a variety of equipment. Multipurpose sprayers have plunger or piston pumps that can deliver approximately 3–8 gallons per minute (gpm) at pressures as high as 800 pounds per square inch (psi). Low- and high-pressure boom sprayers with 15- to 60-ft folding booms are used in forest and nursery pest control. High-pressure hydraulic sprayers with hand-held spray guns are commonly used to apply insecticides to ornamental and shade trees. Generally, appropriate suspensions or emulsion are applied at approximately 35–60 gpm at 650 psi, to the point of runoff (the insecticide runs or drips off the leaves and down the stem of the tree).

Mist blowers, which are a type of airblast sprayer, are also used on shade trees by arborists, landscapers, and nursery managers. The working principle is based on the use of air to transport pesticides to the tree. These economical and effective sprayers vary in size from backpack models to large truck-mounted blowers that apply as much as 60,000 cubic feet per minute (cfm) at nearly 100 mph. Backpack sprayers rely on the airblast to shear the liquid into fine droplets. Larger models provide added shearing by pumping the pressurized liquid through nozzles before the airblast

Table 19.1. Average methoxychlor deposits detected on elm twig crotches after spraying with mistblower, hydraulic sprayer, or helicopter for elm bark beetle control[a]

Type of Equipment	Average μg Methoxychlor/mm² Bark					
	Initial Samples[b]			Final Samples[b]		
	Top	Middle	Bottom	Top	Middle	Bottom
Mist	0.43	1.02	0.95	0.16	0.45	0.54
Hydraulic	0.60	0.91	1.05	0.21	0.44	0.57
Helicopter	0.53	0.24	0.15	0.14	0.10	0.06
Mist versus helicopter	NS[c]	**[d]	—[e]	—	—	—

[a] Modified from Cuthbert *et al.* (1973).
[b] Initial samples taken 6 to 21 days after spraying. Final samples taken 152 to 172 days after spraying.
[c] NS: Not significant at 0.05 level of probability.
[d] **: Significant at 0.01 level of probability.
[e] —: No test completed.

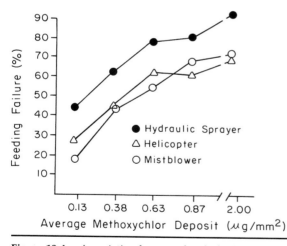

Figure 19.4. Association between beetle feeding and methoxychlor deposits from mistblower (O), helicopter (△), and hydraulic applications (●). Note: In this instance deposits from hydraulic sprayers covered all twigs more uniformly than mistblower or helicopter applications. (From Cuthbert *et al.*, 1973.)

transforms it into a mist. Wind is the major limiting factor in the operation of a mistblower. Winds above 4–5 mph make it difficult to reach the tops of tall trees and cause excessive drift. Generally, highly concentrated emulsions (or sometimes wettable powders) are used in quantities as small as 1 or 2 quarts of concentrate. This quantity is sufficient to reach most large trees. Mistblowers are popular because of their light weight, limited use of water, and ease of operation (Cornell University, 1977; Fronk, 1971).

Granular insecticides and fumigants are most commonly used to control pests in forest nurseries and seed production areas. Use of granular insecticides must be quite precise because incomplete application gives ineffective control and excess application can result in injury to the plants. Soil fumigation may be performed by injection, surface treatment (mixing), or drenching and flooding. Tarpaulin fumigation involves the volatilization of a pesticide under a tentlike tarp.

The vast forest areas that are often infested by insects, the lack of accessibility to these forested areas, and the rapid spread of some outbreak species often make aerial spraying the only adequate method of insecticide application. Although small-scale aerial spraying of arsenicals occurred before the mid-1940s, it was not until the advent of DDT that the full potential of aerial spraying was recognized. Spray technology developed rapidly; by the early and mid-1950s several types of military and civilian aircraft had been adapted for aerial spraying. Concurrently, technological developments provided spray apparatus that could deliver uniform emissions of various formulations of insecticides at desired atomization. Similar advances in calibration procedures and aircraft guidance systems were also developed to provide complete and controlled coverage of insecticidal material (Randall, 1975).

Like any other technique, aerial spraying has its advantages and disadvantages. Most obvious among the advantages are the speed of coverage, the acreage that can be treated per unit time, the ability to treat ordinarily inaccessible areas, and the reduction of labor requirements. However, the potential for environmental risk is greater as the size of treatment areas increases and as the volume of toxicant in the environment increases. In addition, effectiveness is influenced by the ability to thoroughly treat the entire canopy without skipping areas.

Factors Influencing The Effectiveness of Insecticides

The ultimate effectiveness of an insecticide is determined by a number of factors other than the toxicity of the dosage applied. For example, mechanical and procedural aspects of the application of insecticides and the nature of the environmental conditions in which they are applied are very important. In addition, the physiological state and behavior of the target insect and various ecological interactions can influence the effectiveness of insecticides.

The application of insecticides requires skilled personnel. These personnel often must mix insecticides in the field, calibrate equipment, calculate dosages, and use spraying equipment in specific ways. Spray equipment can be complex, and its improper operation or abuse can have a significant impact on the level of control achieved. Critical parts of spray equipment include tanks, agitators (which mix solutions or maintain suspensions), strainers and filtering screens, pumps, pressure regulators, valves, and nozzles. Thus, a vital limitation in achieving the effective use of insecticides often is improper training or carelessness (Pfadt, 1985; Metcalf, 1975).

Physical Factors

An insecticide applied to a tree begins to lose its effectiveness as soon as it is applied, primarily because of weather. Temperature, humidity, wind, and sunlight all affect the potency of an insecticide; the more extreme a factor is, the quicker the degradation of the insecticide. Some aspects of this process are obvious. Rain washes insecticides off foliage; but if the insecticide dries thoroughly before rainfall, removal is less likely.

Humidity can have an important influence, particularly when fine mists are sprayed. High evaporation rates may alter droplet size. Figure 19.5 indicates the time required for a droplet (like that of an insecticide spray) to settle. Humidity determines, in part, how much of that droplet reaches the surface for which it is intended. Figure 19.6 illustrates the degree of reduction of a 145-μm droplet at various relative humidities. Although a droplet of this size takes one minute to settle at 100% relative humidity (R.H.), at 85% R.H. it is reduced to a droplet diameter of 94-μm, thereby reducing the time it takes to settle. All the factors affecting the sizes of droplets and the distances they travel are important because they determine not only drifting patterns but whether the droplets of insecticides reach the canopy or evaporate. The use of additives in the formulation of insecticides reduces evaporation and extends the life of the droplets (Armstrong, 1975).

The nature of the canopy influences the penetration of insecticides. For example, the more dense the canopy, the greater the wind speed reduction in the canopy. In any given stand, wind speeds vary vertically. The stability of the air and the potential for turbulent mixing of air are functions of the vertical temperature strata and the distances between them. These vary both diurnally and as a function of existing weather conditions. These patterns of air movement and turbulence can determine the pattern of droplet deposition.

The movement of air usually causes some of the applied material to move away from the intended target area (Fig. 19.7). The amount of drift varies in response to a variety of factors. Wind is one of the more important of those factors. Wind above 5 mph disrupts and carries away spray or dust particles (Armstrong, 1975).

Components of sunlight, such as heat and ultraviolet radiation, reduce the potency of insecticides. Most insecticides do not perform efficiently at temperatures below 50°F (10°C) or above 95°F (35°C). Physiological changes in the insect are also temperature dependent. The rate of

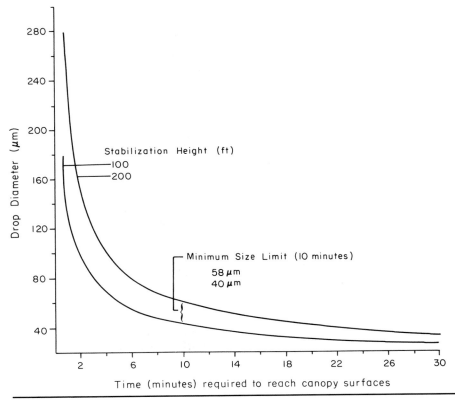

Figure 19.5. Time for water drops of various diameters to fall to the surface from release heights of 100 and 200 ft. (From Barry *et al.*, 1975.)

detoxification by insects is positively proportional to temperature. High temperatures are best for the effective penetration of the toxicant; however, insects may be able to detoxify the chemical readily at these temperatures. Conversely, at low temperatures, the rate of detoxification by the insect is slow, but limited penetration of the toxicant reduces its effectiveness. These relationships may vary with the insecticide. For example, DDT and methoxychlor are most effective at continual low temperatures (Pfadt, 1985).

Drift

Wind and air currents can have a major influence on the effectiveness of aerial spraying and the degree of contamination of peripheral areas. The movement of spray droplets in an insecticidal cloud due to wind through-out the target area and adjacent areas is referred to as drift. Drift and drifting distances vary, depending on wind speed, particle size, height of release, rate of fall of particles, and other related factors (Fig. 19.8).

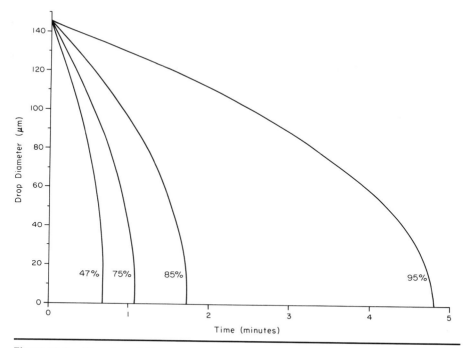

Figure 19.6.   Time for a 145-μm water drop to evaporate in still air for various relative humidities. (From Barry *et al.*, 1975.)

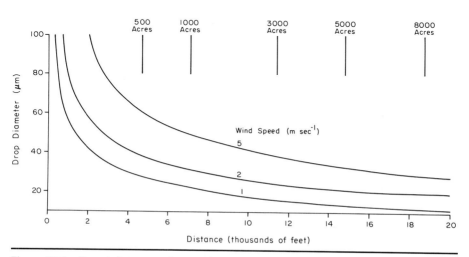

Figure 19.7.   Travel distances of water drops of various diameters for a release height of 100 ft and wind speeds of 1 m/s, 2 m/s, and 5 m/s. (From Barry *et al.*, 1975.)

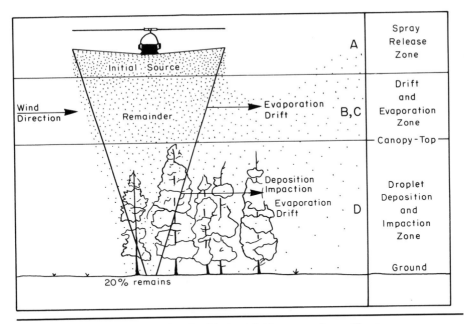

Figure 19.8.   Physical process involved in insecticide spray release. (From Grim and Barry, 1975.)

Predicting spray patterns, particularly with regard to how much toxicant will be deposited on the target insect, is a difficult task. Laboratory simulations using spray towers provide close approximations; however, the vagaries of weather under field conditions may easily negate preliminary evaluations. Generally, early morning and just prior to sunset are the times of the day when most spraying is conducted because of the low wind velocity and minimum air turbulence during these periods. The cooler air near the ground surface, relative to upper air layers, reduces upward air currents (Prebble, 1975).

Certain aerial spraying protocols attempt to take advantage of the tendency of the spray cloud to drift. The Porton method requires that aircraft fly across major wind movements, so that the larger particles drop to the ground immediately and smaller droplets drift downward. Although swath width depends on weather conditions, this procedure results in a wide swath of toxicant deposition (Prebble, 1975).

The shorter the time required for droplets to reach the canopy, the less the droplets drift and the less likely temperature, humidity, or wind will change sufficiently to alter spray patterns. Figure 19.5 shows the relationship between the size of drops (diameter of water droplets) and the time required to reach the target (surface) when released from two heights

(100 and 200 ft). A 100-μm droplet intersects the 100-ft curve at approximately 1.8 min. An 80-μm droplet takes approximately 3 min to settle to the surface when released from 100 ft and 5.5 min from 200 ft.

As previously mentioned, wind speed has a major effect on the distances droplets travel. Figure 19.7 illustrates the distance droplets of various sizes will travel when exposed to wind speeds of 1 m/s, 2 m/s and 5 m/s. The vertical lines across the top of the figure represent the length of one side of a spray block ranging from 500 to 8000 acres in size. Thus, a 40-μm droplet would travel approximately 4000 ft at a wind speed of 2 m/s, which represents a little less than the length of a 500-acre spray block.

Selectivity of Insecticides

Many of the synthetic chemical insecticides used today are broad-spectrum insecticides (i.e., they are effective against a wide range of insects). This characteristic can play a major part in many of the problems associated with chemical insect control. General toxicity to a large number of life forms, low biodegradability, and excessive and inappropriate use of these chemicals may result in an unfavorable environmental impact. Thus, insect pest management emphasizes the selective and limited use of insecticides. The selective use of chemicals is guided by a principle of pest management that states that pest numbers need only be reduced to economic threshold levels, while maximizing natural controls. The selective use of insecticides can include physiological, ecological, and behavioral selectivity (Metcalf, 1975).

A thorough understanding of the strong and weak links in the life cycles of insects is important to the comprehension of ecological selectivity. Alternate host trees and overwintering life stages or highly exposed life stages may provide the most appropriate target for regulation. Knowledge of the timing of life-cycle events and of the role of natural enemies can help determine the most appropriate time to spray the target species. Proper timing is important to avoid spraying natural enemies when they are likely to be most effective.

Another important concept that aids in the effective use of insecticides is behavioral selectivity. If the behavioral patterns of the target organism are known, the timing of applications can be adapted to avoid use of insecticides when they cannot be effective. Insects often exhibit a rhythmicity of behavior that regulates feeding, locomotion, mating, and oviposition. Knowledge of these behavioral cycles may reveal a time of maximum vulnerability. Application of insecticides may produce different mortality levels, depending on whether the larvae are resting in bark flaps or litter or feeding within a bud or freely on the foliage. All of these behaviors affect the exposure of the insect to insecticidal sprays and thus can influence the degree of control achieved (Metcalf, 1975).

The physiological selectivity of insecticides refers to the differential

Table 19.2. Comparative toxicity of carbaryl and diazinon to various instars of the gypsy moth[a]

	LD$_{95}$ Ratio[c]			
	Carbaryl		Diazinon	
Instar[b]	Per Larva	Per g Larva	Per Larva	Per g Larva
Second	1.0	1.0	1.0	1.0
Third	1.1	0.5	5.3	2.5
Fourth	6.1	1.1	9.2	2.8
Fifth	25.1	1.9	50.2	4.5

[a] From Ahmad and Forgash (1975).
[b] Larval age was 1 to 2 days from preceding ecdysis.
[c] $$LD_{95} \text{ ratio} = \frac{LD_{gs} \ (\mu \text{ Carbaryl or Diazinon}) \text{ for third, fourth, or fifth instar}}{LD_{gs} \ (\mu g \text{ Carbaryl or Diazinon}) \text{ for second instar}}$$

effect of toxicants on insects or insect stages that are physiologically different (i.e., they differ in their capacity to detoxify chemical insecticides). One example is provided by the effects of insecticide on gypsy moth larvae. Late instars are more capable of detoxifying insecticides than early instars (Table 19.2).

Assessment of Insecticide Deposition

When forest acreages are sprayed, the assessment of insecticide deposition is critical to the evaluation of application procedures and goals. The information that is essential in this assessment is the volume of material being deposited per unit area and the pattern of distribution of the droplets (i.e., the number of droplets per unit area) (Table 19.3 and 19.4). The method of assessment usually consists of using glass plates or cardboard squares with specially treated surfaces so that residual spray drops show up on the card and can be tabulated. Depending on the formulation, drops are sometimes difficult to see. Dyed sprays may be used so that more visible traces remain on squares. The residues may be further evaluated using colorimetry or other techniques such as fluorometry or gas chromatography. The relationship between drop-stain diameter and the original drop diameter (stain/drop ratio) is a function of many factors, including drop size, formulation ingredients, and type of paper used for sampling cards. Surface treatment of sampling cards can be sophisticated enough to allow measurement of droplets well below 100 μm in diameter.

Recently developed sampling cards are coated with microencapsulated ink. Spray drops with the appropriate solvent make very distinct markings when they fall on the cards. Ultraviolet lights can also be used to

Table 19.3.   Quantity of mexacarbate (Zectran) spray deposit recovered from foliage taken from the four cardinal directions of the tree crowns in a pilot-control study of the control of western spruce budworm (*Choristoneura occidentalis*)[a]

Tree Crown Direction	No. of Samples	Mean Deposit (μg/75 Needles)	Deposit Range (μg/75 Needles)
North	96	3.314 ± 0.857	0–17.577
East	96	2.906 ± 1.051	0–38.455
South	96	1.894 ± 0.620	0–16.389
West	96	2.384 ± 0.742	0–21.975

[a] From Maksymiuk *et al.* (1975).

highlight fluorescent dyes incorporated into insecticides as spray markers. Although the total amount of material deposited is ultimately determined by a variety of meteorological factors, the average measured deposition of material for any given area ranges from 1/4 to 1/3 of the intended application rate (Haliburton *et al.*, 1975).

Regulation of Pesticide Use

Many aspects of the production and use of pesticides in the United States are regulated by state and federal laws. Perhaps the most important legislation that controls pesticide use is the Federal Insecticide, Fungicide and Rodenticide Act (FIFRA) passed in 1947 and amended in 1972, 1975, and 1978. The Environmental Protection Agency (EPA) is responsible for administrating the regulations of FIFRA. Two important provisions of FIFRA that relate to pesticide application are that all pesticides are to be classified for either general use or restricted use and that restricted-use pesticides are to be applied only by certified pesticide applicators. Individual states have established pesticide certification programs, using federal guidelines, to train pesticide applicators. Forestry is one of several categories of pesticide application established by EPA. Though the pesticide laws vary from state to state, it is unlawful for any person to purchase or apply a restricted-use pesticide without the proper state pesticide certification credentials. Be sure to check local laws before applying any pesticide.

Another important aspect of FIFRA is the regulation of pesticide labeling. The information found on a pesticide label is strictly regulated and constitutes a legal document. Information on the label includes the brand name, formulation, chemical name, statement of active and inert ingredients, net contents, name and address of the manufacturer, registration number, signal word (statement of toxicity), special safety warnings,

Table 19.4. Quantity of spray deposit recovered in tree canopies and at ground level under the same trees and in the adjacent small forest openings in a pilot-control study of the effectiveness of mexacarbate (Zectran) against the western spruce budworm (*Choristoneura occidentalis*)[a]

Sampling Station	On Canopy	Under Canopy			In Openings		
	μg/75 Needles	Gal/acre[b]	Liter/hectare	Drops/cm²	Gal/acre	Liter/hectare	Drops/cm²
1	1.094	0.030	0.281	9.8	0.209	1.955	21.2
2	9.846	0.010	0.094	6.8	0.027	0.253	14.8
3	0.333	0.007	0.066	3.0	0.006	0.056	1.2
4	0.534	0.006	0.056	1.0	0.010	0.094	6.5
5	3.602	0.005	0.047	4.0	0.043	0.402	6.2
6	2.556	0.252	2.357	25.8	0.089	0.832	12.5
7	2.774	0.030	0.281	14.2	0.175	1.637	42.8
8	3.504	0.106	0.992	8.5	0.128	1.197	37.5
9	1.112	0.038	0.355	5.8	0.082	0.767	27.5
10	0.426	0.011	0.103	5.2	0.012	0.112	3.2
11	2.503	0.024	0.224	4.0	0.143	1.338	13.2
12	1.302	0.028	0.262	7.2	0.080	0.748	11.8
Mean	2.466	0.046	0.426	7.9	0.084	0.782	16.5
	±1.653	±0.045	±0.419	±4.2	±0.043	±0.405	±8.4

[a] From Maksymiuk *et al.* (1975).
[b] 1 gallon per acre = 9.354 liters per hectare.

and directions for use. Label information is provided to minimize the misuse of pesticides. It is unlawful to use a pesticide in a manner that is inconsistent with the label instructions. The best advice to all would-be pesticide users is "read the label."

Environmental Impact of Insecticide Use

The concern about the impact of insecticidal treatments on the forest ecosystem is based to a great degree on the assumption that a "natural" forest maintains itself as a complex of interacting communities that exhibit relative numerical stability. This stability is maintained by the pressures provided by the interactions of faunal and floral species, which may include symbiosis as well as competition. Divergences from the biotic central tendency of any given forest ecosystem, as might occur due to insecticidal applications, are the focus of environmental impact assessments by forest managers. Implicit in these assessments is an understanding of ecosystem integrity, which may be described in such terms as species diversity, population equilibrium, nutrient cycling, and species stability. Herein lies a major difficulty in the assessment of the environmental impact of chemical control.

The risks inherent in the spraying of forests, plantations, and nurseries are determined by many factors. Some of the variables include the type and formulation of the insecticide, the number of applications (Table 19.5), the method of application, the timing of application, the size of the area being treated, the degree of human habitation in the treatment area, and the nature and age of the treated stand. The lack of uniform and repeated chemical applications in the same area is probably the primary reason for the relative absence of widespread ecological imbalance in forest ecosystems compared with agroecosystems (Varty, 1975). Imbalances can occur in agroecosystem-like forestry operations such as seed orchards, where extensive use of chemicals can create scale and aphid problems.

The technical and scientific challenge of assessment, analysis, and interpretation of ecological impacts is significant and difficult. Analysis of the effects of pesticides on various biotic components of the forest environment generally consists of (a) toxicological studies of the chemical on a variety of target and nontarget organisms, (b) evaluation of field mortality of nontarget organisms, and (c) assessment of long-term changes in populations, whether they be single species populations, functional groupings such as parasitoids, or specific taxa.

Persistence of Insecticide

An important characteristic of an insecticide is persistence. A persistent insecticide remains in the environment as an effective insecticide for a long period of time; thus, fewer additional applications are needed. However, this same trait provides opportunity for the development of resistance by

Table 19.5. Western spruce budworm-infested forests treated with insecticides since 1952 in the northern Rocky Mountains[a]

Year	State	Insecticide	Dosage (lb/acre)	Acres Treated
1952	Montana	DDT	1.0	12,000
1953	Montana	DDT	1.0	117,140
1953	Idaho	DDT	1.0	16,070
1953	Wyoming	DDT	1.0	2,000
1955	Montana	DDT	1.0	246,530
1955	Idaho	DDT	1.0	70,710
1955	Wyoming	DDT	1.0	55,410
1956	Montana	DDT	1.0	765,880
1956	Idaho	DDT	1.0	119,370
1957	Montana	DDT	1.0	715,380
1957	Wyoming	DDT	1.0	69,300
1958	Montana	DDT-Genite	0.5–1.0	18,200
1959	Montana	DDT	1.0	126,880
1960	Montana	DDT	0.5–1.0	117,850
1962	Montana	DDT	0.5	451,760
1963	Montana	DDT	0.5–1.0	415,120
1963	Montana	Malathion	0.5–1.0	13,550
1963	Montana	Phosphamidon	1.0	5,000
1964	Montana	Malathion	0.5–0.75	157,860
1965	Montana	Malathion	0.5–0.8	3,550
1965	Montana	Naled	0.4	1,160
1965	Montana	Mexacarbate	0.15	1,080
1966	Montana	Malathion	0.80	83,030
1966	Montana	Mexacarbate	0.15	5,360
1968	Montana	Mexacarbate	0.06	6,080
1969	Idaho	Mexacarbate	0.15	6,000
1971	Idaho	Mexacarbate	0.15	9,000
1972	Montana	Mexacarbate	0.15	500
1975	Montana	Carbaryl	1.0	3,500
1975	Montana	Trichlorfon	1.0	3,800
1975	Montana	Bt (Dipel WP)	7.28 B.I.U.	3,500
1976	Montana	Acephate	1.0	3,000
1976	Montana	Trichlorfon	1.0	3,000
1977–79	Montana	Carbaryl	1.0	1,400
1979	Montana	Imidan	0.12–0.50	750
1979	Montana	Permethrin	0.24–1.4	750
1979	Montana	Carbaryl	0.25–1.0	750
1979	Montana	S.I.R. 8514	0.12	150
1981	Montana	Bt (Thuricide 16B)	8 B.I.U.	8,150
1981	Montana	Bt (Dipel 4L)	8 B.I.U.	8,600

[a] Modified from Fellin (1983).

insects and for the movement of chemicals throughout and beyond the target area via organisms, soil, water, and air. Insecticides such as DDT have been recovered from organisms at considerable distances from areas of application. However, many pesticides produce residues that persist for only a few days.

In addition, the fate of a pesticide in the environment varies considerably. Some may degrade into pesticides many times more toxic than the parent compound, whereas others change into relatively innocuous substances. Finally, degradation rates differ in different parts of the environment. For example, residues of fenitrothion, used for spruce budworm control in Canada, are relatively short-lived in forest soil and water, but these same residues in coniferous foliage exhibit persistence and stability (Yule, 1975).

Insect Resistance

A critical limitation to the effective use of insecticides is the development of resistance in the target species. Often, insects that have been exposed to repeated insecticide treatments lose their susceptibility. The development of resistance has been noted since the days when only inorganic materials were used. Development of resistance to one group of insecticides does not necessarily provide immunity (cross-resistance) to chemicals in other groups. Resistant populations result from a selection process whereby preadapted genotypes survive in increasing numbers as a result of differential mortality. In general, individuals with alleles for resistance have a functional detoxification mechanism.

Resistance results whenever insects are exposed for long periods of time to selecting levels of an insecticide that causes less than 100% mortality. Residual insecticides are more effective inducers of resistance than nonresidual insecticides because they persist in the environment at functional levels. The relatively limited use of insecticides in forest ecosystems, particularly when compared with their use in agrecosystems, has made the development of insecticide resistance uncommon among forest insects.

Effects on Nontarget Organisms

Even during the early period of use of synthetic organic insecticides the mortality of nontarget organisms was observed. The same characteristic that made materials such as DDT attractive (broad-spectrum activity) also increased the likelihood of nontarget mortality. In addition, high lipid (fat) solubility and other biochemical characteristics resulted in concentration of these insecticides in fatty and nervous tissues of vertebrates at several trophic levels. Physiological abnormalities and declines in insect (including parasites and predators), bird, fish, and mammal populations were also noted.

Early studies of the survival of nontarget arthropods in the woodlands of northeastern United States that had been sprayed with DDT, reported heavy losses of beetles, lacewings, and hemipterous insects (Hoffman *et al.*, 1949). Other studies conducted since then show that syrphids, coccinellids, pentatomids, spiders, and other important predators suffer significant mortality after chemical spraying. In addition to mortality, abnormalities of the reproductive physiology of vertebrates (i.e., thin eggshells and reduced vitality and survival of progeny) as a result of insecticide applications have been noted. A variety of behavioral changes also have been demonstrated in mammals and fish as a result of sublethal dosages of insecticides (Pimentel, 1971; Prebble, 1975).

An important indirect effect of insecticide use is the alteration of the composition and numbers of forest insects that may result after major spraying programs. These changes, although difficult to assess, may in turn cause major changes in the forest ecosystem. Mortality of target and nontarget organisms may reduce the food available to invertebrate and vertebrate predators. Perhaps more importantly, drastic reductions of natural enemies may allow the resurgence or emergence of major and formerly minor species, respectively, as significant pests. For example, phytophagous mites, aphids, and scales may reach epidemic levels after spray operations geared toward other major pests (see Long-Range Changes in Forest Stands in Chapter 7).

In contrast, many other studies fail to demonstrate any deleterious alterations in forest ecosystems as a result of insecticide spraying. These studies include evaluations of the effects of insecticide applications on natural control agents (Blais, 1960; MacDonald and Webb, 1963; Carolin and Coulter, 1971). Other studies demonstrate that, after an initial population decline, insect species recover their numbers in the years following insecticide spraying (Varty, 1975).

Phytotoxicity

Limitations on insecticide use may also develop as a result of their effects on the host trees of the target species. Various insecticides are capable of causing plant injury (phytotoxicity). Symptoms of phytotoxicity include dead leaves and burnlike or scorched spots on leaves. Because these symptoms frequently appear several days after treatment, they are often confused with plant diseases, insect activity, nutritional deficiency, water stress, weather-related injury, or pollution. Whether or not phytotoxicity occurs varies with the type of formulation, concentration of the toxicant, method of application, plant health, and type of insecticide used. Although each situation is different, organophosphates, oils, and carbamates are more likely to cause injury. Oils, even if refined, are often phytotoxic. Certain tree species such as sugar and Japanese maple (*Acer palmatum*), beech, hickory, walnut, Douglas-fir, and spruce are often injured by oil sprays. Applications of oil to Colorado blue spruce (*Picea pungens*) removes the color from the needles. Even in those species that are generally tolerant, oils must be used within a certain temperature range. Use of dormant oils is recommended only between 40 and 75°F (4 and 24°C). Similarly, it is not recommended that any liquid concentrate be sprayed above 85°F (29°C). Phytotoxicity can also be avoided if chlorinated hydrocarbons are not used in hot, dry weather.

As suggested, certain formulations can increase the possibility of phytotoxicity. Emulsifiable concentrates, because of their solvents, are more likely to cause injury than wettable powders. Pesticide mixtures are more likely to cause plant injury. Finally, insecticides vary in their phytotoxicity. Thus, dimethoate tends to cause injury to elms, some hollies, honey locust, dogwood, crab apple, and maple, whereas endosulfan may injure white birch and redbud, and malathion may injure junipers.

Although a great deal of research has been conducted on the environmental hazards of insecticides, it is often difficult to obtain clear patterns or conclusive evidence of direct or indirect effects of any given chemical in any given forest site. One reason for conflicting results is that target insects, nontarget organisms, and forest habitats are dynamic, complex entities whose interactions result in a multitude of ecological permutations. In addition, although the scientific methods used to evaluate impact are designed to enhance objectivity and repeatability, they are still subject to methodological limitations and human frailties (including bias). In the final analysis, many of the hazards of insecticide use in forests may be reduced or avoided by a well-thought-out analysis of available data on target and nontarget organisms, their habitats, and the chemicals and equipment being used. An accurate estimation must also be made of the relative balance between potential risks and benefits (Southwood and Norton, 1972; Pimentel *et al.*, 1978).

Benefit–Cost Analysis

The use of insecticides cannot simply reflect the need to reduce the abundance of an insect pest. Even after an accurate analysis of the insects' effects on host trees has been made, other economic factors must be considered. The development of a benefit–cost analysis produces a balance sheet of the cost of control in relation to added profits resulting from increased yields (or other objectives) associated with control activities.

The value of this type of analysis is based on its thoroughness and objectivity and on the accuracy or validity of economic data. The component parts of a benefit–cost analysis may include only a few items or a wide variety of factors, depending on the size of the infested area. Some of the components that may be considered include the economic value of the total timber volume to be saved and of the watershed areas, cost of extra fire control due to defoliation of a stand, detrimental or beneficial effects of spraying on wildlife and wildlife habitat, economic and aesthetic values of recreational areas, and cost of labor, materials, and equipment necessary to complete the control project. Clearly, the development of an objective benefit–cost ratio is a difficult and important task.

Chapter Summary

The use of chemicals to control forest insect pests is still an important part of the management of dendrophagous species. Insecticides are divided into major groups based on their structure and mode of action. A variety of insecticide formulations have been developed and enhance the utility of these products in forestry. Systemic insecticides are very useful in forestry because of their ability to move throughout the plant and protect individuals from insect damage; phytotoxicity can be a problem with systemics. Specific equipment has been developed to apply chemicals in forest and urban environments. Aerial application via helicopter or fixed-wing aircraft is the most common method in natural forests. Physical factors such as temperature, humidity, wind, and sunlight can significantly influence the effectiveness of chemicals. Federal and state laws govern the use of pesticides in the United States, and applicators are required to obtain training and abide by appropriate laws. Considerable controversy still exists regarding the environmental impact of insecticide use. Persistence, development of resistance, and effects on nontarget organisms are real concerns with insecticide use and must be considered when deciding to apply pesticides as part of the management of forest pests.

CHAPTER 20

Direct Control of Forest Insects

Introduction

Direct control can be defined as tactics or procedures designed to achieve immediate numerical suppression of insect pests. Chemical insect control is a direct control, although it is generally discussed as a separate category (see Chapter 19). Indeed, many of the procedures discussed here include the use of insecticides.

A variety of features characterize most direct control measures. As with other control strategies or procedures, direct control has its advantages and disadvantages. It is used most efficiently when it is part of an overall management scheme that incorporates a number of control methodologies, each designed to meet specific objectives. Direct control procedures are aimed at the prevention of numerical buildup or the reduction of economically injurious levels of insects. Direct control is achieved by various methods including destruction of habitat, exclusion, desynchronization, sterilization, and killing or trapping insect pests.

Various procedures in direct control can be very effective; however, they are often labor intensive. Direct control procedures generally have an immediate impact on their target species, although they do not always result in control of species that have a high reproductive potential. For those species, other methods must be used, or the direct control may have to be repeated. Similarly, asynchrony in the life cycles of individuals of a population may require that repeated applications of the procedure be made. For some direct control measures, such repetition may be biologically or economically unfeasible.

Whereas some procedures are applicable to forest trees, others are feasible only for limited, small-scale use as is typical in urban forests or on individual shade trees. Finally, direct control tactics may represent emergency actions in the face of existing or imminent loss. One example of such a circumstance is illustrated by the use of early harvest or salvage removal as a control tactic.

Insect Removal and Habitat Destruction

Among the simplest but often most effective procedures is the removal of the pest insect, most frequently in the egg stage but also in the larval, pupal, or adult stage. This procedure is perhaps most effective against species attacking shade trees. The removal of webs or tents during the early developmental stages of species such as the eastern tent caterpillar, fall webworm, and pine webworm can be quite effective. Other insect feeding shelters can also be easily removed. Adult females such as white pine weevils can be picked off leaders of isolated trees before any extensive oviposition occurs. For those insect species that bore into buds or twigs, selective pruning may effectively reduce damage or the potential increase of pest populations. Populations of aphid species that attack linden, elm, and ash can be successfully reduced by pruning branches from the preferred portion of host trees (Olkowski *et al.*, 1974). Egg masses, webs, and tents can be destroyed by burning before any significant defoliation occurs.

Prescribed or controlled burning can be an effective direct control method. Recommendations for the control of *Hylobius warreni*, a weevil causing root collar injury to pines, includes prescribed burning. To reduce the hazard from pest buildup in residual trees after a clearcut, it is generally recommended that residual strips and blocks of land should not remain fallow longer than 2–4 yr. In addition, scarification or prescribed burning to remove the duff layer should be instituted soon after cutting to hasten and increase mortality of the weevil larvae and pupae in stumps and the forest floor (Cerezke, 1973). Another example is the burning of leaf litter in the spring, which resulted in as much as 87.5% mortality of maple leaf cutter (*Paraclemensia acerifoliella*) pupae in a Michigan woodlot (Simmons, *et al.*, 1977). Prescribed burning is a promising tool for controlling other insects that overwinter in the forest floor and thus are vulnerable to burning. Preliminary evidence suggests that control burning of pine and herbaceous litter protects some crops in seed production areas from insect damage (Miller, 1978).

Mechanically dislodging or disrupting insects may provide an effective control. Washing off eggs from needles with water under pressure and raking and destroying litter and debris under pinyon pine trees reduces pinyon needle scale (*Matsucoccus acalyptus*) populations by 84–86%. Proper timing of the washing treatment (i.e., before crawlers emerge) is crucial (Flake and Jennings, 1974). In California, high-pressure (up to 600 psi) sprayers have been used to spray water on large (30ft) trees to remove shade tree pests (Olkowski *et al.*, 1974). Applications of tree (water) washes with handgun-type sprayers can be used to remove pear psylla (*Psylla*

Figure 20.1. (a, b) Windthrown trees and logging slash are breeding sites for bark beetles. (From Werner *et al.*, 1977.)

pyricola) honeydew from trees; washes containing a wetting agent increase mortality of pear psylla nymphs (Brunner and Burts, 1981).

The removal of logging residue is a tactic used to prevent the buildup of pest insects (particularly tree-feeding beetles) and is based on certain key concepts and assumptions. These include the concept that most beetle species of economic importance prefer to attack weakened or newly killed host timber. Some species (mostly in the genus *Dendroctonus*) are able to attack and kill living trees, although even these may prefer low-vigor or weakened trees. The basic assumption of slash removal is that logging slash and wind-thrown or weakened trees attract beetles and provide ideal breeding material for wood and bark beetles (Fig. 20.1) and result in an increase of beetles in healthy trees (Lejeune *et al.*, 1961). Thus, control recommendations for effective prevention and reduction of injurious pest population buildups include prompt removal or destruction of beetle broods in infested material (e.g., slash, stumps, trap logs, and wind-thrown trees) to prevent the emergence of new adults.

There has always been some degree of controversy over the effective-ness of sanitation as a control method. Some of the arguments against the use of this technique are economic, that is, the cost–benefit relationships of this approach. However, this can be and should be a criticism of all potential pest management tactics. Other objections question the under-lying biological assumptions of the procedure. Studies conducted in the early 1900s concluded that in northeastern forests, slash burning was not effective as a factor in forest insect control because very few injurious species were found breeding in slash. Species that were considered to be relatively harmless at low population densities, but injurious at high densi-ties, were found to prefer large pieces of slash (Keen *et al.*, 1927). Similarly, stumps furnished ideal breeding places for many insects, particularly the injurious species (Graham, 1922). Specific relationships between slash size and insect development are demonstrated in Table 20.1. The most favor-able breeding material for the Douglas fir beetle consists of freshly felled logs over 8 in. in diameter or similar-sized stumps. Stumps produce more beetles per unit area of bark than other slash types in many North Amer-ican stands. Although beetles may attack slash that is less than 8 in. in diameter, the material tends to dry out quickly and does not allow a sufficient period for brood development (Lejeune *et al.*, 1961).

Stumps left after logging operations serve as breeding sites for a number of economically important species. In North America, species in the tribe Hylobini usually live in stumps of coniferous trees and are bene-ficial scavengers. However, when areas are cleared and opened for the establishment of plantations, these insects can become destructive to planted seedlings and older trees. Adults feed on bark and may girdle trees. Complete removal of stumps in and around plantations or nurseries reduces the damage caused by many of these insects.

Table 20.1.   Differential use and survival of *Trypodendron lineatum* in fir, hemlock, and balsam slash[a,b]

Type of Slash[c]	Bark Surface (ft^2)	Holes per ft^2	Beetles Emerged per Hole	Beetles Emerged per ft^2	Beetles Emerged per Acre
Logs	4109	6.42	9.90	63.6	261,000
Chunks	1767	3.63	6.36	23.1	41,000
Tops	2147	1.91	7.03	13.4	29,000
Stumps	1444	2.88	8.49	24.5	35,000

[a] From Dyer (1963).
[b] Based on data from 4521 square-foot samples and 943 galleries, including those that failed to produce brood.
[c] *Logs* are pieces larger than 6 ft in length and 12 in. in diameter at the mid point of the length. *Chunks* are shorter than 6 ft and greater than 6 in. in diameter. *Tops* are longer than 6 ft, between 6 and 12 in. in diameter, and with more than 3/4 of their surface undamaged. *Stumps* are greater than 6 in. in diameter.

Larvae of *Hylobius warreni* feed in the root collar of healthy lodgepole pine and cause open wounds, resinosis, and partial or complete girdling. This damage ultimately causes growth loss and some tree mortality. Many larvae are able to complete their development in stumps 1 or 2 yr after clear cutting despite a high mortality rate of larval stages. Because adults survive 3–4 yr and are reproductive during at least three summers, adult survivors remaining in clearcuts are a reservoir population that can infest new trees. Indeed, areas adjacent to such clearcuts have had large and significant buildups of this weevil species (Cerezke, 1973).

Studies evaluating the effects of slash removal on the level of beetle attack on residual trees have shown varying results; the beetle population decreased after slash removal in some stands, but it remained unchanged in other stands. Noncommercial thinnings (a 30% thinning in a 16-yr-old stand and a 50% thinning in a 28-yr-old stand) of both plantations and natural stands of jack pine in the north-central United States have been used to determine the importance of leaving slash in a thinned stand. No economic losses could be attributed to the presence of slash in the thinned stands (Schenk *et al.*, 1957). Conversely, other studies have demonstrated that sanitation-salvage procedures provide effective protection from insects within a favorable cost–benefit ratio (Hall, 1958). Effective sanitation can be crucial in the control of insect-related tree diseases. Sanitation or the annual removal of potential elm bark beetle breeding material (e.g., newly dead elm wood) remains an important aspect of the control of Dutch elm disease (Fig. 20.2).

An alternative to the removal of slash or logging debris is the destruction or use of these materials. Commonly procedures include piling and

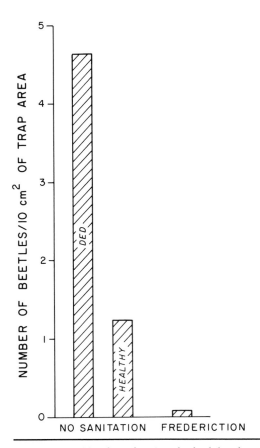

Figure 20.2.   Number of native elm bark beetles trapped per unit of trap area on diseased (DED) and healthy elms in areas of no sanitation and on healthy elms within the Fredericton (Canada) area, where sanitation was practiced (annual removal of potential elm bark beetle breeding material). (From Sterner, 1976. Reproduced with permission of the Minister of Supply and Services Canada.)

burning, peeling the bark, or treating the slash with chemical insecticides (Fig. 20.3). Burning destroys the bark, and peeling exposes the brood to biotic and physical factors. Barking (removal of bark) is also an effective deterrent to oviposition by adults. Oil solutions and water emulsion of various insecticides are also frequently used to destroy broods of bark beetles. Chemicals, such as lindane, are commonly used to treat slash. In some circumstances, storage of attacked logs in water prior to milling destroys the portion of the brood that is underwater (Lejeune *et al.*, 1961).

    In summary, the role of slash removal in preventing the buildup of dendrophagous species and reducing the level of injury to standing trees

Figure 20.3.   Chemical treatment of stacked logs. (Courtesy Shade Tree Laboratory, University of Massachusetts.)

requires further quantitative investigation. It is difficult to assess the validity of the conclusions made in previous studies because many fail to give enough detail on the size of logging debris examined or other relevant environmental factors. Nevertheless, it is clear that the removal of slash in the broad sense (including all sizes of logging debris) has some merit as a control tactic and can be an effective tactic in managing forest insect pests.

The level of success in insect control will increase as we gain further understanding of the biotic and physical factors affecting the attraction and breeding of pest insects in slash. These factors include, but are not limited to, the effect of microenvironmental factors on target insects and slash; the influence of slash type, size, and structure; the role of site differences and species composition; the size of the surrounding target insect population; and the specifics of host selection behavior. An example of the latter point may be helpful. Even though some of the early studies minimized the importance of slash as potential breeding material, they did recognize the distinction between breeding and host selection. Thus, these studies conceded that slash might be important because of its ability to attract and concentrate pest populations from surrounding areas, even when it did not serve as a site for breeding. Little is known about the long-term effects of slash removal or destruction.

Habitat Manipulation

An understanding of the biotic and environmental limitations in the habitat of a species can provide a means of direct control. Habitat manipulation provides a number of good examples of the usefulness of that understanding. The manipulation, rather than the destruction, of logging debris is an example of direct control through habitat manipulation. This approach requires a detailed understanding of the biology and life history of insect pest species. For example, although species of *Ips* breed successfully under a given set of environmental conditions, development of certain *Dendroctonus* species under the same conditions can be suboptimal. Insects such as the ambrosia beetles depend on moisture for successful development and can be easily killed when host material is spread out and exposed to the sun to dry. The high temperatures that occur under bark exposed to direct sunlight is sufficient, particularly at southern latitudes, to kill many bark beetles and wood borers. This phenomenon is especially important for dendrophagous species with thin-bark hosts. Complete drying also can be achieved by stacking host material in open rack or crib formations.

In British Columbia, Canada, log decks are misted with water to protect the stored wood from damage by ambrosia beetles. This procedure may involve the use of a permanent sprinkling system that provides a continuous mist of water (Fig. 20.4). Often the system is operated continuously during the mass adult flight period of ambrosia beetles such as *Trypodendron lineatum*. Wet bark appears to be the factor that impedes beetle attack. Although searching movements can be observed in beetles on the bark, none or few are seen burrowing in the wet bark (Richmond and Nijholt, 1972). Cost–benefit analyses indicate that over a period of 5 yr the benefits of this beetle control method outweigh the costs by a factor varying from four to six (Nijholt, 1978).

Physical Barriers to Insect Attack

The placement of bands or barriers around the tree trunks is a method of control that has been used for many years. It is primarily aimed at insects that actively wander in trees and from tree to tree or species whose wingless females (e.g., cankerworms) climb the trunk to oviposit. Bands are often made of cloth such as burlap and tied around the bole to form two flaps. Bands of sticky resin are the most effective barriers but may be messy to work with. These materials may be applied in a band or applied to paper or cloth bands. Inverted funnels constructed of wire screening or tin also have been used as barriers. Whatever the type, these barriers act by deterring insects from crossing the point of attachment or by trapping crawling individuals. All of these procedures require frequent attention, including the harvesting and killing of accumulated insects. Although these techniques may be useful in light infestations, their effectiveness

Figure 20.4.   Water misting as a direct control. (From Richmond and Nijholt, 1972.)

during severe outbreaks of insects such as the cankerworm is limited. Banding provides little protection during outbreaks of species that are wind disseminated in their early instars. Wind-borne larvae do not crawl to or up a new host and therefore never contact the physical barrier (Hartzell and Louden, 1935).

Ants often have a symbiotic relationship with aphids and scales, in which the ants protect the aphids and scales against predators and parasitoids while the aphids or scales provide honeydew for the ants. This association can enhance numerical increases of aphid or scale populations. In Berkeley, California, the Argentine ant (*Iridomyrmex humilis*) is associated with many pest aphid species. These associations may lead to excessive populations of both ant and aphid. Sticky, adhesive bands (1/8 in. thick) placed around tree trunks at approximately 5 ft above the ground exclude ants and reduce aphid populations (Olkowski *et al.*, 1974).

Quarantine

A variation on the barrier concept is the establishment of a quarantine. This is generally an attempt to limit the spread of a designated pest beyond its normal distribution. The two major tactics of quarantines are inspection and education. Nurseries, plantations, and seed orchards are periodically inspected to insure that material does not harbor potential pests and that material is not shipped that will spread insects. In the case of an insect such as the gypsy moth, state and federal agencies collaborate to prevent its spread from the northeast via campers, tents, and trailers. Larvae, pupae, adults, and particularly egg masses are often found attached to these items (Fig. 20.5). Inspection stations in Arizona and California are attempting to limit the accidental introduction of known pests into these states. Lumber, stone, and quarry products are also inspected. Quarantine activities also may include pheromone trapping of male moths to identify infestations and to determine where control methods are needed.

Early Harvest or Salvage

The use of early harvest as a control for insect pests requires a method of assessing which trees are in imminent danger of attack or are currently infested to such an extent that they should be removed to protect other trees. These methods of assessment are generally called hazard-rating systems (discussed in chapter 17).

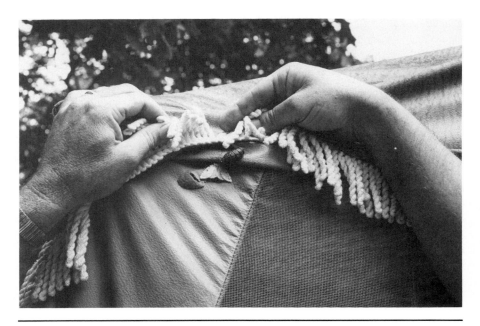

Figure 20.5.   Adult female, pupal case, and egg mass of the gypsy moth (*Lymantria dispar*) on a piece of camping equipment. (Courtesy USDA; photograph by Jim Strawser.)

Early harvest is not always a last-ditch method of control. This approach can also be part of a carefully planned cutting schedule and is thus a useful silvicultural tool to prevent the accumulation of vast stands of overmature trees. In forest stands in the coastal range of southwestern Oregon, prompt salvage of trees has helped to prevent the buildup of Douglas fir and other bark beetle populations. The timing and urgency of salvage operations may vary. In clearcut areas, felled timber is rapidly colonized by beetles, but their broods are not always successful due to the effect of direct exposure to the sun. Conversely, many trees cut for road right-of-ways during logging operations are shaded and provide ideal host material.

In many areas of North America, logging operations are planned to remove felled timber between October and May to eliminate the material most likely to produce a new brood. Timber felled from May to October generally dries out and provides unfavorable host material for emerging spring adults. Another category of host material that should be removed to avoid brood emergence is blowdown trees (Smyth, 1959; Johnson and Zingg, 1969). Another example of accurate timing of harvest illustrated by the susceptibility of host logs to attack by ambrosia beetles of the genus *Trypodendron*. In British Columbia, Canada, logs felled during spring are relatively immune to attack, whereas those felled the previous fall and winter are preferred by the beetles.

Kiln Processing

Direct insect control can often be achieved by exposing logs to kiln temperatures of approximately 180°F (71°C) for 2–6 hr, depending on their size. This type of treatment is effective for various insect species including pine sawyer beetles and ambrosia beetles such as *Trypodendron lineatum, Gnathotrichus sulcatus,* and *Xyleborinus* sp. (Hopping and Jenkins, 1933; French and Johnstone, 1968; Ostaff and Cech, 1978). Kiln drying is also effective against insects that attack wood as it seasons. Starch retention in the sapwood of hardwood species increases their susceptibility to attack by insects such as powder-post beetles. Relatively low temperatures used for slow drying of hardwoods enhance starch depletion and reduce the susceptibility of wood to insect attack (Table 20.2). Although kiln drying is not always more effective than air drying, appropriate procedures usually provide greater protection (Table 20.3).

The infestation of seeds by insects can have a major impact on the success of tree-breeding and reforestation programs. The viability of seeds of certain tree species is often dramatically reduced by insects such as seed chalcids. Seeds can be dried to kill chalcid larvae and prevent the release and spread of the insect into uninfested areas. Temperature treatment at 45°C of Douglas fir seed infested with the Douglas fir seed chalcid (*Megastigmus spermotrophus*) will kill the larvae, but uninfested seeds are still capable of germinating (Ruth and Hedlin, 1974).

Table 20.2. Starch depletion and *Lyctus* attack in air-seasoned and kiln-seasoned 1-in. European oak boards.[a]

Method of Seasoning	No. of Billets	No. of Quarter-sawn boards	Average Starch Grade		Starch Reduction (%)	*Lyctus* Attack			
			Green	Seasoned		No. Boards Infested	No. Exit Holes	Avg. Holes per Board	Avg. Holes per Billet
Air	25	72	5	3.63	27.3	71	2902	40.3	116.1
Kiln	15	41	5	1.86	63.0	21	233	5.6	15.5

[a] From Harris (1961).

Table 20.3. Mortality of insects in kiln-runs[a,b]

Length of Treatment (hr)	Number of Experimental Runs	Kiln Temperature (°F)	Insects Living	Insects Dead	Percent Dead
1	1	160	24	101	80.8
1 1/2	2	160	—	382	100.0
2	2	160	—	257	100.0
3	1	160	—	612	100.0
Controls	—	—	940	1	0.1
2	1	120	50	6	5.4
3	1	120	6	12	66.7
4	1	120	8	71	89.9
6	1	120	4	18	81.8
7	1	120	2	53	96.4
8	1	120	—	78	100.0
9	1	120	—	311	100.0
Controls	—	—	227	5	2.1

[a] From Hopping and Jenkins (1933).
[b] Mortality primarily of *Trypodendron lineatum* but also of some *Gnathotrichus sulcatus* and *Xyleborinus* species.

Trap Trees and Trap Logs

The same behavioral response of bark beetles that may lead to economic damage can be used to control these insects. The attraction of bark beetles to weakened, dying, or felled trees serves as the basis for the trap-tree or trap-log method of direct control. The simplest form of the trap-log method is produced by felling a tree, which subsequently absorbs the beetle population in the immediate vicinity (Parker and Stevens, 1979). Once this is achieved, broods can be destroyed before development is completed. Live standing trees can also be specifically prepared as trap trees by various methods. The trap-tree method has been used in Europe to control insects such as the spruce bark beetle (*Ips typographus*) for more than 200 yr (Bakke and Strand, 1981). Trees are incised (cut) around the bole at breast height, ring-barked (a ring of bark is removed at breast height), topped (the crown is removed), or pruned (the branches are removed). Each technique produces a trap tree that is slightly different in moisture content and distribution. Various insect species may be attracted by each type of trap tree. Incised trees develop centers of moisture accumulation and usually attract moisture-loving insect species. Ring-barked stems produce dry trap trees, which attract insects that prefer dry conditions. Topped and pruned trees attract dry-loving species to the upper stem and moisture-loving species to the lower stem.

Procedural considerations, such as the number of trap trees and their spacing per unit area, the amount of competing host material that occurs in the area, and the density of the beetle population, can have a significant impact on the effectiveness of the trap-tree technique (Doane *et al.*, 1936). Only a few of these interactions have been evaluated. A study in a Canadian spruce forest demonstrates that to compete effectively with naturally occurring sources of attraction for the spruce beetle, approximately 34 pheromone-baited trees are required for each windthrown tree in the target area (Dyer and Safranyik, 1977).

The attraction and subsequent colonization of trees by large numbers of beetles generally involve a complex pheromone system. Both host terpenes and beetle-produced chemicals appear to play significant roles in attraction and colonization of host trees. These phenomena, although not thoroughly understood, are frequently used in the design of trap trees (Ringold *et al.*, 1975) (Table 20.4). Pheromones and host terpene components may be added to trap trees to enhance their attractiveness (Borden *et al.*, 1983a,b). Indeed, in some cases the use of pheromones appears to concentrate surrounding beetle populations in one site, above and beyond the levels that would normally be found there. Many believe that the future of pheromones is not in the use of attractants, but in the use of so-called antiaggregation pheromones. Unlike attractive pheromones, which induce mass attack of host material, antiaggregation pheromones, as the name suggests, inhibit further attack by beetles. The multifunctional

Table 20.4 Selected examples of the use of pheromones, insecticides, and silvicides in the preparation of trap trees.

Species (Insect/Host)	Treatment	Result	Source
Mountain pine beetle (*D. ponderosae*)/western white pine (*Pinus monticola*)	*Trans*-verbenol and α pinene + lindane	Attracted in-flight beetles resulting in heavy attack of baited trees. Lindane did not prevent attack.	Pitman (1971)
Douglas-fir beetle (*D. pseudotsugae*)/ Douglas-fir (*Pseudotsuga menziesii*)	Frontalin, camphene, and α-pinene	Induced attack of 100% of 157 baited compared with 8% of 154 unbaited trees.	Knopf and Pitman (1972)
Southern pine beetle (*D. frontalis*)/ loblolly pine (*Pinus taeda*)	Frontalure + cacodylic acid	In cacodylic acid-treated trees, produced a 3.5-fold increase in aborted attacks and 59.9% brood reduction.	Copony and Morris (1972)
Spruce beetle (*Dendroctonus rufipennis*)/white spruce (*Picea glauca*) and Engelmann spruce (*P. engelmanii*)	Frontalin and cacodylic acid	An 80-in. wide block with treated trees successfully concentrated the attack of the beetles (97–98% in endemic populations).	Dyer (1975)

pheromone MCH (3-methyl-2-cyclohexen-1-one) represses the attractiveness of host material to a variety of beetle species. Attraction of the spruce beetle to natural and synthetic chemical lures can be almost totally suppressed by MCH (Furniss *et al.*, 1976). Controlled-release granular formulations of MCH and several chemical analogs of MCH have also been tested against the Douglas-fir beetle and have reduced Douglas-fir beetle attack and brood density (Rudinsky *et al.*, 1975; Furniss *et al.*, 1977; McGregor *et al.*, 1984). Further research is necessary to make this approach feasible. Information that is needed before this method can be made operational includes data on the nature and effectiveness of antiaggregation pheromones against a variety of species, various formulations of pheromones, and of application technology (Kline *et al.*, 1974; Furniss *et al.*, 1974, Rudinsky *et al.*, 1974a,b). Another consideration is, of course, the cost of synthesizing the chemical. Sex pheromones of various species of forest insects have been either suggested or tested for possible use in trapping-out programs. At present, many of these pheromones are practical and economically feasible for monitoring pest populations.

An interesting blend of attraction and mortality is provided by the use of trap trees and insecticide sprays. Lodgepole pine under attack by mountain pine beetle can be sprayed with lindane emulsions and can act as trap trees for the beetles. The key to the success of this procedure (often referred to as the lethal trap-tree method) lies in the use of appropriate insecticide concentrations that kill a substantial proportion of individuals but still allow the survival of enough attacking beetles to maintain the tree as an attractive center (Smith, 1976). This procedure has been tested against *Dendroctonus rufipennis* and a few other species (Table 20.4). Pheromones can be added to the trap to further enhance attraction. One drawback of this procedure is the apparent detrimental effect of insecticides on some predators and parasitoids of bark beetles, which are often attracted by bark beetle pheromone components (Bakke and Kramme, 1978; Bedard *et al.*, 1980) (Table 20.5).

In recent years additional refinements have been made to the trap-tree technique. Although many of these innovations are experimental, they hold great promise. The use of cacodylic acid (dimethylarsenic acid) is one interesting variation in the development of new types of trap trees. Cacodylic trap trees have been evaluated for use as both a post and a pre-attack control. Although originally used as an herbicide to kill trees, the ability of cacodylic acid to induce brood mortality has renewed interest in the use of the trap-tree method. Cacodylic acid applied to living loblolly pine just before the initiation of attack by the southern pine beetle effectively reduces average population density (Coulson *et al.*, 1975). Other species of bark beetle have been similarly affected (Frye *et al.*, 1977) (Figs. 20.6 and Fig. 20.7).

Cacodylic acid appears to affect the inner bark of the tree; however, the exact mode of action on bark beetles is not known. A current

Table 20.5.   Numbers of *Dendroctonus rufipennis* and *Thanasimus undatulus* caught at insecticide-sprayed and unsprayed trees baited with frontalin[a]

	No. of Trees Sampled	Mean No. per Tree	Std. Error of Mean	% of Total Caught
Dendroctonus caught at sprayed trees	13	130.5	44.68	40.1 ± 0.2
Thanasimus caught at sprayed trees	13	73.7	19.64	41.1 ± 0.2
Estimated no. of *Dendroctonus* in attacked unsprayed trees	25	43.5	2.38	

[a] From Dyer *et al.* (1975).

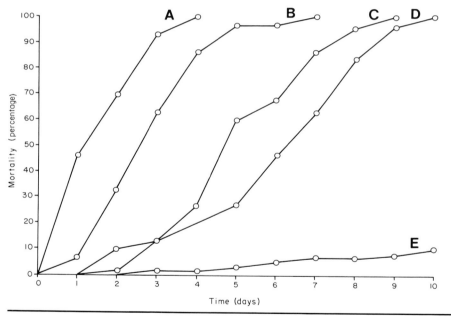

Figure 20.6.   Effect of cacodylic acid at various concentrations on oak bark beetles (*Pseudopityophthorus* sp.). Curve A, 21,000 ppm; curve B, 5200 ppm; curve C, 1300 ppm; curve D, 900 ppm; and curve E, control. (From Rexrode and Lockyer, 1974.)

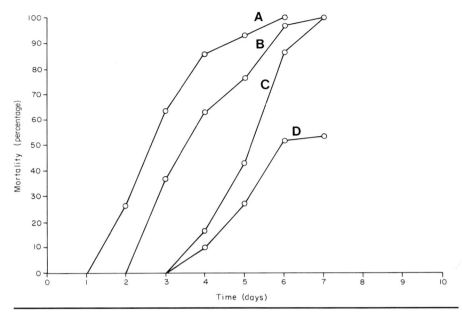

Figure 20.7.   Effect of various concentrations of cacodylic acid on elm bark beetles (*Scolytus multistriatus*). Curve A, 28,000 ppm; curve B, 7000 ppm; curve C, 1000 ppm; and curve D, control. (From Rexrode and Lockyer, 1974.)

hypothesis suggests that the moisture content of the inner bark is increased and maintained at an unsuitable level for beetle development (Coulson *et al.*, 1975).

    The introduction of cacodylic acid into the outer xylem of infested trees (post-attack method) produces variable results in some species. A number of important factors such as timing and subsequent translocation of the chemical affect the success of the procedure. Timing is crucial because the development of blue-stain fungi in xylem inhibits translocation. Nevertheless, injection of cacodylic acid during egg hatch can result in significant reduction of brood in beetle-infested Engelmann spruce (Frye *et al.*, 1977). Similarly, studies on the use of cacodylic acid against mountain pine beetles infesting ponderosa pine suggest that treatment is effective if applied before larval galleries exceed 0.5 in. in length (Stevens *et al.*, 1974a). Another aspect of timing is the length of time between injection of trees and felling. Injection of Engelmann spruce infested with spruce beetles in late August and felling a month later produces the most effective mortality (Lister *et al.*, 1976).

Mass Trapping

Mechanical or physical traps are also used for the direct control of forest and shade insects. Many traps use a sticky resinous material to capture

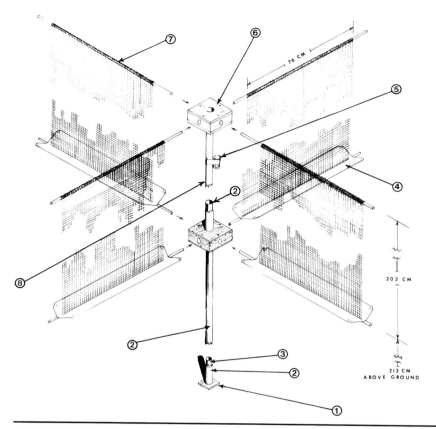

Figure 20.8. Bark beetle vane trap: (1) ground plate, (2) EMT standpipe, (3) iron pipe, (4) aluminum catch trough, (5) pheromone evaporation device, (6) vane block, (7) upper vane dowel, and (8) PVC vane support. (From Browne, 1978.)

insects. To enhance capture, various attractants are added to the trap (Figs. 20.8 and 20.9). Experiments on ambrosia beetles such as *Gnathotrichus sulcatus* and *Trypodendron lineatum* demonstrate that wire-mesh sticky traps baited with the aggregation pheromone sulcatol (6-methyl-5-hepten-2-ol) in a commercial sawmill can capture a significant proportion of beetles attracted to sawn lumber. Because lumber near traps is often heavily attacked by ambrosia beetles, recommendations for the future use of this method include the placement of piles of slabwood near the traps in the mill. After infestation, this material can be converted to pulp chips (McLean and Borden, 1977). A similar attempt to reduce the incidence of Dutch Elm disease through mass trapping is currently being evaluated in urban and suburban areas throughout the United States and shows a great deal of promise. The technique calls for the use of traps made of squares

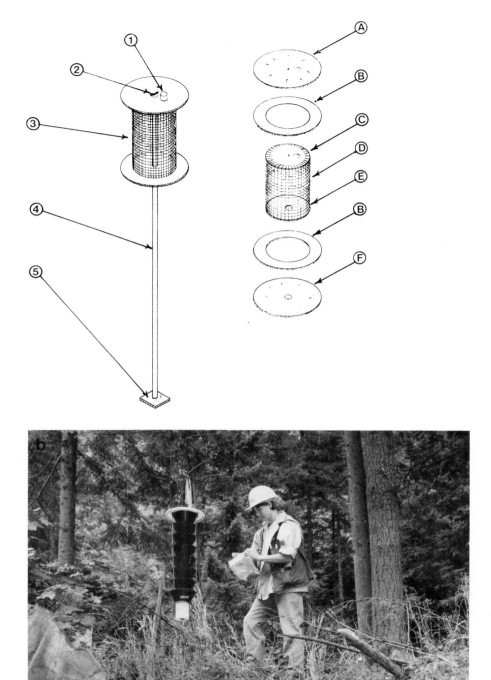

Figure 20.9. Top, Bark beetle cylindrical trap: (1) pheromone evaporation device, (2) wingnut and fender washer fastener, (3) sticky coated veil, (4) PVC trap standard, and (5) ground plate; (A–F) exploded view showing veil and endplates. (From Browne, 1978.) Bottom, Lindgren funnel trap for ambrosia and bark beetles. Trap can be collapsed for easy transport. (From Lindgren, 1983.)

coated with sticky material and baited with the synthetic pheromone Multilure (Cuthbert and Peacock, 1975; Lanier and Jones, 1985; Cuthbert *et al.*, 1977).

In summary, many of the available direct control methodologies are designed to work most effectively in the remedial short-term elimination of an insect pest. They are perhaps most effective at endemic or moderate population levels. The labor-intensive aspects of most of the techniques described here suggest their use in urban forests where the high value of individual trees merits high-cost remedies or in forests where widespread chemical spraying is more costly than preventative or small-scale direct controls.

Chapter Summary

Direct control is defined as a tactic designed to directly reduce insect numbers. Chemical control (Chapter 19) is a kind of direct control, but we discuss it separately. Direct control generally involves removing insects or modifying their habitat to destroy the pests. Physical removal means collecting insects individually, such as removing insect colonies or removing the insects' habitat; sanitation in urban environments to control beetles is an example. In many cases, direct control is the most effective and environmentally appropriate method to manage insects in the urban environment. Physical barriers such as tree bands and trap trees are examples of direct methods that are routinely used to manage pest insects, especially shade tree pests. A menagerie of traps, including those baited with pheromones, is currently being used to mass-trap insects in high-value situations.

CHAPTER 21

Silvicultural Control of Forest Insects

Introduction

Silvicultural control is unusual among the various strategies and procedures for the control of forest and shade tree insects because it is an indirect and often long-term solution. It relies heavily on knowledge of conditions that favor or inhibit the survival of the insect pest. Silvicultural control is any modification of the method of growing or harvesting a forest crop or shade tree that reduces insect damage by avoiding conditions favorable to the target insect. In general, satisfactory control can be achieved only where the optimum conditions of growth for the insect differ from those of the host.

The use of forest cultural methods as well as the conservation of natural enemies of forest insects have been the central core of insect pest control in European forestry for many years. When broad-spectrum insecticides became available, they were used with care in Europe. In the past, North American foresters had an overabundance of forest resources and were relatively unconcerned with the losses due to insects and diseases. When forest resources became less abundant in the late 19th and early 20th centuries, technology and the apparent panacea of insecticidal control began to play an important role in forest management. Although some early forest entomologists promulgated concepts and strategies based on cultural and other harmonious control procedures, these tenets did not immediately become part of the mainstream of forest insect control.

The success of silvicultural controls is limited by (a) the accuracy of data on the factors affecting survival and numerical increase of pests, (b) the feasibility of necessary environmental manipulations, (c) the costs of suggested procedures, (d) the degree to which suggested manipulations interfere with or affect the ultimate objectives of forest management, and (e) the length of time required before the effects of procedures are manifested. The underlying assumptions of most silvicultural controls are that the key factors determining the survival and numerical increase of target insects are food (host quality and availability), microclimate, and natural enemies. Silvicultural procedures tend to minimize host availability, maximize host resistance and natural enemy effectiveness, and/or insure suboptimal microclimatic conditions for insect pests.

Silvicultural control recommendations, although shown to be effective in avoiding the conditions conducive to damage, may be unacceptable. On occasion, recommendations may conflict with short-term goals of commercial concerns, may not be cost-effective, may reduce the effectiveness of harvesting techniques, or may result in undesirable side effects. Some of these conflicts will be discussed throughout this section and, a few illustrations noted. The development of pure mature, even-aged stands may enhance the mechanical harvesting of a crop but may also enhance the potential for the development of pest outbreaks. Any attempts at diversification, asynchronization of growth, or similar changes may reduce harvesting efficiency. Similarly, most of the commercially important hardwoods in the oak–hickory and eastern mixed hardwood forests of the United States range from shade-intolerant to moderately shade-tolerant species and do not grow and reproduce well under the heavy shade characteristic of uneven-aged stands. Thus, with the exception of the northern hardwoods, most of the eastern hardwoods are managed under even-aged management systems (Sander and Phares, 1976).

Finally, the costs of silvicultural procedures such as thinning must be recouped through the sale of the cut trees as poles and pulpwood. Failure to do so may make it impossible to use the practice because it would not be cost-effective.

Host Species and Stand Manipulation

Mixed Stands

Diversification in the types of trees that grow in a forest increases the stability of stands. Increased stability results from an increase in the number of interweaving interactions among resident animal and plant species. In general, catastrophic insect outbreaks occur less often in a forest with a diversity of tree species than in a less diverse or single-species forest (Graham, 1963). Forest management should include the maintenance of a diverse host tree species composition, which in turn will result in a diverse community of predators and parasites. A comparison of the species diversity of insect communities in a white pine watershed with those of a

Table 21.1. Numbers of individuals and species in two insect orders collected (from July 9, 1969 to Sept. 27, 1969) in a mixed hardwood coppice and white pine watershed in North Carolina[a]

Order	Number Collected in Hardwood Coppice			Number Collected in White Pine Area		
	Total Species	Species/ Sample (Mean)[b]	Individuals/ Sample (Mean)[b]	Total Species	Species/ Sample (Mean)[b]	Individuals/ Sample (Mean)[b]
Coleoptera	165	42.0 ± 9.0	395 ± 137	35	8.5 ± 1.5	58 ± 20
Hymenoptera	107	36.6 ± 4.4	91 ± 16	43	12.8 ± 1.8	24 ± 5
Coleoptera + Hymenoptera	272	78.7 ± 12.5	486 ± 148	78	21.3 ± 2.4	82 ± 18

[a] From Crossley *et al.* (1973).
[b] Mean ± SE.

coppice (mixed hardwood regenerated from stump growth) watershed shows marked differences in the two stand types. The mean number of species and individuals per sample are higher in the coppice watershed than in the white pine watershed (Crossley *et al.*, 1973) (Table 21.1). Although the principle that links species diversity and community stability was developed from a number of theoretical and empirical analyses, this axiom is not accepted by all ecologists.

Another important type of diversification can be manifested in differences in the age of trees. Thus, although a given species may predominate in an area, age diversification may reduce the potential for extensive insect damage. This is particularly important in insect species whose numerical increase is enhanced by the presence of mature individuals. For example, extensive monocultures or areas dominated by mature trees of a preferred host are important in the outbreak development of species such as the forest tent caterpillar, spruce budworm, hemlock looper, and larch sawfly. Conversely, large areas dominated by young hosts are vulnerable to outbreaks of aphids and root-feeders (Schowalter *et al.*, 1981; Witcosky *et al.*, 1986b).

Selective cutting can be undertaken to maximize variation in tree species. A mixture of tree species can be achieved by cutting in groups or strips or by creating clearings that allow shade-intolerant species to become established. Similarly, cutting adjacent areas at intervals of more than 15 yr may aid in producing mixed-age stands. In plantations, a mixed planting of white pine and other conifers such as Scots pine offers some degree of protection against the white pine weevil. Scots pines appear to act as trap trees, thereby dramatically reducing the severity and duration of damage to white pine (Belyea, 1923). Hosts of different ages often differ in their susceptibility to attack and injury. The susceptibility of overmature stands to insect attack has been previously discussed. Another example of host age as a factor in susceptibility is the relationship between the white pine weevil and its host, white pine. The first trees attacked by the weevil are usually 5–7 years old. At first the infestation level is light, but then it builds to a maximum when trees are approximately 15 years old. From that point the infestation declines until attacks cease between the 25th and 30th years (Graham, 1918).

When insect species exhibit a strong preference for a certain host species, selective elimination of that species and encouragement of a variety of nonhost species may reduce the potential for injury to the primary host species. Studies in the early 1900s and subsequent field observations of gypsy moth feeding (defoliation) in relation to the available species in New England forests led to the recommendation that growth of less-favored hosts such as pines be encouraged at the expense of the favored oaks (Behre, 1939; Behre and Reineke, 1943). We know now that the presence of favored or nonfavored hosts is not the only factor responsible for the numerical increase of this species; nevertheless, host species composition does play a role in the increase of gypsy moth populations (see section on site conditions).

Alternate Host and Ground Cover Manipulation

The complexity of the life cycle of many dendrophilous species results in their becoming pests, but may provide an opportunity to regulate their populations by manipulating so-called weak links in the cycle. One of those potentially weak links is the need for alternate hosts.

The black army cutworm (*Actebia fennica*) is a pest of herbaceous agricultural crops. Herbaceous foliage is preferred, but when numbers increase and foliage is sparse, they readily defoliate conifer seedlings. Consequently, *A. fennica* has become a pest of planted conifer seedlings in several areas in North America (Ross and Ilnytzky, 1977).

The feeding habits of a group of tortricids that affect young conifer plantations provide an excellent example of the role of alternate hosts and the impact of ground cover manipulations on injury levels. Plantations in the United States and Canada have suffered serious losses from a group of tip-infesting tortricid species. These insects, the most common of which are *Tortrix alleniana* and *T. pallorana*, feed primarily on herbaceous ground cover plants but may also feed on the new shoots of seedlings less than 3 ft in height. They cause injury to leaders and cause growth deformities. These insects feed on more than 20 species of plants, including white sweet clover (*Melilotus alba*), alfalfa (*Medicago sativa*), red clover (*Trifolium pratense*), dandelion (*Taraxacum officinale*), goldenrod (*Solidago* spp.), and dogwood (*Cornus stolonifera*). However, they oviposit only on white sweet

Table 21.2. Effect of ground cover composition on the degree of infestation of Scots pine by *Tortrix pallorana*[a]

Ground Cover Species (Exclusive of Grasses)	Block 1 4% Trees Injured		Block 2 26% Trees Injured	
	Avg. No. Specimens per 100 ft	Frequency[b] (%)	Avg. No. Specimens per 100 ft	Frequency[b] (%)
White sweet clover (*Melilotus alba*)	49	88	1	6
Bladder campion (*Silene cucubalus*)	10	30	20	56
Yellow goat's beard (*Tragopogon pratensis*)	9	56	7	36
Goldenrod (*Solidago* spp.)	6	30	12	6

[a] From Martin (1958).
[b] Frequency is the number of 10-ft lengths in which a species occurs in relation to the total number of 10-ft lengths surveyed, expressed on a percentage basis.

clover and alfalfa. The larvae tend to remain on certain herbaceous hosts for their complete life cycle, though some larvae move to young pines, spruce, and other shrubs and trees during the penultimate and ultimate instars. In many plantations, herbaceous weeds are found only in the furrows along which trees are planted, whereas the spaces between furrows are populated by pure grass complexes. A comparison can be made between adjacent areas that differ in the occurrence and frequency of preferred ground cover species such as alfalfa or white sweet clover. Thus, where white sweet clover is abundant and broadly distributed, there is little movement of larvae from the clover to trees and less injury compared with sites where clover is relatively rare (Martin, 1958) (Table 21.2). Clearly, an understanding of the relationship between ground cover species and tree injury is critical to protection by silvicultural control.

The occurrence and abundance of certain dendrophilous species are often influenced by the occurrence and abundance of alternate hosts. The *Adelgidae* include a number of major pests and are characterized by their need for an alternate host. All adelgids (or chermids) feed on conifers; several species induce galls on trees such as spruce (Fig. 10.8), and species of *Pinus*, *Abies*, *Tsuga*, *Larix*, and *Pseudotsuga* serve as alternate hosts. A common example of this behavior is that of the Cooley's spruce gall aphid, which forms conelike galls on spruce. In late summer or early fall, winged

Table 21.3. Common alternate hosts of the Saratoga spittlebug in Maine[a]

Host Category	Common Name	Scientific Name
Preferred host	Sweetfern	*Comptonia peregrina*
Secondary host[b]	Blueberry	*Vaccinium angustifolium*
	Goldenrod	*Solidago* spp.
	Lambkill	*Kalmia angustifolia*
	Blackberry, raspberry	*Rubus* spp.
	Orange hawkweed	*Hieracium aurantiacum*
	Wintergreen	*Gaultheria procumbens*
	Strawberry	*Fragaria virginiana*
	Wild raisin	*Viburnum cassinoides*
	Gray birch	*Betula populifolia*
	Black chokeberry	*Pyrus melanacarpa*

[a] From Linnane and Osgood (1976).
[b] Includes a wide variety of herbaceous and woody plants.

females disperse from the galls to Douglas fir trees where they lay eggs. Thus, damage by this species can be minimized by not planting Douglas fir near spruce (Wood, 1977). Alternate hosts form an integral part of the life cycle of various other species such as the pine leaf chermid (*Pineus pinifoliae*), *P. floccus*, which uses white pine as an alternate host, and *Adelges larciatus*, which uses tamarack.

Another example of the importance of the ability to control pests by manipulating alternate hosts is that of the Saratoga spittlebug, a pest of young red pine and jack pine plantations. Although adults exhibit a strong preference for pine, nymphs feed on herbs and shrubs (Table 21.3). When eggs laid on red pine hatch, nymphs crawl to the ground and actively seek suitable host plants. They can be found on approximately 30 alternate hosts that grow in plantations. The most commonly reported alternate hosts are sweetfern (*Comptonia peregrina*) and *Rubus* spp. The presence of alternate hosts in a plantation affects the likelihood of an infestation. Plantations that do not contain woody alternate hosts, especially sweetfern and *Rubus* spp., rarely harbor more than a light infestation. Recommendations for silvicultural control of spittlebugs include (a) not planting red pine in open, burned-over areas that previously supported lush sweetfern growth and (b) planting under over-topping hardwoods or adjusting the density of plantations to exclude shade-tolerant alternate hosts such as sweetfern.

Thus, silvicultural control provides effective techniques for protecting susceptible pines from serious injury. Although these techniques rarely result in rapid, drastic reductions of pest numbers, they do substantially reduce the likelihood that a given plantation will support an epidemic population.

Pest problems often result from necessary manipulations of a habitat. Larvae of pales weevil hatch from eggs deposited in freshly cut logs or large roots of stumps and feed just beneath the bark. Upon emergence, adults feed on the tender bark and buds of coniferous seedlings and can cause severe injury (Fig. 21.1) Depending on the circumstances, silvicultural actions can be taken to reduce the potential for injury by this species. Its impact can be reduced by not storing freshly cut white, red, or pitch pine logs in the field or by delaying the planting of conifer seedlings for 2–4 yr after harvest (Peirson, 1937; Nord *et al.*, 1982).

The exploitation of an overabundant tree host, as a result of its cultivation for horticultural, agricultural, or forestry purposes, can also lead to pest problems originating from closely related tree taxa. For many years ornamental varieties of honey locust were considered to have no significant insect pests. However, a number of insect and one mite species have recently been reported to cause serious injury. Some pest species such as those in the family Miridae occurred on native locusts but were

ADULT PALES WEEVIL
AND FEEDING DAMAGE

Figure 21.1.   Adult pales weevil (*Hylobius pales*) and the typical feeding damage it inflicts. Actual length of weevil $\frac{1}{4} - \frac{1}{2}$ in. (Courtesy Dr. John A. Davidson).

rarely mentioned in the literature. Many of these pests have now become injurious because the new honey locust varieties have been widely established as shade trees (Wheeler and Henry, 1976).

Manipulation of Tree Growth and Density

Host Vigor and Insect Problems

Discussions in Chapter 3 provided details on the importance of host vigor to insect injury, the changes exhibited by nonvigorous hosts, and the differential success of pest insects on vigorous and nonvigorous tree species. In general, the more vigorous (or faster growing) the species, the less injury it suffers. Growing conditions that are favorable for a given tree species should be detrimental to its insect enemies. However, there are a few exceptions to this generalization. Certain species prefer or do best on vigorous hosts. An example is the association of the larch sawfly and larch trees; the most favorable growing conditions or sites for larch also favor larch sawfly buildup (Muldrew, 1956).

Thinning

Practices that regulate the density of trees are a key method of enhancing tree vigor and thus reducing pest problems (Table 21.4). General recommendations are difficult to make and are often inappropriate because the kind, quality, amount, and timing of thinning depend on a number of factors including tree species, pest species, site conditions, tree

Table 21.4. Relation of tree mortality due to mountain pine beetle (*Dendroctonus ponderosae*) to stand density in eight stands of ponderosa pine pole timber in Oregon[a]

Stand Density (BA/acre)	Mortality by Stand (%)							
	Site Class V[b]				Site Class IV		Site Class III	
	1	2	3	4	5	6	7	8
100	16.8	14.2	22.0	10.1	—	0.4	—	1.1
200	60.1	55.3	59.1	35.9	20.8	17.1	8.2	8.2
300	—	96.5	96.2	61.6	41.9	33.9	16.9	15.4
400	—	—	—	87.4	63.0	50.3	25.6	22.6

[a] From Sartwell (1971).
[b] Site class V is poorer than site class III.

age, size of area, and cost/benefit ratios. For example, in sites where there is an abnormal degree of competition for moisture or where trees are infested with dwarf mistletoe, recommended spacing would differ from that recommended for similar sites in the absence of moisture stress.

In the silvicultural control of mountain pine beetle-infested ponderosa pine, the beneficial effects of thinning may be reduced if the thinned area is small and surrounded by unmanaged forests in which there exists an unchecked infestation. The appropriate density to reduce the likelihood of severe insect injury, aside from being influenced by site, climate, and size of area, depends on the tree species. In the north-central states the success of sucker reproduction of aspen depends, initially, on the establishment of a very dense stand of sprouts after logging. When this occurs, injury by *Saperda* spp. and *Agrilus* spp. borers, the willow shoot sawfly (*Janus abbreviatus*), and others is light, if it occurs at all. After approximately 5 yr, thinning is required to avoid infestation by insects, particularly if trees have begun to exhibit the effects of competition (Graham, 1959). On a per tree basis, larch casebearer populations are generally higher on trees in less dense stands than in heavily stocked stands. However, casebearer populations in low-density stands appear to be more susceptible to mortality factors such as harsh climate compared with those in unthinned stands (Schmidt, 1978). Root-feeding beetles are attracted to thinned stands of Douglas-fir in Oregon (Witcosky *et al.*, 1986b).

In many circumstances thinning can be quite useful in forest protection. Thinning of pole-sized ponderosa pine that is subject to mountain pine beetle injury substantially reduces tree mortality and results in positive net stand growth (Table 21.5). In Rocky Mountain pine sites,

Table 21.5. Cumulative mortality and net growth 5 yr after thinning of ponderosa pine subject to mountain pine beetle (*Dendroctonus ponderosae*) and pine engraver (*Ips pini*) attack[a]

Thinning Treatment	Stand Density		Mortality				Net Growth
	1967	1972	M.P.B.[b]	Ips[c]	Other	Total	
Stem basal area (ft²/acre)							
Unthinned	173.2	152.5	11.8	4.0	5.2	21.0	−20.7
12 × 12 ft[d]	116.8	113.5	3.2	2.3	0.5	6.0	−3.3
15 × 15 ft	85.8	89.0	0.2	0.0	0.3	0.5	3.2
18 × 18 ft	61.8	64.8	0.0	0.5	0.3	0.8	3.0
21 × 21 ft	35.0	37.2	0.0	0.0	0.8	0.8	2.2

[a] From Sartwell and Stevens (1975).
[b] Mountain pine beetle.
[c] Pine engraver (*Ips pini*)
[d] Spacing between residual trees achieved by cutting the smaller trees.

Table 21.6. Simulation of basal area of typical 90-yr-old Black Hills ponderosa pine stands (site index 60) at 10, 20, and 30 yr after thinning[a]

	Original Stand		Thinning Level		
Year after Thinning	dbh (in.)	Basal Area (ft²/acre)	60	80 (ft²/acre)	100
	8.0	200			
10			74	95	114
20			89	112	130
30			104	128	145
	8.0	160			
10			74	95	114
20			89	112	130
30			104	128	145
	10.0	200			
10			72	94	114
20			85	108	130
30			97	122	144
	10.0	160			
10			73	95	115
20			86	109	129
30			98	123	144
	12.0	200			
10			70	92	114
20			81	104	127
30			91	116	139
	12.0	160			
10			71	93	114
20			82	105	128
30			93	116	140

[a] From Sartwell and Stevens (1975).

outbreaks usually develop in stands with basal areas greater than 150 ft²/acre and in which most trees are greater than 8 in. in dbh (Stevens *et al.*, 1974b). Thinning to maintain stand density well below 150 ft² basal area/acre has been effective in suppressing bark beetle outbreaks. Predictive computer growth models for ponderosa pine indicate that typical stands thinned to 60 to 80 ft² of basal area/acre do not reach 150 ft² even after 30 yr (Table 21.6) (Stevens *et al.*, 1974b; Sartwell and Stevens, 1975; Sartwell and Dolph, 1976).

A number of hypotheses are proposed for the relationship between thinning and insect injury, particularly in bark beetle–tree associations. Improved growth rates are correlated with thinning. In addition, trees remaining after thinning have increased resistance to insect attack (Table 21.7). Still another hypothesis suggests that the increased distance

Table 21.7. Selected examples of silvicultural recommendations based on the relationship between host vigor and insect injury

Insect	Host	Relevant Characteristics	Recommendations	Source
Mountain pine beetle (*Dendroctonus ponderosae*)	Lodgepole pine (*Pinus contorta*)	Beetle brood production is correlated positively with phloem thickness, and phloem thickness is positively correlated to tree diameter.	Management for small tree-size classes through knowledge of pine stand structure (diameter classes) and phloem distribution among diameter classes (e.g., the percentage of trees with 0.11 phloem thickness per class).	Cole and Cahill (1976)
Spruce budworm (*Choristoneura fumiferana*)	Balsam fir (*Abies balsamea*)	Lake states (USA) spruce–fir stands. Most balsam fir found in blocks of about 5–100 acres separated by nonhost types. Thus, small blocks can be logged in a single season, thus reducing the chance of invasion from adjacent untreated areas.	Partial cutting (removing 30% of the basal area of host species) will result in a reduction in the amount of defoliation on the original stand. Commercial clearcutting (removing all host trees having two or more 8-ft bolts) will result in greater reduction in the amount of defoliation.	Batzer (1967)

between trees reduces the probability that mass attack of one tree will result in attack of nearby trees (Gara and Coster, 1968).

Thinning may result in no net change in the level of injury or in an increase in damage (Schenk *et al.*, 1957). Indeed, partially cut (thinned) stands are often subject to excessive windthrow (Batzer, 1967). Thinning may also lead to dramatic changes in the stand microclimate, particularly to an increase in moisture stress. This may stress trees and increase their susceptibility to secondary pests such as bark beetles or various species of *Pissodes* and *Agrilus*. For a number of plantation pests such as Saratoga spittlebug, pine shoot borers, and Nantucket pine tip moth, damage declines sharply as the crown begins to close, an event that is not encouraged by thinning. It is not always clear whether this phenomenon is directly or indirectly related to the spacing of trees.

A close planting of seedlings (e.g., 1200–1500 trees/acre) rather than thinning is one method of protecting white pine from attack by the white pine weevil (Fig. 21.2). Among the several reasons why this approach provides some success is that, unlike most insects, the white pine weevil prefers the most rapidly growing trees. Other factors such as the unfavorable microclimate created by a closed-canopy stand also are important. However, although slowing the growth of trees may reduce injury, other factors such as yield per unit time may be negatively affected (Graham, 1918).

Conflicting results of thinning experiments may be associated with aspects of the insect–tree relationship that are not fully understood. The incidence of damage by the western spruce budworm, for example, is not consistently related to stand density. Occasionally, when damage is relatively light, the percentage of dominant trees with leader damage due to the budworm is high in plots with the fewest trees. This relationship, however, is not evident during subsequent years as budworm damage increases (Fellin, 1976).

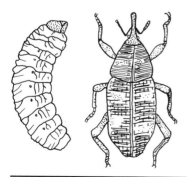

Figure 21.2.   Larva and adult of white pine weevil (*Pissodes strobi*). (Courtesy of Dr. John A. Davidson).

Although other circumstances in forestry are perfectly suited for silvicultural controls such as thinning, we may lack the key information necessary to make reasonable recommendations. One of the best examples is the potential for the silvicultural control of oak borers in the eastern hardwood region of the United States. The economic flexibility in this system is limited. Because of the relatively high costs associated with sales, administration, marketing, and logging road renovation, chemical control measures and many direct control procedures are likely to be too expensive. Additionally, the life history of borers requires that the timing of chemical application be exact, coinciding with adult emergence or egg hatch. The use of direct measures that rely on sex attractants also requires exact timing. Biological control might be feasible if the natural control agents are self-maintaining and do not require periodic augmentation. Alternatively, the silvicultural control of borers only requires the application of ordinary silvicultural procedures that could provide more permanent, long-term results compared, for example, with chemical treatment. However, as noted, the data essential for the extensive implementation of such a program are lacking. Needed information includes the indentification of tree characteristics that enhance susceptibility to borer attack, such as crown class, species, bark texture or thickness, age, and size. The influence of site quality, host defense mechanisms, and optimum stand densities must also be evaluated (Sander and Phares, 1976). Nevertheless, the general principles of silvicultural control can still be applied.

The Right Species in the Right Place

The principle of fitting the tree species to the proper site and soil conditions is often disregarded in plantations, nurseries, and urban/suburban situations. The result of such action is poor development, decreased vigor, and enhanced potential for injury from insects and diseases. This sequence of events is particularly likely to occur if the trees are planted in pure stands.

Planting Requirements and Procedures

Planting tree species within their normal range and on sites where soil can provide adequate nutrition and moisture, and air lacks phytotoxic pollutants and other detrimental stresses, ensures tree vigor. Planting species beyond their normal range results in tree stresses that can increase susceptibility to insect pests and decrease resistance to attack and ability to recover from injury; soil and climatic conditions in these sites may be different enough to alter normal growth patterns and thus the vigor of trees. *Adelges piceae*, for example, is endemic in Europe in even-aged, mature stands of silver fir (*Abies alba*) but is often epidemic in the mixed-aged plantings outside its natural range (Turnock and Muldrew, 1971).

Frequently, dendrophagous species associated with transplanted host trees are transported to new areas that lack significant numbers of the natural enemies that keep them in check. Another good example of improper planting is the damage to red pine by the European pine shoot moth when this tree species is planted south of its normal range.

As indicated above, a corollary to the issue of planting outside the natural range of a tree species is planting in soil that is inappropriate for a tree species. White and red pine are subject to greater pest problems when planted on soils with a high pH and high calcium content. Norway spruce is similarly predisposed when planted on shallow soils (Westveld *et al.*, 1950). Planting species under conditions not representing the proper successional stage for that species can also increase pest damage.

The principle of the right species in the right site also includes the concept that a given area or region has a variety of environmental characteristics and a microclimate that reflects certain potentials as well as limitations to vigorous growth. Although burning, clearcutting, browsing, or cultivating may change environmental conditions, in general, the species that occur in a region as subclimax or climax species are likely to avoid the reduced vigor that often makes them susceptible to attack. Thus, in urban forests, exotic ornamental varieties should be planted only if they are suited to the local conditions or they are ecological analogs of indigenous subclimax (or climax) species (Westveld *et al.*, 1950).

Site Conditions

Forest, shade tree, nursery, and plantation managers are often involved in preparing sites for tree planting. This may include selection of the proper location and extensive preparation of recently cut-over sites. Frequently, little is known of the influence of these preparations on the frequency or severity of any subsequent pest infestations. Site preparation for certain pines may include (in increasing intensity) no burn, burn, burn-and-harrow, and burn, harrow, and bed (where soil is mounded approximately 25.4-cm high in rows approximately 3.1 m apart). In general, tip moths (*Rhyacionia* spp.) infest more loblolly and slash pine trees on intensely prepared sites where tree growth is best. The differences in infestation are not reflected in the trees until the second growing season. Site preparation prior to planting slash pines also has a major impact on levels of infestation by the pine webworm (*Tetralopha robustella*) by the end of the second growing season (Hertel and Benjamin, 1977). Obviously, the influence of site preparation varies depending on the unique aspects of the particular insect–tree relationships.

Establishment of new plantations, nurseries, or reforestation projects following cutting is often a precarious proposition. The removal of trees and the presence of associated residual materials provide conditions that can easily lead to the buildup of a number of pest species and can make the establishment of new seedlings extremely difficult. Weevils, other beetles, and soil-dwelling lepidopterans can pose a significant threat. The weevil *Steremnius carinatus* is a menace to plantations in the west coast area of Canada. Severe damage occurs in plantations established 2–3 yr after timber cutting, a time period that reflects the one or two generations required to develop significant populations of the insect. Recommended remedial silvicultural measures include (1) planting immediately after logging and removing slash to allow seedlings to become established before the broods of weevils emerge from stumps, (b) using older, less attractive seedlings where there is evidence of the presence of a large weevil population, and (c) spring planting rather than autumn planting in hazardous areas to avoid autumn weevil feeding (Condrashoff, 1968). Similar recommendations concerning the time of planting, age of trees to be planted, and preparation of site comprise the silvicultural control strategies for a variety of similar insect pests.

In forests the establishment of new plantings, whether as part of a reforestation program or to maintain stand diversification, often requires specific preparation of sites to ensure tree establishment and survival. Particularly in older stands, successional changes have created site conditions, such as soil and duff characteristics, that make seedling establishment difficult. In some circumstances, various practices such as scarification (the exposure of mineral soil for a seedbed) may be required before seedlings will become established.

Even when trees are planted within their normal range, the selected site must provide all necessary requirements for growth and vigor. Urban shade trees may have to cope with high concentrations of pollutants in the air, hard-packed soils, low soil moisture, and restricted root space. Often species such as red maple, which typically prefer moist or even swamplike conditions, are planted in urban sites like those described.

Inadequate or poor forest site conditions have frequently been associated with increased injury and pest epidemics. As previously noted, in the 1920s and 1930s strong recommendations were made for the silvicultural cultural control of the gypsy moth, consisting primarily of the removal of certain favored food species. These recommendations were based on the theory that severe defoliation was unlikely if the volume of favored tree foliage was substantially reduced (by 50%). Although the concept of favored food species is accurate to a degree, the strategy failed to explain why two or more stands in the same region and similar in composition differed widely in their susceptibility to defoliation. Examination of defoliated areas showed that poorly managed sites, which lacked leaf litter and were characterized by certain indicator species such as gray birch, aspen, lowbush blueberry, and sweetfern, were attacked most severely. Similarly, oaks growing on thin soils, sandy soils, or rocky ridges were defoliated far more heavily than oaks on better sites. With further research it became clear that certain site characteristics could be associated with susceptibility to gypsy moth attack (Bess *et al.*, 1947) (Table 21.8).

Table 21.8. Characteristics of sites that are resistant or susceptible to gypsy moth defoliation[a]

Site Characteristic	Resistant	Susceptible
Soil	Loamy soil	Light sand or gravel soils
Disturbance	Undisturbed sites	Disturbed sites
Moisture	Mesic	Xeric
Stand conditions	Dense crowns	Canopy area less than half the ground area

[a] Based on data from Spurr *et al.* (1947) and Houston (1979).

Recent research has confirmed the importance of site to the severity of gypsy moth attack. These studies point out that gypsy moth survival is highest when larvae remain on trees and away from ground and litter-inhabiting predators. Stands of favored host trees that provide numerous above-ground resting places and scarce leaf-litter habitat for predators support high pest populations and high defoliation levels. Trees in susceptible sites often exhibit external structural features used as resting sites for the gypsy moth larvae, including an abundance of bark flaps, holes, wounds, dead branches, and dead sprout stubs (Houston and Valentine, 1977).

Many similar correlations between poor site conditions and increased damage or population levels have been reported for other dendrophagous species. Indeed, a great deal of research on bark beetles has demonstrated clear correlations between site characteristics of poor quality soils and low moisture and outbreaks of pest insects. Typically, problem pine-beetle areas are closely associated with site and soil conditions characterized by shallow A horizons, a condition that reflects severe land use and abuse. These poor soils are often characterized by poor permeability and aeration and result in restricted root development. Invariably these conditions cause trees to become physiologically stressed (Belanger *et al.*, 1977) (for other examples see Chapter 3: Tree Vigor and Insect Herbivores). The relationship between site and insect damage can be summarized as follows: Poor growing conditions for trees are favorable for dendrophagous species. This interaction may be based on structural features of stressed trees, changes in growth rates, decreased vigor, or other changes induced by marginal or poor site conditions.

In some instances the relationship between site quality and the level of damage is based on more direct influences of site condition on dendrophagous insects. *Hylobius piceus*, a curculionid of North American forests, attacks spruce, pine, and larch and causes serious mechanical injury, leaf lesions, ports of entry for wood-rotting fungi, and the death of many trees. The most significant damage is caused by larvae, which concentrate their feeding on roots and root collars. The degree of damage is

Table 21.9.   Damage due to *Hylobius piceus* and its relation to the moisture of the humus in its habitat.[a]

Year	Loca-tion	Damage Indices: Moisture Classes (Number of Trees)[b]				
		A	B	C	D	E
1952	I	0.8 (20)	1.7 (20)	2.5 (20)	3.7 (10)	3.6 (10)
(max. damage	II	—	1.5 (10)	2.3 (10)	3.8 (10)	
index = 5)	III	—	1.5 (8)	2.6 (12)		
	IV	0.0 (20)				
1953	V	—	0.6 (10)	3.1 (10)	3.9 (10)	
(max. damage	VI	—	0.3 (11)	3.0 (11)		
index = 10)						

[a] Modified from: The effect of some site factors on the abundance of *Hypomolyx piceus* (Coleoptera: Curculionidae). G. L. Warren, *Ecology* **37**: 132–139 (1956). Copyright © 1956 by the Ecological Society of America. Reprinted by permission.
[b] A–E denote very dry, dry, intermediate, wet, and very wet, respectively.

correlated to differences in moisture content of sites as well as associated differences in the physical and chemical properties of the humus layer. In general, damage increases from dry to wet sites (Table 21.9). The conditions provided by sites having moist humus layers are presumed requisites for the survival and/or successful feeding of the larvae of this species. However, it is also possible that the trees' defense mechanisms against this type of damage may break down with increasing moisture (Warren, 1956).

Manipulation of Physical Factors

The development, feeding, reproduction, and survival of all dendrophilous insects are controlled, in large part, by constraints imposed by physical factors of their environment; that is, temperature, moisture, quantity and quality of light, air movement, and the quality of the substrate can determine the ability of a species to survive and to inflict damage. The influence of these environmental factors is altered by many of the various procedures already mentioned, including thinning, tree species regulation, stand and age distribution, ground cover manipulation, and planting procedures. The principle of environmental or physical factor manipulation is difficult to apply. It must be based on a thorough understanding of the pest insect and the microclimate of its habitat.

The use of this approach to control insects may be limited by many of the same factors that affect other approaches, for example, an unfavorable cost–benefit relationship. Depending on the circumstances in which each is used, many of the required manipulations may not be feasible. An example is the manipulation of ground litter or crown closure to alter temperature or moisture conditions, which are often critical for the hatch and/or development of dendrophilous species. A second limitation of physical manipulation involves procedures aimed at producing detrimental environmental conditions for the insect. These manipulations may also be detrimental to the tree.

Nevertheless, the possibility of microclimatological control by regulation of stand density, age distribution, and various other silvicultural procedures can work in appropriate circumstances and thus merits consideration. An example is damage by white pine weevils, which is greater in open-growing stands of white pine than in mixed stands in which pines are shaded by other trees. Copulation, feeding, and particularly oviposition are strongly affected by bark temperatures and atmospheric moisture. Oviposition behavior is most closely governed by these physical factors; optimal activity occurs between 25 and 29°C and 20 to 35% R.H. In mixed conifer-hardwood stands, the period of oviposition and feeding coincides with the period of canopy development. Depending on the point in the season, temperature differences between exposed and shaded sites range from approximately 3 to 11°C. Although the microenvironment created does not entirely prohibit feeding or oviposition, weevil damage is negligible in shaded stands. Conditions favorable to weevil egg laying commonly occur in open-grown pine stands on clear, calm days in spring and early summer (Belyea and Sullivan, 1956; Sullivan, 1961). Cold-water misting is used in British Columbia seed orchards to disrupt synchrony of tree flowering and cone gall midge oviposition phenology.

Resistance to Insects and Selective Breeding

The high degree of variation of physical, chemical, and biological characteristics of trees is noted in previous discussions. This natural variation can serve as a source for the selection of those trees that are better able to cope with insect pests.

Nursery-grown jack pine seedlings from 90 provenance sources throughout the species' range exhibit significant variation in the incidence of eastern pine shoot borer. Borer incidence in jack pine is under strong genetic control and is correlated with the time of shoot-growth initiation and the terminal shoot length in mid-May (i.e., the same time as shoot borer oviposition). In general, trees from intermediate-flushing seed sources have the greatest terminal shoot lengths in mid-May and the greatest shoot borer incidence of the provenances tested. Trees from late-flushing sources have shorter terminal shoots in mid-May and lower shoot borer incidence. Late-flushing trees have greater height growth than intermediate flushing trees (Jeffers, 1978). Douglas-fir clones and progeny show differential susceptibility to cone gall midges and seed chalcids in Oregon (Schowalter *et al.*, 1986b).

An analysis of 433 wild Douglas-fir trees in 87 northwestern localities indicates that geographic variation between coastal and Rocky Mountain varieties is reflected in relative percentages of β-pinene, the terpinene-sabinene group of terpenes, the camphene group, and perhaps limonene (von Rudloff and Rehfeldt 1980). Total phenol content of the foliage of *Picea abies* trees that are resistant to eastern spruce gall aphid is higher than that of susceptible trees. In addition, an unknown phenol occurs consistently in resistant trees but is absent in susceptible trees (Tjia and Houston, 1975).

Although not strictly a form of silvicultural control, reduction of damage by using trees that have natural or artifically selected resistance to certain dendrophagous species shares some basic underlying ecological principles with traditional silvicultural control and can be viewed as a corollary to the maintenance of vigor. It is an approach that relies on the inherent vigor of a tree (i.e., genetically fixed characteristics that allow the tree to resist and/or

Table 21.10. Selected examples in which coniferous and deciduous trees have exhibited intra- and interspecific resistance to monophagous or polyphagous insects that attack fruit, leaves, bark, and wood[a]

Insect	Tree
Argyresthia laevigatella	*Larix decidua*
Contarinia inouyei	*Cryptomeria japonica*
Adelges abietis	*Picea abies*
Adelges cooleyi	*Pseudotsuga menziesii*
Adelges spp.	*Larix leptolepsis, L. decidua*
Choristoneura fumiferana	*Abies balsamea*
Choristoneura pinus	*Pinus banksiana*
Cossus cossus	*Alnus incana, A. glutinosa*
Cryptorhynchus lapathi	*Populus grandidentata*
	P. tremuloides
Curculio elphas	*Castanea* spp.
Cylindroctoptorus eatoni	*Pinus ponderosa*
	P. strobus
	P. monticola
	P. jeffreyi
	P. coulteri hybrids
Dendroctonus monticolae	*Pinus ponderosa*
Dendroctonus ponderosae	*Pinus ponderosa*
Dendroctonus pseudotsugae	*Pseudotsuga menziesii*
Dryocosmus kuriphilus	*Castanea mollisima*
	C. crenata, hybrids
Hylobius abietis	*Picea abies*
Lymantria monacha	*Picea abies*

(continued)

Table 21.10. (*continued*)

Insect	Tree
Megacyllene robiniae	*Robinia pseudoacacia*
Nematus abietum	*Picea abies*
Neodiprion lecontei	*Pinus banksiana*
Panolis flammea	*Pinus sylvestris*
Pissodes strobi	*Picea abies*
	Pinus armandi
	P. ayacahuite
	P. banksiana
	P. cembra
	P. flexilis
	P. griffithii
	P. koraiensis
	P. monticola
	P. parviflora
	P. peuce
	P. strobiformis
	P. strobus
	P. sylvestris
Pityopthorus spp.	*Pinus banksiana*
Popillia japonica	*Populus* hybrids
Pristiphora erichsonii	*Larix decidua*
Pristiphora abietina	*Picea abies*
Rhabdophaga heterobia	*Salix* varieties
Rhagoletis completa	*Juglans regia*
Rhyacionia buoliana	*Pinus banksiana*
	P. contorta
	P. caribaea
	P. densiflora
	P. heldreichii
	P. mugo
	P. palustris
	P. pinaster
	P. tabulaeformis
	P. nigra
	P. resinosa
	P. sylvestris
Rhyacionia frustrana	*Pinus banksiana*
	P. caribea
	P. densiflora
	P. heldreichii
	P. mugo
	P. nigra
	P. palustris
	P. pinaster
	P. ponderosa
	P. resinosa
	P. sylvestris
	P. tabulaeformis
	P. thunbergeii
Taeniothrips laricivorus	*Larix decidua, L. leptolepis*
Tortrix viridana	*Quercus* spp.
Toumeyella numismaticum	*Pinus banksiana*

[a]From Gerhold (1966). Reprinted with permission from Pergamon Press, copyright 1966.

recover from attack). In rural or urban forests the use of insect-resistant trees has great potential. Resistance is a phenomenon that occurs naturally, in varying degrees, in all species. It certainly has been demonstrated in a large number of tree species and a variety of dendrophagous species (Table 21.10).

Simply stated, the objective of pest-resistance control is to systematically take advantage of inherent qualities by selecting or breeding trees that are sufficiently immune to attack to provide a profitable harvest and/or an aesthetically pleasing tree. Implicit in this principle is that resistance is rarely absolute and that minimum standards of resistance are thus required for each host–pest complex, which in turn are determined by the objectives of tree or forest management.

The bases of resistance are quite varied. Resistance to insects can be based on differences in (a) morphological structures and tissue strength, (b) exudation pressure and the quantity and quality (e.g., toxicity) of resins or gums, (c) presence, quality, and relative availability of nutrients, (d) hormonally active protective compounds in the plant, (e) chemical behavior modifiers (e.g., antifeedants), (f) protective substances that act on insect symbionts, and (g) the degree of hypersensitivity to injury (i.e., an active metabolic process involving the degeneration of cells surrounding the site of injury or involving the release of terpenes, polyphenols, and other toxic or inhibitory compounds) (Berryman, 1972; Smelyanets, 1977a,b). Factors involved in resistance in specific host-tree relationships are illustrated in Table 21.11.

Clonal resistance to insects is demonstrable in a variety of instances, including the resistance of *Pseudotsuga menziesii* to attack by *Adelges cooleyi* and the resistance of *Robinia pseudoacacia* to attack by the locust borer (Schreiner, 1960). Among coniferous hosts of bark beetles, considerable variation exists in the toxicity and behavior-modifying capacity of resin vapors and thus in the resistance of hosts and nonhosts of insect pests. Resistance may be a temporal factor. The time of highest susceptibility and the duration of susceptibility of elms to Dutch elm disease (vectored by bark beetles) vary during the growing season, depending on elm species, and between seed sources of a given species (Smalley and Kais, 1964). Mountain pine beetles tend to prefer ponderosa pine trees with low concentrations of limonene and high concentrations of alpha-pinene (Sturgeon, 1979).

Dramatic illustrations of intraspecific variation in resistance, even in individual trees, suggests a bright future for resistant stock. An infestation of black pineleaf scale on ponderosa pine is often characterized by variation in scale density from one pine to the next. Thus, scale-free pines are found standing beside a heavily infested tree even when branches of the hosts are intertwined. Even more fascinating is that when scales are reciprocally transferred from adjacent infested and uninfested trees, survival and reproduction of transferred scales are very poor on the new host. These transfers demonstrate intraspecific heterogeneity in resistance to insect attack and the possible basis for an effective silvicultural control method.

Table 21.11. Selected factors involved in forest tree resistance to insect injury[a]

Target	Resistance Factor
Adelges abietis	Buds hypersensitive
Cryptorhynchus lapathi	Type of bark
Cylindrocoptorus eatoni	Resin ducts, cells
	Preference, antibiosis, tolerance
Dendroctonus brevicomis	Resin flow, resin pressure
	Resin toxicity
	Volatile attractants from interaction
Dendroctonus frontalis	Resin flow
Dendroctonus jeffreyi	Resin toxicity
Dendroctonus ponderosae	Tree vigor, resin flow
	Xylem resin ducts, resin toxicity
Dendroctonus pseudotsugae	Monoterpenes
Dendrotonus spp.	Oleoresins
Diprion simile	Host preference
Dryocosmus kuriphilus	Larval mortality
Elatobium abietinum	Supercooling point of needles and overwintering survival of aphids
Exoteleia pinifoliella	Vigor, needle resin canals
Hylobius abietis	Resin flow, turgor pressure
Ips paraconfusus	Resin flow
	Volatile attractants from insect–host interaction
Ips calligraphus	Volatile attractants from insect–host interaction
Lymantria monacha	Late flushing, terpenes
Panolis flammea	Resin flow
Phenacaspis pinifoliae	Blue needle color
Pissodes strobi	Resin flow
	Bark thickness
	Terminal diameter
	Depth of resin ducts
Taeniothrips laricivorus	Branching form affects recovery
Tortrix viridana	Late-flushing trees
	Early- or late-flushing trees
Xyleborus pavulus	Corky outer bark
Pristiphora abietina	Phenology
	Early-flushing trees
Cecidomyia sp.	Nonviscid
Rhagoletis completa	Time of husk ripening, hardness of husk
Rhyacionia buoliana	Differential larval survival
	Resin flow
	Differential rate of parasitism and oviposition, more buds
Rhyacionia frustrana	Adventitious buds and resin crystallization
Rhyacionia rigidana	Resin crystallization

[a] Modified from Smith (1966), Gerhold (1966), and Parry and Powell (1977).

Integration of Control Strategies

Although various approaches to silvicultural control have been dealt with separately, forest management often uses a variety of these procedures in concert. Any given pest species has preferences for specific hosts (whether primary or alternate) and climatic conditions and is affected in a particular fashion by enhanced vigor of its host. An understanding of host–pest interactions can allow for a complete silvicultural approach. For example, an understanding of the ecology and biology of the redheaded pine sawfly in the south-central United States and the results of interactions with its hosts results in a series of silvicultural control recommendations that include (a) not planting hard pines (e.g., jack, shortleaf, slash, longleaf, red, pitch (*Pinus rigida*), loblolly, and Virginia pines) under hardwoods, (b) not planting hard pines within approximately 25 ft of all hardwood borders, (c) replanting areas of poor survival so as to prevent widely spaced, open plantations, and (d) promoting early crown closures in plantations by using a 6 × 6 in. (or less) spacing. In the north-central states similar or modified procedures are recommended as follows: (a) avoid planting hard pines (particularly jack pine) under hardwood overstories, (b) avoid planting hard pines within approximately 25 ft of hardwood borders or residual hardwood crowns, (c) remove or girdle residual hardwoods prior to planting hard pines, and (d) substitute red pine for jack pine wherever possible (Benjamin, 1955).

For the spruce budworm, certain silvicultural controls or practices are suggested by various investigators. Although these include a variety of recommendations, each attempts to minimize the impact or frequency of outbreaks. The adopted procedures should (a) promote vigorous stands on better soils, (b) include partial cuttings on short (15–20 yr) cutting cycles or clearcut on a 50-yr cycle, (c) break up susceptible cover types into small areas separated by nonsusceptible host types, (d) maintain mixed species composition, (e) identify overmature stands and accelerate cutting, and (f) prevent stands from reaching maturity simultaneously over large areas (Batzer, 1976).

A variety of silvicultural procedures may be used in conjunction with direct and indirect control methods. An example of such an integrated approach in forest insect pest managment is that of the turpentine beetle (*Dendroctonus terebrans*). This insect often attacks and develops epidemic populations after thinning due to its attraction to stumps and residual trees injured by harvesting equipment. Infestations can be substantially reduced by minimizing injury and by removing residual stumps. Where this is not possible or feasible, losses can be reduced by selectively spraying injured trees with insecticides (Feduccia and Mann, 1975).

Use of an integrated program of direct and silvicultural controls can reduce losses due to the mountain pine beetle if treatment is applied over a large enough area to minimize emigration. Such a program might include direct methods such as felling infested trees, piling and burning (where possible), and insecticidal treatment. The aim of these procedures is to destroy beetles before they can emerge and attack green trees. The use of insecticides is frequently considered to prevent successful attacks on high-value trees in campgrounds, recreational areas, and summer home sites. In other circumstances salvage logging or harvesting of infested trees may be preferable. Beetles infesting these trees often are destroyed in the milling process.

Silvicultural recommendations for minimizing damage from insect borers of deciduous trees provides a final illustration. A simple but applicable practice is to fit tree species to the right site. This principle involves the avoidance of low-quality sites that consist of heavy clay soils and hardpan flats where borer damage is heavy. Second, tree species that are least damaged by borers should be favored. Borers exhibit clear preferences; for example, red oak borers (*Ennaphalodes rufulus*) prefer black oak (*Quercus velutina*) over northern red and scarlet oaks, and carpenter worms prefer overcup oak (*Quercus lycata*) over Nuttall oak (*Quercus nuttallii*, an intermediate preference) and willow oak (*Quercus phellos*, a low preference). Third, trees should be kept in vigorous condition, possibly by fertilization or irrigation, and brood trees, which harbor borers, should be identified and removed to reduce spread to other crop trees. Several major borers such as the carpenter worm reinfest trees (i.e., deposit eggs on the same tree in which they developed). Control of vines on tree trunks in high-value stands should also be considered. These create moist, favorable microhabitats for borers and provide protection from birds and other predators. Physiologically mature trees should be harvested because they have a high risk of mortality due to their slow growth, low vigor, and thus increased susceptibility to boring insects.

Injuries during harvesting and other operations should be minimized. The oak timberworm (*Arrhenodes minutus*), for example, attacks almost exclusively at wound sites. Carpenter worms, clearwing borers, and some ambrosia beetles show similar behavior. Close surveillance of timber stands should be maintained following periods of severe tree stress. Serious borer damage may follow prolonged drought, flooding, fire, or storm damage and requires prompt salvage harvesting or direct control (Solomon, 1978).

The use of these procedures is at present the most reliable means of silvicultural control of insect pests. As further research is available, these and other control strategies, including the use of predators, pathogens, parasites, pheromones, repellants, and antifeedants, may become more reliable and compatible with other forest managment practices and may be incorporated into effective forest pest management programs (Stevens *et al.*, 1974b).

Chapter Summary

Silvicultural control of insects is accomplished by modifying forest stands or methods of growing and harvesting trees to reduce damage. This control method requires understanding in detail the conditions in forests that are favorable to pests and then changing these conditions by manipulating species composition, alternate hosts, associated vegetation, stand structure, stand age, and overall vigor in order to create suboptimal conditions for pests. Matching tree species to planting sites is one of the most fundamental methods to minimize pests through silvicultural means. Use of genetically resistant host genotypes is also valuable approach and is used extensively in urban tree plantings. Because they are a normal part of the forest management process, silvicultural manipulations, are often the most effective approaches to manage forest pest insects.

Synopsis of General Entomology

What Is an Insect?

Insects belong to the phylum Arthropoda and are more common than any other form of animal. Most adult insects have certain characteristics that differ from those of other arthropods (Fig. A.1). Insects and some of their close relatives such as the mites are functionally, behaviorally, and morphologically highly variable animals. Thus, they are among the best-adapted, cosmopolitan species of animals on earth. Insects have evolved a high degree of specialization of labor by developing a three-segmented body (Fig. A.2). Each segment has a concentration of organs and structures

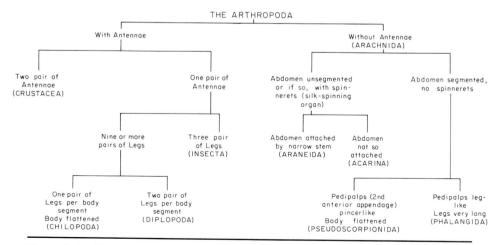

Figure A.1. Characteristics differentiating selected arthropod classes and orders.

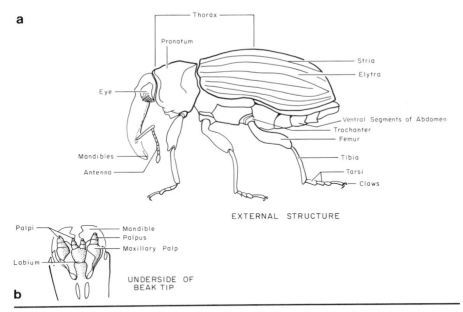

Figure A.2.   Adult weevil, illustrating the three body segments (head, thorax, and abdomen) and selected structures of insect body segments. (a) External structure and (b) underside of beak tip. (Drawing by A. Bartlett.)

adapted for specific critical functions: Feeding structures and major sensory receptors are found in the head; the legs and wings are attached to the thorax and provide locomotion; and maintenance and reproductive systems are in the abdomen. Similar subdivisions occur in many larval insects. The variation that exists in larval forms often reflects adaptation to their environment (Fig. A.3).

The Exoskeleton

Unlike many other animals, insects have an external skeleton rather than an exterior skin. The exoskeleton provides a protective shell and an attachment for muscles. Sclerotization, a hardening process, makes the integument rigid. The molecular structure of the outermost layer of the integument, or epicuticle, provides protection from desiccation. Water loss is the major obstacle to survival for terrestrial animals such as insects.

The exoskeleton also provides some protection against harsh chemicals and pathogens. The multitude of colors found on insects is often due to pigments in the cuticle or is a result of the effects of incident light on cuticular structures. Arising from the integument are various cuticular structures, many of which are sensory. Hairlike setae occur commonly on

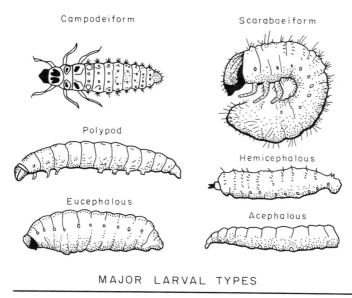

Campodeiform
Scarabaeiform
Polypod
Hemicephalous
Eucephalous
Acephalous

MAJOR LARVAL TYPES

Figure A.3.   Selected examples of major larval types.

larval and adult insects. These and similar structures may respond to vibration (mechanoreceptors) and chemicals (chemoreceptors).

External Structures

Head

Because the head is the first part of the insect to confront most environmental factors, it has a concentration of sensory structures that assist in the coordination of bodily functions.

Vision

The major photoreceptive organ is the compound eye (Fig. A.4). the degree of variation in this structure is considerably less than in the other structures on the head. The compound eye, the image former, consists of a number of separate receptors called ommatidia. The number of these individual receptors determines, in part, the visual acuity of the eye. Predatory insects such as dragonflies may have several thousand more receptors than scavengers or plant feeders, which fly only occasionally. Up to three so-called simple eyes or ocelli are also found on the head. Although they do not form images, they perceive drastic changes in illumination. Caterpillars have eyes that are intermediate between these two

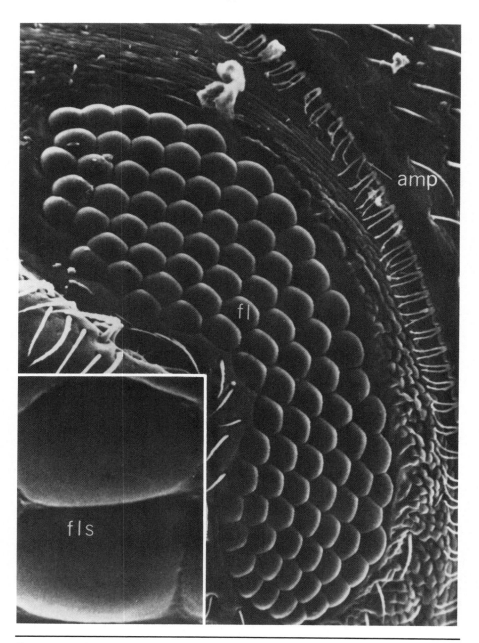

Figure A.4.   Compound eye of the scolytid beetle (*Xyleborus ferrugineus*) showing hexagonal ommatidia. Insert illustrates the smooth facet lens surface. Important structures are facet lens (fl) and facet lens surface (fls). (From Chu *et al.*, 1975.)

receptor types and are called stemmata. Twelve or fewer stemmata are found in two groups on the head and provide limited image resolution.

Antennae

The paired segmented appendages called antennae are major sensor structures. The fascinating polymorphism exhibited by the antennae of insects reflects major and sometimes subtle adaptations to their mosaic environments (Fig. A.5). This variation is particularly useful in the iden-

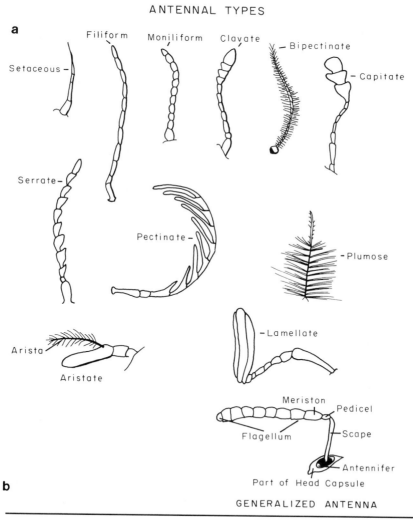

Figure A.5. (a) Selected examples of the types of antennae found in insects; (b) generalized antenna.

tification of insects. Receptors on the antennae respond to vibrations, odors, humidity, temperature, and other stimuli (Fig. A.6). The shape of the antennae of males and females of the same species often are different. One sex may have feathery antennae, which are better suited for the perception of a sex pheromone. In larval forms the antennae may be reduced.

Mouthparts

The basic plan of insect mouthparts is typified by chewing (or mandibulate type) mouthparts. These consist of an upper lip (labrum), a lower lip (labium), a pair of maxillae, and a pair of mandibles (Fig. A.7). The hard (sclerotized) and often large mandibles are adapted for grinding and tearing food. The segmented maxillae are composed of a pair of small sclerotized jawlike structures and a pair of sensory palps.

The haustellate mouthparts, a second major type, are adapted for sucking (Fig. A.8). Modifications of this pattern may include the piercing-sucking needlelike stylets for the intake of plant or animal fluids (as in the Homoptera, biting flies, or fleas), the rasping-sucking mouthparts of thrips, and the sponging-lapping mouthparts of higher Diptera such as blow flies and fruit flies. The mouthparts of adult Lepidoptera consist primarily of a coiled, strawlike tube (proboscis) formed by parts of the maxillae. The labrum is greatly reduced, and the chewing mandibles and hypopharynx (tongue) are lacking. Larval mouthparts are primarily designed for chewing and may be reduced. All these variations and others have evolved to cope with the equally variable food upon which insects feed.

Thorax

The three-segmented thorax is the point of attachment for most locomotory appendages. When present, wings are attached only on the last two segments. In adults there is a pair of legs on each segment. Both legs and wings show substantial variation among species. Although some species have two pairs of wings, others have only one pair, and still others have lost both pairs of wings through natural selection. In beetles, the forewings form a hard, protective shell called the elytra. The hindwings are membranous and folded under the modified forewings. In the Hemiptera the anterior portion of the forewings may be partially hardened, whereas the posterior portion is membranous. These modified wings are called hemelytra. In grasshoppers, crickets, and cockroaches the forewings are leathery to provide protection. The most dramatic modification of the hindwing occurs in flies. In the true flies the hindwings are clubshaped and important as stabilizers during flight. These modified hindwings are called halteres. In many insects having two pair of fully formed wings, a coupling mechanism allows the wings to function as one structure.

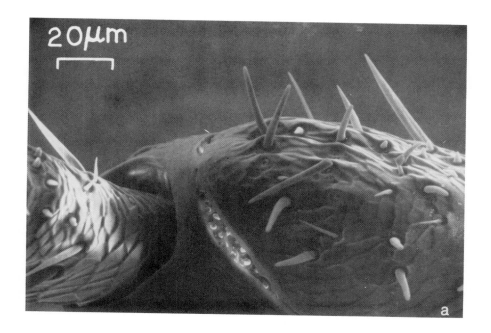

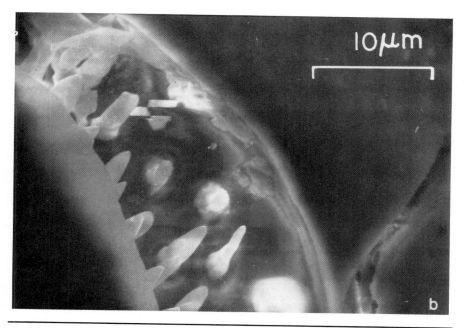

Figure A.6. (a) Chemoreceptors probably involved in olfaction, located in a slit organ on the fourth antennal annulus of a female flatheaded wood borer (*Melanophila drummondi*); (b) the sensory structures (sensilla basiconica), seen in the lumen of the slit organ. (From Scott and Gara, 1975.)

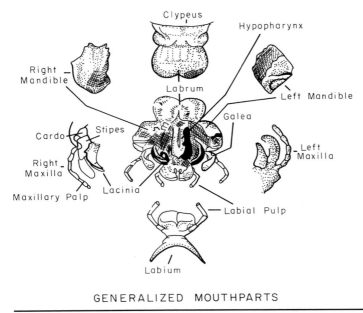

GENERALIZED MOUTHPARTS

Figure A.7.   Basic mouthparts of insects. (Redrawn from Metcalf *et al.*, 1962.)

Minute tubelike structures run between the two layers of a wing, joining them together to form the flat wing. These so-called veins form distinct patterns that are similar among related groups and differ among unrelated insects. Thus, vein characteristics are one of many useful aids in the identification of insects. The wings of most of these insects are covered with minute hairs. Some of these hairs have been modified into flat scales, typically seen in insects such as the Lepidoptera. The coloration of many adult insects results from the alignment of the many rows of scales and how they affect the incident light that falls on them. Coloration also may be due to the pigmentation of the scales.

The structures of insect legs are related to the function they must serve or the nature of the habitat in which the insect lives (Fig. A.9). For example; the femur of the hind pair of legs of grasshoppers is enlarged to enable them to hop; the long, thin legs of predaceous ground beetles, wood roaches, and long-horned beetles are adapted to walking or running; the wide, shovellike front legs of mole crickets or immature cicada, both of which live in the soil, are adaptations for digging; and the front legs of insects such as the praying mantid, with its large spiny projections and muscle-filled enlargements (e.g., the coxa and trochanters), are well suited for grasping prey. Many similar adaptations can be found in terrestrial and aquatic insects.

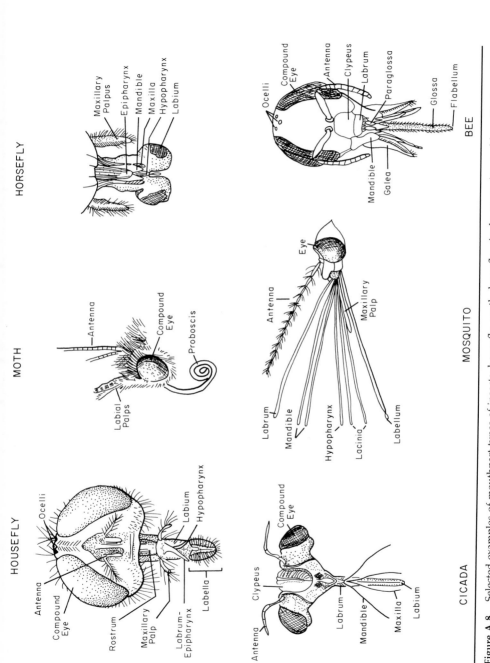

Figure A.8. Selected examples of mouthpart types of insects: housefly, moth, horsefly, cicada, mosquito, and bee. (Redrawn from Ross, 1965; Metcalf *et al.*, 1962; and Waldbauer 1962.

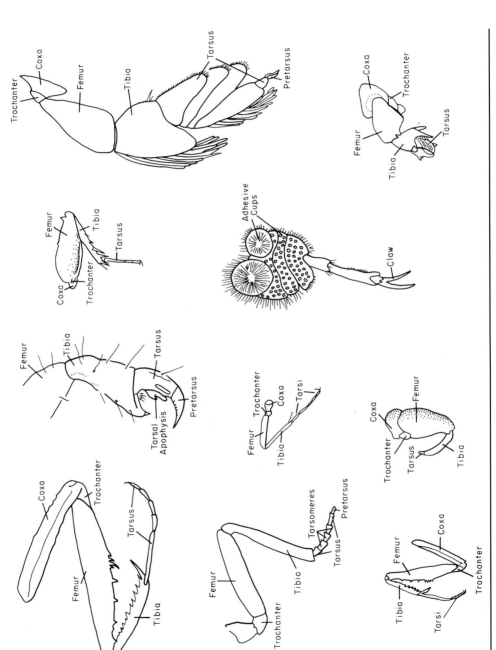

Figure A.9. Selected examples of the diverse leg adaptations found in insects. (Redrawn from Fernald, 1921; Chapman, 1969.)

Figure A.10.   A wasp parasitoid (*Coccygomimus* sp.) shown depositing its egg in a gypsy moth pupa. (Courtesy USDA, photo by Murray Lemmon.)

Abdomen

The abdominal segments of adults exhibit less external variation than the other body areas. The most obvious structures include those involved in reproduction and respiration. A pair of spiracles (the external openings of the respiratory system) are found along the sides of the first eight abdominal segments. At the distal end of the abdomen are the modified appendages that form the ovipositor of the female (Fig. A10). The terminal segment may also bear a pair of appendages referred to as *cerci*, which are believed to be sensory structures. In aquatic insects such as mayflies and stoneflies, gills are very apparent structures, occurring on both sides or at the distal end of the abdomen. The larvae of many Lepidoptera and most sawflies bear nonjointed, leglike paired prolegs on the abdomen. In Lepidoptera larvae, three to five pair of prolegs bear rows of minute hooked claws that aid in substrate attachment and locomotion, and sawfly larvae bear six to eight pair of prolegs without hooks.

Internal Structures

Alimentary System

The organs of the alimentary system are responsible for breaking down and transporting ingested food, digesting and absorbing assimilated

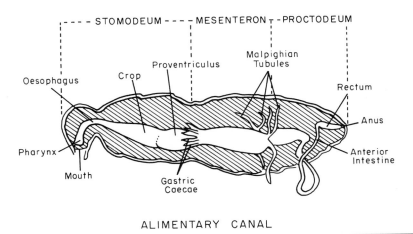

ALIMENTARY CANAL

Figure A.11.   Illustration of the typical organs of the alimentary system of insects.

food, and egesting waste products. The alimentary canal is characterized by three regions differing in structure and function: foregut, midgut, and hindgut (Fig. A.11). The foregut and hindgut are lined with a protective chitinous intima, which is a continuation of the cuticle of the integument. The foregut may be divided into distinct sections. The pharynx is the first section, posterior to the mouth. Although its structure varies widely among insects, with the aid of attached musculature the pharynx often forms a strong pump to draw up fluids or to pass food back to the esophagus, the next section. The esophagus may be enlarged to form a crop or several blind sacs (diverticula), which often function as temporary storage areas. The esophagus may also have a muscular valve known as a proventriculus.

Food moves through the esophagus via muscular undulations known as peristalsis. Little or no digestion or absorption occurs in the anterior portion of the canal. It is in the middle portion of the gut (the mesenteron, or ventriculus) that enzymes are liberated and food is broken down and absorbed. At the anterior end of the midgut, lateral projections called the gastric caecae provide additional surface area. In addition to these diverticula, there are a variable number of elongate excretory structures known as malpighian tubules.

A number of dendrophagous insects have piercing-sucking mouthparts and subsist on plant fluids. Accordingly, the digestive systems of these species are modified to cope with liquid diets. The major problem for these species is excess water in the food. One modification developed by some species is the presence of a crop adjacent to the esophagus for liquid storage. Plant-sap feeders often have a sort of filter chamber that short circuits water between the anterior and posterior sections of the gut. Thus,

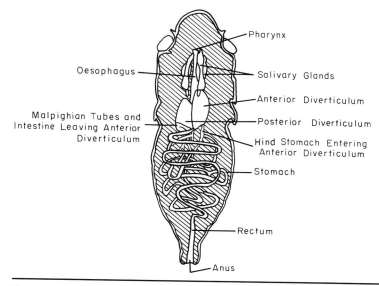

Figure A.12. Diagrammatic dorsal view of the alimentary canal of a
spittlebug nymph similar to those that are occasional feeders on Scots
pine in northeastern United States.

the concentration of food in the water is appropriate for efficient digestion
in the midgut (Fig. A.12).

   A number of other diet-related modifications can be found among
insects, including toxins in carnivores and anticoagulants in blood feeders.
Many of these modifications involve the labial gland, which normally acts
as the salivary gland. The secretions from this paired structure, which lies
ventral to the foregut, are carried along ducts and empty between the base
of the insect's tongue and the base of the labium. Depending on the nature
of the feeding for given species, salivary secretions serve a variety of
functions. These functions include the preliminary digestion of tissues and
the production of anticoagulants, silk, adhesives, or toxins.

   The digestion of food may not be possible in many cases without the
aid of microorganisms that live in the bodies of insects. These microflora
synthesize enzymes that enable the insect to break down a food type, or
they may provide a nutritive substance directly.

   Normally, digestion in insects occurs in the alimentary canal with the
aid of a number of enzymes that act to break down proteins, carbohydrates, and lipids. In some cases specialized enzymes such as cellulases
and pectinases break down substances that are not easily digested. At the
posterior end of the midgut, the pyloric valve regulates the movement of
food into the hindgut. As fecal matter moves through the hindgut (the
proctodeum), it becomes dehydrated as the lining of the gut absorbs water
and salts.

Circulatory System

Since hemolymph (insect blood) can comprise as much as 40% of the body, the circulatory system is, to say the least, an important organ system. It is referred to as an open system because the blood is enclosed in the body cavity (the hemocoel) and is not completely restricted to tubes or vessels. A longitudinal dorsal vessel consists of a series of pumping chambers (hearts) along the abdomen and an aorta along the anterior end to the brain region. Blood enters at the chambers through small openings called ostia. Nutrients are picked up from the midgut and distributed throughout the body; hormones may also be transported. Blood cells do not carry oxygen but may be phagocytic (actively ingesting foreign particles) or may encapsulate (become layered around) large entities such as parasites; they also detoxify toxic metabolites, aid in coagulation or accumulate at wounds, and aid in healing. Excretory products such as uric acid and allantoin are carried in the blood and eventually excreted by the malpighian tubules.

Insect blood is not compressible. Thus, hydraulic presure, developed by constricting one portion of the body, may act to cause a pressure change in another portion of the body or the extension of an appendage. The extension of the wings of a newly emerged adult may be accomplished partially by the hydraulic pressure provided by the hemolymph.

Respiratory System

The respiratory (ventilatory) system functions to exchange air and carbon dioxide. The spiracles, tracheae, and air sacs are important structures in the ventilation process. Spiracles are paired openings, generally in the last two thoracic segments and in eight abdominal segments, which are regulated by valves. Internally from each spiracle, branches (respiratory tubes known as tracheae) radiate dorsally to the muscles and dorsal vessels, medially to the gut and gonads, and ventrally to the muscles and nerve cord. Tracheae branch into smaller tracheae forming minute tubules (tracheoles). Tracheal dilations (air sacs) are formed in parts of the tracheae but are not as well braced as the rest of the tracheal system. These structures vary in size, number, and location and may serve (a) to increase an insect's volume (e.g., to provide increased pressure during molting), (b) to increase buoyancy for flight, (c) to store oxygen, or (d) to help conserve heat. A reduction in the number of spiracles has evolved in some insect groups. In still others, such as some aquatic or endoparasitic species, a closed system has been evolved in which gaseous exchange with the tracheal system occurs across the integument.

Muscular System

Tissues and cells of the muscular organs have an ability to contract or shorten in length. Contraction, depending on the attachment of the muscles, their location, and the strength of the stimuli, results in movements reflected in all aspects of an insect's life. These movements include a range of activities from movement of viscera and maintenance of posture to locomotion. Although some insects have as many muscles as humans (approximately 800), others have as many as 4000. The power of these muscles also varies. In general, because of their small size, insect muscles are relatively more powerful than those of larger animals.

Unlike warm-blooded animals, insects have a body temperature that closely follows that of the environment. Although not having to maintain a constant body temperature has certain advantages, it may limit the rate of metabolic processes and the functioning of certain tissues. For example, the temperature of flight muscles generally must approach 30–35°C for flight to occur. Although this requirement restricts flight in some species, others have evolved behaviors that maximize exposure to the sun, which then provides the needed increase in body temperature.

Excretory System

The primary functions of the organs and tissues of the excretory system are to eliminate metabolic wastes and to regulate salt and water. The major nitrogenous waste product in insects is uric acid. Malpighian tubules, which vary in number from 2 to 250, depending on the species, are usually free in the body cavity and bathed in hemolymph. These organs are the major organs of excretion. Both the malpighian tubules and the rectum, located in the hindgut, function to solve the problems of salt and water balance. Depending on the water content of food (the usual source of water), insects are evolutionarily adapted for maintaining the appropriate proportion of body water. Similar maintenance of the appropriate salt balance insures that the insect has needed ions.

Nervous System

The building blocks of the nervous system are cells called neurons, which have become specialized to transport the responses to stimuli and to coordinate the informational aspects of the stimuli leading to a given behavior. The information is carried in the form of electrical impulses from receptor sensory cells to effectors (such as glands or muscles). The type and location of stimulated sensory cells determine the nature of the response. Neurons are not continuous with each other, but the fine terminal branchings of one neuron are extremely close to the branchings of another neuron. The extremely small distance between them is called a synapse. Specific chemicals (neurotransmitters) help to bridge the gap and initiate the impulse in the following neuron. Neurons may form aggregations of neural material known as ganglia. Finally, bundles of long, very thin fibers (axons) from the neuron cell bodies constitute a nerve.

The central nervous system of insects is composed of a ladderlike double chain of ganglia connected with longitudinal connectives. The three

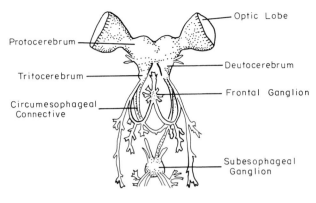

Protocerebrum

Tritocerebrum

Circumesophageal
Connective

Optic Lobe

Deutocerebrum

Frontal Ganglion

Subesophageal
Ganglion

BRAIN AND STOMODAEL NERVOUS SYSTEM
OF THE GRASSHOPPER

Figure A.13.   Typical brain and stomodael nervous system of insects.
(Redrawn from "The Science of Entomology," 2nd Ed., by W.S. Romoser.
Copyright © 1973, 1981 by William S. Romoser.)

anterior-most ganglion pairs are fused to form the brain (Fig. A.13). The
fourth to sixth pairs form the subesophageal ganglion. This ganglion inner-
vates sense organs and muscles of the mouthparts, salivary glands, and
the neck area. It may also have a general influence on an insect's internal
physiological state. Posterior to the subesophageal ganglion are three thoracic ganglia, each controlling sensory and motor activities of its respective
segment. In certain insect groups these may all be fused longitudinally.
The visceral nervous system of insects is comprised of three subsystems.
The stomodael system exerts control over the movements of the gut and
heart and possibly the labial muscles, mandibular muscles, and salivary
ducts. The ventral sympathetic system is associated with the spiracles and
the ganglia of the ventral nerve cord. Finally, the nerves of the caudal
sympathetic system supply the posterior portions of the hindgut and
internal sexual structures.

Endocrine System

The endocrine system includes tissues and organs that secrete chemicals (i.e., hormones), which are produced by one tissue and exert their
influence on another, target tissue or organ. The hormone production
complements the functioning of the nervous system and is often closely
associated with nervous tissues. Indeed, certain neurosecretory cells are
found in insect ganglia. These modified neurons are found, for example, in
parts of the brain, the subesophageal and thoracic ganglia, the corpora
allata, and the corpus cardiaca (Fig. A.14).

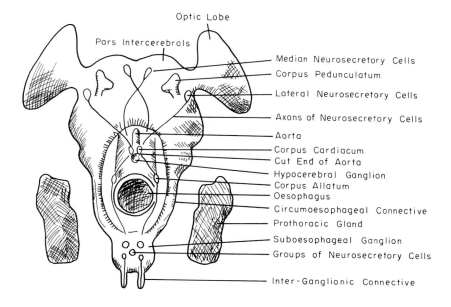

Optic Lobe

Pars Intercerebrals

Median Neurosecretory Cells
Corpus Pedunculatum
Lateral Neurosecretory Cells
Axons of Neurosecretory Cells
Aorta
Corpus Cardiacum
Cut End of Aorta
Hypocerebral Ganglion
Corpus Allatum
Oesophagus
Circumoesophageal Connective
Prothoracic Gland
Suboesophageal Ganglion
Groups of Neurosecretory Cells
Inter-Ganglionic Connective

MAIN ENDOCRINE ORGANS

Figure A.14. Primary endocrine organs of insects. (Redrawn from Chapman, 1969, by permission of the publisher, from "The Insects. Structure and Function," by R. F. Chapman. Copyright 1969 by Elsevier North Holland, Inc.)

The corpora allata secretes juvenile hormone, which stimulates larval development while retarding or preventing adult characteristics. The corpora cardiaca is the organ that accumulates material called brain hormone, which arises from the brain. The prothoracic glands located in the prothorax secrete molting hormone, or ecdysone, which initiates the growth- and molting-related activities of cells. The secretion of the prothoracic gland is stimulated by brain hormone. Various other hormones in insects are responsible for influencing color changes, controlling heartbeat rate and amplitude, synthesizing carbohydrate and protein, melanizing the cuticle, and regulating the excretion of water, as well as other vital functions.

Reproductive System

The male reproductive system consists of paired testes that provide the sperm, which are transferred to the female via a penis (aedeagus) (Fig. A.15). The female vagina opens to the exterior and receives the

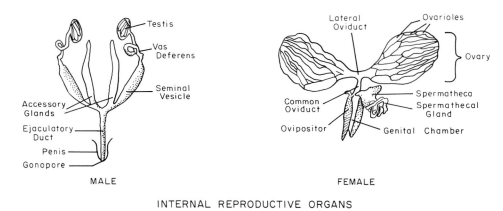

INTERNAL REPRODUCTIVE ORGANS

Figure A.15. Typical internal organs of reproduction of (a) male and (b) female insects. (Redrawn from Snodgrass, 1935; Van Derwerker *et al.*, 1978.)

aedeagus. Ovaries consist of a definite number of tubes or ovarioles joined at a lateral oviduct. The lateral oviducts join to form a median oviduct, which leads to the vagina, or genital chamber. The bursa copulatrix is a saclike expansion of the vagina. An outpocketing of the bursa, vagina, or common oviduct serves as a storage area for spermatozoa prior to fertilization. This structure is called the spermatheca. Accessory glands opening into the bursa copulatrix vary in structure and function. One such function is the production of secretions used to cement eggs together or to a substrate. Eggs move down the oviducts by peristaltic waves and may be deposited singly or in groups. The deposition of eggs is accomplished either by a well-developed organ (i.e., the ovipositor) or a modified abdomen that telescopes into a long tube and acts as an ovipositor.

A variety of reproductive strategies are used by insects. Oviparity is the most common type of reproduction among insects and consists of oviposition of fertilized eggs. Sufficient yolk is usually present to permit complete embryonic development. In ovoviviparity, eggs are normally developed and fertilized but are either retained and hatched in the body of the female or are deposited shortly before hatching occurs. Although fewer eggs are usually produced, the favorable environment in the female's body provides substantial protection to the eggs, and when deposited the immatures are ready to feed. In viviparity, development also takes place in the female, but because the yolk content of eggs may be insufficient for complete development, the embryos are nourished by the female.

Asexual reproduction is another evolutionary option that occurs in insects. Parthenogenesis, the capacity of reproducing without fertilization, is the most common form. Parthenogenic species may also be viviparous, as are aphids, or oviparous, as are parasitic Hymenoptera and numerous sawfly species.

A less common form of asexual reproduction, usually restricted to parasitic species, is polyembryony. In this form of reproduction, a single egg divides from two to several thousand times, producing a comparable number of individuals. The advantages of both sexual and asexual reproduction are incorporated in the life history of some species. Many life cycles exhibit metagenesis, or alternation of generations, from sexual to parthenogenesis and back again.

Postembryonic Growth and Development

For most insects (i.e., oviparous species) postembryonic development spans the period from hatch to adult emergence. Although some species undergo some development in the female, few of these insects are of importance in the forest habitat. Because the exoskeleton of insects is, for the most part, a relatively rigid shell within which the insect grows, the process of development represents a phased or stepwise progression of changes of the outer shell.

A key process in development is molting or ecdysis (i.e., the periodic digestion and shedding of an old cuticle and the secretion of a new cuticle). Between each molt the insect is referred to as an instar, and the time period it spends as that instar is called a stadium. In most insects, molting ceases after the adult emerges. The number of instars that occur varies among insect species, but generally ranges from 2 to 20. Even within a species the number of instars may vary. Increases in body weight as insects develop can be quite dramatic. The carpenter worm, when fully grown, has increased its weight by a factor of 72,000. Although this magnitude of increase is not common, an increase of 1000 times a larva's original weight is not uncommon.

Metamorphosis

As an insect molts, it not only increases in size, but also its tissues and organs differentiate. Changes in the insect may be internal and/or external, slight or profound. The developmental process by which a first instar is transformed into an adult is called metamorphosis, of which there are various types (Table A.1). In certain insects, life-history adaptations may result in a complex series of developmental or life-history forms (Table A.2).

The control and regulation of growth and metamorphosis are accomplished by the hormones of the endocrine system. The neurosecretory cells in the brain secrete the brain hormone, which accumulates in the corpora cardiaca and is ultimately released into the hemolymph. The brain hormone stimulates the prothoracic glands which, in turn, produce molting hormone. The juvenile hormone produced by the corpora allata retards

Table A.1. General types of metamorphosis

Type	Characteristics	Life Cycle	Example
Ametabola	The immature forms, called young, resemble the adult except in size. These insects do not have wings (Apterygota).	Egg → young → adult	Silverfish
Hemimetabola (gradual or incomplete)	The immatures, called nymphs, resemble the adult except that they are smaller and possess wing pads in later stages. Adults posses fully developed wings. The nymphs are usually terrestrial.	Egg → nymph → adult	Cricket
	The immatures, called naiads, are aquatic, usually respiring through gills. Later, immatures possess wing pads. The adult has fully developed wings. In many species the adults are terrestrial.	Egg → naiad → adult	Stonefly
Holometabola (complete)	These insects undergo four different life stages. The immatures, called larvae, look nothing like the adult and do not develop wing pads. The pupa is a transition stage between larva and adult.	Egg → larva → pupa → adult	Japanese beetle

Table A.2.   Selected life-history forms of some aphid and adelgid species and their characteristic traits[a]

Life-history Form	Characteristic[b]
	APHIDS
Fundatrix (stem mother)	Apterous, parthenogenetic, viviparous or oviparous female hatching from the overwintering egg of female sexualis and living on primary host.
Spuriae apterae	Parthenogenetic, viviparous females living on primary host.
Spuriae alatae	Parthenogenetic, viviparous females developing on primary host and migrating to secondary host.
Spuriae apterae	Parthenogenetic, viviparous females developing on secondary host.
Sexuparae alatae	Parthenogenetic, viviparous or oviparous females, progeny of the exulis sistens and developing or overwintering on secondary host and migrating to primary host to deposit progeny.
Sexuales	Small apterous males and oviparous females, progeny of sexupara.
	ADELGIDS
Gallicola migrans	Winged parthenogenetic female, progeny of the fundatrix on the primary host. The larva settles at base of a needle on a new shoot. This form completes the formation of the gall. After the fourth molt, the winged adult leaves the gall and flies to the secondary host and oviposits.
Exsulis	Apterous parthenogenetic female that develops from the egg of the gallicola migrans and gives rise to other generations of exules on the secondary host.
Exsulis sistens	Form that has three molts and is characterized by a period of "diapause" in the first stage after inserting its stylets.
Hiemosistens	The overwintering generation of the exulis.
Aestivosistens	The summer generation.
Exsulis progrediens	Form that has four molts and usually develops without a period of rest.
Neosistens or neoprogrediens	The first instars of the exules.

[a] Based on terminology of Patch (1920) and Balch (1952).
[b] Alatae refers to winged forms, and apterae refers to wingless forms.

adult characteristics. Both molting and juvenile hormones are produced during the immature stages. In the last immature instar, juvenile hormone is not produced, and metamorphosis to the adult occurs. In hemimetabolous insects, there is a gradual decrease in the juvenile hormone concentration with a concomitant, slow increase in adult characteristics.

Diapause

Dormancy resulting from the cessation of growth during the postembryonic period is called diapause. In the adult it takes the form of the cessation of reproduction. This phenomenon provides a number of advantages for insects, foremost of which is that it allows them to survive through the nongrowing season. In addition, it synchronizes the life cycle (a) to allow insects to emerge when there is optimal quantity or quality of food, (b) to compensate for slow-growing individuals by ceasing growth at one stage, or (c) to numerically maximize the emergence of males and females, thereby enhancing the opportunity for mating.

Diapause occurs in all life stages of insects. Although diapause is obligatory for many insects (i.e., insects enter diapause regardless of variation in environmental conditions), other species exhibit facultative diapause. The latter form is common among multivoltine (multiple generation) species and is more sensitive to environmental stimuli. These stimuli commonly are photoperiod or temperature but may include indirect cues such as lowered moisture or reduced food quality. The physiological control of diapause is believed to be associated with the endocrine system.

Insect Behavior

The behavior of insects in any situation represents the total response of an individual to a number of concurrent stimuli. The response may be genetically fixed (so-called innate behavior), learned (environmentally influenced or experience-related behavior), or some combination of both.

Perhaps of most frequent occurrence and importance to forest insects are innate behaviors. Some of the simplest innate behaviors are reflexes that may involve the whole body or only a portion of the body. These responses include righting reflexes and maintenance of posture and equilibrium. More complex reflexes result from contact with an appropriate surface or substrate.

Orientation patterns are examples of more complex behavior. The movements of insects are classified on the basis of the mechanisms of orientation they use. Among these are kineses, or undirected locomotory responses to stimuli. Kineses are characterized by the lack of orientation of the long axis of the body relative to the source of the stimuli. An example is the increased activity of an insect in a suboptimal microenvironment compared with those in areas having optimal conditions. Taxis responses are characterized by directed responses of the insect relative to the source of the stimulus, such that the insect moves toward or away from the stimulus source. The nature of the stimulus determines the category of the taxis response. Thus, responses are classified as photo- , geo- , anemo- , thigmo- , or theotaxic, representing reactions to light, gravity, air currents, contact, and water currents, respectively. Other mechanisms of orientation include transverse orientations in which the body is oriented at a fixed angle to the direction of the stimulus. A transverse orientation in which locomotion occurs at a fixed angle to light rays is referred to as a light–compass reaction. A second type of transverse orientation is the dorsal or ventral light reaction, in which long and transverse axes of the body are kept perpendicular to the light source.

Finally, the most complex innate behavior patterns are adaptive species-specific responses that require a certain state of internal readiness before they can occur. The stimuli that act as releasers of the state of readiness or motivation can be internal or external. The stimuli may be hormonal secretions, light, temperature, mechanical contact, or pheromones. Examples of these behavior patterns include silk-spinning behavior of lepidopteran larvae, mate-calling, and food ingestion. When a releaser stimulus increases the state of internal readiness, the insect exhibits an appetitive behavior (e.g., increased locomotion). Although this behavior does not reduce the motivation, it does increase the likelihood of exposure to the releaser stimulus. Such exposure results in the consummatory act, which is a specific behavior pattern involved with the motivation and which lowers the motivation.

Reproductive Behavior

The initial phase of reproduction is mate location. The large number of mechanisms that are evolved in bringing the sexes together can be categorized as visual, olfactory, and auditory. The simplest form of visual stimulus is the movement of one individual in the visual field of another. Thus, a male often will attempt to mate with any moving object within a certain size range that moves into its visual field. More complex interactions occur when individuals respond to visual stimuli comprised of specific markings, shapes, or silhouettes. Often, different types of cues occur in unison or in sequence. Once a general visual stimulus brings individuals of the same species together, olfactory stimuli may differentiate the sexes.

The use of olfactory stimuli is extremely common among insects. Sex attractants can be produced by either males or females (Table A.3.). These pheromones are detectable by insects, although they often are produced in minute quantities. These chemicals are produced in special glands; for example, in many female Lepidoptera the pheromone gland is a modified intersegmental membrane often between the eighth and ninth abdominal segments (Lalanne-Cassou *et al.*, 1977).

Table A.3. Selected sex attractants in forest and shade tree insects

Scientific Name	Common Name	Pheromone
Dendroctonus (general)	Pine beetles	Frontalin [1,5-dimethyl-6,8-dioxabicyclo(3·2·1)octane]
		2-Methyl-2-cyclohexen-1-ol
		2-Methyl-2-cyclohexen-1-one
		trans-Verbenol
		[*trans*-4,6,6-Trimethylbicyclo(3·1·1)-3-hepten-2-ol]
		cis-Verbenol
		[*cis*-4,6,6-Trimethylbicyclo(3·1·1)-3-hepten-2-ol]
		Verbenone[a] [4,6,6-trimethylbicyclo(3·1·1)-3-heptenone]
Dendroctonus brevicomis	Western pine beetle	*endo*-Brevicomin[a]
		[*endo*-6-Ethyl-1-methyl-7,8-dioxabicyclo(3·2·1)octane]
		exo-Brevicomin
		[*exo*-6-Ethyl-1-methyl-7,8-dioxabicyclo(3·2·1)octane]
		Brevilure-33C (33% Frontalin in 3-carene)
		Brevilure-33M (33% Frontalin in myrcene)
Anarsia lineatella	Peach twig borer	(E)-5-Decen-1-ol (or Acetate)
Ips paraconfusus	Pine bark beetle	Ipsdienol [2-methyl-6-methylene-2,7-octadien-4-ol]
		Ipsenol [2-methyl-6-methylene-7-octen-4-ol]
		cis-Verbenol [*cis*-4,6,6-trimethylbicyclo(3·2·1)-3-hepten-2-ol]
Popillia japonica	Japanese beetle	Japonilure
		7:3 mixture of 2-phenylethanol propionate and
		4-allyl-2-methoxyphenol (Eugenol)
Scolytus multistriatus	Elm bark beetle	Blend of the following:
		Multilure I or Multistriatin
		[2,4-Dimethyl-6,8-dioxabicyclo(3·2·1)octane]
		Multilure II [α-cubebene]
		Multilure III [4-methyl-3-heptanol]

(continued)

Table A.3. (*continued*)

Scientific Name	Common Name	Pheromone
Dendroctonus frontalis	Southern pine beetle	Frontalure-33 (33% frontalin in α-pinene)
Dendroctonus pseudotsugae	Douglas-fir beetle	Douglure-33 (33% frontalin in a 1:1 mixture of α-pinene and camphene) 3-Methyl-2-cyclohexen-1-ol 3-Methyl-2-cyclohexen-1-one
Dendroctonus ponderosae	Mountain pine beetle	Pondelure-90 (90% *trans*-verbenol with 10% α-pinene)
Gnathotrichus sulcatus	Ambrosia beetle	Sulcatol [6-methyl-5-hepten-2-ol]
Barbara colfaxiana	Douglas-fir cone moth	*cis*-9-Dodecen-1-ol
Laspeyresia youngana	Spruce seed moth	*trans*-7-Dodecen-1-ol (and up to 10% of cis-isomer)
Ryacionia buoliana	European pine shoot moth	*trans*-9-Dodecenylacetate (blend of isomers)
Prionoxystus robiniae	Carpenter worm	(Z,E)-3,5-Tetradecadiem-1-ol acetate
Archips semiferanus	Oak leaf roller	(Z)-10-Tetradecen-1-ol acetate
Choristoneura fumiferana	Spruce budworm	Blend of the following: Soolure I [(E)-11-tetradecenal] Soolure II [(Z)-11-tetradecenal]
Choristoneura rosaceana	Oblique-banded leaf roller	Riblure [(Z)-11-ettradecen-1-ol acetate]
Orgyia pseudotsugata	Douglas-fir tussock moth	Tussolure [(Z)-6-heneicosen-11-one]
Lymantria dispar	Gypsy moth	Disparlure [2-methyl-cis-7,8-epoxyoctadecane]

[a] Acts as a repellent for some insects.

The sexes of many species are often stimulated by what appears to be the same compound. However, detailed investigations usualy reveal that there are subtle differences in the substances that are most attractive. One isomer of a substance can be more effective than the other, or frequently the most attractive material is a specific blend (or proportion) of each isomer. Other isolating mechanisms may also restrict cross attractancy. These include differences in diel periodicity of adult flight, seasonal occurrence, or hosts exploited by the species.

Acoustic signals are also important to insects. So-called calling songs, which are one part of mate location or courtship, are particularly important in species of the Orthoptera, Homoptera, and Diptera and in a few species of Lepidoptera, Heteroptera, and Coleoptera. These species-specific calling songs are often produced by specialized structures. The sound production is often referred to as stridulation, which is defined as sound production by friction of one specialized body part against another.

Among the Coleoptera, stridulation appears to play an important role in the life history of many Scolytidae, Passalidae, and Cerambycidae, although in the last group its specific function remain unclear. Sound production has been studied in detail in various species of *Dendroctonus*, such as *D. pseudotsugae, D. frontalis, D. valens*, as well as *Ips tridens, I. pini, I. calligraphus, I. paraconfusus. I. concinnus, Leperisinus fraxini, Hylesinus oleiperda*, and others. In these and other species, auditory signals appear to be closely intertwined with chemical signals. Stridulation may function as (a) a signal to the opposite sex to produce antiaggregation pheromone, (b) a rivalry signal associated with aggressive behavior of one male toward another, or (c) a stress signal or chirp produced under stress (see Chapter 4). In general, these acoustic signals act as behavioral primers and releasers (Oester and Rudinsky, 1979).

The problem of placing eggs in the most appropriate habitat is particularly crucial for those species that have specific requirements. The choice of oviposition site is most likely an evolved behavior pattern mediated by a variety of environmental cues and the internal physiological state of females.

Periodicity of Behavior

Many of the activities of insects occur with a certain degree of predictable repetitiveness. That is, the behavior, whether locomotion, oviposition, or flight, shows a recurrence at specific intervals (Table A.4). This periodicity may be circadian (of approximately 24 hr), annual, or even lunar. Long-lasting periodicities also may occur. Each of these intervals may be subdivided; thus, diel activity may be crepuscular (dawn and dusk), diurnal (daytime), or nocturnal (nighttime). Rhythms are periodicities that are controlled by an innate time-measuring sense or biological clock. Exogenous (i.e., environmental) cues regulate periodicities. Many physiological–

Table A.4. Endogenous rhythm illustrated by male catches of Douglas-fir tussock moths on virgin female baited sticky trap panels. Note peak male flight occurred between 1500 and 1800 hr[a]

Time (PST)	Males Caught, by Trap Number[b]										Total
	1	2	3	4	5	6	7	8	9	10	
1100	0	0	0	7	0	2	0	0	0	0	9
1200	0	4	0	1	0	0	0	2	0	0	7
1300	0	4	0	14	0	7	0	5	0	0	30
1400	0	13	0	8	0	12	0	11	0	3	47
1500	0	31	0	54	0	23	0	12	0	6	126
1600	0	36	0	52	0	27	0	40	0	29	184
1700	1	26	0	52	0	62	0	16	0	8	165
1800	0	7	0	22	0	42	0	8	0	6	85
1900	0	0	0	5	0	17	0	3	0	1	26
2000	0	0	0	0	0	2	0	0	0	0	2
2100	0	0	0	0	0	0	0	0	0	0	0

[a] Based on data from Wickman *et al.* (1975).
[b] Even-numbered traps baited with virgin females.

behavioral events exhibit specific periodicities; these include reproductive cycles or events (e.g., mate-calling), developmental events, emergence of adults, and dispersal.

Behavioral Communication

A great variety of other behavior patterns of forest insects are mediated by chemical signals. The type of signal may vary depending on the source of the signal, its species-specificity, and the behavior it elicits (Table A.5). Nevertheless, chemical communication appears to be the primary means for information exchange among insects.

Artifical categories such as chemical, visual, and auditory communication cues are pedagogical conveniences. The reality is more an inseparable mixture of all types of cues involved in the transfer of information.

Insect Relatives

Arachnida

Included in this group are a wide variety of animals such as scorpions, whipscorpions, and sunspiders. More important (to our discussion)

Table A.5.   Definitions of chemical mediators important in chemical interactions among organisms[a]

Hormone   Chemical agent, produced by tissue or endocrine glands, that controls various physiological processes within an organism.

Semiochemical   Chemical involved in the chemical interaction between organisms.

> **Pheromone**   Substance that is secreted by an animal or plant to the outside that causes a specific reaction in a individual of the same species that detects the substance.

> **Allelochemical**   Chemical significant to organisms of a species different from their source, for reasons other than food as such.

> > **Allomone**   Substance, produced or acquired by an organism, that (when it contacts an individual of another species in the natural context) evokes in the receiver a behavioral or physiological reaction adaptively favorable to the emitter but not to the receiver.

> > **Kairomone**   Substance, produced, acquired by, or released as a result of the activities of an organism, that (when it contacts an individual of another species in the natural context) evokes in the receiver a behavior or physiological reaction favorable to the receiver but not to the emitter.

> > **Synomone**   Substance, produced or acquired by an organism, that (when it contacts an individual of another species in the natural context) evokes in the receiver a behavioral or physiological reaction adaptively favorable to both emitter and receiver.

> > **Apneumone**   Substance, emitted by a nonliving material, that evokes a behavioral or physiological reaction adaptively favorable to a receiving organism but detrimental to an organism of another species, which may be found in or on the nonliving material.

[a] From Nordlund (1981). "Semiochemicals: Their Role in Pest Control," D. A. Nordlund, R. L. Jones, and W. J. Lewis eds. Copyright © 1981 John Wiley & Sons, Inc. Reprinted by permission of John Wiley & Sons, Inc.

and more frequently encountered representatives include mites, ticks, pseudoscorpions, spiders, and harvestmen (daddy long-legs). Most have pincerlike (or fanglike) appendages that are modified for feeding (chelicerae) and pedipalps that usually have a sensory function. In addition, they lack antennae and have two body segments (the cephalothorax and the abdomen) and four pairs of legs.

Phalangida (Harvestmen or Daddy Long-legs)

These common arthropods are easily recognized by their extremely long, slender legs which hold their small, oval bodies high over the ground surface. These common woodland organisms feed on the fluids of plants or dead insects; some are predatory.

Mites

Mites rival insects in their diversity of form and life style, as well as the variety of habitats in which they occur. Mites successfully exist as animal parasites, plant feeders, fungus feeders, predators, microorganism filter feeders, scavengers, commensals of vertebrates and invertebrates, and gall-makers. They outnumber all other arthropods in the forest soil and promote soil richness as a result of their role in the breakdown of organic matter.

Initial symptoms of mite damage to tree foliage often appear as bronzed and/or sticky foliage. Aside from chlorosis of needles or leaves, premature needle- or leaf-drop may occur in severe infestations. This injury results when the mites withdraw plant fluids with their needlelike stylets.

Three families of mites are commonly associated with trees: the Phytoseiidae, Tetranychidae, and Eriophyidae. These include both phytophagous and predaceous species. There are hundreds of species in the family Phytoseiidae, most of which are predaceous on other mites, particularly spider mites and eriophyid mites. As adults they are characteristically oval, translucent white, and very active. The most important group in the family Tetranychidae is the spider mites. This large group includes common pests of trees, particularly in the genera *Entetranychus*, *Eurytetranychus*, *Panonychus*, *Eotetranchus*, *Oligonychus*, and *Tetranychus*. Although some species, such as the honey locust mite, are host-specific, many others are polyphagous. Both males and females occur in this species, but reproduction can also be parthenogenetic. Spider mites are so named because of the silk that they produce from silk glands near their mouths. Mites live under a silk mat produced over the surface of the leaves. Brown needles and fine silken webs are indicators of the presence of these mites, which feed on a variety of conifers, including spruce, pine, and cedar, and some deciduous species. Examples include the spruce spider mite (*Oligonychus ununguis*), honey locust spider mite (*Eotetranychus matthyssei*), two-spotted mite (*Tetranychus urticae*), European red mite (*Panonychus ulmi*), linden spider mite (*Eotetranychus tiliarum*), and oak red mite (*Oligonychus bicolor*).

The family Eriophyidae includes rust mites, which feed on surface cells of foliage, bud mites, which feed and live in plant buds, and gall mites, which induce plants to produce growths in which the mites feed and live. The gall mites are usually elongate or wormlike with two pair of legs at the anterior end. Common representatives include the so-called erineum mites (which feed on maples, birch, and other hosts and cause red, green, or yellow feltlike patches of growth on the foliage) and the bladder gall mites (which feed on red maple, elm, linden, and wild cherry (*Prunus emarginata*) and cause red or green growths on upper foliage surfaces). Many other representatives are included in this group (e.g., spindle galls and ash flower galls). Still other Eriophyids produce marginal leaf rolls, within which they feed, live, and reproduce.

Eriophyid mites are very small. Often, mature adults are smaller than spider mite eggs. They are wormlike and cylindrical, and their color varies from translucent white to dark orange.

Free-living mites in all three families are disseminated or transported in a variety of ways. Locomotion, artificial transport on shipped plant material, and windborne transport are three common methods. Hitchhiking on winged insects (phoresy) also is a common means of transport. For example, at least 14 species of mites are phoretic on the southern pine beetle; as much as 89% of emerged brood adults have been found to carry mites. Although the average number of *Trichouropoda australis* mite per infested *D. frontalis* beetle ranged from 4.9 to 5.1, single individuals have been found to carry as many as 64 mites (Moser, 1976a,b; Roton, 1978).

Ticks

These acarines are larger than mites and are parasitic on birds and mammals. Ticks take blood meals from many of the small mammals commonly found in the forest.

Spiders

This large and abundant group includes species occurring in a variety of forest microhabitats. All spiders are predaceous, feed on insects, and can be important in the natural regulation of forest and shade tree insect pests. Spiders capture their prey in a variety of ways, including web building and free-style hunting. In the forests of the northeastern United States, for example, spiders of the families Theridiidae, Linyphiidae, Agelenidae, Hahniidae, Amaurobiidae, and Dictynidae spin webs of various types in litter, under rocks and logs, and in other similar sites. The species in the families Gnaphosidae, Lycosidae, Clubionidae, Salticidae, and Thomisidae may spin retreat-type webs but rely on several types of active hunting paradigms. Spiders constitute a significant component of the forest floor. Estimates of the number of spiders in the forest ground layer range from between 60 to 230 individuals per m^2 of forest floor (Norton, 1973).

Chilopoda

Centipedes are characterized by having two body segments: a head and a trunk that is comprised of many segments. The head has a pair of long antennae, mandibles (the feeding appendages), and two pair of maxillae (sensory feeding structures). Each of the segments on the trunk has a single pair of legs, making this arthropod a rapidly moving predator. Centipedes are nocturnal feeders; they consume slugs, earthworms, and other arthropods.

Diplopoda

Millipedes have two body sections (a head and a truck), a pair of short antennae, a pair of mandibles, and a single pair of fused maxillae. Each segment of the body trunk typically bears two pairs of legs. Millipedes feed on decomposing plants and live in moist habitats, such as leaf litter, moss, rotting logs, soil, and under stones. Many roll up and form tight balls when disturbed.

References

Adams, R. H. 1941. Stratification, diurnal, and seasonal migration in a deciduous forest. *Ecol. Monogr.* **11**: 190–227.

Adams, R. P. 1969. Chemosystematic and numerical studies in natural populations of *Juniperus*. Ph.D. Thesis, Univ. of Texas, Austin. .

Adams, R. P. 1970. Seasonal variation of terpenoid constituents in natural population of *Juniperus pinchotii* Sirdiv. *Phytochem.* **9**: 397–402.

Adams, R. P. 1977. Chemosystematics—analysis of populational differentiation and variability of ancestral and recent populations of *Juniperus ashei*. *Ann. Mo. Bot. Gard.* **64**: 184–209.

Adams, R. P., and A. Hagerman. 1976. A comparison of the volatile oils of mature versus young leaves of *Juniperus scopulorum*: Chemosystematic significance. *Biochem. Syst. Ecol.* **4**: 75–79.

Adams, R. P., and A. Hagerman. 1977. Diurnal variation in the volatile terpenoids of *Juniperus scopulorum* (Cupressaceae). *Am. J. Bot.* **64**: 278–285.

Adlung, K. G. 1966. A critical evaluation of the European research on use of red wood ants (*Formica rufa*. group) for the protection of forests against harmful insects. *Z. Angew. Entomol.* **57**: 167–189.

Ahmad, S., and A. J. Forgash. 1975. Toxicity of carbaryl and diazinon to gypsy moth larval stages in relation to larval growth. *J. Econ. Entomol.* **68**: 803–806.

Alexander, S. A., J. M. Skelly, and R. S. Webb. 1981. Effects of *Heterobasidion annosum* on radial growth in southern pine beetle infested loblolly pine. *Phytopath.* **71**: 479–481.

Alfaro, R. I., A. J. Thomson, and G. A. Van Sickle. 1985. Quantification of Douglas-fir growth losses caused by western spruce budworm defoliation using stem analysis. *Can. J. For. Res.* **15**: 5–19.

All, J. N., and D. M. Benjamin. 1975. Influence of needle maturity on larval feeding preference and survival of *Neodiprion swainei* and *N. rugifrons* on jack pine, *Pinus banksiana*. *Ann. Entomol. Soc. Am.* **68**: 579–584.

All, J. N. and D. M. Benjamin. 1976. Potential of antifeedants to control larval feeding of selected *Neodiprion* sawflies (Hymenoptera: Diprionidae). *Can. Entomol.* **108**: 1137–1144.

Allee, W. C. 1926. Distribution of animals in a tropical rain forest with relation to environmental factors. *Ecology* **7**: 445–468.

Allen, M. D., and I. W. Selman. 1955. Egg-production in the mustard beetle *Phaedon cochleariae* (F.) in relation to diets of mineral-deficient leaves. *Bull. Entomol. Res.* **46**: 393–397.

Allen, T. C. 1947. Suppression of insect damage by means of plant hormones. *J. Econ. Entomol.* **40**: 814–817.

Amman, G. D. 1972. Mountain pine beetle brood production in relation to thickness of lodgepole pine phloem. *J. Econ. Entomol.* **65**: 138–140.

Amman, G. D. 1975. Insects affecting lodgepole pine productivity. *In* "Management of Lodgepole Pine Ecosystems: Symposium Proceedings" (D. M. Baumgartner, ed.), pp. 310–341. Washington State Univ. Coop. Ext. Serv., Pullman.

Amman, G. D., and L. A. Rasmussen. 1974. A comparison of radiographic and bark removal methods for sampling of mountain pine beetle populations. *USDA For. Serv. Res. Pap.* **INT-151**.

Anderson, N. A., M. E. Ostry, and G. W. Anderson. 1976. Hypoxylon canker of aspen associated with *Saperda inornata* galls. *USDA For. Serv. Res Note NC* **NC-214**.

Andrewartha, H. G. and L. C. Birch. 1954. "The Distribution and Abundance of Animals." Univ. of Chicago Press, Chicago, Illinois.

Angus, T. A. 1962. Microbiological control of forest insects as a new approach. *Proc. World For. Congr., 5th, 1960* pp. 1–7.

Anonymous. 1973. Insect and mite control on ornamentals. Coop. Ext. Service, *Bull. Ohio State Univ. Coop. Ext. Serv.* No. 504.

Appleby, J. E. and R. B. Malek. 1982. *Proc. Natl. Pine Wilt Workshop: Ill. Nat. Hist. Survey*.

Armstrong, J. A. 1975. Technology of aerial control: meteorological influences. *In* "Aerial Control of Forest Insects in Canada" (M. L., Prebble, ed.), pp. 56–58. Information Canada, Ottawa.

Arthur, A. P. 1963. Life histories and immature stages of four ichneumonid parasites of the European pine shoot moth, *Rhyacionia buoliana* (Schiff.) in Ontario. *Can. Entomol.* **95**: 1078–1091.

Ashley, M. D. and D. Stark. 1976. Photo field guide for on-the-ground evaluation of spruce budworm damage (*Choristoneura fumiferana* Clem.) on balsam fir (*Abies balsamea* Mill.). *Maine Agric. Exp. Stn. Misc. Publ.*

Askew, R. R. 1971. "Parasitic Insects." Heinemann Educational Books, Ltd., London.

Auchmoody, L. R. 1974. Nutrient composition of blades, petioles, and whole leaves from fertilized and unfertilized yellow poplar. *USDA For. Serv. Res. Note* **NE-198**.

Auchmoody, L. R., and K. P. Hammack, 1975. Foliar nutrient variation in four species of upland oaks. *USDA For Serv. Res. Pap.* **NE-331**.

Aylor, D. E. 1976. Estimating peak concentration of pheromones in the forest. *In* "Perspectives in Forest Entomology" (J. F. Anderson and H. K. Kaya, eds.), pp. 177–188. Academic Press, New York.

Baker, W. L. 1949. Studies on the transmission of a virus causing phloem necrosis of American elm with notes on the biology of its insect vector. *J. Econ. Entomol.* **42**: 729–732.

Baker, W. L. 1972. Eastern Forest Insects. *USDA For. Serv. Misc. Publ.* No. 1175.

Bakke, A. and T. Kramme. 1978. Kairomone response by the predators *Thanasimus formicarius* and *T. rufipes* to the synthetic pheromone of *Ips typographus*. *Norw. J. Entomol.* **25**: 41–43.

Bakke, A. and L. Strand. 1981. Pheromones and traps as part of an integrated control of the spruce bark beetle. Some results from a control program in Norway in 1979 and 1980. *Rapp. Nord. Inst. Skogforsk.* **5**: 1–39.

Balch, R. E. 1946. The spruce budworm and forest management in the Maritime Provinces. *Can. Dep. Agric. Div. Entomol. Processed Publ.* No. 60; See also *For. Chron.* **22** (1951).

Balch, R. E. 1952. Studies of the balsam woolly aphid, *Adelges piceae* (Ratz.), and its effects on balsam fir, *Abies balsamea* (L.). Mill. *Can. Dep. Agric. Sci. Serv. Div. For. Biol. Dom. Entomol. Lab. Publ.* No. 867.

Balch, R. E., and F. T. Bird. 1944. A disease of the European spruce sawfly, *Gilpinia hercyniae* (Hartig), and its place in natural control. *Sci. Agric. (Ottawa)* **25**: 65–80.

Balch, R. E., and J. S. Prebble. 1940. The bronze birch borer and its relation to the dying of birch in New Brunswick forests. *For. Chron.* **16**: 179–201.

Balch, R. E., and G. R. Underwood. 1950. The life-history of *Pineus pinifoliae* (Fitch) (Homoptera : Phylloxeridae) and its effect on white pine. *Can. Entomol.* **82**: 117–123.

Baltensweiler, W. 1964. *Zeiraphera griseana* Hubner (Lepidoptera : Tortricidae) in the European Alps. A contribution to the problem of cycles. *Can. Entomol.* **96**: 792–800.

Barbosa, P. 1978. Host plant exploitation by the gypsy moth, *Lymantria dispar*. *Entomol. Exp. Appl.* **24**: 28–37.

Barbosa, P., and W. Baltensweiler. 1987. Phenotypic plasticity and herbivore outbreaks. *In* "Insect Outbreaks" (P. Barbosa and J. C. Schultz, eds.), pp. 469–503. Academic Press, New York.

Barbosa, P., and J. L. Capinera. 1978. Population quality, dispersal and numerical change in the gypsy moth, *Lymantria dispar* (L). *Oecologia* **36**: 203–209.

Barbosa, P., and E. A. Frongillo, Jr. 1977. Influence of light intensity and temperature on the locomotory and flight activity of *Brachymeria intermedia* (Hym. : Chalcidae), a pupal parasitoid of the gypsy moth. *Entomophaga* **22**: 405–411.

Barbosa, P., and J. Greenblatt. 1979a. Suitability, digestibility and assimilation of various host plants of the gypsy moth, *Lymantria dispar* (L.). *Oecologia* **43**: 111–119.

Barbosa, P., and J. Greenblatt. 1979b. Effects of leaf age and position on larval preferences of the fall webworm, *Hyphantria cunea* (Lepidoptera : Arctiidae). *Can. Entomol.* **111**: 381–383.

Barbosa, P., J. Greenblatt, W. Withers, W. Cranshaw, and E. A. Harrington. 1979. Host-plant preferences and their induction in larvae of the gypsy moth, *Lymantria dispar*. *Entomol. Exp. Appl.* **26**: 180–188.

Barnes, H. F. 1935. Studies of fluctuations in insect populations. VI. Discussion of results of studies I–V. *J. Anim. Ecol.* **4**: 254–263.

Barras, S. J., and J. D. Hodges. 1969. Carbohydrates of inner bark of *Pinus taeda* as affected by *Dendroctonus frontalis* and associated microorganisms. *Can. Entomol.* **101**: 489–493.

Barras, S. J. 1970. Antagonism between *Dendroctonus frontalis* and the fungus *Ceratocystis minor*. *Ann. Entomol. Soc. Am.* **63**: 1187–1190.

Barras, S. J. 1973. Reduction of progeny and development in the southern pine beetle following removal of symbiotic fungi. *Can. Entomol.* **105**: 1295–1299.

Barry, J. W., R. K. Dumbauld, and H. E. Cramer. 1975. Application of meteorological prediction models to forest spray problems. *U.S. Army Tech. Note* **TECOMS-CO-403-000-051**.

Barter, G. W. 1965. Survival and development of the bronze poplar borer *Agrilus liragus* Barter & Brown (Coleoptera : Buprestidae). *Can. Entomol.* **97**: 1063–1068.

Barton, G. M. 1976. Foliage. Part II. Foliage chemicals, their properties and uses. *Appl. Polym. Symp.* No. 28: 465–484.

Basham, J. T., and R. M. Belyea. 1960. Death and deterioration of balsam fir weakened by spruce budworm defoliation in Ontario. Part III. The deterioration of dead trees. *For. Sci.* **6**: 78–96.

Batra, L. R. 1967. Ambrosia fungi: A taxonomic revision, and nutritional studies of some species. *Mycologia* **56**(6): 976–1017.

Batzer, H. O. 1967. Spruce budworm defoliation is reduced most by commercial clear cutting. *USDA For. Serv. Res. Note* **NC-36**.

Batzer, H. O. 1973. Net effect of spruce budworm defoliation on mortality and growth of balsam fir. *J. For.* **71**: 34–37.

Batzer, H. O. 1976. Silvicultural control techniques for the spruce budworm. *In* "Proceedings of a Symposium on the Spruce Budworm." *USDA For. Serv. Misc. Publ.* No. 1327: 110–116.

Beal, J. A. 1942. Mortality of reproduction defoliated by the red-headed pine sawfly (*Neodiprion lecontei* (Fitch)). *J. For.* **40**: 562–563.

Beall, H. W. 1934. The penetration of rainfall through hardwood and softwood forest canopy. *Ecology* **15**: 412–415.

Beckwith, R. C. 1976. Influence of host foliage on the Douglas-fir tussock moth. *Environ. Entomol.* **5**: 73–77.

Beckwith, R. C., and W. P. Kemp. 1984. Shoot growth models for Douglas-fir and grand fir. *For. Sci.* **30**(3): 743–746.

Bedard, W. D. 1965. The biology of *Tomicobia tibialis* (Hymenoptera:Pteromalidae) parasitizing *Ips confusus* (Coleoptera : Scolytidae) in California. *Contrib. Boyce Thompson Inst.* **23**: 77–81.

Bedard, W. D., P. E. Tilden, D. L. Wood, R. M. Silverstein, R. G. Brownlee, and J. O. Rodin. 1969. Western pine beetle: field response to its sex pheromone and a synergistic host terpene, myrcene. *Science* **164**: 1284–1285.

Bedard, W. D., R. M. Silverstein, and D. L. Wood. 1970. Bark beetle pheromones. *Science* 167: 1638–1639.

Bedard, W. D., D. L. Wood, P. E. Tilden, K. Q. Lindahl, R. M. Silverstein, and J. O. Rodin. 1980. Field responses of the western pine beetle and one of its predators to host- and beetle-produced compounds. *J. Chem. Ecol.* **6**: 625–641.

Behre, C. E. 1939. The opportunity for forestry practice in the control of gypsy moth on Massachusetts woodlands. *J. For.* **37**: 546–551.

Behre, C. E., and L. H. Reineke. 1943. The opportunity for silvicultural control of gypsy moth in southwestern Maine. *J. For.* **41**: 811–815.

Belanger, R. P., G. E. Hatchell, and G. E. Moore. 1977. Soil and stand characteristics related to southern pine beetle infestations: A progress report for Georgia and North Carolina. *Proc. South. For. Soils Workshop., 6th South. For. Soils Counc.* pp. 99–107.

Belyea, R. M., and C. R. Sullivan. 1956. The white pine weevil: A review of current knowledge. *For. Chron.* **32**: 58–67.

Benjamin, D. M. 1955. The biology and ecology of the red-headed pine sawfly. *USDA For. Serv. Tech. Bull.* No. 118.

Benjamin, D. M., H. O. Batzer, and H. G. Ewan. 1953. The lateral-terminal elongation growth ratio of red pine as an index of Saratoga spittlebug injury. *J. For.* **51**: 822–823.

Bennett, R. B. and J. H. Borden. 1971. Flight arrestment of tethered *Dendroctonus pseudotsugae* and *Trypodendron lineatum* (Coleoptera : Scolytidae) in response to olfactory stimuli. *Ann. Entomol. Soc. Am.* **64**: 1273–1286.

Berry, F. H. 1945. Effect of site and the locust borer on plantations of black locust in the Duke forest. *J. For.* **43**: 751–754.

Berryman, A. A. 1966. Studies on the behavior and development of *Enoclerus lecontei* (Wolcott), a predator of the western pine beetle. *Can. Entomol.* **98**: 519–526.

Berryman, A. A. 1967. Preservation and augmentation of insect predators of the western pine beetle. *J. For.* **65**: 260–262.

Berryman, A. A. 1969. Responses of *Abies grandis* to attack by *Scolytus ventralis* (Coleoptera: Scolytidae). *Can. Entomol.* **101**: 1033–1041.

Berryman, A. A. 1972. Resistance of conifers to invasion by bark beetle-fungus associations. *BioScience* **22**: 598–602.

Berryman, A. A. 1973. Population dynamics of the fir engraver, *Scolytus ventralis* (Coleoptera : Scolytidae). I. Analysis of population behavior and survival from 1964–1971. *Can. Entomol.* **105**: 1465.

Berryman, A. A. 1982. Population dynamics of bark beetles. *In* "Bark Beetles in North American Conifers: A Study in Evolutionary Biology" (J. B. Mitton and K. B. Sturgeon, eds.), pp. 264–314. Univ. of Texas Press, Austin.

Berryman, A. A. 1986. "Forest Insects: Principles and Practice of Population Management." Plenum, New York.

Berryman, A. A., and M. Ashraf. 1970. Effects of *Abies grandis* resin on the attack behavior and brood survival of *Scolytus ventralis* (Coleoptera : Scolytidae). *Can. Entomol.* **102** : 1229–1236.

Bess, H. A., S. H. Spurr, and E. W. Littlefield. 1947. Forest site conditions and the gypsy moth. *Harv. For. Bull.* No. 22.

Bethlahmy, N. 1975. A Colorado episode: beetle epidemic, ghost forests, and streamflow. *Northwest Sci.* **49**: 95–105.

Betts, M. M. 1955. The food of titmice in oak woodland. *J. Anim. Ecol.* **24**: 282–323.

Bevan, D. 1974. Control of forest insects: There is a porpoise close behind us. *In* "Biology in Pest and Disease Control" (D. P. Jones and M. E. Solomon, eds.) *Symp. Br. Ecol. Soc.*, Vol. 13, pp. 302–312. Blackwell, Oxford.

Billany, D. J., J. H. Borden, and R. M. Brown. 1978. Distribution of *Gilpinia hercyniae* (Hymenoptera:Diprionidae) eggs with Sitka spruce trees. *Forestry* **51**: 67–72.

Billings, R. F., R. I. Gara, and B. F. Nrutfiord. 1976. Influence of ponderosa pine resin volatiles on the response of *Dendroctonus ponderosae* to synthetic trans-verbenol. *Environ. Entomol.* **5**: 171–179.

Birch, L. C. and D. P. Clark. 1953. Forest soil as an ecological community with special reference to the fauna. *Q. Rev. Biol.* **28**: 13–36.

Birch, M. C. 1978. Chemical communication in pine bark beetles. *Am. Sci.* **66**: 409–420.

Birch, M. C., and D. M. Light. 1977. Inhibition of the attractant pheromone response in *Ips pini* and *I. paraconfusus* (Coleoptera : Scolytidae): field evaluation of ipsenol and linalool. *J. Chem. Ecol.* **3**: 257–267.

Birch, M. C., D. M. Light, and K. Mori. 1977. Selective inhibition of response of *Ips pini* to its pheromone by the (S)-(-)-enantiomer of ispenol. *Nature (London)* **270**: 738–739.

Bird, F. T. 1953. The use of a virus disease in the biological control of the European pine sawfly, *Neodiprion sertifer* (Geoffr.). *Can. Entomol.* **85**: 437–446.

Bird, F. T. 1955. Virus diseases of sawflies. *Can. Entomol.* **87**: 124–127.

Bird, F. T. 1961. Transmission of some insect viruses with particular reference to ovarial transmission and its importance to the development of epizootic. *J. Insect Pathol.* **3**: 352–380.

Bird, F. T. 1969. Infection and mortality of spruce budworm, *Choristoneura fumiferana,* and forest tent caterpillar, *Malacosoma disstria,* caused by nuclear and cytoplasmic polyhedrosis viruses. *Can. Entomol.* **101**: 1267–1285.

Bird, F. T., and M. M. Whalen. 1953. A virus disease of the European Pine sawfly, *Neodiprion sertifer* (Geoffr.). *Can. Entomol.* **85**: 433–437.

Blackman, M. W. 1924. The effect of deficiency and excess of rainfall upon the hickory bark beetle. *J. Econ. Entomol.* **17**: 460–470.

Blais, J. R. 1952. The relationship of the spruce budworm (*Choristoneura fumiferana* (Clem.) to the flowering condition of balsam fir (*Abies balsamea* (L.) Mill.). *Can. J. Zool.* **30**: 1–29.

Blais, J. R. 1958a. Effects of defoliation by spruce budworm (*Choristoneura fumiferana* Clem.) on radial growth at breast height of balsam fir (*Abies balsamea* (L.) Mill.) and white spruce (*Picea glauca* (Moench) Voss.). *For. Chron.* **34**: 39–47.

Blais, J. R. 1958b. The vulnerability of balsam fir to spruce budworm attack in northwestern Ontario, with special reference to the physiological age of the tree. *For. Chron.* **34**: 405–422.

Blais, J. R. 1960. Spruce budworm parasite investigations in the lower St. Lawrence and Gaspe regions of Quebec. *Can. Entomol.* **92**: 384–396.

Blais, J. R. 1961. Spruce budworm outbreaks in the lower St. Lawrence and Gaspe regions. *For. Chron.* **37**: 192–202.

Blais, J. R. 1974. The policy of keeping trees alive via spring operations may hasten the recurrence of spruce budworm outbreaks. *For. Chron.* **50**: 19–21.

Blake, E. A., M. R. Wagner, and T. W. Koerber. 1986. Insects destructive to ponderosa pine cone crops in northern Arizona. *In* Proceedings—Conifer Tree Seed in the Inland Mountain West Symposium. *USDA For. Serv. Gen. Tech. Rep.* **INT-203**: 238–242.

Blake, I. H. 1926. A comparison of the animal ecology of coniferous and deciduous forests. *III. Biol. Monogr.* **10**: 5–149.

Blake, I. H. 1931. Further studies on deciduous forest animal communities. *Ecology* **12**: 508–527.

Blaustein, A. R., A. M. Kuris, and J. J. Alio. 1983. Pest and parasite species richness problems. *Am. Nat.* **122**: 352–370.

Bletchly, J. D., and R. H. Farmer. 1959. Some investigations into the susceptibility of Corsican and Scots pines and of European oak to attack by the common furniture beetle, *Anobium punctatum* DeGeer (Col. : Anobiidae). *J. Inst. Wood. Sci.* **3**: 2–20.

Bletchly, J. D., and J. M. Taylor. 1964. Investigations on the susceptibility of home-grown Sitka spruce (*Picea sitchensis*) to attack by the common furniture beetle (*Anobium punctatum* DeGeer). *J. Inst. Wood. Sci.* **12**: 29–43.

Bodenheimer, F. S. 1928 Welche Faktoren regulieren die Individuenzahl einer Insecktenart in der Natur? *Biol. Z. entrallbl.* **48**: 714–739.

Bordasch, R. P. and A. A. Berryman. 1977. Host resistance to the fir engraver beetle, *Scolytus ventralis* (Coleoptera : Scolytidae). 2. Repellency of *Abies grandis* resins and some monoterpenes. *Can. Entomol.* **109**: 95–100.

Borden, J. H. 1974a. Aggregation pheromones in the Scolytidae. *In* "Pheromones" (M. Birch, ed.), pp. 135–160. Am. Elsevier, New York.

Borden, J. H. 1974b. Pheromone mask produced by male *Trypodendron lineatum* (Coleoptera : Scolytidae). *Can. J. Zool.* **52**: 533–536.

Borden, J. H. 1982. Aggregation pheromones. *In* "Bark Beetles in North American Conifers" (J. B. Mitton and K. B. Sturgeon, eds.), pp. 74–139. Univ. of Texas Press, Austin.

Borden, J. H., L. J. Chong, and M. C. Fuchs. 1983a. Application of semiochemicals in post-logging manipulation of the mountain pine beetle, *Dendroctonus ponderosae*. *J. Econ. Entomol.* **76**: 1428–1432.

Borden, J. H., J. E. Conn, L. M. Friskie, B. E. Scott, L. J. Chong, H. D. Pierce, Jr., and A. C. Oehlschlager, 1983b. Semiochemicals for the mountain pine beetle, *Dendroctonus ponderosae*, in British Columbia: baited-tree studies. *Can. J. For. Res.* **13**: 325–333.

Borden, J. H. and C. F. Slater. 1969. Sex pheromone of *Trypodendron lineatum*: production in the female hindgut-malpighian tubule region. *Ann. Entomol. Soc. Am.* **62**: 454–455.

Bornemissza, G. F. 1957. An analysis of arthropod succession in carrion and the effect of its decomposition on the soil fauna. *Aust. J. Zool.* **5**: 1–12.

Borrer, D. J., D. M. DeLong, and C. A. Triplehorn. 1964. "An Introduction to the Study of Insects." Holt, New York.

Boyce, J. S. 1961. "Forest Pathology," 3rd Ed. McGraw-Hill, New York. N. Y.

Bradshaw, W. E. 1974. Phenology and seasonal modeling in insects. *In* "Phenology and Seasonality Modeling" (H. Lieth, ed.), 127–137. Ecol. Studies 8. Springer-Verlag, Berlin and New York.

Brand, J. M., J. W. Bracke, A. J. Barkovetz, D. L. Wool, and L. E. Browne. 1975. Production of verbenol pheromone by a bacterium isolated from bark beetles. *Nature (London)* **254**: 136–137.

Brand, J. M., J. W. Bracke, L. N. Britton, A. J. Markovetz, and S. J. Barras. 1976. Bark beetle pheromones: production of verbenone by a mycangial fungus of *Dendroctonus frontalis J. Chem. Ecol.* **2**: 195–199.

Braun, E. J. and W. A. Sinclair. 1979. Phloem necrosis of elms: symptoms and histopathological observations in tolerant hosts. *Phytopathology* **69**: 354–358.

Bray, J. R. 1964. Primary consumption in three forest canopies. *Ecology* **45**: 165–167.

Brewer, J. W. and P. R. Johnson. 1977. Biology and parasitoids of *Contarinia coloradensis* Felt., a gall midge on ponderosa pine. *Marcellia* **39**: 391–398.

Bridges, J. R. 1982. Effects of juvenile hormone on pheromone synthesis in *Dendroctonus frontalis*. *Environ. Entomol.* **11**: 417–420.

Brooks, M. A. 1963. The microorganisms of healthy insects. *In* "Insect Pathology: An Advanced Treatise" (E. A.Steinhaus, ed.), pp. 215–250. Academic Press, New York.

Brookes, M. H., R. W. Stark, and R. W. Campbell (eds.). 1978. The Douglas-fir tussock moth: a synthesis. *USDA For. Serv. Sci. Educ. Agency Tech. Bull.* No. 1585.

Brown, J. M., C. A. Budelsky, and H. C. Fritts. 1965. A study of photosynthesis and other physiological processes in some drought subjected ponderosa pine. *Bull. Ecol. Soc. Am.* **46**: 92–93.

Brown, N. R., and R. C. Clark. 1956a. Studies of predators of the balsam woolly aphid, *Adelges piceae* (Ratz.) (Homoptera : Adelgidae). I. Field identification of *Neoleucopis obscura* (Hal.), *Leucopina americana* (Mall.), and *Cremifania nigrocellulata* Cz. (Diptera : Chamaemyiidae). *Can. Entomol.* **88**: 272–279.

Brown, N. R., and R. C. Clark. 1956b. Studies of the predators of the balsam woolly aphid, *Adelges piceae* (Ratz.) (Homoptera : Adelgidae). II. An annotated list of the predators associated with balsam woolly aphid in eastern Canada. *Can. Entomol.* **88**: 678–683.

Brown, N. R., and R. C. Clark. 1957. Studies of predators of the balsam woolly aphid, *Adelges piceae* (Ratz.) (Homoptera : Adelgidae). IV. *Neoleucopis obscura* (Hal.) (Diptera : Chamaemyiidae), an introduced predator in eastern Canada. *Can. Entomol.* **99**: 533–546.

Browne, L. E. 1978. A trapping system for the western pine beetle using attractive pheromones. *J. Chem. Ecol.* **4**: 261–275.

Brunner, J. F. and E. C. Burts. 1981. Potential of tree washes as a management tactic against the pear psylla. *J. Econ. Entomol.* **74**: 71–74.

Bruns, H. 1960. The economic importance of birds in forests. *Bird Study* **7**: 193–208.

Buckner, C. H. 1966. The role of vertebrate predators in the biological control of forest insects. *Annu. Rev. Entomol.* **11**: 449–470.

Buckner, C. H. 1967. Avian and mammalian predators of forest insects. *Entomophaga* **12**: 491–501.

Buckner, C. H., and W. J. Turnock. 1965. Avian predation on the larch sawfly, *Pristiphora erichsonii* (Htg.) (Hymenoptera : Tenthredinidae). *Ecology* **46**: 223–236.

Burges, H. D. 1973. Enzootic diseases of insects. *In* "Regulations of Insect Populations by Microorganisms" (L. A. Bulla, Jr., ed.), *Ann. N. Y. Acad. Sci.* **217**: 31–49.

Bush, G. L. 1975. Modes of animal speciation. *Annu. Rev. Ecol. Syst.* **6**: 339–364.

Bushing, R. W., and D. L. Wood. 1964. Rapid measurement of oleoresin exudation pressure in *Pinus ponderosa* Doug. ex Laws. *Can. Entomol.* **96**: 510–513.

Butcher, J. W., and A. C. Hodson. 1949. Biological and ecological studies on some lepidopterous bud and shoot insects of jack pine (Lepidoptera : Olethreutidae). *Can. Entomol.* **81**: 161–173.

Butterick, P. L. 1913. Insect destruction of fire-killed timber in the Black Hills of South Dakota. *J. For.* **11**: 363–371.

Byers, J. A. 1981. Pheromone biosynthesis in the bark beetle *Ips paraconfusus* during feeding or exposure to vapours of host plant precursors. *Insect Biochem.* **11**: 563–569.

Byers, J. A., J. W. Brewer, and D. W. Denna. 1976. Plant growth hormones on pinyon insect galls. *Marcellia* **39**: 125–134.

Byers, J. A., D. L. Wood, L. E. Browne, R. H. Fish, B. Piatek, and L. B. Hendry. 1979. Relationship between a host plant compound, myrcene, and pheromone production in the bark beetle *Ips paraconfusus*. *J. Insect Physiol.* **25**: 477–482.

Cade, S. C., B. F. Hrutiford, and R. I. Gara. 1970. Identification of a primary attraction for *Gnathotrichus sulcatus* isolated from western hemlock logs. *J. Econ. Entomol.* **63**: 1014–1015.

Caird, R. W. 1935. Physiology of pines infected with bark beetles. *Bot. Gaz.* **96**: 709–733.

Callahan, R. Z. 1966. Nature of resistance of pines to bark beetles in breeding pest-resistant trees. *In* "Breeding Pest-Resistant Trees" (H. D. Gerhold, R. McDermott, E. Schreiner, and J. Winiesk, eds.), pp. 197–201. Pergamon, Oxford.

Cameron, D. G., G. A. McDougall, and C. W. Bennett. 1968. Relation of spruce budworm development and balsam fir shoot growth to heat units. *J. Econ. Entomol.* **61**: 857–858.

Campbell, R. W. 1961. Population dynamics of the gypsy moth. Ph.D. Thesis, Univ. of Michigan, Ann Arbor.

Campbell, R. W., and R. J. Sloan. 1976. Influence of behavioral evolution on gypsy moth pupal survival in sparse populations. *Environ. Entomol.* **5**: 1211–1217.

Campbell, R. W., and R. J. Sloan. 1977a. Natural regulation of innocuous gypsy moth populations. *Environ. Entomol.* **6**: 315–322.

Campbell, R. W., and R. J. Sloan. 1977b. Forest stand responses to defoliation by the gypsy moth. *For. Sci. Monogr.* No. 19.

Campbell, R. W., and R. J. Sloan. 1978. Numerical bimodality among north American gypsy moth populations. *Environ. Entomol.* **7**: 641–646.

Campbell, R. W., D. L. Hubbard, and R. J. Sloan. 1975a. Patterns of gypsy moth occurrence within a sparse and numerically stable position. *Environ. Entomol.* **4**: 535–542.

Campbell, R. W., D. L. Hubbard, and R. J. Sloan. 1975b. Location of gypsy moth pupae and subsequent pupal survival in sparse, stable populations. *Environ. Entomol.* **4**: 597–600.

Cannon, W. N., Jr., and L. C. Terriere. 1966. Egg production of the two-spotted spider mite on bean plants supplied nutrient solutions containing various concentrations of iron, manganese, zinc, and cobalt. *J. Econ. Entomol.* **59**: 89–93.

Capinera, J. L., and P. Barbosa. 1975. Transmission of nuclear-polyhedrosis virus to gypsy moth larvae by *Calosoma sycophanta*. *Ann. Entomol. Soc. Am.* **68**: 593–594.

Capinera, J. L., and P. Barbosa. 1976. Dispersal of first instar gypsy moth larvae in relation to population quality. *Oecologia* **26**: 53–64.

Capinera, J. L., and P. Barbosa. 1977. Influence of natural diets and larval density on gypsy moth, *Lymantria dispar* (Lepidoptera : Orgyiidae), egg mass characteristics. *Can. Entomol.* **109**: 1313–1318.

Carolin, V. M., and W. K. Coulter. 1971. Trends of western spruce budworm and associated insects in Pacific Northwest forests sprayed with DDT. *J. Econ. Entomol.* **64**: 291–297.

Carolin, V. M., Jr., and G. E. Daterman. 1974. Hazard of European pine forests. *J. For.* **72**: 136–140.

Carrow, J. R. 1974. Detection of balsam woolly aphid. *Environ. Conserv. For. Serv. Fact Sheet.*

Carter, E. E. 1916. *Hylobius pales* as a factor in the reproduction of conifers in New England. *Proc. Soc. Am. For.* **11**: 297–307.

Carter, W. 1962. "Insects in Relation to Plant Disease," 2nd Ed. John Wiley, New York.

Cerezke, H. F. 1972. Observations on the distribution of the spruce bud midge (*Rhabdophaga swanei* Felt) in black and white spruce crowns and its effect on height growth. *Can. J. For. Res.* **2**: 69–72.

Cerezke, H. F. 1973. Survival of the weevil *Hylobius warreni* Wood in lodgepole pine stumps. *Can. J. For. Res.* **3**: 367–372.

Cerezke, H. F. 1974. Effects of partial girdling on growth in lodgepole pine with application to damage by the weevil *Hylobius warreni* Wood. *Can. J. For. Res.* **4**: 312–320.

Champion, F. J. 1975. Products of American forests. *USDA For. Serv. Misc. Publ.* No. 861.

Chapman, J. A. 1963. Field selection of different log odors by scolytid beetles. *Can. Entomol.* **95**: 673–676.

Chapman, J. A. 1974. Ambrosia beetle. *Can. For. Serv. Pac. For. Res. Cent. [Rep.] BC-X* **BC-X-103**.

Chapman, R. F. 1969. "The Insects. Structure and Function," 1st Ed. Am. Elsevier, New York.

Chater, C. S., and F. W. Holmes. 1973. Insect and Disease Control Guide for Trees and Shrubs. *Univ. Mass. USDA Coop. Ext. Serv.*

Chew, R. M. 1974. Consumers as regulators of ecosystems; an alternative to energetics. *Ohio J. Sci.* **74**(6): 359–370.

Chitty, D. 1957. Self-regulation of numbers through changes in viability. *Cold Spring Harbor Symp. Quant. Biol.* **22**: 277–280.

Chitty, D. 1960. Population processes in the vole and their relevance to general theory. *Can. J. Zool.* **38**: 99–113.

Chitty, D. 1971. The natural selection of self-regulating behavior in animal populations. *In* "Natural Regulations of Animal Populations" (I. A. McLaren, ed.) Atherton Press, New York.

Christian, J. J. 1950. The adreno-pituitary system and population cycles in mammals. *J. Mammal.* **31**: 247–259.

Christian, J. J. 1959. The role of endocrine and behavioral factors in the growth of mammalian populations. *In* "Symposium on Comparative Endocrinology," (A. B. Grobman, ed.), pp. 7–97. Wiley, New York.

Christy, H. R. 1952. Vertical temperature gradients on a beech forest in central Ohio. *Ohio J. Sci.* **52**: 199–209.

Chu, H., D. M. Norris, and S. D. Carlson. 1975. Ultrastructure of the compound eye of the diploid female beetle, *Xyleborus ferrugineus*. *Cell Tissue Res.* **165**: 23–36.

Churchill, G. B., H. H. John, D. P. Duncan, and A. C. Hodson. 1964. Long-term effects of defoliation of aspen by the forest tent caterpillar. *Ecology* **45**: 630–633.

Clark, E. W., J. D. White, and E. L. Bradley. 1975. Seasonal changes in foliar fatty acids and sterols in *Carya glabra* (Mill.) Sweet and *Quercus falcata* Michx, two hosts of the elm spanworm, *Ennomos subsignarius* (Hubner). *Environ. Entomol.* **4**: 935–943.

Clark, R. C., and N. R. Brown. 1958. Studies of predators of the balsam woolly aphid, *Adelges piceae* (Ratz.) (Homoptera: Adelgidae). *V. Laricobius erichsonii* Rosen. (Coleoptera: Aerondontidae), an introduced predator in eastern Canada. *Can. Entomol.* **90**: 657–672.

Clarke, G. L. 1954. "Elements of Ecology." Wiley, New York.

Clausen, C. P. 1956. Biological control of insect pests in the continental United States. *USDA Tech. Bull.* No. 1139.

Cleary, B. D. 1970. The effect of plant moisture stress on physiology and establishment of planted Douglas-fir and ponderosa pine seedlings. Ph.D. Thesis, Oregon State Univ., Corvallis.

Clement, R. C., and I. C. T. Nisbet. 1977. The suburban woodland. Trees and insects in the human environment. *Audubon Conserv. Rep.* No. 2.

Cobb, F. W., Jr., E. Zavarin, and J. Bergot. 1972. Effect of air pollution on the volatile oil from leaves of *Pinus ponderosa*. *Phytochemistry* **11**: 1815–1818.

Cole, W. E. 1960. Sequential sampling in spruce budworm control projects. *For. Sci.* **6**: 51–59.

Cole, W. E., G. D. Amman, and C. E. Jensen. 1976. Mathematical model for the mountain pine beetle-lodgepole pine interaction. *Environ. Entomol.* **5**: 11–19.

Cole, W. E., and D. B. Cahill. 1976. Cutting strategies can reduce probabilities of mountain pine beetle epidemics in lodgepole pine. *J. For.* **74**: 294–297.

Collins, S. 1961. Benefits to understory from canopy defoliation by gypsy moth larvae. *Ecology* **42**: 836–838.

Collis, D. G., and G. A. Van Sickle. 1978. Damage appraisal cruises in spruce budworm defoliated stands of Douglas-fir in 1977. *Can. For. Serv.* **BC-P-19-1978**.

Condrashoff, S. F. 1968. Biology of *Steremnius carinatus* (Coleoptera: Curculionidae), a reforestation pest in coastal British Columbia. *Can. Entomol.* **100**: 386–394.

Conn, J. E., J. H. Borden, D. W. A. Hunt, J. Holman, H. S. Whitney, O. J. Spanier, H. D. Pierce, Jr., and A. C. Oehlschlager. 1984. Pheromone production by axenically reared *Dendroctonus ponderosae* and *Ips paraconfusus* (Coleoptera : Scolytidae). *J. Chem. Ecol.* **10**: 281–290.

Connola, D. P., W. E. Waters, and E. R. Nason. 1959. A sequential sampling plan for red-pine sawfly, *Neodiprion nanulus* Schedl. *J. Econ. Entomol.* **52**: 600–602.

Copony, J. A., and C. L. Morris. 1972. Southern pine beetle suppression with frontalin and cacodylic acid treatments. *J. Econ. Entomol.* **65**: 754–757.

Coppel, H. C. and N. F. Sloan. 1971. Avian predation, an important adjunct in the suppression of larch casebearer and introduced pine sawfly populations in Wisconsin forests. *Tall Timbers Conf., Ecol. Anim. Control Habitat Manage.* **2**: 259–272.

Cornell University. 1977. "Pesticide Applicator Training Manual. Category 2. Forest Pest Control Cooperative Extension Chemicals—Pesticides Program." Ithaca, New York.

Coster, J. E. 1980. Developing integrated management strategies. *In* "The Southern Pine Beetle" (R. C. Thatcher, J. L. Searcy, J. E. Coster, and G. D. Hertel, eds.) *USDA For. Serv. Tech. Bull.* No. 1631: 195–203.

Coster, J. E., and J. P. Vite. 1972. Effect of feeding and mating on pheromone release in the southern pine beetle. *Ann. Entomol. Soc. Am.* **65**: 263–266.

Couch, J. N. 1938. "The Genus Septobasidium." Univ. of North Carolina Press, Chapel Hill.

Coulson, R. N. 1981. Evolution of concepts of integrated pest management in forest. *J. Ga. Entomol. Soc.* **16**: 301–316.

Coulson, R. N., J. L. Foltz, A. M. Mayyasi, and F. P. Hain. 1975. Quantitative evaluation of frontalure and cacodylic acid treatment effects on within-tree populations of the southern pine beetle. *J. Econ. Entomol.* **68**: 671–678.

Coutts, M. P. and J. E. Dolezal. 1969. Emplacement of fungal spores by the woodwasp *Sirex noctilis* during oviposition. *For. Sci.* **15**: 412–416.

Craig, T. P., P. W. Price, and J. K. Itami. 1986. Resource regulation by a stem-galling sawfly on the arroyo willow. *Ecology* **67**: 419–425.

Craighead, F. C. 1925. Relation between mortality of trees attacked by the spruce budworm (*Cacoecia fumiferana* Clem.) and previous growth. *J. Agric. Res.* **30**: 541–555.

Craighead, F. C. 1940. Some effects of artificial defoliation on pine and larch. *J. For.* **38**: 885–888.

Crist, C. R. and D. F. Schoeneweiss. 1975. The influence of controlled stresses on susceptibility of European white birch stems to attack by *Bostryosphaeria dothidea*. *Phytopathology* **65**: 369–373.

Crookston, N. L., and R. W. Stark. 1985. Forest-bark beetle interactions: stand dynamics and prognosis. *In* "Integrated Pest Management in Pine-Bark Beetle Ecosystems" (W. E. Waters, R. W. Stark, and D. L. Wood, eds.), pp. 81–103. Wiley, New York.

Crossley, D. A., Jr., R. N. Coulson, and C. S. Gist. 1973. Trophic level effects on species diversity in arthropod communities of forest canopies. *Environ. Entomol.* **2**: 1097–1100.

Cumming, M. E. P. 1962. The biology of *Pineus similis* (Gill) (Homoptera : Phylloxeridae) on spruce. *Can. Entomol.* **94**: 395–408.

Cuthbert, R. A., and J. W. Peacock. 1975. Attraction of *Scolytus multistriatus* to pheromone-baited traps at different heights. *Environ. Entomol.* **4**: 889–890.

Cuthbert, R. A. and J. W. Peacock. 1978. Response of the elm bark beetle *Scolytus multistriatus* (Coleoptera : Scolytidae) to component mixtures and doses of the pheromone multilure. *J. Chem. Ecol.* **4**: 363–373.

Cuthbert, R. A., J. H. Barger, A. C. Lincoln, and P. A. Reed. 1973. Formulation and application of methoxychlor for elm bark beetle control. *USDA For. Serv. Res. Pap.* NE-283.

Cuthbert, R. A., J. W. Peacock, and W. N. Cannon, Jr. 1977. An estimate of the effectiveness of pheromone-baited traps for the suppression of *Scolytus multistriatus* (Coleoptera : Scolytidae). *J. Chem. Ecol.* **3**: 527–537.

Daterman, G. E. 1974. Synthetic sex pheromone for detection survey of European pine shoot moth. *USDA For. Serv. Res. Pap.* **PNW-180**.

Daterman, G. E., and V. M. Carolin, Jr. 1973. Survival of European pine shoot moth, *Rhyacionia buoliana* (Lepidoptera : Olethreutidae), under caged conditions in a ponderosa pine forest. *Can. Entomol.* **105**: 929–940.

De Barr, G.L. 1969. The damage potential of flower thrips in slash pine seed orchards. *J. For.* **67**: 326–327.

De Barr, G. L. and P. P. Kormanik. 1975. Anatomical basis for conelet abortion on *Pinus echinata* following feeding by *Leptoglossus corculus* (Hemiptera : Coreidae). *Can. Entomol.* **107**: 81–86.

deBary, A. 1879. "Die Erscheinung der Symbiose." Trubner, Strassburg.

De Groat, R. C. 1967. Twig and branch mortality of American beech infested with oystershell scale. *For. Sci.* **13**: 448–455.

De Gryse, J. J. 1934. Quantitative methods in the study of forest insects. *Sci. Agric.* (Ottawa) **14**: 477–495.

Dewey, J. E., W. M. Ciesla, and H. E. Meyer. 1974. Insect defoliation as a predisposing agent to a bark beetle outbreak in eastern Montana. *Environ. Entomol.* **3**: 722.

Dickens, J. C., and T. L. Payne. 1977. Bark beetle olfaction: pheromone receptor system in *Dendroctonus frontalis*. *J. Insect Physiol.* **23**: 481–489.

Dickens, J. C., and T. L. Payne. 1978. Structure and function of the sensilla on the antennal club of the southern pine beetle, *Dendroctonus frontalis* (Zimmerman) (Coleoptera : Scolytidae). *Int. J. Insect Morphol. Embryol.* **7**: 251–265.

Dickson, R. E., and P. R. Larson. 1977. Muka from *Populus* leaves: a high energy feed supplement for livestock. *Tappi* **60**: 95–99.

Dimond, J. B. 1974. Sequential surveys for the pine leaf chermid, *Pineus pinifoliae*. *Univ. Maine Tech. Bull.* No. 68.

Dixon, A. F. G. 1963. Reproductive activity of the sycamore aphid, *Drepanosiphum plalanoides* (Schr.) (Hemiptera : Aphididae). *J. Anim. Ecol.* **42**: 33–48.

Dixon, A. F. G. 1970. Quality and availability of food for a sycamore aphid population. *In*

"Animal Populations in Relation to Their Food Resources." (A. Watson, ed.) Symp. Br. Ecol. Soc. Vol. 10, pp. 271–287. Blackwell, Oxford.

Dixon, A. F. G. 1971a. The role of aphids in wood formation. I. The effect of the sycamore aphid, *Drepanosiphum platanoides* (Schr.) (Aphididae), on the growth of sycamore, *Acer pseudoplatanus* (L.). *J. Appl. Ecol.* **8**: 165–179.

Dixon, A. F. G. 1971b. The role of aphids in wood formation. II. The effect of the lime aphid, *Eucallipterus tiliae* L. (Aphididae), on the growth of lime, *Tilia x vulgaris* Hayne. *J. Appl. Ecol.* **8**: 165–179.

Dixon, A. F. G. 1973. "Biology of Aphids. Studies in Biology." Arnold, London.

Dixon, A. F. G. 1985. "Aphid Ecology." Blackie, Glasgow.

Doane, C. C. 1975. Infectious sources of nuclear polyhedrosis virus persisting in natural habitats of the gypsy moth. *Environ. Entomol.* **4**: 392–394.

Doane, C. C., and D. E. Leonard. 1975. Orientation and dispersal of late-stage larvae of *Porthetria dispar* (Lepidoptera: Lymantriidae). *Can. Entomol.* **107**: 1333–1338.

Doane, C. C., and P. W. Schaefer. 1971. Aerial application of insecticides for control of the gypsy moth, with studies of effects on non-target insects and birds. *Bull. Conn. Agric. Exp. Stn.* No. 724: 23.

Doane, R. W., E. C. Van Dyke, W. J. Chamberlin, and H. E. Burke. 1936. "Forest Insects." McGraw-Hill, New York.

Dochinger, L. S. 1963. Artificial defoliation of eastern white pine duplicates some effects of chlorotic dwarf disease. *USDA For. Serv. Res. Note* **CS-16**.

Doerksen, A. H., and R. G. Mitchell. 1965. Effects of the balsam woolly aphid upon wood anatomy of some western true firs. *For. Sci.* **11**: 181–188.

Donnelly, J. R. 1976. Carbohydrate levels in current-year shoots of sugar maple. *USDA For. Serv. Res. Pap.* **NE-347**.

Doskotch, R. W., S. K. Chatterji, and J. W. Peacock. 1970. Elm bark derived feeding stimulants for the smaller European elm bark beetle. *Science* **167**: 380–382.

Dowden, P. B., H. A. Jaynes, and V. M. Carolin, 1953. The role of birds in a spruce budworm outbreak in Maine. *J. Econ. Entomol.* **46**: 307–312.

Dowdy, W. W. 1947. An ecological study of the arthropoda of an oak-hickory forest, with reference to stratification. *Ecology* **28**: 418–439.

Dowdy, W. W. 1951. Further ecological studies on stratification of the arthropoda. *Ecology* **32**: 37–52.

Drooz, A. T. 1965. Some relationships between host, egg potential, and pupal weight of the elm span worm, *Ennomos subsignarius* (Lepidoptera: Geometridae). *Ann. Entomol. Soc. Am.* **58**: 243–245.

Drooz, A. T. 1966. Intrinsic and extrinsic factors that cause variability in wing length in the elm spanworm, *Ennomos subsignarius*. *Ann. Entomol. Soc. Am.* **59**: 1021–1022.

Drooz, A. T. 1970a. The elm spanworm (Lepidoptera: Geometridae): How several natural diets affect its biology. *Ann. Entomol. Soc. Am.* **63**: 391–397.

Drooz, A. T. 1970b. Rearing the elm spanworm on oak or hickory. *J. Econ. Entomol.* **63**: 1581–1585.

Drooz, A. T., A. E. Bustillo, G. F. Fedde, and V. H. Fedde. 1977. North American egg parasite successfully controls a difficult host genus in South America. *Science* **197**: 390–391.

Duff, C. H., and N. J. Nolan. 1953. Growth and morphogenesis in the Canadian forest species. I. The controls of cambial and apical activity in *Pinus resinosa* Ait. *Can. J. Bot.* **31**: 471–513.

Dunbar, D. M. and G. R. Stephens. 1975. Association of two-lined chestnut borer and shoestring fungus with mortality of defoliated oak in Connecticut. *For. Sci.* **21**: 169–174.

Duncan, D. P., and A. C. Hodson. 1958. Influence of the forest tent caterpillar upon the aspen forests of Minnesota. *For. Sci.* **4**: 71–93.

Durzan, D. J. 1968. Nitrogen metabolism of *Picea glauca*. I. Seasonal changes of free amino acids in buds, shoot apices, and leaves and the metabolism of uniformity labeled HC-L-arginine by buds during the onset of dormancy. *Can. J. Bot.* **46**: 909–919.

Dyer, E. D. A. 1963. Attack and brood production of ambrosia beetles in logging debris. *Can. Entomol.* **95**: 624–631.

Dyer, E. D. A. 1975. Frontalin attractant in stands infested by the spruce beetle, *Dendroctonus rufipennis* (Coleoptera : Scolytidae). *Can. Entomol.* **107**: 979–988.

Dyer, E. D. A., and L. Safranyik. 1977. Assessment of the impact of pheromone baited trees on a spruce beetle population (Coleoptera : Scolytidae). *Can. Entomol.* **109**: 77–80.

Dyer, E. D. A., P. M. Hall, and L. Safranyik. 1975. Numbers of *Dendroctonus rufipennis* (Kirby) and *Thanasimus undatulus* Say at pheromone baited, poisoned and unpoisoned trees. *J. Entomol. Soc. B. C.* **72**: 20–22.

Ebel, B. H. 1961. Thrips injure slash pine female flowers. *J. For.* **59**: 374–375.

Ebell, L. F. 1971. Girdling: its effects on carbohydrate status and on reproductive bud and cone development of Douglas-fir. *Can. J. Bot.* **49**: 453–466.

Edelman, N. M. 1963. Age changes in the physiological condition of certain arbivorous larvae in relation to feeding conditions. *Entomol. Rev.* **42**: 4–9.

Edmonds, R. L. 1976. Atmospheric transport of insects and disease. *In Proc. Natl. Conf. Fire For. Meteorol., 4th,* (D. H. Baker and M. A. Fosberg, tech. coords.), *USDA For. Serv. Gen. Tech. Rep.* **RM-32**: pp. 83–93.

Edmunds, G. F. and R. K. Allen. 1958. Comparison of black pine leaf scale population density on normal ponderosa pine and those weakened by other agents. *Int. Congr. Entomol. Proc.,* 10th **4**: 391–392.

Edwards, C. A. and G. W. Heath. 1963. The role of soil animals in breakdown of leaf material. *In* "Soil organisms" *Proc. Colloq. Soil Fauna, Soil Microflora Their Relat. 1962* (J. Doeksen, and J. Van Der Drift, eds.), pp. 76–84.

Edwards, C. A., D. E. Reichle, and D. A. Crossley. 1970. The role of soil invertebrates in turnover of organic matter and nutrients. *In* "Analysis of Temperate Forest Ecosystems" (D. E. Reichle, ed.), pp. 147–172. Springer-Verlag, Berlin.

Edwards, P. J., S. D. Wratten, and S. Greenwood. 1986. Palatability of British trees to insects: constitutive and induced defences. *Oecologia* **69**: 316–319.

Eichhorn, O. 1968. Problems of the population dynamics of silver fir woolly aphids, genus Adelges (=Dreyfusia), Adelgidae. *Z. Angew. Entomol.* **61**: 157–214.

Eidt, D. C. and C. H. A. Little. 1966. Insect control through induced host-insect asynchrony: a progress report. *J. Econ. Entomol.* **63**: 1966–1968.

Eidt, D. C. and C. H. A. Little. 1968. Insect control by artificially prolonging plant dormancy — a new approach. *Can. Entomol.* **100**: 1278–1279.

Elkinton, J. S., and D. L. Wood. 1980. Feeding and boring behavior of the bark beetle *Ips paraconfusus* (Coleoptera : Scolytidae) on the bark of host and non-host tree species. *Can. Entomol.* **112**: 589–601.

Elliot, K. R. 1960. A history of recent infestations of the spruce budworm in northwestern Ontario, and an estimate of resultant timber losses. *For Chron.* **36**: 61–82.

Elton, C. 1973. The structure of invertebrate populations inside neotropical rain forest. *J. Anim. Ecol.* **42**: 55–104.

Embree, D. G. 1965. The population dynamics of the winter moth in Nova Scotia, 1954–1962. *Mem. Entomol. Soc. Can.* **46**: 1–57.

Embree, D. G. 1967. Effects of the winter moth on growth and mortality of red oak in Nova Scotia. *For. Sci.* **13**: 295–299.

Esterbauer, H., D. Grill, and G. Beck. 1975. Unter-suchungen uber phenole in Nadeln von *Picea abies. Phyton (Horn, Austria)* **17**: 87–99.

Evenden, J. C. 1940. Effects of defoliation by the pine butterfly upon ponderosa pine. *J. For.* **38**: 949–955.

Ewan, H. G. 1961. The Saratoga spittlebug. A destructive pest in red pine plantations. *USDA For. Serv. Tech. Bull.* No. 1250.

Falcon, L. A. 1971. Use of bacteria for microbial control. *In* "Microbial Control of Insects and Mites" (H. D. Burges, and N. W. Hussey, eds.), pp. 67–90. Academic Press, London.

Farris, S. H., and A. Funk. 1965. Repositories of symbiotic fungus in the ambrosia beetle *Platypus wilsoni* Swaine (Coleoptera : Platypodidae). *Can. Entomol.* **97**: 527–532.

Fauss, D. L., and W. R. Pierce. 1969. Stand conditions and spruce budworm damage in a western Montana forest. *J. For.* **67**: 322–329.

Fedde, G. F. 1964. Elm spanworm, a pest of hardwood forests in the southern Appalachians. *J. For.* **62**: 102.

Feduccia, D. P., and W. F. Mann, Jr. 1975. Black turpentine beetle infestations after thinning in a loblolly pine plantation. *USDA For. Serv. Res. Note* **SO-206**.

Feeny, P. 1968. Effect of oak leaf tannins on larval growth of the winter moth, *Operophtera brumata* L. *J. Insect Physiol.* **14**: 805–817.

Feeny, P. 1970. Seasonal changes in oak leaf tannins and nutrients as a cause of spring feeding by winter moth caterpillars. *Ecology* **51**: 565–582.

Feeny, P. P., and H. Bostock. 1968. Seasonal variation in the tannin content of oak leaves. *Phytochemistry* **7**: 871–880.

Feldman, R. M., G. L. Curry, and R. N. Coulson. 1981. A mathematical model of field population dynamics of the southern pine beetle. *Ecol. Modell.* **13**: 261–281.

Fellin, D. G. 1976. Forest practices, silvicultural prescriptions and the western spruce budworm. Proc. Symp. on the Spruce Budworm, 1974. *USDA Misc. Publ.* No. 1327: 117–121.

Fellin, D. G. 1983. Chemical insecticides vs. the western spruce budworm: After three decades, what's the score. *West. Wildlands* **9**: 8–12.

Felt, E. P. 1914. Notes on forest insects. *J. Econ. Entomol.* **7**: 373–375.

Felt, E. P. 1940. "Plant Galls and Gall Makers." Comstock, New York.

Fernald, H. T. 1921. "'Applied Entomology. An Introductory Text-Book of Insects in Their Relations to Man." McGraw-Hill, New York.

Ferrell, G. T. 1974. Moisture stress and fir engraver (Coleoptera : Scolytidae) attack in white fir infected by true mistletoe. *Can. Entomol.* **106**: 315–318.

Fichter, E. 1939. An ecological study of Wyoming spruce-fir arthropods with reference to stratification. *Ecol. Monogr.* **9**: 184–215.

Finnegan, R. J. 1962. The pine root-collar weevil, *Hylobius radicis* Buch., in southern Ontario. *Can. Entomol.* **94**: 11–17.

Finnegan, R. J. 1969. Assessing predation by ants on insects. *Insectes Soc.* **16**: 61–65.

Finnegan, R. J. 1971. An appraisal of indigenous ants as limiting agents of forest pests in Quebec. *Can. Entomol.* **103**: 1489–1493.

Finnegan, R. J. 1973. Diurnal foraging activity of *Formica sublucida*, *F. sanguinea subnuda*, and *F. fossaceps* (Hymenoptera : Formicidae) in Quebec. *Can. Entomol.* **105**: 441–444.

Finnegan, R. J. 1974. Ants as predators of forest pests. *Entomophaga* **7**: 53–59.

Finnegan, R. J. 1975. Introduction of a predaceous red wood ant, *Formica lugubris* (Hymenoptera : Formicidae), from Italy to eastern Canada. *Can. Entomol.* **107**: 1271–1274.

Fisher, R. C., G. H. Thompson, and W. E. Webb. 1953. Ambrosia bettles in forest and sawmill. Part I. Their biology, economic importance, and control. *For. Abstr.* **14**: 381–390.

Fitzgerald, T. D., and E. M. Gallagher. 1976. A chemical trail factor from the silk of eastern tent caterpillar, *Malacosoma americanum* (Lepidoptera : Lasiocampidae). *J. Chem. Ecol.* **2**: 187–193.

Flake, H. W., Jr., and D. T. Jennings. 1974. A cultural control method for pinyon needle scale. *USDA For. Serv. Res. Note* **RM-270**.

Flieger, B. W. 1970. Forest fire and insects: the relation of fire to insect outbreak. *Proc. Annu. Tall Timbers Fire Ecol. Conf.* **10**: 107–120.

Floden, K., and N. Fries. 1978. Studies on volatile compounds from *Pinus silvestris* and their effect on wood decomposing fungi. II. Effects of some volatile compounds on fungal growth. *Eur. J. For. Pathol.* **8**: 300–310.

Fogal, W. H. 1974. Nutritive value of pine foliage for some diprionid sawflies. *Proc. Entomol. Soc. Ont.* **105**: 101–118.

Foltz, J. L., A. M. Mayyasi, F. P. Hain, R. N. Coulson, and W. C. Martin. 1976. Egg-gallery length relationship and within-tree analyses for the southern pine beetle, *Dendroctonus frontalis* (Coleoptera : Scolytidae). *Can. Entomol.* **108**: 341–352.

Fonken, G. S., and R. A. Johnson. 1972. "Chemical Oxidations with Microorganisms." Dekker, New York.

Forbes, A. R., and D. B. Mullick. 1970. The stylets of the balsam woolly aphid, *Adelges piceae* (Homoptera : Adelgidae). *Can. Entomol.* **102**: 1074–1082.

Forbush, E. A., and C. H. Fernald. 1896. "The Gypsy Moth, *Porthetria dispar* (Lann.)." Wright & Potter, Boston, Massachusetts.

Francke, Von W., and V. Heemann. 1976. The odour-bouquet of *Blastophagus piniperda* L. (Col.: Scol). *Z. Angew Entomol.* **82**: 117–119.

Francke-Grosmann, H. 1963. Some new aspects in forest entomology. *Annu. Rev. Entomol.* **8**: 415–438.

Francke-Grosmann, H. 1967. Ectosymbiosis in wood inhabiting insects. *In* "Symbiosis. Vol. 2: Associations of Invertebrates, Birds, Ruminants and Other Biota" (S. M. Henry, ed.), pp. 141–205. Academic Press, New York.

Franklin, E. C., and E. B. Snyder. 1971. Variation and inheritance of monoterpene composition in longleaf pine. *For. Sci.* **17**: 178–179.

Franklin, R. T. 1973. Insect influence on the forest canopy. *In* "Analysis of Temperate Forest Ecosystems" (D. E. Reichle, ed.), pp. 86–99. Springer-Verlag. Berlin and New York.

Franz, J. M. 1961. Biological control of pest insects in Europe. *Annu. Rev. Entomol.* **6**: 183–200.

French, J. R. J., and R. S. Johnstone. 1968. Heat sterilization of block stacked timber in wood destroying insect control. *J. Inst. Wood Sci.* **20**: 42–46.

Friend, R. B. 1931. The European pine shoot moth in red pine plantations. *J. For.* **29**: 551–556.

Fronk, W. D. 1971. Chemical Control. *In* "Fundamentals of Applied Entomology" (R. E. Pfadt, ed.), pp. 191–217. Macmillan, New York.

Frye, R. H., J. M. Schmid, C. K. Lister, and P. E. Buffam. 1977. Post-attack injection of Silvisar 510 (cacodylic acid) in spruce beetle (Coleoptera : Scolytidae) infested trees. *Can. Entomol.* **109**: 1221–1225.

Furniss, M. M., G. E. Daterman, L. N. Kline, M. D. McGregor, G. C. Trostle, L. F. Pettinger, and J. A. Rudinsky. 1974. Effectiveness of the Douglas-fir beetle antiaggregative pheromone methylcyclohexenone at three concentrations and spacings around felled host trees. *Can. Entomol.* **106**: 381–392.

Furniss, M. M., B. H. Baker, and B. B. Hostetter. 1976. Aggregation of spruce beetles (Coleoptera) to seudenol and repression of attraction by methylcyclohexenone in Alaska. *Can. Entomol.* **108**: 1297–1302.

Furniss, M. M., J. W. Young, M. D. McGregor, R. L. Livingston, and D. R. Hamel. 1977. Effectiveness of controlled-release formulations of MCH for preventing Douglas-fir beetle (Coleoptera : Scolytidae) infestation in felled trees. *Can. Entomol.* **109**: 1063–1069.

Gara, R. I., and J. E. Coster. 1968. Studies on the attack behavior of the southern pine beetle. III. Sequence of tree infestation within stands. *Contrib. Boyce Thompson Inst.* **24**: 77–85.

Gardiner, L. M. 1975. Insect attack and value loss in wind-damaged spruce and jack pine stands in northern Ontario. *Can. J. For. Res.* **5**: 387–398.

George, J. L., and R. T. Mitchell. 1948. Calculations on the extent of spruce budworm control by insectivorous birds. *J. For.* **46**: 454–455.

Gerhold, H. D. 1966. In quest of insect-resistant forest trees. *In* "Breeding Pest-Resistant Trees" (H. D. Gerhold, E. J. Schreiner, R. E. McDermott, and J. A. Winieski, eds.), pp. 305–318. Pergamon, New York.

Gerken, B., and P. Hughes. 1976. Hormonal stimulation of sex-specific volatiles in bark beetles. *Z. Angew. Entomol.* **82**: 108–110.

Gerry, R. W., H. E. Young, W. P. Apgar, and H. C. Dickey. 1977. Muka feeding trials with Maine farm livestock. *Univ. Maine Agric. Exp. Stn. Res.* **24**: 1–6.

Giese, R. L., and D. M. Benjamin. 1959. The biology and ecology of the balsam gall midge in Wisconsin. *For. Sci.* **5**: 193–208.

Gilbert, B. L., J. E. Baker, and S. D. M. Norris. 1967. Juglone (5-hydroxy-1,4-napthoquinone) from *Carya ovata*, a deterrent to feeding by *Scolytus multistriatus*. *J. Insect Physiol.* **13**: 1453–1459.

Gilmore, A. R. 1977. Effects of soil moisture stress on monoterpenes in loblolly pine. *J. Chem. Ecol.* **3**: 667–676.

Gobeil, A. R. 1941. *Dendroctonus piceaperda*: A detrimental or beneficial insect? *J. For.* **39**: 632–640.

Godzik, S., and M. M. A. Sassen. 1978. A scanning electron microscope examination of *Aesculus hippocastanum* L. leaves from control and air-polluted areas. *Environ. Pollut.* **17**: 13–18.

Gore, W. E., G. T. Pearce, G. N. Lanier, J. B. Simeone, R. M. Silverstein, J. W. Peacock, and R. A. Cuthbert. 1977. Aggregation attractant of the European elm bark beetle, *Scolytus multistriatus*, production of individual components and related aggregation behavior. *J. Chem. Ecol.* **3**: 429–446.

Goyer, R. A. and L. H. Nachod. 1976. Loblolly pine conelet, cone and seed losses to insects and other factors in a Louisiana seed orchard. *For. Sci.* **22**: 386–391.

Grace, J. R. 1986. The influence of gypsy moth on the composition and nutrient content of litter fall in a Pennsylvania oak forest. *For. Sci.* **32**(4): 855–870.

Graham, K. 1967. Fungal-insect mutualism in trees and timber. *Annu. Rev. Entomol.* **12**: 105–126.

Graham, K. 1968. Anaerobic induction of primary chemical attractancy for ambrosia beetles. *Can. J. Zool.* **46**: 905–908.

Graham, S. A. 1918. The white pine weevil and its relation to second-growth white pine. *J. For.* **16**: 192–202.

Graham, S. A. 1922. Some entomological aspects of the slash disposal problem. *J. For.* **20**: 437–447.

Graham, S. A. 1924. Temperature as a limiting factor in the life of subcortical insects. *J. Econ. Entomol.* **17**: 377–383.

Graham, S. A. 1925. The felled tree trunk as an ecological unit. *Ecology* **6**: 397–411.

Graham, S. A. 1931. The effect of defoliation on tamarack. *J. For.* **29**: 199–206.

Graham, S. A. 1939. Forest insect populations. *Ecol. Monogr.* **9**: 301–310.

Graham, S. A. 1959. Control of insects through silvicultural practices. *J. For.* **57**: 281–283.

Graham, S. A. 1963. Making hardwood forests safe from insects. *J. For.* **61**: 356–359.

Graham, S. A., and L. G. Baumhofer. 1930. Susceptibility of young pines to tipmoth injury. *J. For.* **28**: 54–65.

Green, G. W. 1954a. Some laboratory investigations of the light reactions of the larvae of *Neodiprion americanus banksianae* Roh. and *N. lecontei* (Fitch) (Hymenoptera : Diprionidae). *Can. Entomol.* **86**: 207–222.

Green, G. W. 1954b. Humidity reactions and water balance of larvae of *Neodiprion americanus banksianae* Roh. and *N. lecontei* (Fitch) (Hymenoptera: Diprionidae). *Can. Entomol.* **86**: 261–274.

Green, G. W. 1962. Low winter temperatures and the European pine shoot moth *Rhyacionia buoliana* (Schiff.) in Ontario. *Can. Entomol.* **94**: 314–336.

Green, G. W., and A. S. DeFreitas. 1955. Frass drop studies of larvae of *Neodiprion americanus banksianae* Roh. and *N. lecontei* (Fitch) (Hymenoptera : Diprionidae). *Can. Entomol.* **87**: 427–440.

Greenbank, D. O. 1956. The role of climate and dispersal in the initiation of outbreaks of the spruce budworm in New Brunswick. *Can. J. Zool.* **34**: 453–476.

Greenbank, D. O. 1957. The role of climate and dispersal in the initiation of outbreaks of the spruce budworm in New Brunswick. II. The role of dispersal. *Can. J. Zool.* **35**: 385–403.

Grier, C. C., and R. H. Waring. 1974. Conifer foliage mass related to sapwood area. *For. Sci.* **20**: 205–206.

Grim, B. R., and J. W. Barry. 1975. A canopy penetration model for aerially disseminated insecticide spray released above coniferous forests. *U. S. Army USDA For. Serv. Final Rep. TECOM* No. 5-CO-153-USF-001.

Grimble, D. G., and J. D. Kasile. 1974. A sequential sampling plan for saddled prominent eggs. *Appl. For. Res. Int. Res. Rep.* No. 15.

Grimble, D. G., and R. G. Newell. 1972. Saddled prominent pupal sampling and related studies in New York state. *Appl. For. Res. Inst. Res. Rep.* No. 11.

Grimble, D. G., and C. E. Palm. 1976. Drop traps to estimate the relative abundance of red pine scale crawlers. *Appl. For. Res. Inst. Res. Note* No. 20.

Haack, R. A., and D. M. Benjamin. 1982. The biology and ecology of the twolined chestnut borer, *Agrilus bilineatus* (Coleoptera : Buprestidae), on oaks, *Quercus* spp., in Wisconsin. *Can. Entomol.* **114**: 385–396.

Haack, R. A., and F. Slansky, Jr. 1987. Nutritional ecology of wood-feeding Coleoptera, Lepidoptera, and Hymenoptera. *In* "Nutritional Ecology of Insects, Mites, Spiders, and Related Invertebrates" (F. Slansky, Jr. and J. G. Rodriguez, eds.). Wiley InterScience, New York.

Haliburton, W., W. W. Hopewell, and W. N. Yule. 1975. Deposit assessment, chemical insecticides. *In* "Aerial Control of Forest Insects in Canada" (M. L. Prebble, ed.), pp. 59–67. Information Canada, Ottawa.

Hall, R. C. 1958. Sanitation-salvage controls of bark beetles in southern California recreation areas. *J. For.* **56**: 9–11.

Hamilton, W. J., Jr., and D. B. Cook. 1940. Small mammals and the forest. *J. For.* **38**: 468–473.

Hanover, J. W. 1975. Comparative physiology of eastern and western white pines: oleoresin composition and viscosity. *For. Sci.* **21**: 214–221.

Happ, G. M., C. M. Happ, and S. J. Barras. 1971. Fine structure of the prothoracic mycangium, a chamber for the culture of symbiotic fungi, in the southern pine beetle, *Dendroctonus frontalis. Tissue Cell* **3**: 295–308.

Hard, J. S. 1974. Budworm in coastal Alaska. *J. For.* **72**: 26–31.

Hard, J. S., and T. R. Torgersen. 1975. Field and laboratory techniques for evaluating hemlock sawfly infestations. *USDA For. Serv. Note* PNW-252.

Hardin, J. W. 1979. Patterns of variation in foliar trichomes of eastern North American *Quercus. Am. J. Bot.* **66**: 576–585.

Harding, D. J. L., and R. A. Stuttard. 1974. Microarthropods. *In* "Biology of Plant Litter Decomposition" (C. H. Dickinson and G. J. F. Pugh, eds.) Vol. 2, pp. 489–532. Academic Press, New York.

Hare, R. C. 1966. Physiology of resistance to fungal diseases in plants. *Bot. Rev.* **32**: 95–137.

Harlow, W. M. 1931. The identification of the pines of the United States, native and introduced, by needle structure. *Bull N. Y. State Coll. For. Syracuse Univ. Tech. Publ.* No. 32.

Harper, J. D. 1974. "Forest Insect Control with *Bacillus thuringiensis*. Survey of Current Knowledge" *Agric. Exp. Stn.* Auburn Univ., Auburn, Alabama.

Harring, C. M. 1978. Aggregation pheromones of the European fir engraver beetles *Pityokteines curvidens, P. spinidens*, and *P. vorontzovi* and the role of juvenile hormone in pheromone biosynthesis. *Z. Angew. Entomol.* **85**: 281–317.

Harris, E. C. 1961. Kiln and air drying of European oak: effect on starch depletion and consequent susceptibility to *Lyctus* attack. *J. Inst. Wood Sci.* **7**: 3–14.

Harris, P. 1960. Production of pine resin and its effect on survival of *Rhyacionia buoliana* (Schiff.) (Lepidoptera : Olethreutidae). *Can. J. Zool.* **38**: 121–130.

Harrison, S., and R. Karban. 1986. Effects of an early season folivorous moth on the success of a later-season species, mediated by a change in the quality of the shared host, *Lupinus arboreus* Sims. *Oecologia* **69**: 354–359.

Hart, G., and H. W. Lull. 1963. Some relationships among air, snow, and soil temperatures and soil frost. *USDA For. Serv. Res. Note* NE-3.

Hartzell, A., and W. J. Louden. 1935. Efficiency of banding for the control of cankerworms. *Contrib. Boyce Thompson Inst.* **7**: 365–377.

Harvey, G. T. 1974. Nutritional studies of eastern spruce budworm. I. Soluble sugars. *Can. Entomol.* **106**: 353–365.

Harvey, G. T., and J. M. Burke. 1974. Mortality of spruce budworm on white spruce caused by *Entomophthora sphaerosperma fresenius Can. For. Serv. Bi-Mon. Res. Notes* **30**: 23–24.

Hatcher, R. J. 1964. Spruce budworm damage to balsam fir in immature stands. Quebec *For. Chron.* **40**: 372–383.

Hathway, D. E. 1952. Oak bark tannins. *Biochem. J.* **70**: 34–42.

Haukioja, E., and T. Hakala. 1975. Herbivore cycles and periodic outbreaks. Formulation of a general hypothesis. *Rep. Kevo Subarct. Res. Stn.* **12**: 1–9.

Haukioja, E., and P. Niemela. 1977. Retarded growth of a geometrid larva after mechanical damage to leaves of its host tree. *Ann. Zool. Fenn.* **14**: 48–52.

Haukioja, E., and S. Neuvonen. 1987. Insect population dynamics and induction of plant resistance: The testing of hypotheses. *In* "Insect Outbreaks" (P. Barbosa and J. C. Schultz, eds.), pp. 411–432. Academic Press, New York.

Hawley, L. C., and P. W. Stickle. 1956. "Forest Protection," 2nd Ed. Wiley, New York.

Hedlin, A. F. 1964a. A six-year plot study on Douglas-fir cone insect population fluctuations. *For. Sci.* **10**: 124–128.

Hedlin, A. F. 1964b. Life history and habits of a midge, *Phytophaga thujae* Hedlin (Diptera : Cecidomyiidae), in western red cedar cones. *Can. Entomol.* **96**: 950–957.

Hedlin, A. F., H. O. Yates III, D. C. Tovar, B. H. Ebel, T. W. Koerber, and E. P. Merkel. 1980. Cone and seed insects of North America conifers. *Can. For. Serv.–U. S. For. Serv.–Sec. Agric. Recur. Hidr., Mex.*

Heichel, G. H., and N. C. Turner. 1976. Phenology and leaf growth of defoliated hardwood trees. *In* "Perspectives in Forest Entomology" (J. F. Anderson, and H. K. Kaya, eds.), pp. 31–40. Academic Press, New York.

Heichel, G. H., N. C. Turner, and G. S. Walton. 1972. Anthracnose causes dieback of regrowth on defoliated oak and maple. *USDA Plant Dis. Rep.* **56**: 1046–1047.

Hendry, L. B., B. Piatek, L. E. Browne, D. L. Wood, J. A. Byers, R. H. Fish, and R. A. Hicks. 1980. *In vivo* conversion of a labelled host plant chemical to pheromones of the bark beetle *Ips paraconfusus. Nature (London)* **284**: 485.

Henson, W. R. 1951. Mass flights of the spruce budworm. *Can. Entomol.* **83**: 240.

Henson, W. R. 1958. The effects of radiation on the habitat temperatures of some poplar-inhabiting insects. *Can. J. Zool.* **36**: 463–478.

Henson, W. R. 1962. Laboratory studies on the adult behavior of *Conophthorus coniperda* (Coleoptera : Scolytidae). III. Flight. *Ann. Entomol. Soc. Am.* **55**: 524–430.

Hensen, W. R., and R. F. Shepherd. 1952. The effects of radiation on the habitat temperatures of the lodgepole needleminer, *Recurvaria milleri* Busck (Gelechiidae : Lepidoptera). *Can. J. Zool.* **30**: 144–153.

Heron, R. J. 1951. Notes on the feeding of larvae of the larch sawfly *Pristiphora erichsonii* (Htg.) (Hymenoptera : Tenthredinidae). *Annu. Rep. Entomol. Soc. Ont.* **82**: 67–70.

Heron, R. J. 1965. The role of chemotactic stimuli in the feeding behavior of spruce budworm larvae on white spruce. *Can. J. Zool.* **43**: 247–269.

Hertel, G. D., and D. M. Benjamin. 1977. Intensity of site preparation influences on pine webworm and tip moth infestations of pine seedlings in northcentral Florida. *Environ. Entomol.* **6**: 118–122.

Hetrick, L. A. 1949. Some overlooked relationships of southern pine beetle. *J. Econ. Entomol.* **42**: 466–469.

Heyd, R. L., and L. F. Wilson. 1981. Risk-rating red pine plantations to predict losses from Saratoga spittlebug for management decisions. *In* "Hazard-Rating Systems in Forest Insect Pest Management: Symposium Proceedings" (R. L. Hedden, S. J. Barras, and J. E. Coster, tech. coords.), *USDA For. Serv. Gen. Tech. Rep.* **WO-27**: 93–98.

Hicks, R. R., Jr., J. E. Howard, K. G. Watterston, and J. E. Coster. 1980. Rating forest stand susceptibility to southern pine beetle in east Texas. *For. Ecol. Manage.* **2**: 269–283.

Hierholzer, O. 1950. Ein beitrag zur frage der Orientierung von *Ips curvidens Germ. Z. Tierpsychol.* **7**: 588–620. Cited in Schoonhoven (1972b).

Hillis, W. E., and I. Inoue. 1968. The formation of polyphenols in trees. IV. The polyphenols formed in *Pinus radiata* after *Sirex* attack. *Phytochemistry* 7: 13–22.

Hines, G. S., H. A. Taha, and F. M. Stephen. 1980. Model for predicting southern pine beetle population growth and tree mortality. *In* "Modeling Southern Pine Beetle Populations: Symposium Proceedings" (F. M. Stephen, J. L. Searcy, and G. D. Hertel, eds.), *USDA For. Serv. Tech. Bull.* No. 1630: 4–12.

Hodges, J. D. and P. L. Lorio, Jr. 1969. Carbohydrate and nitrogen fractions of the inner bark of loblolly pines under moisture stress. *Can. Bot.* 47: 1651–1657.

Hodges, J. D., and P. L. Lorio, Jr. 1975. Moisture stress and composition of xylem oleoresin in loblolly pine. *For. Sci.* 21: 283–290.

Hodges, J. D., and L. S. Pickard. 1971. Lightening in the ecology of the southern pine beetle, *Dendroctonus frontalis* (Coleoptera : Scolytidae). *Can. Entomol.* 103: 44–51.

Hodges, J. D., S. J. Barras, and J. K. Mauldin. 1968. Amino acids in inner bark of loblolly pine, as affected by the southern pine beetle and associated microorganisms. *Can. J. Bot.* 46: 1467–1472.

Hodges, J. D., W. W. Elan, W. F. Watson, and T. E. Nebeker. 1979. Oleoresin characteristics and susceptibility of four southern pines to southern pine beetle (Coleoptera : Scolytidae) attacks. *Can. Entomol.* 111: 889–898.

Hodson, A. C., and M. A. Brooks. 1956. The frass of certain defoliators of forest trees in the north central U. S. and Canada. *Can. Entomol.* 88: 62–68.

Hoffard, W. H., and J. E. Coster. 1976. Endoparasitic nematodes of *Ips* bark beetles in eastern Texas. *Environ. Entomol.* 5: 128–132.

Hoffman, C. H., H. K. Townes, H. H. Swift and R. I. Sailer. 1949. Field studies on the effects of airplane spraying of DDT on forest invertebrates. *Ecol. Monogr.* 19: 1–46.

Holling, C. S. 1965. The functional response of predators to prey density and its role in mimicry and population regulation. *Mem. Entomol. Soc. Can.* 45: 1–60.

Hopping, G. R., and J. H. Jenkins. 1933. The effect of kiln temperatures and air-seasoning on ambrosia insects (pinworms). *Can. For. Serv. Circ.* No. 38.

Houston, D. R. 1975. Beech bark disease: the aftermath forests are structured for a new outbreak. *J. For.* 73: 660–663.

Houston, D. R. 1979. Classifying forest susceptibility to gypsy moth defoliation. *U. S. Dep. Agric. Handb.* No. 542.

Houston, D. R., and H. T. Valentine. 1977. Comparing and predicting forest stand susceptibility to gypsy moth. *Can. J. For. Res.* 7: 447–461.

Howden, H. F., and G. B. Vogt. 1951. Insect communities of standing dead pine (*Pinus virginiana* Mill.). *Ann. Entomol. Soc. Am.* 44: 581–595.

Hughes, P. R. 1973a. Effect of a-pinene exposure on trans-verbenol synthesis in *Dendroctonus ponderosae* Hopk. *Naturwissenschaften* 5: 261–262.

Hughes, P. R. 1973b. *Dendroctonus*: production of pheromones and related compounds in response to host monoterpenes. *Z. Angew. Entomol.* 73: 294–312.

Hughes, P. R. 1974. Myrcene: a precursor of pheromones in *Ips* beetles. *J. Insect Physiol.* 20: 1271–1275.

Hughes, P. R. 1975. Pheromones of *Dendroctonus*: origin of a-pinene oxidation products present in emergent adults. *J. Insect Physiol.* 21: 687–691.

Hughes, P. R. 1976. Responses of female southern pine beetles to the aggregation pheromone frontalin. *Z. Angew. Entomol.* 80: 280–284.

Hughes, P. R., and J. A. A. Renwick. 1977a. Hormonal and host factors stimulating pheromone synthesis in female western pine beetles, *Dendroctonus brevicomis*. *Physiol. Entomol.* 2: 289–292.

Hughes, P. R., and J. A. A. Renwick, 1977b. Neural and hormonal control of pheromone biosynthesis in the bark beetle, *Ips paraconfusus*. *Physiol. Entomol.* 2: 117–123.

Hundertmark, A. 1937a. Helligkeits-und Farbenunterscheidungsvermogen der Eiraupe der Nonne (*Lymantria monancha* L.). *Z. Vergl. Physiol.* 24: 42–57. Cited in Schoonhoven (1972a).

Hundertmark, A. 1937b. Das formenunterscheidungsvermogen der Eiraupe der Nonne (*Lymantria monacha* L.). *Z. Vergl. Physiol.* **24**: 563–582. Cited in Schoonhoven (1972a).

Hunt, J. R., and G. M. Barton. 1978. Nutritive value of spruce muka (foliage) for the growing chick. *Anim. Feed Sci. Technol.* **3**: 67–72.

Hynum, B. G., and A. A. Berryman. 1980. *Dendrotonus ponderosae* (Coleoptera : Scolytidae):pre-aggregation landing and gallery intiation on lodgepole pine. *Can. Entomol.* **112**: 185-191.

Ives, W. G. H. 1954. Sequential sampling of insect populations. *For Chron.* **30**: 287–291.

Ives, W. G. H. 1955a. Effect of moisture on the selection of cocooning sites by the larch sawfly, *Pristiphora erichsonii* (Hartig). *Can. Entomol.* **87**: 301–311.

Ives, W. G. H. 1955b. Estimation of egg populations of the larch sawfly, *Pristiphora erichsonii* (Htg.). *Can. J. Zool.* **33**: 370–388.

Ives, W. G. H. 1973. Heat units and outbreaks of the forest tent caterpillar, *Malacosoma disstria* (Lepidoptera : Lasiocampidae). *Can. Entomol.* **105**: 529–543.

Ives, W. G. H., and L. D. Nairn. 1966. Effects of water levels on overwintering survival and emergence of the larch sawfly in a bog habitat. *Can. Entomol.* **98**: 768–777.

Ives, W. G. H., and R. M. Prentice. 1958. A sequential sampling technique for surveys of the larch sawfly. *Can. Entomol.* **90**: 331–338.

Jackson, R. M. and F. Raw. 1966. "Life in the Soil." St. Martin's, New York.

Jacot, A. P. 1936. Soil structure and soil biology. *Ecology* **17**: 359–379.

Jeffers, R. M. 1978. Seed source-related eastern pineshoot borer incidence in jack pine. *Silvae Genet.* **27**: 211–214.

Jennings, D. T. 1975. A sex-lure trap for *Rhyacionia* tip moths. *USDA For. Serv. Res. Note* **RM-284**.

Jennings, D. T. and H. A. Pase, III. 1975. Spiders preying on *Ips* bark beetles. *Southwest. Nat.* **20**: 225–229.

Jensen, G. L. and C. S. Koehler. 1969. Biological studies of *Scythropus californicus* on Monterey pine in northern California. *Ann. Entomol. Soc. Am.* **62**: 117–120.

Jensen, V. 1974. Decomposition of angiosperm tree leaf litter. *In* "Biology of Plant Litter Decomposition" (C. H. Dickinson and G. T. F. Pugh, eds.), pp. 69–104. Academic Press, New York.

Johnson, K. N., D. B. Jones, and B. M. Kent. 1980. Forest Planning Model (FORPLAN): User's Guide and Operations Manual (Draft). "USDA For. Serv. Land Manage. Plann." Fort Collins, Colorado.

Johnson, N. E. 1963. Cone-scale necrosis and seed damage associated with attacks by Douglas-fir cone midges. *For. Sci.* **9**: 44–51.

Johnson, N. E. 1965. Reduced growth associated with infestations of Douglas-fir seedlings by *Cinara* species (Homoptera : Aphididae). *Can. Entomol.* **97**: 113–119.

Johnson, N. E. 1968. "Insect Attack in Relationship to the Physiological Condition of the Host Tree" *Entomol. Limnol. Mimeogr. Rev. No. 1.* State Coll. Agric. Publ., Cornell Univ., Ithaca, New York.

Johnson, N. E., and A. F. Hedlin. 1967. Douglas-fir cone insects and their control. *Can. For. Branch Dep. Publ.* No. 1168.

Johnson, N. E., and H. J. Heikkenen. 1958. Damage to the seed of Douglas-fir by the Douglas-fir cone midge. *For. Sci.* **4**: 274–282.

Johnson, N. E., K. R. Shea, and R. L. Johnsey. 1970. Mortality and deterioration of looper-killed hemlock in western Washington. *J. For.* **68**: 162–163.

Johnson, N. E., and J. G. Zingg. 1969. Transpirational drying of Douglas-fir: effect on log moisture content and insect attack. *J. For.* **67**: 816–819.

Johnson, N. E., R. G. Mitchell, and K. H. Wright. 1963. Mortality and damage to pacific silver fir by the balsam woolly aphid in southwestern Washington. *J. For.* **61**: 854–860.

Johnson, P. C. 1972. Bark beetle risk in mature ponderosa pine forests in western Montana. *USDA For. Serv. Res. Pap.* **INT-119**.

Kapler, J. E., and D. M. Benjamin. 1960. The biology and ecology of the red-pine sawfly in Wisconsin. *For. Sci.* **6**: 253–268.

Karban, R. 1983. Induced responses of cherry trees to periodical cicada oviposition. *Oecologia* **59**: 226–231.

Katagiri, K. 1969. Review on microbial control of insect pests in forests in Japan. *Entomophaga* **14**: 203–214.

Kaupp, W. J., and J. C. Cunningham. 1977. Aerial application of a nuclear polyhedrosis against the red-headed pine sawfly, *Neodiprion lecontei* Fitch. *Can. For. Serv. Rep.* **IP-X-14.**

Kaya, H. K., and J. F. Anderson. 1973. Elm spanworm egg distribution in a Connecticut hardwood forest. *Ann. Entomol. Soc. Am.* **66**: 825–829.

Keays, J. L. 1976. Foliage. Part I. Practical utilization of foliage. *Appl. Polym. Symp.* No. 28: 445–464.

Keen, F. P. 1936. Relative susceptibility of ponderosa pines to bark-beetle attack. *J. For.* **34**: 919–927.

Keen, F. P. 1943. Ponderosa pine tree classes redefined. *J. For.* **41**: 249–253.

Keen, F. P., and K. A. Salman. 1942. Progress in pine beetle control through tree selection. *J. For.* **40**: 854–858.

Keen, F. P., F. C. Craighead, S. A. Graham, J. C. Evenden, W. D. Edmonston, J. E. Patterson, J. M. Maller, H. E. Burke, and H. L. Person. 1927. The relation of insects to slash disposal. *US Dep. Agric. Dep. Circ.* No. 411.

Kegg, J. D. 1971. The impact of gypsy moth: repeated defoliation of oak in New Jersey. *J. For.* **69**: 852–854.

Kegg, J. D. 1973. Oak mortality caused by repeated gypsy moth defoliations in New Jersey. *J. Econ. Entomol.* **66:** 639–641.

Kemp, W. P., D. O. Everson, and W. G. Wellington. 1985. Regional climatic patterns and western spruce budworm outbreaks. *USDA For. Serv. Tech. Bull.* No. 1693.

Kendeigh, S. C. 1947. Bird population studies in the coniferous forest biome during a spruce budworm outbreak. *Ontario Dep. Lands Forests Biol. Bull.* No. 1: 1–100.

Kennedy, P. C., and L. F. Wilson. 1971. Pine root collar weevil damage to red pine plantations in Michigan related to host age, temperature and stand location. *Can. Entomol.* **103**: 1685–1690.

Kevan, D. K. 1962. "Soil Animals." Witherby, London.

Kidd, N. A. C. 1976a. Aggregation in the lime aphid (*Eucallipterus tiliae* L.). 1. leaf vein selection and its effect on distribution on the leaf. *Oecologia* **22**: 299–304.

Kidd, N. A. C. 1976b. Aggregation in the lime aphid (*Eucallipterus tiliae* L.). 2. Social aggregation. *Oecologia* **25**: 175–185.

Kimmins, J. P. 1971. Variations in the foliar amino acid composition of flowering and non-flowering balsam fir (*Abies balsamea* (L.) Mill.) and white spruce (*Picea glauca* (Moench) Voss.) in relation to outbreaks of the spruce budworm (*Choristoneura fumiferana* (Clem.)). *Can. J. Zool.* **49**: 1005–1011.

King, E. W. 1972. Rainfall and epidemics of the southern pine beetle. *Environ. Entomol.* **1**: 279–285.

Kinzer, G. W., A. F. Fentiman, Jr., T. F. Page, Jr., R. L. Foltz, J. P. Vite, and G. B. Pitman. 1969. Bark beetle attractants: identification, synthesis, and field bioassay of a new compound isolated from *Dendroctonus*. *Nature (London)* **221**: 477–478.

Kinzer, H. G. 1976. Biology and behavior of cone beetles of ponderosa pine and southwestern white pine in New Mexico. *Bull. N. M. State Univ. Agric. Exp. Stn.* No. 641.

Klimetzek, D., and W. Francke. 1980. Relationship between the enantiomeric composition of a-pinene in host trees and the production of verbenols in *Ips* species. *Experientia* **36**: 1343–1344.

Kline, L. N., R. F. Schmitz, J. A. Rudinsky, and M. M. Furniss. 1974. Repression of spruce beetle (Coleoptera) attraction by methylcyclohexenone in Idaho. *Can. Entomol.* **106**: 485–491.

Knerer, G., and C. E. Atwood. 1973. Diprionid sawflies: polymorphism and speciation changes in diapause and choice of food plants led to new evolutionary units. *Science* **179**: 1090–1098.

Knight, F. B. 1958. The effects of woodpeckers on populations of the Engelmann spruce beetle. *J. Econ. Entomol.* **51**: 603–607.

Knopf, J. A. E., and G. B. Pitman. 1972. Aggregation pheromone for manipulation of the Douglas-fir beetle, *J. Econ. Entomol.* **65**: 723–726.

Kogan, M. 1977. The role of chemical factors in insect/plant relationships. *Proc. Int. Congr. Entomol., 15th, 1976* pp. 211–227.

Kostaka, S. J., and J. L. Sherald. 1982. An evaluation of electrical resistance as a measure of vigor in eastern white pine. *Can. J. For. Res.* **12**: 463–467.

Kozlowski, T. T. 1949. Light and water in relation to growth and competition of Piedmont forest tree species. *Ecol. Monogr.* **19**: 207–231.

Kozlowski, T. T. 1963. Growth characteristics of forest trees. *J. For.* **61**: 655–662.

Kozlowski, T. T. 1969. Tree physiology and forest pests. *J. For.* **67**: 118–123.

Kozlowski, T. T. 1985. Tree growth in response to environmental stress. *J. Arboric.* **11**(4): 97–111.

Kozlowski, T. T., and J. J. Clausen. 1966. Shoot characteristics of heterophyllous woody plants. *Can. J. Bot.* **44**: 827–843.

Kozlowski, T. T., and T. Keller. 1966. Food relations of woody plants. *Bot. Rev.* **32**: 293–382.

Kozlowski, T. T., and C. H. Winget. 1964. The role of reserves in leaves, branches, stems, and roots on shoot growth of red pine. *Am. J. Bot. Sci.* **522**: 529.

Krebs, C. J. 1972. "Ecology—The Experimental Analysis of Distribution and Abundance." Harper, New York.

Kriebel, H. B. 1954. Bark thickness as a factor in resistance to white pine weevil injury. *J. For.* **52**: 842–845.

Krischik, V. A., P. Barbosa, and C. F. Reichelderfer. 1988. Three trophic level interactions: Allelochemicals, *Manduca sexta* and *Bacillus thuringiensis* var. Kurstaki. *Environ. Entomol.* **17**: 476–482.

Kuc, J. 1982. Induced immunity to plant disease. BioScience 32: 854–860.

Kuc, J., and F. L. Caruso. 1977. Activated coordinated chemical defense against disease in plants. *In* "Host Plant Resistance to Pests. American Chemical Society Symposium Series 62." pp. 78–89.

Kulman, H. M. 1965. Effects of artificial defoliation of pine on subsequent shoot and needle growth. *For. Sci.* **11**: 90–98.

Kulman, H. M. 1971. Effects of insect defoliation on growth and mortality of trees. *Annu. Rev. Entomol.* **16**: 289–324.

Kulman, H. M., and A. C. Hodson. 1961. The jack pine budworm as a pest of other conifers with special reference to red pine. *J. Econ. Entomol.* **54**: 1221–1224.

Kulman, H. M., A. C. Hodson, and D. P. Duncan. 1963. Distribution and effects of jack pine budworm defoliation. *For. Sci.* **9**: 146–157.

Kushmaul, R. J., M. D. Cain, C. E. Rowell, and R. L. Porterfield. 1979. Stand and site conditions related to southern pine beetle susceptibility. *For Sci.* **25**: 656–664.

Kushner, D. J., and G. T. Harvey. 1962. Antibacterial substances in leaves: their possible role in insect resistance to disease. *J. Invert. Pathol.* **4**: 155–184.

Lalanne-Cassou, B., J. Percy, and J. H. MacDonald. 1977. Ultrastructure of sex pheromone gland cells in *Lobesia botrana* Den. & Schiff. (Lepidoptera : Olethreutidae). *Can. J. Zool.* **55**: 672–680.

Lance, D. R., and P. Barbosa. 1981. Host tree influences on the dispersal of late instar gypsy moths, *Lymantria dispar* (L.). *Oikos* **38**: 1–7.

Lance, D., and P. Barbosa. 1981. Host tree influences on the dispersal of first instar gypsy moths *Lymantria dispar* (L.). *Ecol. Entomol.* **6**: 411–416.

Lanier, G. N., and W. E. Burkholder. 1974. Pheromones in speciation of Coleoptera. *In* "Pheromones" (M. C. Birch, ed.), pp. 161–189. North-Holland Publ., Amsterdam.

Lanier, G. N., and A. H. Jones. 1985. Trap trees for elm bark beetles. Augmentation with pheromone baits and chlorpyrifos. *J. Chem. Ecol.* **11**: 11–20.

Larson, P. R. 1963. Stem form development of forest trees. *For. Sci. Monogr.* **5**.

Larson, P. R. 1964. Contribution of different-aged needles to growth and wood formation of young red pines. *For. Sci.* **10**: 224–238.

Lautenschlager, R. A., and J. D. Podgwaite. 1977. Passage of infectious nuclear polyhedrosis virus through the alimentary tracts of two small mammal predators of the gypsy moth, *Lymantria dispar. Environ. Entomol.* **6**: 737–738.

Leach, J. G. 1940. "Insect Transmission of Plant Disease." McGraw-Hill, New York.

Leaf, A. L., C. B. Davey, and G. K. Voigt. 1973. Forest soil organic matter and nutrient element dynamics. *In Proc. Soil Microcommunities Conf., 1st* (D. L. Dindal, ed.) *US AEC Off. Inf. Serv. Tech. Inf. Cent.* **CONF-711076**.

Lejeune, R. R., and B. Filuk. 1947. The effect of water levels on larch sawfly populations. *Can. Entomol.* **79**: 155–160.

Lejeune, R. R., W. H. Fell, and D. P. Burbidge. 1955. The effect of flooding on development and survival of the larch sawfly, *Pristiphora erichsonii* (Tenthredinidae). *Ecology* **36**: 63–70.

Lejeune, R. R., L. H. McMullen, and M. D. Atkins. 1961. The influence of logging on Douglas-fir beetle populations. *For. Chron.* **37**: 308–314.

Leonard, D. E. 1972. Survival in a gypsy moth population exposed to low winter temperatures. *Environ. Entomol.* **1**: 549–554.

Lessard, G., and D. T. Jennings. 1976. Southwestern pine tip moth damage to ponderosa pine reproduction. *USDA For. Serv. Res. Pap.* **RM-168**.

Lester, D. T. 1974. Geographic variation in leaf and twig monoterpenes of balsam fir. *Can. J. For. Res.* **4**: 55–60.

Levin, D. A. 1973. The role of trichomes in plant defense. *Q. Rev. Biol.* **48**: 3–15.

Levinson, A. S., G. Lemoine, and E. C. Smart. 1971. Volatile oil from foliage of *Sequoiadendron giganteum*: change in composition during growth. *Phytochemistry* **10**: 1087–1094.

Lewis, T. 1970. Patterns of distribution of insects near a windbreak of tall trees. *Ann. Appl. Biol.* **65**: 213–220.

Lingren, B. S. 1983. A multiple funnel trap for scolytid beetles (Coleoptera). *Can. Entomol.* **115**: 299–302.

Linnane, J. P., and E. A. Osgood. 1976. Controlling the Saratoga spittlebug in young red pine plantations by the removal of alternate hosts. *Univ. Maine Agric. Exp. Stn. Tech. Bull.* No. 84.

Linsley, E. G. 1959. Ecology of Cerambycidae. *Annu. Rev. Entomol.* **4**: 99–138.

Linzon, S. N. 1958. The effect of artificial defoliation of various ages of leaves upon white pine growth. *For. Chron.* **34**: 50–56.

Lister, C. K., J. M. Schmid, C. D. Minnemeyer, and R. H. Frye. 1976. Refinement of the lethal trap tree method for spruce beetle control. *J. Econ. Entomol.* **69**: 415–418.

Little, C. H. A. 1970. Seasonal changes in carbohydrate and moisture content in needles of balsam fir (*Abies balsamea*). *Can. J. Bot.* **48**: 2021–2028.

Livingston, R. L., and A. A. Berryman. 1972. Fungus transport structures in the fir engraver, *Scolytus ventralis* (Coleoptera : Scolytidae). *Can. Entomol.* **104**: 1793–1800.

Livingston, W. H., A. C. Mangini, H. G. Kinzer, and M. E. Mielke. 1983. Association of root diseases and bark beetles (Coleoptera : Scolytidae) with *Pinus ponderosa* in New Mexico. *Plant Dis.* **67**: 674–676.

Lodhi, M. A. K. 1975. Allelopathic effects of hackberry in a bottomland forest community. *J. Chem. Ecol.* **1**: 171–182.

Lodhi, M. A. K. 1976. Role of allelopathy as expressed by dominating trees in a lowland forest in controlling the productivity and pattern of herbaceous growth. *Am. J. Bot.* **63**: 1–8.

Lodhi, M. A. K. 1978. Allelopathic effects of decaying litter of dominant trees and their associated soil in a lowland forest community. *Am. J. Bot.* **65**: 340–344.

Lorio, P. L. 1968. Soil and stand conditions related to southern pine beetle activity in Hardin County, Texas. *J. Econ. Entomol.* **61**: 565–566.

Lorio, P. L., Jr. 1978. Developing stand risk classes for the southern pine beetle. *USDA For. Serv. Res. Pap.* **SO-144**.

Lorio, P. L., Jr., and J. D. Hodges. 1968. Oleoresin exudation pressure and relative water content of inner bark as indicators of moisture stress in loblolly pines. *For. Sci.* **14**: 392–398.

Luck, R. F., and D. L. Dahlsten. 1967. Douglas-fir tussock moth (*Hemerocampa pseudotsugata*) egg-mass distribution on white fir in northeastern California. *Can. Entomol.* **99**: 1193–1203.

Lunderstadt, J., U. Postel, and A. Reymers. 1977. As to food quality of spruce needles for forest damaging insects. 7. Studies on the constituents of the body of larvae of *Gilpinia hercyniae* Htg. (Hym.: Diprionidae). *Z. Angew. Entomol.* **82**: 380–394.

Lyons, L. A. 1956. Insects affecting seed production in red pine. Part I. *Conophthorus resinosae* Hopk. (Coleoptera: Scolytidae). *Can. Entomol.* **88**: 599–608.

Lyons, L. A. 1964. The spatial distribution of two pine sawflies and methods of sampling for the study of population dynamics. *Can. Entomol.* **96**: 1373–1407.

MacDonald, D. R., and F. E. Webb. 1963. Insecticides and the spruce budworm. In "The dynamics of epidemic spruce budworm populations" *Mem. Entomol. Soc. Can.* **31**: 288–310.

Maddox, J. V. 1975. Use of diseases in pest management. In "Introduction to Insect Pest Management" (R. L. Metcalf and W. H. Luckmann, eds.), pp. 189–233. Wiley, New York.

Maddox, D. M., and M. Rhyne. 1975. Effects of induced host-plant mineral deficiencies on attraction, feeding and fecundity of the alligatorweed flea beetle. *Environ. Entomol.* **4**: 682–686.

Magnoler, A. 1974. Effects of a cytoplasmic polyhedrosis on larval and postlarval stages of the gypsy moth, *Porthetria dispar J. Invert. Pathol.* **23**: 263–274.

Magnoler, A. 1975. Bioassay of nucleopolyhedrosis virus against larval instars of *Malacosoma neustria. J. Invert. Pathol.* **25**: 343–348.

Maksymiuk, B. 1979. Microbial control of insects. In "Forest Insect Survey and Control" (J. A. Rudinsky, ed.), pp. 135–171. Oregon State Univ. Book Stores, Corvallis.

Maksymiuk, B., J. Neisess, R. A. Waite, and R. D. Orchard. 1975. Distribution of aerially applied mexacarbate in a coniferous forest and correlation with mortality of *Choristoneura occidentalis* (Lepid.: Tortricidae). *Z. Angew. Entomol.* **79**: 194–204.

Mansingh, A. 1972. Developmental response of *Antheraea pernyi* to seasonal changes in oak leaves from two localities. *J. Insect Physiol.* **18**: 1395–1401.

Marshall, H. G. W. 1967. The effect of woodpecker predation on wood-boring larvae of families Siricidae (Hymenoptera) and Melandryidae (Coleoptera). *Can. Entomol.* **99**: 978–985.

Martignoni, M. E., and P. J. Iwai. 1986. A catalog of viral diseases of insects, mites, and ticks. *USDA For. Ser. Gen. Tech. Rep.* **PNW-195**.

Martin, J. L. 1958. Observations on the biology of certain tortricids in young coniferous plantations in southern Ontario. *Can. Entomol.* **90**: 44–53.

Martin, J. L. 1966. The insect ecology of red pine plantations in central Ontario. IV. The crown fauna. *Can. Entomol.* **98**: 10–27.

Martin, M. M. 1979. Biochemical implications of insect mycophagy. *Biol. Rev.* **54**: 1–21.

Mason, R. R. 1969. Sequential sampling of Douglas-fir tussock moth populations. *USDA For. Serv. Res. Note PNW* **PNW-102**.

Mason, R. R. 1970. Development of sampling methods for the Douglas-fir tussock moth, *Hemerocampa pseudotsugata* (Lepidoptera: Lymantriidae). *Can. Entomol.* **102**: 836–845.

Mason, R. R. 1971. Soil moisture and stand density affect oleoresin exudation flow in a loblolly pine plantation. *For. Sci.* **17**: 170–177.

Mason, R. R. 1974. Population change in an outbreak of the Douglas-fir tussock moth, *Orgyia pseudotsugata* (Lepidoptera: Lymantriidae), in central Arizona. *Can. Entomol.* **106**: 1171–1174.

Mason, R. R. 1976. Life tables for a declining population of the Douglas-fir tussock moth in northeastern Oregon. *Ann. Entomol. Soc. Am.* **69**: 948–958.

Mason, R. R. 1978. Detecting suboutbreak populations of the Douglas-fir tussock moth by sequential sampling of early larvae in the lower tree crown. *USDA For. Serv. Res. Pap.* **PNW-238**.

Mason, R. R., and J. W. Baxter. 1970. Food preference in a natural population of the Douglas-fir tussock moth. *J. Econ. Entomol.* **63**: 1257–1259.

Mason, R. R., and T. C. Tigner. 1972. Forest-site relationships within an outbreak of lodgepole needle-miner in central Oregon. *USDA For. Serv. Res. Pap.* **PNW-146**.

Massey, C. L. 1966. The influence of nematode parasites and associates on bark beetles in the United States. *Bull. Entomol. Soc. Am.* **12**: 384–386.

Mattson, W. J., Jr. 1971. Relationship between cone crop size and cone damage by insects in red pine seed production areas. *Can. Entomol.* **103**: 617–621.

Mattson, W. J. 1978. The role of insects in the dynamics of cone production of red pine. *Oecologia* **33**: 327–349.

Mattson, W. J. 1980. Herbivory in relation to plant nitrogen content. *Annu. Rev. Ecol. Syst.* **11**: 119–161.

Mattson, W. J., and N. D. Addy. 1975. Phytophagous insects as regulators of forest primary production. *Science* **190**: 515–522.

Mattson, W. J., and R. A. Haack. 1987. The role of drought in outbreaks of plant-eating insects. *BioScience* **37**: 110–118.

Mattson, W. J., F. B. Knight, D. C. Allen, and J. L. Foltz. 1968. Vertebrate predation on the jack-pine budworm in Michigan. *J. Econ. Entomol.* **61**: 229–234.

Mazanec, Z. 1968. Influence of defoliation by the phasmatid *Didymuria violescens* on seasonal diameter growth and the pattern of growth rings in alpine ash. *Aust. For.* **32**: 3–14.

McCambridge, W. F. 1974. Influence of low temperature on attack, oviposition and larval development of mountain pine beetle, *Dendroctonus ponderosae* (Coleoptera : Scolytidae). *Can. Entomol.* **106**: 979–984.

McCarthy, J., and C. E. Olson, Jr. 1983. Assessing spruce budworm damage with small-format aerial photographs. *Can. J. For. Res.* **13**: 395–399.

McClure, M. S. 1977. Dispersal of the scale *Fiorinia externa* (Homoptera : Diaspididae) and effects of edaphic factors on its establishment on hemlock. *Environ. Entomol.* **6**: 539–544.

McCullough, D. G., and M. R. Wagner. 1987. Evaluation of four techniques to assess vigor of water-stressed ponderosa pine. *Can. J. For. Res.* **17**: 138–145.

McDonogh, R. S. 1939. The habitat, distribution and dispersal of the psychid moth, *Luffia erchaultella*, in England and Wales. *J. Anim. Ecol.* **8**: 10–28.

McFarland, W. N., and F. W. Munz. 1975. The visible spectrum during twilight and its implications to vision. *In* "Light as an Ecological Factor (II)" (G. C. Evans, R. Bainbridge, and O. Rackham, eds.), pp. 249–270. Blackwell, Oxford.

McFarlane, R. W. 1976. Birds as agents of biological control. *Biologist* **58**: 123–140.

McGee, C. E. 1975. Change in forest canopy affects phenology and development of northern red and scarlet oak seedlings. *For. Sci.* **21**: 175–179.

McGregor, M. D., M. M. Furniss, R. D. Oaks, K. E. Gibson, and H. E. Meyer. 1984. MCH pheromone for preventing Douglas-fir beetle infestation in windthrown trees. *J. For.* **82**: 613–616.

McKenzie, H. L. 1941. *Matsucoccus bisetosus* Morrison, a potential enemy of California pines. *J. Econ. Entomol.* **34**: 783–785.

McKenzie, H. L., L. S. Gill, and D. E. Ellis. 1948. The Prescott scale (*Maltucoccus verillorum*) and associated organisms that cause flagging injury to ponderosa pine in the Southwest. *J. Agric. Res.* **76**: 33–51.

McLaughlin, R. E. 1973. Protozoa as microbial control agents. *Misc. Publ. Entomol. Soc. Am.* **9**: 95–98.

McLean, J. A., and J. H. Borden. 1975. Survey for *Gnathotrichus sulcatus* (Coleoptera : Scolytidae) in a commercial sawmill with the pheromone sulcatol. *Can. J. For. Res.* **5**: 586–591.

McLean, J. A., and J. H. Borden. 1977. Suppression of *Gnathotrichus sulcatus* with sulcatol-baited traps in a commercial sawmill and notes on the occurrence of *G. retusus* and *Trypodendron lineatum*. *Can. J. For. Res.* **7**: 348–356.

McLeod, J. M. 1970. The epidemiology of the Swaine jack-pine sawfly, *Neodiprion swainei* Midl. *For. Chron.* **46**: 126–133.

McLeod, J. M., and L. Daviault. 1963. Notes on the life history and habits of the spruce cone worm, *Dioryctria renicullella* (Grt.) (Lepidoptera : Pyralidae). *Can. Entomol.* **95**: 309–316.

McLintock, T. F. 1949. Mapping vulnerability of spruce-fir stands in the northeast to spruce budworm attack. *USDA For. Serv. Northeast. For. Exp. Stn. Pap.* No. 21.

McManus, M. L. 1976. Weather influences on insect populations. *In*: Proc. IV Nat'l. Conf. Fire and For. Meteor. *USDA For. Serv. Gen. Tech. Rep.* **RM-32**: 94–100.

McManus, M. L., and R. L. Giese. 1968. The Columbian timber beetle, *Corthylus columbianus*. VII. The effect of climatic integrants on historic density fluctuations. *For. Sci.* **14**: 242–253.

McNamee, P. J. 1977. A process model for eastern black-headed budworm. *Inst. Res. Ecol., Univ. B.C. Work. Pap.* **W-23**.

McNaughton, S. J. 1983. Compensatory plant growth as a response to herbivory. *Okios* **40**: 329–336.

Menzel, R. 1975. Polarized light sensitivity in arthropods. *In* "Light as an Ecological Factor (II)" (G. C. Evans, R. Bainbridge, and O. Rackham, eds.), pp. 289–303. Blackwell, Oxford.

Metacalf, C. L., W. P. Flint, and R. L. Metcalf. 1962. "Destructive and Useful Insects. Their Habits and Control. McGraw-Hill, New York.

Metcalf, R. L. 1975. Insecticides in pest management. *In* "Introduction to Insect Pest Management" (R. L. Metcalf, and W. Luckmann, eds.), pp. 235–273. Wiley, New York.

Metz, L. J., and M. H. Farrier. 1969. Acarina associated with decomposing forest litter in the North Carolina piedmont. *Proc. Int. Congr. Acarol., 2nd* pp. 43–52.

Meyer, H. J., and D. M. Norris. 1967. Behavioral responses by *Scolytus multistriatus* (Coleoptera : Scolytidae) to host- (Ulmus) and beetle-associated chemotactic stimuli. *Ann. Entomol. Soc. Am.* **60**: 642–647.

Miles, P. W. 1968. Insect secretions in plants. *Annu. Rev. Phytopath.* 6: 137–164.

Millar, C. S. 1974. Decomposition of coniferous leaf litter. *In* "Biology of Plant Litter Decomposition" (C. H. Dickinson and G. J. F. Pugh, eds.), Vol. 1, New York. pp. 105–128. Academic Press.

Miller, C. A. 1958. The measurement of spruce budworm populations and mortality during the first and second larval instars. *Can. J. Zool.* **36**: 409–422.

Miller, C. A. 1966. The black-headed budworm in eastern Canada. *Can. Entomol.* **98**: 592–613.

Miller, J. M. 1931. High and low lethal temperatures for the western pine beetle. *J. Agric. Res.* **43**: 303–321.

Miller, J. M., and F. P. Keen. 1960. Biology and control of the western pine beetle. *USDA For. Serv. Misc. Publ.* No. 800.

Miller, J. M., and J. E. Patterson. 1927. Preliminary studies on the relation of fire injury to bark beetle attack in western yellow pine. *J. Agric. Res.* **34**: 597–613.

Miller, K. K., and M. R. Wagner. 1984. Factors influencing pupal distribution of the Pandora moth (Lepidoptera : Saturnidae) and their relationship to prescribed burning. *Environ. Entomol.* **13**: 430–431.

Miller, W. E. 1977. Weights of *Polia grandis* pupae reared at two constant temperatures (Lepidoptera : Noctuidae). *Great Lakes Entomol.* **10**: 47–49.

Miller, W. E. 1978. Use of prescribed burning in seed production areas to control red pine cone beetle. *Environ. Entomol.* **7**: 698–702.

Milne, A. 1957. The natural control of insect populations. *Can. Entomol.* **89**: 193–213.

Milne, A. 1962. On a theory of natural control of insect populations. *J. Theor. Biol.* **3**: 19–50.

Mitchell, R. G. 1967. Abnormal ray tissue in three true firs infested by the balsam woolly aphid. *For. Sci.* **13**: 327–332.

Mitchell, R. G., R. H. Waring, and G. B. Pitman. 1983. Thinning lodgepole pine increases tree vigor and resistance to mountain pine beetle. *For. Sci.* **29**: 204–211.

Mitchell, R. T. 1952. Consumption of spruce budworms by birds in a Maine spruce-fir forest. *J. For.* **50**: 387–389.

Mittler, T. E. 1958. Studies on the feeding and nutrition of *Tuberolachnus salignus* (Gmelin)

(Homoptera: Aphididae). II. The nitrogen and sugar composition of ingested phloem sap and excreted honeydew. *J. Exp. Biol.* **35**: 74–84.

Moeck, H. A. 1970. Ethanol as the primary attractant for the ambrosia beetle, *Trypodendron lineatum* (Coleoptera: Scolytidae). *Can. Entomol.* **102**: 985–995.

Moeck, H. A., D. L. Wood, and K. Q. Lindahl, Jr. 1981. Host selection behavior of bark beetles (Coleoptera: Scolytidae) attacking *Pinus ponderosa*, with special emphasis on the western pine beetle, *Dendroctonus brevicomis*. *J. Chem. Ecol.* **7**: 49–83.

Moericke, V. 1969. Host plant-specific colour behavior by *Hyalopterus pruni* (Aphidiae). *Entomol. Exp. Appl.* **12**: 524–534.

Moore, G. E. 1971. Mortality factors caused by pathogenic bacteria and fungi of the southern pine beetle in North Carolina. *J. Invert. Pathol.* **17**: 28–37.

Morris, O. N. 1976. A two-year study of the efficacy of *Bacillus thuringiensis*–chitinase combinations in spruce budworm (*Choristoneura fumiferana*) control. *Can. Entomol.* **108**: 225–233.

Morris, O. N. and J. A. Armstrong. 1975. Preliminary field trials with *Bacillus thuringiensis*-chemical insecticide combinations in the integrated control of the spruce budworm, *Choristoneura fumiferana* (Lepidoptera: Tortricidae). *Can. Entomol.* **107**: 1281–1288.

Morris, O. N., J. A. Armstrong, G. M. Howse, and J. C. Cunningham. 1974. A two-year study of virus-chemical insecticide combination in the integrated control of the spruce budworm, *Choristoneura fumiferana* (Tortricidae: Lepidoptera). *Can. Entomol.* **106**: 813–824.

Morris, R. F. 1949. Frass-drop measurement in studies of the European spruce sawfly. *Univ. Mich. Sch. For. Conserv. Bull.* No. 12.

Morris, R. F. 1951a. The effects of flowering in the foliage production and growth of balsam fir. *For. Chron.* **27**: 40–57.

Morris, R. F. 1951b. The importance of insect control on a forest management program. *Can. Entomol.* **83**: 176–181.

Morris, R. F. 1955. The development of sampling techniques for forest insect defoliators, with particular reference to the spruce budworm. *Can. J. Zool.* **33**: 107–223.

Morris, R. F. 1967. Influence of parental food quality on the survival of *Hyphantria cunea*. *Can. Entomol.* **99**: 24–33.

Morris, R. F. 1971a. Observed and simulated changes in genetic quality in natural populations of *Hyphantria cunea*. *Can. Entomol.* **103**: 892–906.

Morris, R. F. 1971b. The influence of land use and vegetation on the population density of *Hyphantria cunea*. *Can. Entomol.* **103**: 1525–1536.

Morris, R. F., and C. W. Bennett. 1967. Seasonal population trends and extensive census methods for *Hyphantria cunea*. *Can. Entomol.* **99**: 9–17.

Morris, R. F., W. F. Cheshire, C. A. Miller, and D. G. Mott. 1958. The numerical response of avian and mammalian predators during a gradation of the spruce budworm. *Ecology* **38**: 487–494.

Morris, R. F., and W. A. Reeks. 1954. A larval population technique for the winter moth (*Operophtera brumata* (Lann.) (Lepidoptera: Geometridae)). *Can. Entomol.* **86**: 433–438.

Morrow, R. R. 1965. Height loss from white pine weevil. *J. For.* **63**: 201–203.

Moser, J. C. 1975. Mite predators of the southern pine beetle. *Ann. Entomol. Soc. Am.* **68**: 1113–1116.

Moser, J. C. 1976a. Phoretic carrying capacity of flying southern pine beetles (Coleoptera: Scolytidae). *Can. Entomol.* **108**: 807–808.

Moser, J. C. 1976b. Surveying mites (Acarina) phoretic on the southern pine beetle (Coleoptera: Scolytidae) with sticky traps. *Can. Entomol.* **108**: 809–813.

Mosher, F. G. 1915. Food plants of the gypsy moth in America. *USDA Bull.* No. 250.

Mott, D. G. 1963. The forest and the spruce budworm. *In*: The dynamics of epidemic spruce budworm populations. *Mem. Entomol. Soc. Can.* No. 31: 189–201.

Mott, D. G., L. D. Nairn, and J. A. Cook. 1957. Radial growth in forest trees and effects of insect defoliation. *For. Sci.* **3**: 286–304.

Moulder, B. C., and D. E. Reichle. 1972. Significance of spider predation in the energy dynamics of forest-floor arthropod communities. *Ecol. Monogr.* **42**: 473–498.

Mounts, J. 1976. 1974 Douglas-fir tussock moth control project. *J. For.* **74**: 82–86.

Muldrew, J. A. 1956. Some problems in the protection of tamarack against the larch sawfly, *Pristiphora erichsonii* (Htg.). *For. Chron.* **32**: 20–29.

Murphy, P. W. 1955. Ecology of the fauna of forest soils. *In* "Soil Zoology" (D. K. McE. Kevan, ed.), pp. 99–124. Butterworths, London.

Nakashima, T. 1975. Several types of the mycetangia found in platypodid ambrosia beetles (Coleoptera : Platypodidae). *Insecta Matsumurana* **7**: 1–69.

Nakashima, T. 1978. Ambrosia beetles-fungus growers. *Insectarium* **15**: 14–22.

Neilson, M. M. 1965. Effects of cytoplasmic polyhedrosis on adult Lepidoptera. *J. Invert. Path.* **7**: 306–314.

Neumann, F. G., J. L. Morey, and R. J. McKimm. 1987. The sirex wasp in Victoria. *Conserv. For. Lands Bull.* No. 29. Victoria, Australia.

Nickle, W. R. 1963. Observations on the effect of nematodes on *Ips confusus* (LeConte) and other bark beetles. *J. Insect Pathol.* **5**: 386–389.

Nickle, W. R. 1971. Behavior of the shothole borer, *Scolytus rugulosus*, altered by the nematode parasite *Neoparasitylenchus scoli*. *Ann. Entomol. Soc. Am.* **64**: 751.

Niemann, G. J. 1976. Phenolics from *Larix* needles. XII. Seasonal variation of main flavonoids in leaves of *L. leptolepsis*. *Acta Bot. Neerl.* **25**: 349–359.

Niemann, G. J. 1979. Some aspects of the chemistry of pinaceae needles. *Acta Bot. Neerl.* **28**: 73–88.

Niemela, P., E-M. Aro, and E. Haukioja. 1979. Birch leaves as a resource for herbivores: damage-induced increase in leaf phenols with trypsin inhibiting effects. *Rep. Kevo. Subarct. Res. Stn.* **15**: 37–40.

Nijholt, W. W. 1973. The effect of male *Trypodendron lineatum* (Coleoptera : Scolytidae) on the response of field populations to secondary attraction. *Can. Entomol.* **105**: 583–590.

Nijholt, W. W. 1978. Evaluation of operational water misting for log protection from ambrosia beetle damage. *Can. For. Serv. Pac. For. Res. Cent. Inf. Rep.* **BC-P-22-78**.

Noel, A. R. A. 1970. The girdled tree. *Bot. Rev.* **36**: 162–195.

Nord, J. C., J. H. Ghent, H. A. Thomas, and C. A. Doggett. 1982. Control of pales and pitch-eating weevils in the South. *USDA For. Serv. Reprint* **SA-FR 21**.

Nordlund, D. A. 1981. Semiochemical: A review of the terminology. *In* "Semiochemicals. Their Role in Pest Control" (D. A. Nordlund, R. L. Jones, and W. J. Lewis, eds.), pp. 13–28. Wiley, New York.

Norris, D. M. 1967. Systemic insecticides in trees. *Annu. Rev. Entomol.* **12**: 127–148.

Norton, R. A. 1973. Ecology of soil and litter spiders. *In Proc. Soil Microcommunities Conf., 1st*, (D. L. Dindal, ed.), pp. 138–156. *Off Int. Serv.* US AEC *Tech. Inf. Cent.*

O'Brien, R. D. 1967. "Insecticides: Action and Metabolism." Academic Press, New York.

Odell, T. M., P. A. Godwin, and W. B. White. 1974. Radiographing puparia of tachinid parasites of the gypsy moth and application in parasite-release programs. *USDA For. Serv. Res. Note* **NE-194**.

Odum, E. P. 1971. "Fundamentals of Ecology," 3rd Ed. Saunders, Philadelphia, Pennsylvania.

Oester, P. T., and J. A. Rudinsky. 1979. Acoustic behavior of three sympatric species of *Ips* (Coleoptera : Scolytidae) co-inhabiting Sitka spruce. *Z. Angew. Entomol.* **87**: 398–412.

Ohigashi, H., M. R. Wagner, F. Matsumura, and D. M. Benjamin. 1981. Chemical basis of differential feeding behavior of the larch sawfly, *Pristiphora erichsonii* (Hartig). *J. Chem. Ecol.* **7**: 599–614.

Ohmart, C. P., L. G. Stewart, and J. R. Thomas. 1983. Leaf consumption by insects in three *Eucalyptus* forest types in southeastern Australia and their role in short-term nutrient cycling. *Oecologia* **59**: 322–330.

Oksanen, H., V. Perttunen, and E. Kangas. 1970. Studies on the chemical factors involved in the olfactory orientation of *Blastophagus piniperda* (Coleoptera : Scolytidae). *Contrib. Boyce Thompson Inst.* **24**: 299–304.

Oku, H., T. Shiraishi, S. Ouchi, S. Kurozumi, and H. Ohta. 1980. Pine wilt toxin, the metabolite of a bacterium associated with a nematode. *Naturwissenschaffen* **67**: 198–199.

Oliver, C. D., and F. P. Stephens. 1977. Reconstruction of a mixed species forest in central New England. *Ecology* **58**: 562–572.

Olkowski, W., C. Pinnock, W. Toney, G. Mosher, W. Neasbitt, R. Van Den Bosch, and H. Olkowski. 1974. An integrated insect control program for street trees. *Calif. Agric.* **28**: 3–4.

O'Neil, L. C. D. 1963. The suppression of growth rings in jack pine in relation to defoliation by the Swaine jack-pine sawfly. *Can. J. Bot.* **41**: 227–235.

Osborne, D. J. 1972. Mutual regulation of growth and development in plants and insects. *In* "Insect/Plant Relationships" (H. F. Van Emden, ed.) *Symp. R. Entomol, Soc. London* No. 6, pp. 33–42. Blackwell, Oxford.

Ostaff, D., and M. Y. Cech. 1978. Heat sterilization of spruce-pine-fir lumber containing sawyer beetle larvae (Coleoptera : Cerambycidae, *Monochramus* sp.). *Fish Environ. Can. East. For. Prod. Lab. Rep.* **OP-X-200 E**.

Ostry, M. E., and N. A. Anderson. 1979. Infection of *Populus tremuloides* by *Hypoxylon mammatum* at oviposition sites of cicadas (*Magicicada septendecim* (L.)). *Phytopathology* **69**: 1041.

Otvos, I. S. 1965. Studies on avian predators of *Dendroctonus brevicomis* Le Conte (Coleoptera: Scolytidae) with special reference to Picidae. *Can. Entomol.* **97**: 1184–1199.

Page, G. 1975. The impact of balsam woolly aphid damage on balsam fir stands in Newfoundland. *Can. J. For. Res.* **5**: 195–209.

Paine, T. D., F. M. Stephen, and R. G. Cates. 1985. Induced defenses against *Dendroctonus frontalis* and associate fungi: variation in loblolly pine resistance. *In* "Integrated Pest Management Research Symposium: The Proceedings pp. 169–176. *USDA For. Serv. Southern For. Exper. Stn. Gen. Tech. Rep.* **50–56**.

Papageorgis, C. 1975. Mimicry in neotropical butterflies. *Am. Sci.* **63**: 522–532.

Papp, R. P., J. T. Kliejunas, R. S. Smith, Jr., and R. F. Scharp. 1979. Association of *Plagithmysus bilineatus* (Coleoptera : Cerambycidae) and *Phytophthora cinnamomi* with the decline of Ohi a-lehua forests on the island of Hawaii. *For. Sci.* **25**: 187–196.

Paquet, G. 1943. Bioclimatic study of the European spruce sawfly in Quebec. *Dep. Lands For. Que. Can. Entomol. Serv. Contrib.* No. 20.

Parker, D. L., and R. E. Stevens. 1979. Mountain pine beetle infestation characteristics in ponderosa pine, Kaibab Plateau, Arizona, 1975–1977. *USDA For. Serv. Res. Note* **RM-367**.

Parker, J. 1977. Phenolics in black oak bark and leaves. *J. Chem. Ecol.* **3**: 489–496.

Parkin, E. A. 1942. Symbiosis and siricid woodwasps. *Ann. Appl. Biol.* **29**: 268–274.

Parr, T. 1937. Notes on the golden or pit-making oak scale. *J. For.* **35**: 51–58.

Parry, W. H. 1971. Differences in the probing behavior of *Elatobium abietinum* feeding on Sitka and Norway spruces. *Ann. Appl. Biol.* **69**: 177–185.

Parry, W. H. 1974. The effects of nitrogen levels in Sitka needles on *Elatobiuim abietinum* (Walker) populations in north-eastern Scotland. *Oecologia* **15**: 305–320.

Parry, W. H. 1976. The effect of needle age on the acceptability of Sitka spruce needles to the aphid *Elatobium abietinum* (Walker). *Oecologia* **23**: 297–313.

Parry, W. H. 1978. Studies on the factors affecting the population levels of *Adelges cooleyi* (Gillette) on Douglas-fir. 1. Sistentes on mature needles. *Z. Angew. Entomol.* **85**: 365–378.

Parry, W. H. and W. Powell. 1977. A comparison of *Elatobium abietinum* populations on Sitka spruce trees differing in needle retention during aphid outbreaks. *Oecologia* **27**: 239–252.

Patch, E. M. 1920. The life cycle of aphids and coccids. *Ann. Entomol. Soc. Am.* **13**: 156–167.

Patterson, J. E. 1921. Life history of *Recurvaria milleri* Busck, the lodgepole pine needleminer, in the Yosemite National Park, California. *J. Agric. Res.* **21**: 127–143.

Payne, J. A. 1965. A summer carrion study of the baby pig *Sus scrofa* L. *Ecology* **46**: 592–602.

Payne, T. L. 1974. Pheromone perception. *In* "Pheromones" (M. C. Birch, ed.), pp. 35–61. North-Holland Publ. Amsterdam.

Peacock, J. W., Silverstein, A. C. Lincoln, and J. B. Simeone. 1973. Laboratory investigations of the frass of *Scolytus multistriatus* (Coleoptera : Scolytidae) as a source of pheromone. *Environ. Entomol.* **23**: 355–359.

Peirson, H. B. 1922. Mound-building ants in forest plantations. *J. For.* **20**: 325–336.

Peirson, H. B. 1937. The pales weevil (*Hylobus pales* Herbst). *Mass. For. Park Assoc. Tree Pest Leafl.* No. 13.

Percy, J. E., and J. Weatherston. 1974. Gland structure and pheromone production in insects. *In* "Pheromones" (M. C. Birch, ed.), pp. 11–34. North-Holland Publ., Amsterdam.

Percy, K. E., and R. T. Riding. 1978. The epicuticular waxes of *Pinus strobus* subjected to air pollutants. *Can. J. For. Res.* **8**: 474–477.

Person, H. L. 1931. Theory in explanation of the selection of certain trees by the western pine beetle. *J. For.* **29**: 696–699.

Perttunen, V., H. Oksanen, and E. Kangas. 1970. Aspects of the external and internal factors affecting the olfactory orientation of *Blastophagus piniperda* L. (Col. : Scol.). *Contrib. Boyce Thompson Inst.* **24**: 293–298.

Peterman, R. M. 1978. The ecological role of mountain pine beetle in lodgepole pine forests. *In* "Theory and Practice of Mountain Pine Beetle Management in Lodgepole Pine Forests" (A. A. Berryman, G. D. Amman, R. W. Stark, and D. L. Kibbee, eds.), pp. 16–26. For., Wildl. Range Exp. Stn., Univ, of Idaho, Moscow.

Pfadt, R. E. (ed.). 1985. "Fundamentals of Applied Entomology," 4th Ed. Macmillan, New York.

Piene, H., and K. van Cleve. 1978. Weight loss of litter and cellulose bags in a thinned white spruce forest in interior Alaska. *Can. J. For. Res.* **8**: 42–46.

Pimentel, D. 1961. Animal population regulation by the genetic feedback mechanism. *Am. Nat.* **95**: 65–79.

Pimentel, D. 1963. Introducing parasites and predators to control native pests. *Can. Entomol.* **95**: 785–792.

Pimentel, D. 1971. "Ecological Effect of Pesticides on Non-Target Species." Exec. Off. Pres., Off. Sci. and Technol. Washington, D. C.

Pimentel, D., J. Krummel, D. Gallahan, J. Hough, A. Merrill, I. Schreiner, P. Vattum, F. Koziol, E. Back, O. Yen, and S. Fiance. 1978. Benefits and costs of pesticide use in U. S. food production. *BioScience* **28**: 772–784.

Pitman, G. B. 1966. Studies on the pheromone of *Ips confusus* (Le Conte). III. The influence of host material on pheromone production. *Contrib. Boyce Thompson Inst.* **23**: 147–157.

Pitman, G. B. 1969. Pheromone response in pine bark beetles: influence of host volatiles. *Science* **166**: 905–906.

Pitman, G. B. 1971. Trans-verbenol and alpha-pinene: their utility in manipulation of the mountain pine beetle. *J. Econ. Entomol.* **64**: 426–430.

Pitman, G. B., and J. P. Vite. 1969. Aggregation behavior of *Dendroctonus ponderosa* (Coleoptera : Scolytidae) in response to chemical messengers. *Can. Entomol.* **101**: 143–149.

Pitman, G. B., R. A. Kliefoth, and J. P. Vite. 1965. Studies on the pheromone of *Ips confusus* (LeConte). II. Further observations on the site of production. *Contrib. Boyce Thompson Inst.* **23**: 13–17.

Pittendrigh, C. S. 1950. The ecoclimatic divergence of *Anopheles bellator* and *A. homonculus*. *Evolution* **4**: 43–63.

Poinar, G. O., Jr. 1971. Use of nematodes for microbial control of insects. *In* "Microbial Controls of Insects and Mites" (H. D. Burges and N. W. Hussey, eds.),p. 181–201. Academic Press, London.

Pollard, D. F. W. 1972. Estimating woody dry matter loss resulting from defoliation. *For. Sci.* **18**: 135–138.

Polster, H. 1963. Photosynthesis and growth of a five-year-old stand of poplar trees in relation to water economy of the site. *In* "The Water Relations of Plants" (A. J. Rutter and F. H. Whitehead, eds.), pp. 257–271. Blackwell, Oxford.

Pomeroy, M. K., D. Siminovitch, and F. Wightman. 1970. Seasonal biochemical changes in the living bark and needles of red pine (*Pinus resinosa*) in relation to adaptation to freezing. *Can. J. Bot.* **48**: 953–967.

Potts, S. F. 1938. The weight of foliage from different crown levels of trees and its relation to insect control. *J. Econ. Entomol.* **31**: 631–632.

Powell, R. A., and R. P. Adams. 1973. Seasonal variation in the volatile terpenoids of *Juniperus scopulorum* (Cupressaceae). *Am. J. Bot.* **60**: 1041–1050.

Prebble, M. L. 1975. "Aerial Control of Forest Insects in Canada." *Information Canada, Dep. Environ.*, Ottawa. Canada.

Prebble, M. L., and K. Graham. 1957. Studies of attack of ambrosia beetles in softwood logs on Vancouver Island, British Columbia. *For. Sci.* **3**: 90–112.

Price, P. W. 1975. "Insect Ecology." Wiley, New York.

Pschorn-Walcher, H. 1977. Biological control of forest insects. *Annu. Rev. Entomol.* **22**: 1–22.

Puritch, G. S. 1971. Water permeability of the wood of grand fir (*Abies grandis* (Doug.) Lindl.) in relation to infestation by the balsam woolly aphid, *Adelges piceae* (Ratz.). *J. Exp. Bot.* **22**: 936–945.

Puritch, G. S., and R. P. C. Johnson. 1971. Effects of infestation by the balsam woolly aphid, *Adelges piceae* (Ratz.), on the ultrastructure of bordered-pit membranes of grand fir, *Abies grandis* (Doug.) Lindl. *J. Exp. Bot.* **22**: 953–958.

Puritch, G. S. and J. A. Petty. 1971. Effect of balsam woolly aphid, *Adelges piceae* (Ratz.), infestation on the xylem of *Abies grandis* (Doug.) Lindl. *J. Exp. Bot.* **22**: 946–952.

Puritch, G. S. and M. Talmon-De L'Armee. 1971. Effect of balsam woolly aphid, *Adelges piceae*, infestation on the food reserves of grand fir, *Abies grandis*. *Can. J. Bot.* **49**: 1219–1223.

Radwan, M. A. and W. D. Ellis. 1975. Clonal variation in monoterpene hydrocarbons of vapors of Douglas-fir foliage. *For. Sci.* **21**: 63–67.

Rafes, P. M. 1970. Estimation of the effects of phytophagous insects on forest production. *In* "Analysis of Temperate Forest Ecosystems," (D. E. Reichle, ed.), pp. 100–106. Springer-Verlag, Berlin and New York.

Rafes, P. M. 1971. Pests and the damage which they cause to forests. *In* "Productivity of Forest Ecosystems," (P. Duvigneaud, ed.) UNESCO, Paris. pp. 357–367.

Rafes, P. M., and Y. I. Ginenko. 1973. The survival of leaf-eating caterpillars (Lepidoptera) as related to their behavior. *Entomol. Rev.* **52**: 204–211.

Rafes, P. M., and V. K. Sokolov. 1976. Intersection of background foliage pests with food trees. *Dokl. Biol. Sci.* (Engl. Transl.) **228**: 179–180.

Rafes, P. M., Y. I. Ginenko, and V. K. Sokolov. 1972. The interaction between trees and leaf eating insects. *Byull. MOIP otd. biol.* **77**: 8–19.

Raimo, B., R. C. Reardon, and J. D. Podgwaite. 1977. Vectoring gypsy moth nuclear polyhedrosis virus by *Apanteles melanoscelus* (Hym. : Braconidae). *Entomophaga* **22**: 207–215.

Randall, A. P. 1975. Technology of aerial control: application technology. *In* "Aerial Control of Forest Insects in Canada" (M. L. Prebble, ed.), pp. 34–55. Information Canada, Ottawa.

Randall, C. F., and M. F. Heisley. 1933. Our forests: what they are and what they mean to us. *USDA Misc. Publ.* No. 162.

Raske, A. G., and D. G. Bryant. 1976. Distribution of overwintering birch casebearer larvae, *Coleophora fuscedinella*, on white birch (Lepidoptera : Coleophoridae). *Can. Entomol.* **108**: 407–414.

Readshaw, J. L. 1965. A theory of phasmid outbreak release. *Aust. J. Zool.* **13**: 475–490.

Reardon, R. C. 1981. Rearing, evaluating, and attempting to establish exotic species and redistributing established species. pp. 348–351. *In* "The Gypsy Moth: Research Toward Integrated Pest Management" *USDA For. Serv. Tech. Bull. No.* 1584.

Reardon, R. C., and J. D. Podgwaite. 1976. Disease parasitoid relationships in natural populations of *Lymantria dispar* (Lep. : Lymantriidae) in the northeastern United States. *Entomophaga* **21**: 333–341.

Redmond, D. R. 1959. Mortality of rootlets in balsam fir defoliated by the spruce budworm. *For. Sci.* **5**: 64–69.

Reeks, W. A. 1956. Sequential sampling for larvae of the winter moth, *Operophtera brumata* (L.). *Can. Entomol.* **88**: 241–246.

Reeks, W. A., and G. W. Barter. 1951. Growth reduction and mortality of spruce caused by the European spruce sawfly, *Gilpinia hercyniae* (Htg.) (Hymenoptera : Diprionidae). *For. Chron.* **27**: 140–156.

Reichle, D. E., and D. A. Crossley, Jr. 1967. Investigation on heterotrophic productivity in forest insect communities. *In* "Secondary Productivity of Terrestrial Ecosystems" (K. Petrusewicz, ed.), pp. 563–587. Pa'nstwowe Wydawn, Warsaw-Krakow, Poland.

Reichle, D. E., R. A. Goldstein, R. I. Van Hook, Jr., and G. J. Hodson. 1973. Analysis of insect consumption in a forest canopy. *Ecology* **54**: 1076–1084.

Reid, R. W. 1961. Moisture changes in lodgepole pine before and after attack by the mountain pine beetle. *For. Chron.* **37**: 368–375.

Reid, R. W., H. S. Whitney, and J. A. Watson. 1967. Reactions of lodgepole pine to attack by *Dendroctonus ponderosae* Hopkins and blue stain fungi. *Can. J. Bot.* **45**: 1115–1126.

Reil, W. O., and J. A. Beutel. 1976. A pressure machine for injecting trees. *Calif. Agric.* **30**: 4–5.

Renwick, J. A. A. 1970. Chemical aspects of bark beetle aggregation. *Contrib. Boyce Thompson Inst.* **24**: 337–342.

Renwick, J. A. A., P. R. Hughes, and I. S. Krull. 1976. Selective production of *cis*- and *trans*-verbenol from (-)- and (-)-α-pinene by a bark beetle. *Science* **191**: 199–201.

Renwick, J. A. A., and J. P. Vite. 1968. Isolation of the population aggregating pheromone of the southern pine beetle. *Contrib. Boyce Thompson Inst.* **24**: 65–68.

Renwick, J. A. A., and J. P. Vite. 1969. Bark beetle attractants: mechanism of colonization by *Dendroctonus frontalis*. *Nature (London)* **224**: 1222–1223.

Renwick, J. A. A., and J. P. Vite. 1970. Systems of chemical communication in *Dendroctonus*. *Contrib. Boyce Thompson Inst.* **24**: 283–292.

Renwick, J. A. A., and J. P. Vite. 1972. Pheromones and host volatiles that govern aggregation of the six-spined engraver beetle, *Ips calligraphus*. *J. Insect Physiol.* **18**: 1215–1219.

Renwick, J. A. A., P. R. Hughes, and De J. Ty. Tanletin. 1973. Oxidation products of pinene in the bark beetle, *Dendroctonus frontalis*. *J. Insect Physiol.* **19**: 1735–1740.

Renwick, J. A. A., P. R. Hughes, and J. P. Vite. 1975. The aggregation pheromone system of a *Dendroctonus* bark beetle in Guatemala. *J. Insect Physiol.* **21**: 1097–1100.

Rexrode, C. O. 1968. Tree-wounding insects as vectors of the oak wilt fungus. *For. Sci.* **14**: 181–189.

Rexrode, C. O., and T. W. Jones. 1970. Oak bark beetle-important vectors of oak wilt. *J. For.* **68**: 294–297.

Rexrode, C. O., and J. W. Lockyer. 1974. Laboratory assay of cacodylic acid and meta-systox-R on *Scolytus multistriatus* and *Pseudopityophthorus* spp. *USDA For. Serv. Res. Note* **NE-190**.

Rhoades, D. F. 1979. Evolution of plant chemical defense against herbivores. *In* "Herbivores: Their Interaction with Secondary Plant Metabolites" (G. A. Rosenthal and D. H. Janzen, eds.), pp. 3–54. Academic Press, New York.

Richmond, H. A., and R. R. Lejeune. 1945. The deterioration of fire-killed white spruce by wood-boring insects in northern Saskatchewan. *For. Chron.* **21**: 168–192.

Richmond, H. A., and W. W. Nijholt. 1972. Water misting for log protection from ambrosia beetles in B. C. *Can. For. Serv. Pac. For. Res. Cent. Inf. Rep.* **BC-P-4-72**.

Ringold, G. B., P. J. Gravelle, D. Miller, M. M. Furniss, and M. D. McGregor. 1975. Characteristics of Douglas-fir beetle infestation in northern Idaho resulting from treatment with Douglure. *USDA For. Serv. Res. Note* **INT-189**. 10 p.

Ritchie, G. A., and T. M. Hinckley. 1975. The pressure chamber as an instrument for ecological research. *Adv. Ecol. Res.* **9**: 165–254.

Romoser, W. S. 1973. "The Science of Entomology." Macmillan, New York.

Rose, A. H. 1958. The effect of defoliation on foliage production and radial growth of quaking aspen. *For. Sci.* **4**: 335–342.

Rosenthal, S. S. and C. S. Koehler. 1971. Intertree distributions of some cynipid (Hymenoptera) galls on *Quercus lobata*. *Ann. Entomol. Soc. Am.* **64**: 571–574.

Ross, D. A. 1960. Damage by long-horned wood borers in fire-killed white spruce, central British Columbia. *For. Chron.* **36**: 355–361.

Ross, D. A. and S. Ilnytzky. 1977. The black army cutworm, *Acetebia fennica* (Tauscher), in British Columbia. *Can. For. Serv. Pac. For. Res. Cent. Rep.* **BC-X-154**.

Ross, H. H. 1965. "A Textbook of Entomology." Wiley, New York.

Roton, L. M. 1978. Mites phoretic on the southern pine beetle: when and where they attack. *Can. Entomol.* **110**: 557–558.

Rudinsky, J. A. 1962. Ecology of Scolytidae. *Annu. Rev. Entomol.* 7: 327–348.

Rudinsky, J. A. 1966a. Host selection and invasion by the Douglas-fir beetle, *Dendroctonus pseudotsugae* Hopkins, in coastal Douglas-fir forests. *Can. Entomol.* **98**: 98–111.

Rudinsky, J. A. 1966b. Scolytid beetles associated with Douglas-fir: response to terpenes. *Science* **152**: 218–219.

Rudinsky, J. A. 1968. Pheromone-mask by the female *Dendroctonus pseudotsugae* Hopk.: an attraction regulator. *Pan-Pac. Entomol.* **44**: 248–250.

Rukinsky, J. A. 1969. Masking of the aggregating pheromone in *Dendroctonus pseudotsugae* Hopk. *Science* **166**: 884–885.

Rudinsky, J. A. 1973. Multiple functions of the southern pine beetle pheromone verbenone. *Environ. Entomol.* **2**: 511–514.

Rudinsky, J. A., and I. Schneider. 1969. Effects of light intensity on the flight pattern of two *Gnathotrichus* (Coleoptera : Scolytidae) species. *Can. Entomol.* **101**: 1248–1255.

Rudinsky, J. A., M. E. Morgan, L. M. Libbey, and T. B. Putnam. 1974a. Additional components of the Douglas-fir beetle (Col. : Scolytidae) aggregative pheromone and their possible utility in pest control. *Z. Angew. Entomol.* **76**: 65–77.

Rudinsky, J. A., M. E. Morgan, L. M. Libbey, and T. B. Putnam. 1974b. Antiaggregative rivalry pheromone of the mountain pine beetle, and a new arrestant of the southern pine beetle. *Environ. Entomol.* **3**: 90–98.

Rudinsky, J. A., L. N. Kline, and J. D. Diekman. 1975. Response inhibition by four analogues of MCH, an antiaggregative pheromone of the Douglas-fir beetle. *J. Econ. Entomol.* **68**: 527–528.

Ruesink, W. G., and M. Kogan. 1975. The quantitative basis of pest management: sampling and measuring. *In* "Introduction to Insect Pest Management" (R. L. Metcalf and W. H. Luckmann, eds.), pp. 309–351. Wiley-Interscience, New York.

Ruth, D. S., and A. F. Hedlin. 1974. Temperature of Douglas-fir seeds to control the seed chalcid *Megastigmus spermotrophus* Wachtl. *Can. J. For. Res.* **4**: 441–445.

Safranyik, L., D. M. Shrimpton, and H. S. Whitney. 1974. Management of lodgepole pine to reduce losses from the mountain pine beetle *Can. For. Serv. Techn. Rep.* No. 1.

Sakai, T., and M. Sakai. 1972. Essential oil of the genus *Pseudotsuga*. 2. High boiling constituents in the wood oil of Douglas-fir. *In Abstr. Pap. Symp. Terpene Perfum. Essent. Oil Chem., 16th,* 1971. (J. A. Rudinsky, ed.), pp. 26–28.

Salman, K. A., and J. W. Bongberg. 1942. Logging high risk trees to control insects in the pine stands of northeastern California. *J. For.* **40**: 533–539.

Salonius, P. O., C. C. Smith, H. Prene, and M. K. Mahandrappa. 1977. Relationships between air and ground temperatures in spruce and fir forests. *Can. For. Serv. Dep. Fish Environ. Inf.* **M-X-77**.

Sander, I. L., and R. E. Phares. 1976. Reducing the impact of oak borers by silvicultural methods. *In Proc. Res. Coord. Meet.: Res. Insect Borers Hardwoods, Curr. Status, Needs, Appl.* pp. 69–73.

Sanders, C. J. 1975. Factors affecting adult emergence and mating behavior of the eastern spruce budworm, *Choristoneura fumiferana* (Lepidoptera : Tortricidae). *Can. Entomol.* **107**: 967–977.

Sanders, C. J. 1978. Evaluation of sex attractant traps for monitoring spruce budworm populations (Lepidoptera : Tortricidae). *Can. Entomol.* **110**: 43–50.

Sanders, C. J., and G. S. Lucuik. 1975. Effects of photoperiod and size on flight activity and oviposition in the eastern spruce budworm (Lepidoptera: Tortricidae). *Can. Entomol.* **107**: 1289–1299.

Santamour, F. S., Jr. 1965a. Insect-induced crystallization of white pine resins. I. White-pine weevil. *USDA For. Serv. Res. Note* **NE-38**.

Santamour, F. S., Jr. 1965b. Insect-induced crystallization of white pine resins. II. White pine cone beetle. *USDA For. Serv. Res. Note* **NE-39**.

Santamour, F. S., Jr. 1980. Bagworms eat (almost) everything. *Arboriculture* **6**: 291–293.

Sartwell, C. 1971. Thinning ponderosa pine to prevent outbreaks of mountain pine beetle. *In* "Proc. Precommercial Thinning of Coastal and Intermountain Forests in the Pacific Northwest" (D. Baumgartner, ed.), pp. 41–52. Washington State Univ. Coop. Ext. Serv., Pullman.

Sartwell, C., and R. E. Dolph, Jr. 1976. Silvicultural and direct control of mountain pine beetle in second-growth ponderosa pine. *USDA For. Serv. Res. Note* **PNW-268**.

Sartwell, C., and R. E. Stevens. 1975. Mountain pine beetles in ponderosa pine, prospects for silvicultural control in second-growth stands. *J. For.* **73**: 136–140.

Satchell, J. E., and M. D. Mountford. 1962. A method of assessing caterpillar populations on large forest trees, using a systemic insecticide. *Ann. Appl. Biol.* **50**: 443–450.

Satoo, T. 1970. A synthesis of studies by the harvest method: primary production relations in the temperate deciduous forests of Japan. *In* "Analysis of Temperate Forest Ecosystems" (D. E. Reichle, ed.), pp. 55–72. Springer-Verlag. New York.

Satoo, T., and N. Negisi. 1961. Experiments on the effect of soil moisture on photosynthesis of conifer seedlings. *In* "Recent Advances in Botany," pp. 1317–1321. Univ. of Toronto Press, Toronto.

Saunders, J. L., and D. M. Norris, Jr. 1961. Nematode parasites and associates of the smaller European elm bark beetle, *Scolytus multistriatus* (Marsham). *Ann. Entomol. Soc. Am.* **54**: 792–797.

Saxena, K. N., and P. Khattar. 1977. Orientation of *Papilio demoleus* larvae in relation to size, distance and combination pattern of visual stimuli. *J. Insect Physiol.* **24**: 1421–1428.

Schabel, H. G. 1976. Green muscardine disease of *Hylobius pales* (Herbert) (Coleoptera : Curculionidae). *Z. Angew. Entomol.* **81**: 413–421.

Schaefer, D. A., and W. G. Whitford. 1981. Nutrient cycling by the subterranean termite *Gnathamitermes tubiformans* in a Chihuahuan Desert ecosystem. *Oecologia* **48**: 277–283.

Schaller, F. 1968. "Soil Animals." Univ. of Michigan Press, Ann Arbor.

Schedl, W. 1962. Ein beitrag zur kenntnis der pizubertragungsweise bei xylomycetophagen Scolytiden (Coleoptera). *Sitzungs ber. Oesterr. Akad. Wiss. Math.-Naturwiss. K1, Abt. 1* **171**: 363–387.

Schenk, J. A., R. C. Dosen, and D. M. Benjamin. 1957. Noncommercial thinning of stagnated jack pine stands and losses attributable to bark beetles. *J. For.* **55**: 838–841.

Schenk, J. A., J. A. Moore, D. L. Adams, and R. L. Mahoney. 1977. A preliminary hazard rating of grand fir stands for mortality by the fir engraver. *For. Sci.* **23**: 103–110.

Schimitschek, Von E., and E. Wienke. 1966. Untersuchungen uber die Befallsbereitachaft von baumarten fur sekundarschadlinge, III. Teil. *Z. Angew. Entomol.* **58**: 398–441.

Schmid, J. M. 1970. *Enoclerus sphegeus* (Coleoptera : Cleridae), a predator of *Dendroctonus ponderosae* (Coleoptera : Scolytidae) in the Black Hills. *Can. Entomol.* **102**: 969–977.

Schmid, J. M. 1976. Temperatures, growth, and fall of needles of Engelmann spruce infested by spruce beetles. *USDA For. Serv. Res. Note* **RM-331**: 1–4.

Schmid, J. M., and R. H. Frye. 1976. Stand rating for spruce beetles. *USDA For. Serv. Res. Note* **RM-309**.

Schmid, J. M., J. C. Mitchell, K. D. Carlin, and M. R. Wagner. 1984. Insect damage, cone dimensions, and seed production in crown levels of ponderosa pines. *Great Basin Nat.* **44**: 575–578.

Schmidt, W. C. 1978. Some biological and physical responses to forest stand density. *In Proc. World For. Congr., 8th*, No. 25.

Schmitt, D. M., D. G. Grimble, and J. L. Searcy. 1984. Managing the Spruce Budworm in Eastern North America. *US Dep. Agric. Handb.* No. 620.

Schneider, I., and J. A. Rudinsky. 1969. The site of pheromone production in *Trypodendron lineatum* (Coleoptera : Scolytidae): bio-assay and histological studies of the hindgut. *Can. Entomol.* **101**: 1181–1186.

Schoeneweiss, D. F. 1975. Predisposition, stress and plant disease. *Annu. Rev. Phytopath.* **13**: 193–211.

Scholander, P. F., H. T. Hammel, E. D. Bradstreet, and E. A. Hemmingsen. 1965. Sap pressure in vascular plants. *Science* **148**: 339–346.

Schooley, H. O. 1976. Recovery of young balsam fir trees damaged by balsam woolly aphid. *For. Chron.* **52**: 1–2.

Schooley, H. O. 1978. Effects of spruce budworm on cone production by balsam fir. *For. Chron.* **54**: 298–301.

Schoonhoven, L. M. 1972a. Plant recognition by lepidopterous larvae. *In* "Insect/Plant Relationships" (H. F. Van Emden, ed.), pp. 87–99. *Symp. R. Entomol. Soc. London*, No. 6, Blackwell, Oxford.

Schoonhoven, L. M. 1972b. Some aspects of host selection and feeding in phytophagous insects. *In* "Insect and Mite Nutrition" (J. G. Rodriguez, ed.), pp. 557–566. North-Holland Publ., Amsterdam.

Schoonhoven, L. M. 1977. Chemosensory systems and feeding behaviour in phytophagous insects. *Colloq. Int. C.N.R.S.* No. 265: 391–398.

Schopmeyer, C. S. 1974. Seeds of woody plants in the United States. US *Dep. Agric. Handb.* No. 450.

Schowalter, T. D. 1985. Adaptations of insects to disturbance. *In* "The Ecology of Natural Disturbance and Patch Dynamics" (S. T. A. Pickett and P. S. White, eds.), pp. 235–252. Academic Press, New York.

Schowalter, T. D. 1986a. *Lepesoma lecontei* (Coleoptera : Curculionidae): an agent of cone abortion in a Douglas-fir seed orchard in western Oregon. *J. Econ. Entomol.* **79**: 843–846.

Schowalter, T. D. 1986b. Ecological strategies of forest insects: the need for a community-level approach to reforestation. *New Forests* **1**: 57–66.

Schowalter, T. D., and D. A. Crossley, Jr. 1983. Forest canopy arthropods as sodium, potassium, magnesium, and calcium sinks in forests. *For. Ecol. Manage.* **7**: 143–148.

Schowalter, T. D., and D. A. Crossley, Jr. 1987. Canopy arthropods and their response to forest disturbance. *In* "Forest Hydrology and Ecology at Coweeta" (D. A. Crossley, Jr., and W. T. Swank, eds.) pp. 207–218. Springer-Verlag, New York.

Schowalter, T. D., J. W. Webb, and D. A. Crossley, Jr. 1981a. Community structure and nutrient content of canopy arthropods in clearcut and uncut forest ecosystems. *Ecology* **62**: 1010–1019.

Schowalter, T. D., R. N. Coulson, and D. A. Crossley, Jr. 1981b. Role of southern pine beetle and fire in maintenance of structure and function of the southeastern coniferous forest. *Environ. Entomol.* **10**: 821–825.

Schowalter, T. D., W. W. Hargrove, and D. A. Crossley, Jr. 1986a. Herbivory in forested ecosystems. *Annu. Rev. Entomol.* **31**: 177–196.

Schowalter, T. D., D. L. Overhulser, A. Kanaskie, J. D. Stein, and J. Sexton. 1986b. *Lygus hesperus* as an agent of apical bud abortion in Douglas-fir nurseries in western Oregon. *New Forests* **1**: 5–15.

Schreiner, E. J. 1960. Objectives of pest-resistance improvement in forest trees and their possible attainment. *Proc. World For. Congr., 5th*, pp. 1–8.

Schultz, J. C., P. J. Nothnagle, and I. T. Baldwin. 1982. Seasonal and individual variation in leaf quality of two northern hardwood tree species. *Am. J. Bot.* **69**: 753–759.

Schwenke, W. 1968. New indications of the dependence of population increases of leaf and needle feeding forest insects on the sugar content of their diet. *Z. Angew. Entomol.* **61**: 365–369.

Schwerdtfeger, F. 1941. Uber die ursachen des massenwechsels der insekten. *Z. Angew. Entomol.* **28**: 254–303.

Scott, D. W. and R. I. Gara. 1975. Antennal sensory organs of two melanophila species (Coleoptera : Buprestidae). *Ann. Entomol. Soc. Am.* **68**: 842–846.

Scriber, J. M. 1975. Comparative nutritional ecology of herbivorous insects. Generalized and specialized feeding strategies in the Papilionidae and Saturniidae (Lepidoptera). Ph.D. Thesis. Cornell Univ., Ithaca, New York.

Scriber, J. M. 1979. Post-ingestive utilization of plant biomass and nitrogen by Lepidoptera legume feeding by the southern armyworm. *J. N.Y. Entomol. Soc.* **87**: 141–153.

Scriber, J. M. 1982. Nitrogen nutrition of plants and insect invasion. *In* "Nitrogen in Crop Protection" (R. D. Hauck, ed.), pp. 441–460. *Am. Soc. Agron.*

Seastedt, T. R. and D. A. Crossley, Jr. 1981. Microarthropod response following cable logging and clear-cutting in the southern Appalachians. *Ecology* **62**: 126–135.

Seastedt, T. R. and C. M. Tate. 1981. Decomposition rates and nutrient contents of arthropod remains in forest litter. *Ecology* **62**: 13–19.

Sen-Sarma, P. K. 1982. Insect vectors of sandal spike disease. *Eur. J. For. Pathol.* **12**: 297–299.

Sheehan, K. A., N. L. Crookston, W. P. Kemp, and J. J. Colbert. 1986. Modeling budworm and its hosts. *In* Western spruce budworm. *US Dep. Agric. Tech. Bull.* No. 1694: 96–116.

Shepherd, R. F. 1959. Phytosociological and environmental characteristics of outbreak and non-outbreak areas of the two year cycle spruce budworm, *Choristoneura fumiferana.* *Ecology* **40**: 608–620.

Shepherd, R. F. and C. E. Brown. 1971. Sequential egg-band sampling and probability methods of predicting defoliation by *Malacosoma disstria* (Lasiocampidae : Lepidoptera). *Can. Entomol.* **103**: 1371–1379.

Shigo, A. L. 1973. Insect and disease control: forest fertilization relations. *In* Forest Fertility Symp. Proc. *USDA For. Serv. Gen. Tech. Rep.* **NE-3**: 117–121.

Shigo, A. L. and G. Yelenosky. 1963. Fungus and insect injury to yellow birch seeds and seedlings. *USDA For. Serv. Res. Pap.* **NE-11**.

Shirley, H. L. 1929. Light requirements and silvicultural practice. *J. For.* **27**: 535–538.

Shook, R. S. and P. H. Baldwin. 1970. Woodpecker predation on bark beetles in Engelmann spruce logs as related to stand density. *Can. Entomol.* **102**: 1345–1354.

Silver, G. T. 1968. Studies on the Sitka spruce weevil, *Pissodes sitchensis,* in British Columbia. *Can. Entomol.* **100**: 93–110.

Simkover, H. G. and R. D. Shenefelt. 1952. Phytotoxicity of some insecticides to coniferous seedlings with particular reference to benzene-hexachloride. *J. Econ. Entomol.* **45**: 11–15.

Simmons, G. A., J. Mahar, M. K. Kennedy, and J. Ball. 1977. Preliminary test of prescribed burning for control of maple leaf cutter (Lepidoptera : Incurvariidae). *Great Lakes Entomol.* **10**: 209–210.

Sinclair, W. A., and R. J. Campana, eds. 1978. Dutch elm disease perspectives after 60 years. *Cornell Univ. Agric. Exp. Stn. Search Agric.* **8**: 52.

Sloan, N. F., and H. C. Coppel. 1968. Ecological implications of bird predators on the larch casebearer in Wisconsin. *J. Econ. Entomol.* **61**: 1067–1070.

Smalley, E. B., and A. G. Kais. 1964. Seasonal variations in the resistance of various elm species to Dutch elm disease. *In* "Breeding Pest-Resistant Trees" (H. D. Gerhold, E.J. Schreiner, R. E. McDermott, and J. A. Winieski, eds.), pp. 279–287. Pergamon Press, New York.

Smelyanets, V. P. 1977a. Mechanisms of plant resistance in pine trees, *Pinus sylvestris* 1. Indicators of physiological state in interacting plant–insect populations. *Z. Angew. Entomol.* **83**: 225–233.

Smelyanets, V. P. 1977b. Mechanisms of plant resistance in Scotch pine (*Pinus sylvestris*) 4. Influence of food quality on physiological state of pine pests (trophic preferendum). *Z. Angew. Entomol.* **84**: 232–241.

Smirnoff, W. A. 1959. Predators of *Neodiprion swainei* Midd. (Hymenoptera : Tenthredinidae), larval vectors of virus diseases. *Can. Entomol.* **91**: 246–248.

Smirnoff, W. A. 1961. Predators of larvae of *Neodiprion swainei* Midd. (Hymenoptera : Tenthredinidae). *Can. Entomol.* **93**: 272–275.

Smirnoff, W. A. 1974a. Three years of aerial field experiments with *Bacillus thuringiensis* plus chitinase formulation against the spruce budworm. *J. Invert. Pathol.* **24**: 344–348.

Smirnoff, W. A. 1974b. Sensibilite de *Lambdina fiscellaria fiscellaria* (Lepidoptera : Geometridae), a l'infection par *Bacillus thuringiensis* Berliner seul ou en presence de chitinase. *Can. Entomol.* **106**: 429–432.

Smirnoff, W. A., J. J. Fettes, and R. Desaulniers. 1973. Aerial spraying of a *Bacillus thuringiensis*-chitinase formulation for control of the spruce budworm (Lepidoptera: Tortricidae). *Can. Entomol.* **105**: 1535–1544.

Smirnov, O. V. 1976. Mixed virus infections on insects. *Entomol. Rev.* **55**: 145–150.

Smith, C. C. 1952. The life history and galls of a spruce gall midge, *Phytophaga picene* Felt (Diptera : Cecidomyiidae). *Can. Entomol.* **84**: 272–275.

Smith, F. F., and R. G. Linderman. 1974. Damage to ornamental trees and shrubs resulting from oviposition by periodical cicada. *Environ. Entomol.* **3**: 725–732.

Smith, H. S. 1935. The role of biotic factors in the determination of population densities. *J. Econ. Entomol.* **28**: 873–898.

Smith, L. S. 1966. "Ecology and Field Biology." Harper, New York.

Smith, R. H. 1960. Resistance of pines to the pine reproduction weevil, *Cylindrocapturus eatoni*. *J. Econ. Entomol.* **53**: 1044–1048.

Smith, R. H. 1963. Toxicity of pine resin vapors to three species of *Dendroctonus* bark beetles. *J. Econ. Entomol.* **56**: 827–831.

Smith, R. H. 1964. Variation in the monoterpenes of *Pinus ponderosa* Laws. *Science* **143**: 1337–1338.

Smith, R. H. 1966a. The monoterpene composition of *Pinus ponderosa* xylem resin and of *Dendroctonus brevicomis* pitch tubes. *For. Sci.* **12**: 63–68.

Smith, R. H. 1966b. Resin quality as a factor on the resistance of pines to bark beetles. *In* "Breeding Pest-Resistant Trees" (H. D. Gerhold, E. J. Schreiner, R. E. McDermott, and J. A. Winieski, eds.), pp. 186–196. Pergamon, New York.

Smith, R. H. 1972. Xylem resin in the resistance of the *Pinaceae* to bark beetles. *USDA For. Serv. Gen. Tech. Rep.* **PSW-1**.

Smith, R. H. 1976. Low concentration of lindane plus induced attraction traps mountain pine beetle. *USDA For. Serv. Note* **PSW-316.**

Smith, R. H., B. E. Wickman, R. C. Hall, C. J. DeMars, and G. T. Ferrell. 1981. The California pine risk-rating system: its development, use, and relationship to other systems. *In* "Hazard-Rating Systems" (R. L. Hedden, S. J. Barras, and J. E. Coster, tech. coords.) *USDA For. Serv. Gen. Tech. Rep.* **WO-27**: 53–69.

Smith, V. G. 1928. Animal communities of the deciduous forest succession. *Ecology* **9**: 479–500.

Smith, V. G. 1930. The tree layer society of maple-red oak climax forest. *Ecology* **11**: 601–606.

Smyth, A. V. 1959. The Douglas-fir bark beetle epidemic in the Millicoma forest: methods used for control and salvage. *J. For.* **57**: 278–280.

Snodgrass, R. E. 1935. "Principles of Insect Morphology." McGraw-Hill, New York.

Solomon, J. D. 1969. Woodpecker predation on insect borers in living hardwoods. *Ann. Entomol. Soc. Am.* **62**: 1214–1215.

Solomon, J. D. 1977. Frass characteristics for identifying insect borers in living hardwoods. *Can. Entomol.* **109**: 295–303.

Solomon, J. D. 1978. Impact and control of insect borers in southern hardwoods. *Proc. Symp. South East. Hardwods, 2nd*, pp. 94–100.

Southwood, T. R. E. 1960. The abundance of the Hawaiian trees and the number of their associated insect species. *Proc. Hawaii Entomol. Soc.* **17**: 299–303.

Southwood, T. R. E. 1961. The number of species of insects associated with various trees. *J. Anim. Ecol.* **30**: 1–8.

Southwood, T. R. E. 1966. "Ecological Methods, with Particular Reference to the Study of Insect Populations." Wiley, New York.

Southwood, T. R. E. 1978. "Ecological Methods with Particular Reference to the Study of Insect Population," 2nd Ed. Chapman & Hall. London.

Southwood, T. R. E., and G. A. Norton. 1972. Economic aspect of pest management strategies and decisions. *In* "Insects: Studies in Population Management" (P. W. Geier, L. R. Clark, D. J. Anderson, and N. A. Nix, eds.), *Ecol. Soc. Aust. Mem.* **1**: 168–181.

Speers, C. F. 1958. Pales weevil rapidly becoming serious pest of pine reproduction in the South. *J. For.* **56**: 723–726.

Spurr, S. H., and R. B. Friend. 1941. Compression wood in weeviled northern white pine. *J. For.* **39**: 1005–1006.

Spurr, S. H., E. W. Littlefield, and H. A. Bess. 1947. Ecology of susceptible and resistant forest types and stands. *Hawaii For. Bull.* No. 22: 27–39.

Squillace, A. E. 1971. Inheritance of monoterpene composition in cortical oleoresin of slash pine. *For. Sci.* **17**: 381–387.

Squillace, A. E. 1977. Use of monoterpene composition in forest genetics research with slash pine. *Proc. South. For. Tree Improv. Conf., 14th*, pp. 227–235.

Stage, A. R. 1973. Prognosis model for stand development. *USDA For. Serv. Res. Pap.* **INT-137**.

Stairs, G. R. 1971. Uses of viruses for microbial control of insects. *In* "Microbial Control of Insects and Mites" (H. D. Burges, and N. W. Hussey, eds.), pp. 97–124. Academic Press, London.

Stairs, G. R. 1973. Means for regulation: viruses. *In* "Regulation of Insect Populations by Microorganisms" (L. A. Bulla, Jr., ed.) *Ann. N.Y. Acad. Sci.* **217**: 58–64.

Stark, R. W. 1952. Sequential sampling of the lodgepole needle miner. *For. Chron.* **28**: 57–60.

Stark, R. W. 1965. Recent trends in forest entomology. *Annu. Rev. Entomol.* **10**: 303–324.

Stark, R. W., and J. A. Cook. 1957. The effects of defoliation by the lodgepole needle miner. *For. Sci.* **3**: 376–396.

Stark, R. W., and W. E. Waters. 1985. Concept and structure of a forest pest management system. *In* "Integrated Pest Management in Pine-Bark Beetle Ecosystems" (W. G. Waters, R. W. Stark and D. L. Wood, eds.), pp. 49–60. Wiley Inter-Science, New York.

Stark, R. W., P. R. Miller, E. W. Cobb, D. L. Wood, and J. R. Parmeter, Jr. 1968. Incidence of bark beetle infestation on injured trees. *In* "Photochemical oxidant injury and bark beetle (Coleoptera : Scolytidae) infestation of ponderosa pine." *Hilgardia* **39**: 121–123.

Stein, J. D., and D. J. Doran. 1975. A nondestructive method of whole-tree sampling for spring cankerworm. *USDA For. Serv. Res. Note* **RM-290**.

Stelzer, M. J., J. Neisess, and C. G. Thompson. 1975. Aerial applications of a nucleopolyhedrosis virus and *Bacillus thuringiensis* against the Douglas-fir tussock moth. *J. Econ. Entomol.* **68**: 269–272.

Stepanov, E. V. 1976a. Volatile terpenic metabolites of Siberian fir and the resistance of fir to injurious factors. *Sov. J. Ecol.* (Engl. Trans.) **7**: 408–414.

Stepanov, E. V. 1976b. Volatile terpenic metabolites of various organs of forest forming coniferous species. *Sov. J. Ecol.* (Engl. Trans.) **7**: 414–419.

Stephen, F. M. and G. W. Wallis. 1978. Tip damage as an index of Nantucket pine tip moth population density (Lepidoptera : Tortricidae). *Can. Entomol.* **110**: 917–920.

Stephens, G. R. 1971. The relation of insect defoliation to mortality in Connecticut forests. *Bull. Conn. Agric. Exp. Stn., New Haven*, No. 723.

Stephens, G. R., N. C. Turner, and H. C. De Roo. 1972. Some effects of defoliation by gypsy moth (*Porthetria dispar* L.) and elm spanworm (*Ennomos subsignarius* Hbn.) on water balance and growth of deciduous forest trees. *For. Sci.* **18**: 326–330.

Sterner, T. E. 1976. Dutch elm disease vector populations are low within Fredericton, N. B., sanitation area. *Can. For. Serv., Bi-Mon. Res. Notes* **43**(4): 20.

Stevens, R. E., and F. G. Hawksworth. 1984. Insect-dwarf mistletoe associations: an update. *In* "Proceedings of biology of dwarf mistletoe." *USDA For. Serv. Gen. Tech. Rep.* **RM-111**: 94–100.

Stevens, R. E., D. B. Cahill, K. Lister, and G. E. Metcalf. 1974a. Timing cacodylic acid treatments for control of mountain pine beetles in infested ponderosa pines. *USDA For. Serv. Res. Note* **RM-262**.

Stevens, R. E., C. A. Myers, W. F. McCambridge, G. L. Downing, and J. G. Laut. 1974b. Mountain pine beetle in front range ponderosa pine: what it's doing and how to control it. *USDA For. Serv. Gen. Tech. Rep.* **RM-7**.

St. George, R. A. 1930. Drought affected and injured trees attractive to bark beetles. *J. Econ. Entomol.* **23**: 825–828.

Stillwell, M. A. 1960. Decay associated with woodwasps in balsam fir weakened by insect attack. *For. Sci.* **6**: 225–231.

St. Julian, G., L. A. Bulla, Jr., E. S. Sharpe, and G. L. Adams. 1973. Bacteria, spirochetes, and rickettsia as insecticides. *In* "Regulation of insect populations by microorganisms" (L. A. Bulla, Jr., ed.) *Ann. N.Y. Acad. Sci.* **217**: 65–75.

Stoetzel, M. B., and J. A. Davidson. 1973. Life history variations of the obscure scale (Homoptera : Diaspididae) on pin oak and white oak in Maryland. *Ann. Entomol. Soc. Am.* **66**: 308–311.

Strong, D. R. 1974a. Nonasymptotic species richness models and the insects of British trees. *Proc. Natl. Acad. Sci. U.S.A.* **73**: 2766–2769.

Strong, D. R. 1974b. Rapid asymptotic species accumulation in phytophagous insect communities: the pests of cacao. *Science* **185**: 1064–1066.

Strong, D. R., Jr., E. D. McCoy, and J. R. Rey. 1977. Time and the number of herbivore species: the pests of sugarcane. *Ecology* **58**: 167–175.

Sturgeon, K. B. 1979. Monoterpene variation in ponderosa pine xylem resin related to western pine beetle predation. *Evolution* **33**: 803–814.

Sukachev, V., and N. Dylis. 1964. "Fundamentals of Forest Biogeocoenology." Oliver & Boyd, Edinburgh.

Sullivan, C. R. 1961. The effect of weather and the physical attributes of white pine leaders on the behaviour and survival of the white pine weevil, *Pissodes strobi* Peck, in mixed stands. *Can. Entomol.* **93**: 721–741.

Sullivan, C. R., and W. G. Wellington. 1953. The light reactions of larvae of the tent caterpillars, *Malacosoma disstria* Hbn., *M. americanum* (Fab.), and *M. pluviale* (Dyar) (Lepidoptera: Lasiocampidae). *Can. Entomol.* **85**: 297–310.

Sumimoto, M., T. Kondo, and Y. Kamiyama. 1974. Attractants for the scolytid beetle *C. fulvus*. *J. Insect Physiol.* **20**: 2071–2077.

Summers, J. N. 1922. Effect of low temperature on the hatching of gypsy-moth eggs. U. S. *Dep. Agric. Bull.* No. 1080.

Sutton, R. F., and E. L. Stone, Jr. 1974. White grubs: a description for foresters, and an evaluation of their silvicultural significance. *Can. For. Serv. Dep. Environ. Inf. Rep.* **O-X-212**.

Swaine, J. M. 1933. The relation of insect activities to forest development as exemplified in the forests of eastern North America. *Sci. Agric.* **14**: 8–31.

Swetnam, T. W., M. A. Thompson, and E. K. Sutherland. 1985. Using dendrochronology to measure radial growth of defoliated trees. U. S. *Dep. Agric. For. Serv. Agric. Handb.* No. 639.

Tadaki, Y. 1966. *Bull. Gov. For. Exp. Stn. Tokyo* **184**: 135.

Tagestad, A. D. 1975. Spruce budworm detection in North Dakota shelterbelts and nurseries with a synthetic sex attractant. *N.D. Farm Res.* **33**: 17–19.

Takekawa, J. Y. and E. O. Garton. 1984. How much is an evening grosbeak worth? *J. For.* **82**: 426–428.

Tanada, Y. 1959. Microbial control of insect pests. *Annu. Rev. Entomol.* **4**: 277–302.

Tanada, Y. 1967. Microbial pesticides. *In* "Pest Control. Biological, Physical and Selected Chemical Methods" (W. W. Kilgore and R. L. Doutt, eds.), pp. 31–88. Academic Press, New York.

Tanada, Y. 1973. Environmental factors external to the host. *In* "Regulation of insect populations by microorganisms" (L. A. Bulla. Jr., ed.) *Ann. N. Y. Acad. Sci.* **217**: 120–130.

Tanada, Y. 1976. Ecology of insect viruses. *In* "Perspectives in Forest Entomology" (J. F. Anderson and H. K. Kaya, eds.) pp. 215–283. Academic Press, New York.

Tappeiner, J. C. 1969. Effect of cone production on branch, needle and xylem ring growth of Sierra-Nevada Douglas-fir. *For. Sci.* **15**: 171–174.

Tatro, V. E., R. W. Scora, F. C. Vasek, and J. Kumamoto. 1973. Variation in the leaf oils of three species of *Juniperus*. *Am. J. Bot.* **60**: 236–241.

Tauber, M. J., and C. A. Tauber. 1976. Insect seasonality: diapause maintenance, termination and postdiapause development. *Annu. Rev. Entomol.* **21**: 81–107.

Taylor, G. S., and R. E. B. Moore. 1979. A canker of red maple associated with oviposition by the narrow-winged tree cricket. *Phytopathology* **69**: 236–239.

Taylor, L. F., J. W. Apple, and K. C. Berger. 1953. Response of certain insects to plants grown on varying fertility levels. *J. Econ. Entomol.* **61**: 843–848.

Thatcher, R. C. 1960. Influence of the pitch eating weevil on pine regeneration in east Texas. *For. Sci.* **6**: 354–361.

Thatcher, T. O. 1961. "Forest Entomology." Burgess, Minneapolis, Minnesota.

Thofelt, L. 1975. Studies on leaf temperature recorded by direct measurement and by thermography. Ph.D. Thesis Uppsala Univ., Uppsala, Sweden.

Thompson, W. R. 1929. On natural control. *Parasitology* **21**: 269–281.

Thomson, A. J., R. F. Shepherd, J. W. E. Harris, and R. H. Silverside. 1984. Relating weather to outbreaks of western spruce budworm, *Choristoneura occidentalis* (Lepidoptera: Tortrcidae) in British Columbia. *Can. Entomol.* **116**: 375–381.

Thomson, M. G. 1958. Egg sampling for the western hemlock looper. *For. Chron.* **34**: 248–256.

Ticehurst, M. 1976. Comparison of radiographic and classical techniques to determine the condition of gypsy moth pupae. *J. Econ. Entomol.* **69**: 451–455.

Tinbergen, L. 1960. The dynamics of insect and bird populations in pine woods. *Arch. Neerl. Zool.* **13**: 265.

Tjia, B., and D. B. Houston. 1975. Phenolic constituents of Norway spruce resistant or susceptible to the eastern spruce gall aphid. *For. Sci.* **21**: 180–184.

Tostowaryk, W., and J. M. McLeod. 1972. Sequential sampling for egg clusters of the Swaine jack pine sawfly, *Neodiprion swainei* (Hymenoptera: Diprionidae). *Can. Entomol.* **104**: 1343–1347.

Trager, W. 1953. Nutrition. *In* "Insect Physiology" (K. D. Roeder, ed.), pp. 350–386. Wiley, New York.

Tripp, H. A., and A. F. Hedlin. 1956. An ecological study and damage appraisal of white spruce cone insects. *For. Chron.* **32**: 400–410.

Turnock, W. J., and J. A. Muldrew. 1971. Parasites. *In* "Toward integrated control." *USDA For. Serv. Res. Pap.* **NE-184**: 59–87.

Uvarov, B. P. 1931. Insects and climate. *Trans. Entomol. Soc. London* **79**: 1–247.

Valentine, H. T., and D. R. Houston. 1981. Stand susceptibility to gypsy moth defoliation. *In* "Hazard-Rating Systems in Forest Insect Pest Management: Symposium Proceedings" (R. L. Hedden, S. J. Barras, and J. E. Coster, tech. coords.) 137–144. *USDA For. Serv. Gen. Tech. Rep.* **WO-27**.

Valentine, H. T., W. E. Wallner, and P. M. Wargo. 1983. Nutritional changes in host foliage during and after defoliation and their relationships to the weight of gypsy moth pupae. *Oecologia* **57**: 298–302.

van Buijtenen, J. P., and F. S. Santamour, Jr. 1972. Resin crystallization related to weevil resistance in white pine (*Pinus strobus*). *Can. Entomol.* **104**: 215–219.

Vandenberg, J. D., and R. S. Soper. 1975. Isolation and identification of *Entomophthora* spp. Fres. (Phycomycetes: Entomophthoralis) from the spruce budworm *Choristoneura fumiferana* Clem. (Lepidoptera: Tortricidae). *J. N.Y. Entomol. Soc.* **83**: 254–255.

Van den Driessche, R., and J. E. Webber. 1975. Total and soluble nitrogen in Douglas-fir in relation to plant nitrogen status. *Can. J. For. Res.* **5**: 580–585.

Vander Sar, T. J. D., and J. H. Borden. 1977a. Aspects of host selection behaviour of *Pissodes strobi* (Coleoptera: Curculionidae) as revealed in laboratory feeding bioassays. *Can. J. Zool.* **55**: 405–414.

Vander Sar, T. J. D., and J. H. Borden. 1977b. Visual orientation of *Pissodes strobi* Peck

(Coleoptera : Curculionidae) in relation to host selection behaviour. *Can. J. Zool.* **55**: 2042–2049.

Van Derwerker, G. K., E. F. Cook, and H. M. Kulman. 1978. Morphology of the adult reproductive systems of the yellow headed spruce sawfly, *Pikonema alaskensis*. *Ann. Entomol. Soc. Am.* **71**: 615–618.

van Lenteren, J. C., and K. Bakker. 1976. Functional responses in invertebrates. *Neth. J. Zool.* **26**: 567–572.

van Lenteren, J. C., and K. Bakker. 1978. Behavioral aspects of the functional responses of a parasite (*Pseudocoila bochei* Weld) to its host (*Drosophila melanogaster*). *Neth. J. Zool.* **28**: 213–233.

Varty, I. W. 1963. A survey of the sucking insects of the birches in the maritime provinces. *Can. Entomol.* **95**: 1097–1106.

Varty, I. W. 1975. Side effects of pest control projects on terrestrial arthropods other than the target species. *In* "Aerial Control of Forest Insects in Canada", pp. 266–275. Information Canada. Ottawa.

Vasechko, G. I. 1978. Host selection by some bark beetles (Col. : Scolytidae). I. Studies of primary attraction with chemical stimuli. *Z. Angew. Entomol.* **85**: 66–76.

Vite, J. 1961. The influence of water supply on oleoresin exudation pressure and resistance to bark beetle attack in *Pinus ponderosa*. *Contrib. Boyce Thompson Inst.* **21**: 37–66.

Vite, J. P., and A. Bakke. 1979. Synergism between chemical and physical stimuli in host colonization by an ambrosia beetle. *Naturwissenschaften* **66**: 528–529.

Vite, J. P., and R. G. Crozier. 1968. Studies on the attack behavior of the southern pine beetle. IV. Influence of host condition on aggregation pattern. *Contrib. Boyce Thompson Inst.* **24**: 87–93.

Vite, J. P., and W. Francke. 1976. The aggregation pheromones of bark beetles: progress and problems. *Naturwissenschaften* **63**: 550–555.

Vite, J. P., and R. I. Gara. 1962. Volatile attractants from ponderosa pine attacked by bark beetles (Coleoptera : Scolytidae). *Contrib. Boyce Thompson Inst.* **21**: 251–273.

Vite, J. P., and G. B. Pitman. 1968. Bark beetle aggregation: effects of feeding on the release of pheromones in *Dendroctonus* and *Ips*. *Nature* (London) **218**: 169–170.

Vite, J. P., and J. A. A. Renwick. 1976. Applicability of bark beetle pheromones: configuration and consequences. *Z. Angew. Entomol.* **82**: 112–116.

Vite, J. P., A. Bakke, and J. A. A. Renwick. 1972. Pheromones in *Ips* (Coleoptera : Scolytidae): occurrence and production. *Can. Entomol.* **104**: 1967–1975.

Vite, J. P., D. Klimetzek, G. Loskant, R. Hedden, and K. Mori. 1976. Chirality of insect pheromones: response interruption by inactive antipodes. *Naturwissenschaften* **63**: 582–583.

Vogel, S. 1968. "Sun leaves" and "shade leaves." Differences in convective heat dissipation. *Ecology* **49**: 1203–1204.

Voigt, G. K. 1954. The effect of fungicides, herbicides and insecticides on the accumulation of phosphorus by *Pinus radiata* as determined by us of P_{32}. *J. Agronomy*. **46**: 511–513.

Von Rudloff, E. 1962. Gas-liquid chromatography of terpenes. Part V. The volatile oils of the leaves of black, white and Colorado spruce. *Tappi* **45**: 181–184.

Von Rudloff, E. 1972. Seasonal variation in the composition of the volatile oil of the leaves, buds and twigs of white spruce (*Picea glauca*). *Can. J. Bot.* **50**: 1595–1603.

Von Rudloff, E., and G. E. Rehfeldt. 1980. Chemosystematic studies in the genus *Pseudotsuga*. IV. Inheritance and geographical variations of the leaf oil terpenes of Douglas-fir from the Pacific Northwest. *Can. J. Bot.* **58**: 546–556.

Wagner, M. R. 1988. Induced defenses in ponderosa pine to defoliating insects. *In* "Mechanisms of Woody Plant Resistance to Insects and Pathogens" (W. Mattson, C. Bernard-Dagan and J. Levieux, eds.) Springer-Verlag, New York.

Wagner, M. R., T. Ikeda, D. M. Benjamin, and F. Matsumura. 1979. Host derived chemicals: the basis for preferential feeding behavior of the larch sawfly, *Pristiphora erichsonii* (Hymenoptera : Tenthredinidae) on tamarack *Larix laricina*. *Can. Entomol.* **111**: 165–169.

Wagner, T. L., and D. E. Leonard. 1979. The effects of parental and progeny diet on development, weight gain and survival of pre-diapause larvae of the satin moth, *Leucoma salicis* (Lepidoptera : Lymantriidae). *Can. Entomol.* **111**: 721–729.

Wallner, W. E., and G. S. Walton. 1979. Host defoliation: a possible determinant of gypsy moth population quality. *Ann. Entomol. Soc. Am.* **72**: 62–67.

Walter, J. A., H. M. Kulman, and A. C. Hodson. 1972. Life tables for the forest tent caterpillar. *Ann. Entomol. Soc. Am.* **65**: 25–31.

Wargo, P. M. 1975. Estimating starch content in roots of deciduous trees-a visual technique. *USDA For. Serv. Res. Pap.* **NE-313**.

Wargo, P. M., and H. R. Skutt. 1975. Resistance to pulsed electric current: an indicator of stress in forest trees. *Can. J. For. Res.* **5**: 557–561.

Wargo, P. M., J. Parker, and D. R. Houston. 1972. Starch content in roots of defoliated sugar maple. *For. Sci.* **18**: 203–204.

Waring, R. H., and B. D. Cleary. 1967. Plant moisture stress: evaluation by pressure bomb. *Science* **155**: 1248–1254.

Waring, R. H., W. G. Thies, and D. Muscato. 1980. Stem growth per unit of leaf area: A measure of tree vigor. *For. Sci.* **26**: 112–117.

Warren, G. L. 1956. The effect of some site factors on the abundance of *Hypomolyx piceus* (Coleoptera : Curculionidae). *Ecology* **37**: 132–139.

Waters, W. E. 1959. A quantitative measure of aggregation in insects. *J. Econ. Entomol.* **52**: 1180–1184.

Waters, W. E. 1974. Systems approach to managing bark beetles. *In* "Proceedings of the Southern Pine Beetle Symposium" (T. L. Payne, R. N. Coulson, and R. C. Thatcher, eds.), pp. 12–14. Tex. Agric. Exp. Stn., Texas A&M Univ., College Station.

Waters, W. E., T. McIntyre, and D. Crosby. 1955. Loss in volume of white pine in New Hampshire caused by the white pine weevil. *J. For.* **53**: 271–274.

Waters, W. E., and R. W. Stark, 1980. Forest pest management: concept and reality. *Annu. Rev. Entomol.* **25**: 479–509.

Watson, W. Y., G. R. Underwood, and J. Reid. 1960. Notes on *Matsucocus macrocicatrices* Richards (Homoptera : Margarodidae) and its association with *Septobasidium pinicola* Snell in eastern Canada. *Can. Entomol.* **92**: 662–667.

Watt, K. E. F 1968. "Ecology and Resource Management." McGraw-Hill, New York.

Way, M. J. and M. Cammell. 1970. Aggregation behaviour in relation to food utilization by aphids. *In* "Animal populations in relation to their food resources" (A. Watson, ed.), pp. 229–247, *Symp. Br. Ecol. Soc.*, Vol. 10. Blackwell, Oxford.

Weary, G. C. and H. G. Merriam. 1978. Litter decomposition in a red maple woodlot under natural conditions and under insecticide treatment. *Ecology* **59**: 180–184.

Webb, F. E. 1959. Aerial chemical control of forest insects with reference to the Canadian situation. *Can. Fish Cult.* **24**: 1–14.

Webb, W. L., and J. J. Karchesy. 1977. Starch content of Douglas-fir defoliated by the tussock moth. *Can. J. For. Res.* **7**: 186–188.

Weese, A. O. 1924. Animal ecology of an Illinois elm-maple forest. *Biol. Monogr.* **9**: 1–93.

Wellington, W. G. 1948. The light reactions of the spruce budworm, *Choristoneura fumiferana* Clemens (Lepidoptera : Tortricidae). *Can. Entomol.* **80**: 56–82.

Wellington, W. G. 1949a. The effects of temperature and moisture upon the behaviour of the spruce budworm, *Choristoneura fumiferana* Clemens (Lepidoptera : Tortricidae). I. The relative importance of graded temperatures and rates of evaporation in producing aggregations of larvae. *Sci. Agric. (Ottawa)* **29**: 201–215.

Wellington, W. G. 1949b. The effects of temperature and moisture upon the behaviour of the spruce budworm, *Choristoneura fumiferana* Clemens (Lepidoptera : Tortricidae). II. The responses of larvae to gradients of evaporation. *Sci. Agric. (Ottawa)* **29**: 216–229.

Wellington, W. G. 1950. Effects of radiation on the temperatures of insectan habitats. *Sci. Agric. (Ottawa)* **30**: 209–234.

Wellington, W. G. 1952. Air-mass climatology of Ontario north of Lake Huron and Lake

Superior before outbreaks of the spruce budworm, *Choristoneura fumiferana* (Clem.), and the forest tent caterpillar, *Malacosoma disstria* Hbn. (Lepidoptera : Tortricidae : Lasiocampidae). *Can. J. Zool.* **30**: 114–127.

Wellington, W. G. 1954a. Atmospheric circulation processes and insect ecology. *Can. Entomol.* **86**: 213–333.

Wellington, W. G. 1954b. Weather and climate in forest entomology. *Meteorol. Monogr.* **2**: 11–18.

Wellington, W. G. 1955. Solar heat and plane polarized light versus the light compass reaction in the orientation of insects on the ground. *Ann. Entomol. Soc. Am.* **48**: 67–76.

Wellington, W. G. 1957. Individual differences as a factor in population dynamics: the development of a problem. *Can. J. Zool.* **35**: 293–323.

Wellington, W. G. 1960. Qualitative changes in natural populations during changes in abundance. *Can. J. Zool.* **38**: 298–314.

Wellington, W. G. 1974. Changes in mosquito flight associated with natural changes in polarized light. *Can. Entomol.* **106**: 941–948.

Wellington, W. G., and W. R. Henson. 1947. Notes on the effects of physical factors on the spruce budworm, *Choristoneura fumiferana* (Clem.). *Can. Entomol.* **79**: 168–170.

Wellington, W. G., and R. M. Trimble. 1984. Weather. *In* "Ecological Entomology." (C. B. Huffaker and R. L. Rabb, eds.), pp. 399–425. Wiley, New York.

Wellington, W. G., J. J. Fethes, K. B. Turner, and R. M. Belyea. 1950. Physical and biological indicators of the development of outbreaks of the spruce budworm, *Choristoneura fumiferana* (Clem.) (Lepidoptera : Tortricidae). *Can. J. Res.* **28**: 308–330.

Wellington, W. G., C. R. Sullivan, and G. W. Green. 1951. Polarized light and body temperature level as orientation factors in the light reactions of some hymenopterous and lepidopterous larvae. *Can. J. Zool.* **29**: 339–351.

Wendt, R., J. Weidhass, G. J. Griffin, and J. R. Elkins. 1983. Association of *Endothia parasitica* with mites isolated from cankers on American chestnut trees. *Plant Dis.* **67**: 757–758.

Werner, R. A. 1964. White spruce seed loss caused by insects in interior Alaska. *Can. Entomol.* **96**: 1462–1464.

Werner, R. A. 1972a. Aggregation behavior of the beetle *Ips grandicollis* in response to host-produced attractants. *J. Insect Physiol.* **18**: 425–427.

Werner, R. A. 1972b. Response of the beetle *Ips grandicollis* to combinations of host and insect produced attractants. *J. Insect Physiol.* **18**: 1403–1412.

Werner, R. A. 1979. Influence of host foliage on development, survival, fecundity and oviposition of the spear-marked black moth, *Rheumaptera hastata* (Lepidoptera : Geometridae). *Can. Entomol.* **111**: 317–322.

Werner, R. A., B. H. Baker, and P. A. Rush. 1977. The spruce beetle in white spruce forests of Alaska. *USDA For. Serv. Gen. Tech. Rep.* **PNW-61.**

Weseloh, R. M. 1976. Behavioral responses of the parasite *Apanteles melanoscelus* to gypsy moth silk. *Environ. Entomol.* **5**: 1128–1132.

West, A. S., Jr. 1936. Winter mortality of larvae of the European pine shoot moth, *Rhyacionia buoliana* Schiff., in Connecticut. *Ann. Entomol. Soc. Am.* **29**: 438–448.

Westveld, M. 1945. A suggested method for rating the vulnerability of spruce-fir stands to budworm attack. *USDA For. Serv. Northeast. For. Exp. Stn.*

Westveld, M. 1954. A budworm vigor-resistance classification for spruce and balsam fir. *J. For.* **52**: 11–24.

Westveld, M., H. J. MacAloney, and J. R. Hansbrough. 1950. Forest crop security: the right tree on the right site. *For. Chron.* **26**: 144–151.

Weyh, R., and U. Maschwitz. 1978. Trail substance in larvae of *Eriogaster lanestris* L. *Naturwissenschaften* **65**: 64.

Wheeler, A. G., Jr., and T. J. Henry. 1976. Biology of the honeylocust plant bug, *Diaphnocoris chlorionis*, and other mirids associated with ornamental honeylocust. *Ann. Entomol. Soc. Am.* **69**: 1095–1104.

White, J. 1981. Flagging: hosts defenses versus oviposition strategies in periodical cicadas. *Can. Entomol.* **113**: 727–738.

White, M. G. 1962. The effect of blue stain in Scots pine (*Pinus sylvestris* L.) on growth of larvae of the house longhorn beetle (*Hylotrupes bajulus* L.). *J. Inst. Wood Sci.* **9**: 27–31.

White, T. C. R. 1970. Some aspects of the life history, host selection, dispersal and oviposition of adult *Cardiaspina densitexta* (Homoptera: Psyllidae). *Aust. J. Zool.* **18**: 105–117.

White, T. C. R. 1974. A hypothesis to explain outbreaks of looper caterpillars, with special reference to populations of *Selidosema suavis* in a plantation of *Pinus radiata* in New Zealand. *Oecologia* **16**: 279–301.

White, T. C. R. 1976. Weather, food, and plagues of locusts. *Oecologia* **22**: 119–134.

Whitham, T. G., and S. Mopper. 1985. Chronic herbivory: impacts on architecture and sex expression of pinyon pine. *Science* **228**: 1089–1091.

Whitney, H. S. 1982. Relationships between bark beetles and symbiotic organisms. In "Bark Beetles in North American Conifers. A System for the Study of Evolutionary Biology" (Jeffry B. Mitton and Kareen B. Sturgeon, eds.), pp. 183–211. Univ. of Texas Press, Austin.

Whittaker, R. H. 1952. A study of summer foliage insect communities in the Great Smokey Mountains. *Ecol. Monogr.* **22**: 1–44.

Wickman, B. E. 1976a. Douglas-fir tussock moth egg hatch and larval development in relation to phenology of grand fir and Douglas-fir in northeastern Oregon. *USDA For. Serv. Res. Pap.* **PNW-206**.

Wickman, B. E. 1976b. Phenology of white fir and Douglas-fir tussock moth egg hatch and larval development in California. *Environ. Entomol.* **5**: 316–322.

Wickman, B. E. 1977. Douglas fir tussock moth egg hatch and larval development in relation to phenology of white fir in southern Oregon. *USDA For. Serv. Res. Note* **PNW-295**.

Wickman, B. E. 1980. Increased growth of white fir after a Douglas-fir tussock moth outbreak. *J. Forestry* **78**: 31–33.

Wickman, B. E., R. R. Mason, and H. G. Paul. 1975. Flight, attraction and mating behavior of the Douglas-fir tussock moth in Oregon. *Environ. Entomol.* **4**: 405–408.

Wilford, B. H. 1937. The spruce gall aphid (*Adelges abietis* L.) on southern Michigan. *Univ. Mich. Sch. For. Conserv. Circ.* No. 2: 1–34.

Wilkinson, R. C., G. C. Becker, and D. M. Benjamin. 1966. The biology of *Neodiprion rugifrons*, a sawfly infesting jack pine in Wisconsin. *Ann. Entomol. Soc. Am.* **59**: 786–792.

Williams, C. B. 1967. Third generation pesticides. *Sci. Am.* **217**: 13–17.

Williams, L. H. and J. K. Mauldin. 1974. Anobiid beetle, *Xyletinus pellatus* (Coleoptera : Anobiidae) oviposition on various woods. *Can. Entomol.* **106**: 949–955.

Williamson, D. L., J. A. Schenk, and W. F. Barr. 1966. The biology of *Conophthorus monticolae* in northern Idaho. *For. Sci.* **12**: 234–240.

Wilson, L. F. 1959. Branch "tip" sampling for determining abundance of spruce budworm egg masses. *J. Econ. Entomol.* **52**: 618–621.

Wilson, L.F. 1966. Effects of different population levels of European pine sawfly on young Scotch pine trees. *J. Econ. Entomol.* **59**: 1043–1049.

Wilson, L. F. 1968a. Life history and habits of *Rhabdophaga* sp. (Diptera : Cecidomyiidae), a gall midge attacking willow in Michigan. *Can. Entomol.* **100**: 184–189.

Wilson, L. F. 1968b. Life history and habits of the willow beaked gall midge, *Mayetiola rigidae* (Diptera : Cecidomyiidae), in Michigan. *Can. Entomol.* **100**: 202–206.

Witcosky, J. J., T. D. Schowalter, and E. M. Hansen. 1986a. *Hylastes nigrinus* (Coleoptera: Scolytidae) and *Pissodes fasciatus* and *Steremnius carinatus* (Coleoptera: Curculionidae)as vectors of black-stain root disease of Douglas-fir. *Environ. Entomol.* **15**: 1090–1095.

Witcosky, J. J., T. D. Schowalter, and E. M. Hansen. 1986b. The influence of time of precommercial thinning on the colonization of Douglas-fir by three species of root-colonizing insects. *Can. J. For. Res.* **16**: 745–749.

Witkamp, M., and D. A. Crossley, Jr. 1966. The role of arthropods and microflora in breakdown of white oak litter. *Pedobiologia* **6**: 293–303.

Witter, J. A., and A. M. Lynch. 1985. Rating spruce-fir stands for spruce budworm damage in eastern North America. *USDA For. Serv. Coop. Stn. Res. Serv. Agric. Handbk.* No. 636.

Witter, J. A., H. M., Kulman, and A. C. Hodson. 1972. Life tables for the forest tent caterpillar. *Ann. Entomol. Soc. Am.* **65**: 25–31.

Wollam, J. D., and W. G. Yendol. 1976. Evaluation of *Bacillus thuringiensis* and a parasitoid for suppression of the gypsy moth. *J. Econ. Entomol.* **69**: 113–118.

Wood, C. 1977. Cooley spruce gall aphid. *Can. For. Serv. Pest Leafl.* **FPL-6**.

Wood, D. L. 1963. Studies on host selection by *Ips confusus* (Le Conte) (Coleoptera : Scolytidae) with special reference to Hopkins' host selection principle. *Univ. Calif. Publ. Entomol.* **27**: 241–282.

Wood, D. L. 1972. Selection and colonization of ponderosa pine by bark beetles. *In* "Insect/Plant Relationships" (H. F. Van Emden, ed.) pp. 101–117. *Symp. R. Entomol. Soc. London.* No. 6, Blackwell, Oxford.

Woodwell, G. M., R. H. Whittaker, and R. A. Houghton. 1975. Nutrient concentrations in plants in the Brookhaven oak–pine forest. *Ecology* **56**: 318–332.

Wygant, N. D. 1941. An infestation of the pandora moth, *Coloradia pandora* Blake, in lodgepole pine in Colorado. *J. Econ. Entomol.* **34**(5): 697–702.

Wynne-Edwards, V. C. 1962. "Animal Dispersion in Relation to Social Behaviour." Oliver & Boyd, Edinburgh.

Yendol, W. G. and R. A. Hamlen. 1973. Ecology of entomogenous viruses and fungi. *In* "Regulation of insect populations by microorganisms" (L. A. Bulla, Jr., ed.) *Ann. N.Y. Acad. Sci.* **217**: 18–30.

Yule, W. N. 1975. Persistence and dispersal of insecticide residues. *In* "Aerial Control of Forest Insects in Canada" (M. L. Prebble, ed.), pp. 263–265. Ottawa Dep. of Environ., Ottawa.

Zach, R. and J. B. Falls. 1975. Response of the ovenbird (Aves : Parulidae) to an outbreak of the spruce budworm. *Can. J. Zool.* **53**: 1669–1672.

Zavarin, E. 1975. The nature, variability, and biological significance of volatile secondary metabolites from Pinaceae. *Phytochem. Bull.* **8**: 6–15.

Zavarin, E., F. W. Cobb, J. Bergot, and H. W. Barber. 1971. Variation of the *Pinus ponderosa* needle oil with season and needle age. *Phytochemistry* **10**: 3107–3114.

Zavarin, E., K. Snyberg, and P. Senter. 1979. Analysis of terpenoids from seedcoats as a means of identifying seed origin. *For. Sci.* **25**: 20–24.

Zerillo, R. T. 1975. A photographic technique for estimating egg density of the white pine weevil, *Pissodes strobi* (Peck). *USDA For. Serv. Res. Pap.* **NE-318**.

Zethner-Moller, O., and J. A. Rudinsky. 1967. Studies on the site of sex pheromone production in *Dendroctonus pseudotsugae* (Coleoptera : Scolytidae). *Ann. Entomol. Soc. Am.* **60**: 575–582.

INDEX OF
SCIENTIFIC NAMES

Insects

Plants

Animals

Fungi

Bacteria

Other

INDEX